# Volcanoes in the Sea

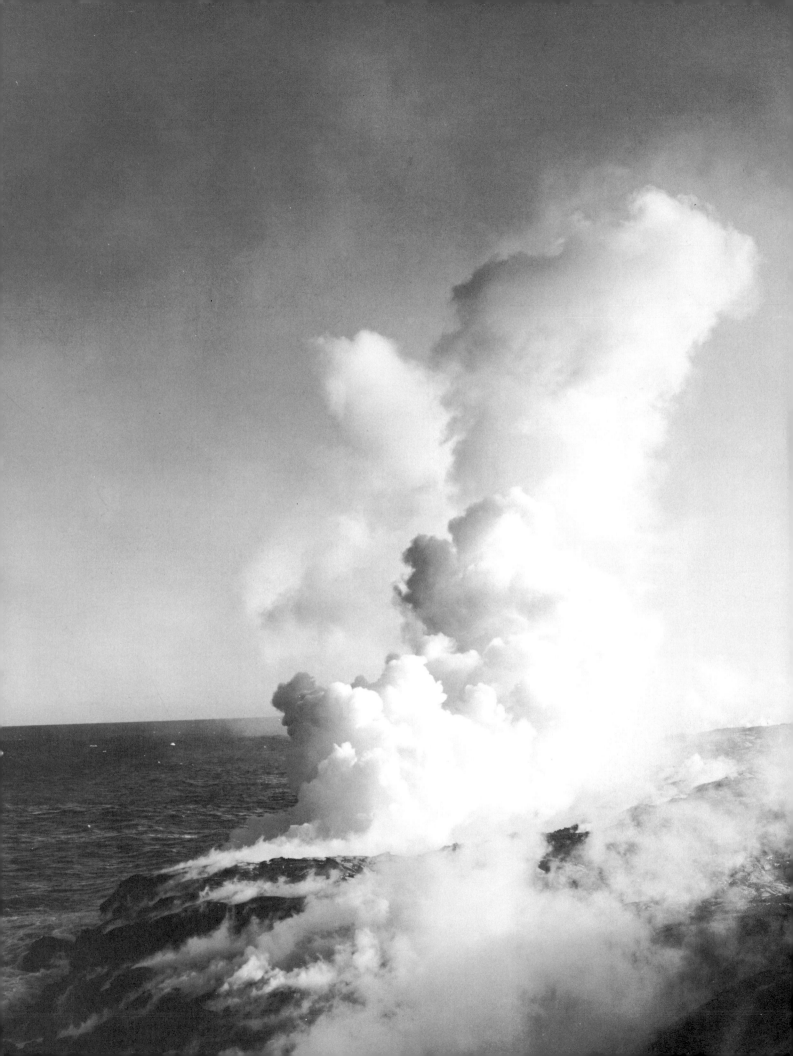

# Volcanoes in the Sea

## THE GEOLOGY OF HAWAII

*Gordon A. Macdonald* and *Agatin T. Abbott*

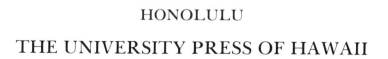

HONOLULU

THE UNIVERSITY PRESS OF HAWAII

FIRST PRINTING 1970
SECOND PRINTING 1971
THIRD PRINTING 1974

*To the people of Hawaii*
*with the hope that*
*they may better understand how the islands*
*came to be, and so more fully enjoy them*
*as they are today*

# Contents

# Preface

SINCE THE DAYS OF THE EARLY EXPLORERS AND WHAL-
ers, the beauty and mystery of the Hawaiian Islands and other islands
of the mid-Pacific have beckoned the traveler and charmed the
visitor. We who make our homes in the islands tend to become
complacent in our surroundings, giving little thought to the grandeur
of the scenery around us and to the dramatic processes that brought
it into being. Thus we lose much of the pleasure that should be ours.
But we can easily recapture the thrill of the islands if we take time to
look again at them and to consider the forces which have built them
up from the sea and sculptured the magnificent cliffs and valleys and
mountain ranges that we see today. These processes are closely
related to such great unsolved mysteries as the nature of the interior
of the earth, and the nature and origin of the tremendous forces that
have formed the present surface features of the earth—mysteries fully
as worthy of study, and in need of exploration, as those of outer
space.

Furthermore, in these days of increasing concern over the changes
we are making in our environment, it becomes apparent that we must
learn to foresee the effects of our actions and plan intelligently to
preserve an environment worth living in. But to do so, we must
understand the processes at work in nature. Man acts, not in an
otherwise static environment, but in a dynamic one which is of itself
undergoing constant and inevitable change. We must correlate our
activities with the natural processes, to guard against unnecessarily
rapid and perhaps disastrous change.

How were the great Hawaiian mountains formed? And what
processes have modeled the ridges and valleys of today? These are
the questions which this book attempts to answer. They are, of
course, the basic problems relating to all parts of the earth that are
dealt with in the science of geology. But the answers to all of the
geological problems of the Hawaiian Islands will not be found in this
book, or in any other. Some of the most fundamental problems still
await solution; the answers to others are as yet tentative and
probably will be modified as our knowledge increases. Geology is a
rapidly growing science, and we hope that some of our readers will
be inspired to take part in its development.

The book is not a highly technical treatise, written for specialists in the field. Rather, it is intended primarily for persons with little or no previous training in geology. On the other hand, although they are presented for the most part in nontechnical language, the geological facts and interpretations are stated as accurately as is possible, and the coverage is as thorough, detailed, and up-to-date (to April 1970) as it can be made in a general survey of this type. Consequently it is hoped that visiting geologists also will find the book useful, particularly as most of the original sources of technical information on the Hawaiian Islands are now out of print.

Summaries of the geology of the islands on a somewhat different level, and a guide to points of geologic interest along the highways, have been published by H. T. Stearns (1946, 1966a, 1966b). These, particularly the guidebook, will be useful supplements.

For those wishing to delve still further into Hawaiian geology, suggestions for additional reading are given at the end of each chapter. In addition, a fairly extensive reference list includes complete citations for all published work referred to in the text and for all references in the suggested reading lists. Still other sources can be found in bibliographies by N. D. Stearns (1935) and Macdonald (1946). Additional information on the operation of the various geologic processes can be found in any good text on physical geology, such as those by Holmes (1966), Gilluly, Waters, and Woodford (1968), and Leet and Judson (1965).

The greater part of the text of this book was written by Macdonald, and most of the photographs were taken by Abbott. However, the work of many persons has contributed to the present knowledge of the geology of the Hawaiian Islands. Oustanding is that of T. A. Jaggar on the volcanoes, and of H. T. Stearns and C. K. Wentworth on the areal geology and ground water. Others are mentioned in appropriate places in the text, but still others remain unnamed because of limited space. Particularly to our colleagues at the University of Hawaii and the U.S. Geological Survey, we are indebted for much information and for profitable discussion of problems over the years.

The directors of the Hawaii Division of Land and Water Development (successor to the Hawaii Division of Hydrography), the U.S. Geological Survey, and the Hawaii Institute of Geophysics have generously allowed us to reproduce photographs from earlier reports and from their official collections. Individual credits for these and for all other photographs used in the book are given on page 431. Other illustrations (line drawings and maps) from earlier books and technical papers are acknowledged in the individual figure titles. To all of the friends, coworkers, and publishers who have thus helped us, we express our appreciation.

The illustrations taken directly from the bulletins of the Hawaii Division of Hydrography were drawn by James Y. Nitta, whom we wish to thank for his superb draftsmanship. Special thanks go also to Ruth Macdonald and Paula Abbott for their critical reading of the manuscript.

# Volcanoes in the Sea

*Haleakala and
Molokini Islet*

# CHAPTER ONE

# Introduction

ACROSS THE FACE OF THE CENTRAL PACIFIC OCEAN, 2,000 miles from the nearest continent and 500 miles from any other land, lies the row of islands known as the Hawaiian Archipelago. From Kure Island at the northwest, the archipelago forms a southeast-trending line of islands 1,500 miles long (fig. 1). Each island is the summit of a great mountain rising from the floor of the ocean. The islands in the northwestern part of the chain are low, barely projecting above sea level, but even the mountains beneath these can hardly be considered puny, for they rise some 15,000 feet above their bases. Those forming the southeastern part of the chain are greater yet. Mauna Kea and Mauna Loa, two of the five mountains that make up the southernmost island, Hawaii, stand more than 30,000 feet above the adjacent ocean floor, and rise higher above their bases than any other mountain on earth.

The southeastern 400 miles of the chain consists of eight major islands, projecting 1,200 to nearly 14,000 feet above sea level (table 1). The island of Hawaii is among the world's largest volcanic islands, second in size only to Iceland. Its area of 4,030 square miles is greater than that of all the other Hawaiian islands put together, and is more than three times that of the state of Rhode Island. Around the edges of some of the islands are narrow flatlands near sea level, but on all of them high central land masses are notched by deep, steep-sided canyons and studded with sharp peaks—mountain scenery unsurpassed in beauty and grandeur.

The great mountain range that constitutes the Hawaiian Archipelago has been built almost wholly by volcanic activity. Each island is the top of an enormous volcanic mountain (fig. 2), modified by stream and wave erosion and minor amounts of organic growth.

At some time many millions of years ago, but probably very late in the history of the earth, a series of fissures opened along a narrow, northwest-trending zone in the ocean floor. Along those fissures molten rock issued at intervals from the interior of the earth, hardened, and gradually piled up, layer upon layer, to build the mountains. Each mountain contains literally thousands of these thin layers. Above sea level, where erosion has cut into them, almost all of the layers are seen to be lava flows, each a sheet of once-molten rock

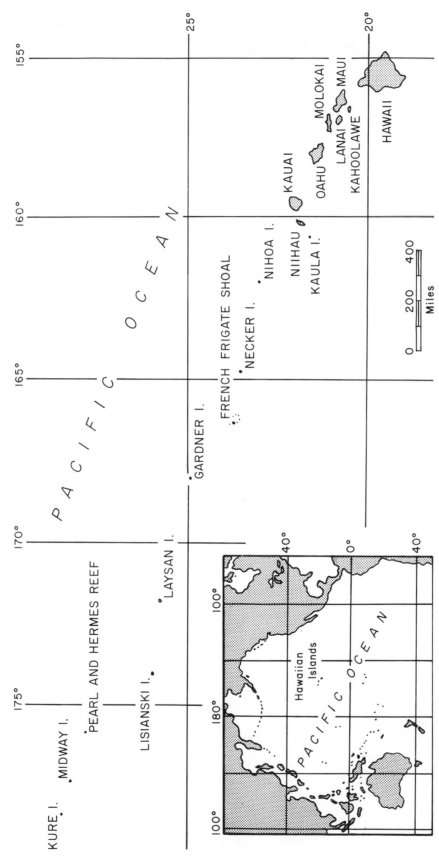

Figure 1. Map of the Hawaiian Archipelago. The small inset shows the location of the islands in the Pacific Ocean.

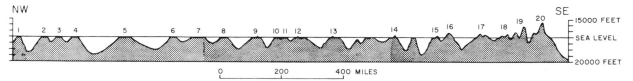

Figure 2. Profile along the Hawaiian Archipelago. The vertical scale is nearly 20 times the horizontal, making the mountains appear much steeper than they actually are. The features indicated by numbers are as follows: *1,* Kure Island; *2,* Midway Islands; *3,* Gambia Shoal; *4,* Pearl and Hermes Reef; *5,* Fisher Reef and Neva Shoal; *6,* Laysan Island; *7,* Maro Reef; *8,* Raita Bank; *9,* Gardner Island; *10,* St. Rogatien Bank; *11,* Brooks Bank; *12,* French Frigate Shoals and La Perouse Rocks; *13,* Necker Island; *14,* Nihoa Island; *15,* Niihau; *16,* Kauai; *17,* Oahu; *18,* Molokai; *19,* Maui; *20,* Hawaii. (After Stearns, 1946.)

that was poured out rather quietly onto the surface. Material thrown out by explosions is almost absent in the visible part of the mountains. At shallow depths in the ocean, just before the summit of the mountain reached sea level, contact of the hot lava with seawater may have resulted in violent steam explosions, and the part of the mountain built under those conditions may contain a greater proportion of exploded ("pyroclastic") material. This is wholly hypothetical, however, because since that time the mountains have sunk, carrying that zone below sea level and out of our range of vision. There is one bit of evidence that supports the hypothesis: the average slopes of the volcanoes are a little steeper below sea level than they are above, suggesting some change in the building process. At great depths in the

Table 1. Geographical data on the Hawaiian Islands

| Island | Length (miles) | Width (miles) | Area (square miles) | Length of shoreline (miles) | Highest points | |
| | | | | | Name of mountain | Altitude (feet above sea level) |
|---|---|---|---|---|---|---|
| Hawaii | 93 | 76 | 4,038 | 313 | Mauna Kea | 13,796 |
| | | | | | Mauna Loa | 13,677 |
| | | | | | Hualalai | 8,271 |
| | | | | | Kohala Mountain | 5,480 |
| | | | | | Kilauea | 4,090 |
| Maui | 48 | 26 | 729 | 149 | Red Hill (Haleakala) | 10,023 |
| | | | | | Puu Kukui (West Maui) | 5,788 |
| Oahu | 44 | 30 | 608 | 209 | Kaala (Waianae Range) | 4,025 |
| | | | | | Puu Konahuanui (Koolau Range) | 3,150 |
| Kauai | 33 | 25 | 553 | 110 | Kawaikini | 5,243 |
| Molokai | 38 | 10 | 261 | 106 | Kamakou (East Molokai) | 4,970 |
| | | | | | Puu Nana (West Molokai) | 1,381 |
| Lanai | 18 | 13 | 140 | 52 | Lanaihale | 3,370 |
| Niihau | 18 | 6 | 73 | 50 | Paniau | 1,281 |
| Kahoolawe | 11 | 6 | 45 | 36 | Lua Makika | 1,477 |
| TOTAL | — | — | 6,447 | 1,025 | — | — |

5

ocean the weight of the overlying water almost surely would prevent explosion, and the bases of the mountains probably consist wholly of lava flows, in the bulbous and billowy forms known as pillow lavas. To some unknown extent the lavas probably were granulated by contact with water to form masses of fragmental, sandy "hyaloclastite." The presence of any appreciable quantity of this fragmental material, or of the fragmental products of steam explosions at shallower levels, would help explain the rather low velocities of earthquake waves traveling through the lower parts of the mountains, but to date very little evidence of the existence of either material has been found by dredging or in undersea photographs. A mass of material of this sort, which is plastered against the southern slope of the island of Hawaii, may well have been formed by the granulation of recent lava flows erupted above sea level as they poured into the ocean.

For convenience the archipelago is commonly divided into two parts: the Windward Islands, comprising the major islands of the chain, from Niihau and Kauai to Hawaii; and the small, widely scattered Leeward Islands, which continue the chain for more than a thousand miles beyond Niihau. In general, the islands to the northwest are older than those to the southeast. The southern part of the island of Hawaii is still growing and is almost untouched by erosion, but on the more northwesterly islands there has been time for streams to cut deep canyons and for waves to remove a large proportion of the original volcanoes. To be sure, in comparatively recent times, a brief revival of volcanic activity occurred on the more northerly of the major islands, particularly on parts of Oahu, Kauai, and Niihau, and erosion has made little headway on these younger rocks. The main masses of the northern islands are several millions of years old, whereas the age of the oldest visible rocks in the oldest part of Hawaii Island is less than 1 million years. The volcanic islands of the Leeward group are still older and more eroded than Kauai and Niihau, and the islands beyond them are coral reefs resting on still older truncated volcanoes that have been submerged beneath the sea. Actually, we may be dealing with volcanoes of two distinctly different geologic ages—an older group that underlies the coralline islands, and a younger group, most of which lie farther southeast. Ancient shallow-water fossils dredged from a submerged shelf off Honolulu suggest, however, that the older volcanic chain extended at least as far southeast as Oahu.

The deep ocean basins, in the largest of which the Hawaiian Islands are located, are fundamentally different geologically from the continents. The outstanding difference is the absence in the ocean basins of the layer of light "granitic" rock which characterizes the continents, and which, because it is relatively light, appears to make possible the existence of the high-standing continental platforms. The other differences certainly are related, either directly or indirectly, to the presence or absence of the granitic layer. They extend to the character of the volcanic activity—predominantly gentle extrusion of fluid basalt lava flows in the ocean basins as compared with abundant explosion in many continental areas. Within the ocean basins, there is a general uniformity of rock types and geological processes. To be sure, there are local differences. Thus, for instance, the rocks of the Society Islands contain, on the average, considerably more potassium than do those of Hawaii. But these differences are relatively minor. The entire mid-Pacific province is essentially uniform geologically. What we learn of the Hawaiian Islands can be applied, for the most part, directly to the other islands farther south, so long as we remain within the boundaries of the true oceanic basin.

Nearly all the geological processes that are known elsewhere on earth have operated, at least to some extent, in Hawaii. Thus a book on geology of the Hawaiian Islands comes close to being one on geological processes in general. Even glaciation, which seems so foreign to the usual concept of subtropical Hawaii, has operated to modify the summit region of Mauna Kea on the island of Hawaii. Only paleontology, with its record of the development of life through past ages, is conspicuously lacking.

The rocks of the islands are all very young, and the fossil record in them is meager. Some interesting details are starting to emerge from the study of very recent fossils in coral reefs left stranded above water by changes of sea level during the most recent epoch of geologic time, but for the broad picture of the evolution of life the reader must turn to other parts of the world, and to other books. In discussing the various geological processes as observed in Hawaii, we have thought it best first to review the nature and operation of each process as it has been worked out elsewhere in the world as well as in Hawaii, and then to describe more specifically the features it has produced in Hawaii.

Throughout, volcanism has been the process of basic importance in the development of the islands. Other processes have sculptured the land surface into the forms we now see, but their work could not begin until a land mass had been built. For that reason, in the chapters that follow, we will consider first the volcanoes.

*Lava fountains
in Halemaumau,
1967*

# CHAPTER TWO

# Hawaiian Volcanic Activity

FOR MONTHS THE PRESSURE HAD BEEN SLOWLY INCREAS-ing beneath Kilauea and Mauna Loa volcanoes. Twenty miles to the east, at the edge of Kilauea Crater, the ground had been tilting gently eastward as the summit of Mauna Loa was pushed up by the accumulation of molten rock beneath it. Then suddenly, at 9:25 on the evening of June 1, 1950, visitors at the Volcano House at Kilauea saw an orange-red glow appear near the top of Mauna Loa. The volcano was in eruption.

Actually the eruption probably had already been in progress for about 20 minutes, but because of its remoteness and the poor visibility it was not immediately detected. At about 9:10, residents of the small town of Naalehu, on the south flank of Mauna Loa, heard a deep rumbling sound from the direction of the top of the mountain; and even earlier, at 9:04, seismographs had started recording a steady slight trembling of the ground, known as volcanic tremor, that is caused by the movement of molten rock through the feeding channels of the volcano.

At first, two columns of gas, glowing bright orange-red from the reflection of molten lava flows beneath, rose from a point on the southwest flank of the mountain about 12,600 feet above sea level. The eruption point was located on a zone of cracks that extends down the southwest side of the mountain (the so-called southwest rift zone), and during the next few minutes the erupting cracks opened farther and farther downslope. Within 15 minutes the line of erupting cracks was 2.5 miles long, and from it spurted fountains of molten rock (lava) several hundred feet high. The gas cloud rose in a narrow column about 2 miles into the air, then spread out to form a mushroom-shaped cloud brightly lighted by the orange glare of the incandescent lava beneath it. A flood of very fluid, gas-rich lava poured from the crack and westward down the mountainside, forming many short streams and one main flow that traveled downslope about 5 miles.

At about 10:15 P.M. another puff of gas was seen to rise from a point much lower down on the zone of fractures. This point was 8 miles southwest of the erupting crack near the top of the mountain and about 8,250 feet above sea level. Ten minutes later a bright glow

9

Figure 3. The "curtain of fire"—a row of lava fountains 600 feet high on the southwest rift zone of Mauna Loa, June 2, 1950.

appeared on the lower gas cloud, showing that orange-hot lava was pouring out at the lower site also. The glowing gas cloud grew rapidly broader as the erupting fissure beneath it continued to open both up and down the mountainside. By 4 o'clock the next morning a line of lava fountains 8 miles long (fig. 3) was shooting from the opened fissures to a height perhaps as much as 1,000 feet above the ground.

From the base of the fountains a flood of molten lava poured down the mountainside southward and westward from the crest of the broad ridge along which the erupting cracks were located. By daybreak on June 2 one lava stream on the south slope was 10 miles long. Even faster was the advance of a flow on the west slope. Shortly after midnight the flow was burning its way through the forest not far above the highway, and at 1:05 it plunged into the sea. This flow had traveled 15 miles at an average speed of 5.8 miles an hour. En route it had wiped out most of a small village—all the villagers had escaped, though some by a narrow margin.

By noon on June 2 these early flows had ended, but two other flows were pouring down the west slope (fig. 4). One entered the ocean just after noon, destroying two groups of ranch buildings along the way. The other reached the shore at 3:30 P.M. (fig. 55). This one became the principal flow of the eruption. It gradually spread laterally, destroying as it did so a new restaurant and nearby buildings.

During the next few days thousands of people visited the flow where it crossed the highway. Vendors moved back and forth along the road selling ice cream and soft drinks, and a police detail was kept busy directing the parking of cars and preventing visitors, fascinated with the hot lava, from becoming too careless.

The eruption continued for two weeks. More than 600,000,000 cubic yards of lava were poured out, covering nearly 35 square miles of country above sea level. Buildings and pastureland were destroyed, but no one was injured.

This eruption was only one of more than 70 that have taken place in the Hawaiian Islands since the beginning of the 19th century. During that interval Mauna Loa and Kilauea, on the island of Hawaii, have been among the earth's most active volcanoes. But just what is a volcano?

### WHAT IS A VOLCANO?

A volcano may be defined as a place where molten rock (*magma*) and/or gas from within the earth issue at the surface. The term is also used for the hill or mountain built by the rock that is poured or blown out. Using the word in the latter sense, small volcanoes may be hillocks only a few feet high, while large ones such as the Hawaiian mountains may be many hundreds of cubic miles in volume. The bulk of Mauna Loa is estimated to be about 10,000 cubic miles, and it rises nearly 30,000 feet above its base at the ocean floor. The truly enormous size of this mountain (probably the

Figure 4. Lava flow descending through the forest on the west slope of Mauna Loa, June 2, 1950. In the center of the picture the flow is nearly a mile wide. Note the brilliant incandescence of the rapidly advancing flow front in the foreground.

11

largest single mountain on earth) can be appreciated when it is compared with the great volcanic mountains, Shasta and Fuji, each of which has a volume of about 100 cubic miles. Mauna Loa is one hundred times larger than either of them.

Sometimes the word *crater* is used incorrectly to mean a volcanic hill or mountain. Craters are bowl-shaped or funnel-shaped depressions—not hills. Volcanic craters are commonly found at the summit of a volcano, but also often on its flanks; or they may occur even in nonvolcanic rocks completely apart from any volcanic mountain. So long as they are formed by volcanic action they are volcanic craters, no matter what the nature of the surrounding rocks.

Some volcanoes erupt only once; others erupt repeatedly, and many volcanic mountains are built of the products of dozens or hundreds of eruptions. Volcanic eruptions may differ widely in character. Some are violently explosive; others are gentle. Eruptions of differing character may take place at the same volcano. For the most part, however, each volcano has a rather uniform and characteristic "habit" of eruption. Exceptions include the occasional explosions that occur at normally quiet volcanoes as a result of water coming in contact with molten or very hot rock. The explosions of Kilauea volcano in 1924 were of that sort. It is also recognized that volcanoes commonly change as they grow older, becoming more explosive toward the end of their life spans.

The explosiveness of an eruption depends largely on two factors: the amount of gas in proportion to liquid rock reaching the surface, and the viscosity, or "stickiness," of the liquid lava. Thinly fluid lava allows moderate amounts of gas to bubble out with little more than minor spattering; but in very viscous lava it is difficult for the gas to work its way upward and break through the surface of the liquid, and it accumulates until the pressure is sufficiently high to allow it to burst free. With high enough gas pressure, a large enough amount of gas, and viscous enough lava, a major explosion or series of explosions may

result. There are many other complicating factors, of course. On the whole, however, the amount of gas present seems to be more important than the viscosity of the magma. Since an explosion results from the rapid expansion of gas, no very big one can occur unless a large amount of gas is present. But a moderately large amount of gas rushing out of even very fluid lava produces a moderately explosive eruption (this is the case in some of the late-stage Hawaiian eruptions discussed in a later chapter).

## CHARACTERISTICS OF HAWAIIAN-TYPE ERUPTIONS

In the course of the development of each individual Hawaiian volcano, its characteristic pattern of eruption changes. However, by far the largest part of the building of the Hawaiian volcanic mountains is of the sort exemplified by present-day Kilauea and Mauna Loa, of which the 1950 eruption already described was an unusually large example. Indeed, Hawaii has given its name to this sort of volcanic activity wherever it may occur. It is known by volcanologists as *Hawaiian-type eruption.*

Kilauea and Mauna Loa release very fluid lava containing only a relatively small amount of gas. The gas escaping at the vent during an eruption forms an impressively huge cloud that rises thousands of feet into the atmosphere, but actually the cloud is much diluted by air, and a very small amount of gas (in terms of weight percent of the magma reaching the surface) occupies a very large volume in the atmosphere. Calculations based on estimates of the amount of gas given off during the 1940 eruption of Mauna Loa and the 1952 eruption of Kilauea indicate that the gas forms only about 1 percent of the weight of the erupting magma. The outstanding characteristic of Hawaiian-type eruptions, as compared with other types, is their gentleness. Almost no explosion is involved; this unquestionably is because of both the fluidity of the magma and the small amount of gas it contains. To be sure, jets of liquid lava shoot into the air, forming lava fountains (fig. 5) that may continue

12

uninterrupted for many days and commonly reach heights of several hundred feet above the vent opening. In 1949 the great fountain at the summit of Mauna Loa probably was more than 1,000 feet high, and the tremendously spectacular fountain of the 1959 eruption in Kilauea Iki Crater reached a height of at least 1,500 feet. But these jets, impressive as they are, are essentially nonexplosive. The rise of magma into the air is aided by the expansion of the gas within it, but for the most part the fountains are simply streams of molten rock under pressure, shooting into the air much like a stream of water from a hose or a fountain of water in a park.

SOURCE OF THE MAGMA

It is generally recognized today that the earth consists of three principal parts (see chap. 18). At the center is the core, some 2,200 miles in radius; on the outside is a very thin crust, only a few miles thick; and between them is the earth's mantle, nearly 1,800 miles thick (fig. 166). Evidence, discussed in chapter 18, indicates that the earth is essentially solid down to the boundary of the core; yet there is also evidence that the magma erupted by volcanoes is formed at comparatively shallow depths within the earth. In Hawaii seismographs record groups of earthquakes that come from depths of about 30 to 40 miles below the summits of the active volcanoes, and we believe that these earthquakes mark the zone of origin of the magma (Eaton and Murata, 1960). They are accompanied by a peculiar trembling of the ground (volcanic tremor) which we have come to associate with the movement of molten rock beneath the surface. Recent studies by Russian scientists indicate that beneath the Kamchatkan volcanoes, magma probably originates at about the same depth. In Kamchatka this is close to the boundary between the earth's crust and mantle, or in the very uppermost part of the mantle. Beneath Hawaii, where the base of the crust is only 9 to 12 miles below sea level,

Figure 5. Lava fountain about 150 feet high on the east rift zone of Kilauea, March 14, 1955, just starting to build a spatter cone at its base. Note the irregular forms of the bomb-size shreds of spatter in the air.

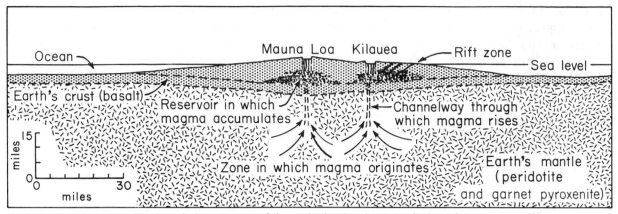

Figure 6. Vertical cross section through the outer part of the earth, showing the place of origin of the magma of Kilauea and Mauna Loa volcanoes, and the shallow magma reservoir in which it accumulates before eruptions. (Modified after Macdonald, 1961.)

and in some places even less, this depth lies well within the upper part of the mantle (fig. 6). There appears to be little question that at least in these two regions, and probably nearly everywhere, magma of the sort erupted in Hawaii (basaltic magma: see chap. 4) is formed by melting of previously solid rock in the outer part of the earth's mantle.

In deep mines and wells, wherever measurements have been made, we find that the temperature increases downward within the earth. The rate of increase varies somewhat from place to place, but it averages about 1° C. for every 100 feet of depth. Under surface conditions most lava rocks begin to melt near 1100° C. and are fully fluid at 1200° C.; with dissolved gases present to act as fluxes they should melt at even lower temperatures. Very fluid lavas erupted at Kilauea have temperatures between 1100° and 1200° C. (very roughly 2000° to 2200° F.). Thus, if the temperature continues to increase at the same rate deep within the earth as it does near the surface, temperatures high enough to melt rocks should be reached at a depth of only a little more than 20 miles. But if this is so, why is the earth's mantle essentially solid? Apparently the pressure, which also increases downward within the earth due to the increasing weight of the thicker mass of overlying rock, raises the melting point of the rocks enough to keep them solid.

In the mantle, then, at depths below 20 miles, the temperature is high enough to melt the rock if it were under the same pressure that prevails at the surface, but because of the very

high pressure, it remains solid. Local melting of the rock can then result either from a rise of temperature to a level above the melting temperature even under pressure, or from a local reduction of pressure which would lower the melting temperature of the rock. Exactly what brings about the melting we do not know, but there can be no question that it happens, because the molten rock emerges at the surface for all to see.

Once molten, the liquid rock starts its journey toward the surface of the earth, along opening fissures, pressed upward because it is lighter than the surrounding solid rocks, and perhaps also by compressional forces of other origin.

COMPOSITION OF THE MAGMA

The magma reaching the surface is quite different in chemical composition from the peridotite that we believe constitutes the mantle (see chap. 18). Some of this difference results from chemical processes within the magma as it rises, and some perhaps from melting or dissolving of rocks of the earth's crust with which it comes in contact. We believe, however, that the original magma, as it starts to rise, is already considerably different from the peridotite, and that here in Hawaii it corresponds quite closely with that erupted at Mauna Loa or Kilauea, with a good deal more silicon and less iron and magnesium than the peridotite (see table 6, chap. 4). This is probably because not all the minerals in the peridotite melt at the same temperature. Those

14

with the lowest melting temperatures melt first, supplying the material that forms the magma, leaving behind as a solid residuum the components that melt at higher temperatures.

Part of the magma consists of volatiles —substances which, at the surface, tend to separate out as gases and escape into the atmosphere. The exact composition of the gas dissolved in the magma when it starts its rise from depth is very hard to determine, because the gases react with each other and with the other components of the magma as the conditions of temperature and pressure change during the rise, and because an unknown amount of water almost certainly is picked up by the magma from the rocks bordering the conduit near the surface. Chemical analyses of the best collections of gas show (Jaggar, 1940) that they contain, on the average, about 70 percent water vapor, 14 percent carbon dioxide, 5 percent nitrogen, 6 percent sulfur dioxide (the choking gas one smells from a burning sulfur match or fumigating candle), 2 percent sulfur trioxide, less than 1 percent each of carbon monoxide, hydrogen, and argon, and a bare trace of chlorine. Fluorine, a common gas at continental volcanoes, has seldom been found in Hawaii, though it was present in gases collected from the 1959 lavas in Kilauea Iki Crater (Murata, Ault, and White, 1964).

When it starts toward the surface the magma probably is usually wholly liquid, with at the most only a very few solid grains suspended in it. However, as it rises it enters regions of progressively lower temperature and pressure. Analysis of the pattern of swelling and shrinking of the volcanic mountain (p. 34), coupled with earthquake evidence, seems to indicate that beneath Kilauea the magma pauses in its rise in a chamber 2 or 3 miles below the summit of the volcano (fig. 6), and no doubt the same is true of other volcanoes. In these cooler regions the magma loses heat to the surrounding rocks, and as it becomes cooler solid crystals of minerals start to form in it. The first to form generally are olivines—the glassy green crystals commonly seen in Hawaiian lava rocks (see chap. 4). Growing slowly, these crystals become fairly large, often as much as a quarter of an inch across. If at

that stage the magma is brought out onto the earth's surface by eruption, the rest of the liquid cools quickly and forms a fine-grained matrix of gray to black crystalline and glassy material enclosing the larger crystals. (The large crystals in a finer matrix are called *phenocrysts.*) Rocks of this sort are widespread in all the Hawaiian mountains. Magma that has issued onto the surface of the earth is called *lava,* and when solidified it is still referred to as lava, or sometimes as lava rock.

As the magma enters the zone of low pressure very close to the earth's surface, still another process begins. Gas starts to come out of solution and forms bubbles in the magma, just as it does in soda pop when the internal pressure in the bottle is relieved by removing the cap. At this stage the magma consists of three physical phases: a liquid, solid crystals, and bubbles of gas. This is the condition in which most magmas are erupted. Solidifying of the liquid rock perpetuates the bubble holes, which are then known as *vesicles* (fig. 7). Bubbly solidified rock froth is called *scoria,* or the rock is said to be *scoriaceous.* Extremely inflated, very light rock froth is *pumice,* which often is so light it will float on water. Hawaiian pumice (reticulite) commonly is lighter even than the pumice of other regions, but it usually sinks quickly in water because it is so inflated that the holes interconnect and allow it quickly to become waterlogged.

## PYROCLASTIC MATERIALS AND STRUCTURES

### Types of Ejecta

Volcanic gas escaping at the earth's surface carries up with it into the air fragments of the magma, and sometimes also fragments of solidified lava or of old rocks that form the walls of the volcanic conduit. If the gas rushes out violently it may carry large amounts of material high into the air, but if it escapes gently it may cause nothing more than a weak spattering. In any case, the fragments thrown into the air fall back to the ground, either close around the vent or at a greater distance, and form what are known as *pyroclastic* ("fire-broken") *rocks.*

15

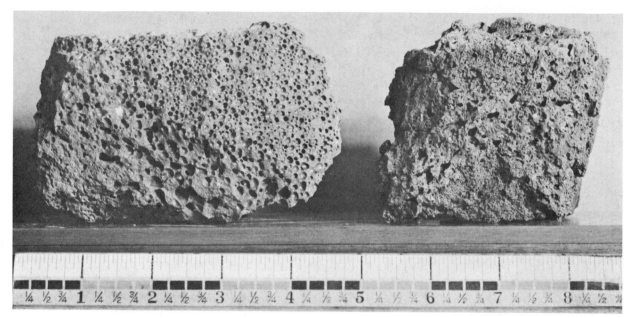

Figure 7. Specimens of pahoehoe *(left)* and aa *(right)*, showing the contrast in vesicle shapes. The scale is in inches.

Recently, the Icelandic volcanologist Sigurdur Thorarinsson has revived an ancient term used by Aristotle, and proposes to call pyroclastic rocks by the shorter designation, *tephra*. Although this brief collective designation is very useful and worthy of general adoption, it has been accepted as yet by only a few writers on volcanoes. All fragments thrown up (ejected) by volcanic explosion are also designated by the term *volcanic ejecta*.

Ejecta are classified primarily according to size, but the larger fragments also according to their fluidity when they are thrown out. Fragments larger than 1.5 inches in diameter are called *bombs* or *blocks,* depending on their shape, which in turn depends on their fluidity when ejected. Bombs were thrown out in a fluid state. In the air, many of them take on a rounded shape. Some are nearly spherical; others are drawn out at the ends to a form resembling the mass of yarn on the spindle of an old-fashioned spinning wheel (fig. 8 *A, B, C*) and are known as fusiform or spindle bombs. Long thin bombs, which are merely frozen "squirts" of fluid lava, are ribbon bombs.

16 Spherical, spindle, and ribbon bombs have solidifed in the air sufficiently to retain their shape when they strike the ground, but others are still so fluid that they flatten out (fig. 8 *C, D*), or even splash, when they hit. These are called pancake bombs, or even more descrip-

tively, cow-dung bombs. They are quite characteristic of Hawaiian eruptions. Some bombs, particularly those of spherical shape, are formed when liquid lava adheres to a fragment of already solid rock, completely enclosing it like the chocolate surrounding the cream center of a bonbon. These are cored bombs. Broken open, they reveal the angular core, which may be a fragment of older rock, sometimes even of nonvolcanic nature, or a piece of spatter or cinder formed in an earlier stage of the same eruption. Although regular-shaped bombs are most conspicuous, many are very irregular—fragments of liquid froth torn apart by expanding gas and frozen in the air.

Some bombs are dense; others are moderately to highly scoriaceous. Frothy fragments of this sort, commonly very irregular in out-

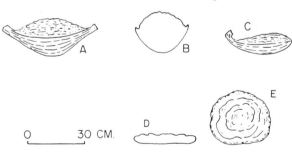

Figure 8. Sketches and cross sections of volcanic bombs. *A,* Spindle bomb with projections at both ends; *B,* cross section of *A; C,* spindle bomb with projection at only one end; *D,* cross section of cow-dung bomb; *E,* cow-dung bomb. (Modified after Macdonald, 1967.)

line, are called *cinder* or *scoria*. Pumice is simply an extremely light form of cinder. Many cinder and pumice fragments are too small to be called bombs, and fall into the classes of lapilli or ash, described in later paragraphs.

Blocks are angular (fig. 13), having been either solid at the time of ejection or so viscous that they could not take on a rounded form in the air. The solid blocks may be fragments of older volcanic rocks, or even nonvolcanic rocks. Koko Head, on Oahu, for instance, is full of blocks of limestone torn from the coral reef beneath by the explosions that built the cone. Other blocks are fragments of a solidified crust that had formed on the erupting lava in the crater and was torn apart by the outrushing gas.

Masses of angular blocks are called volcanic *breccia*. The term is a general one, used by geologists to designate any aggregate of angular fragments. Even volcanic breccias may form in other ways than by explosion. Great heaps of angular rock fragments accumulate at the foot of cliffs, forming taluses (see chap. 9), which may be buried by lava flows and preserved in the structure of the volcano. Many of these talus breccias, exposed by erosion, have been found in the older Hawaiian volcanoes. Commonly they mark the position of the cliffs that once bounded the caldera of the volcano. Other breccias are formed by mudflows. It is useful, therefore, to designate those formed by explosion as explosion breccias. Beds of explosion breccia can be seen in the highway cut in the side of Koko Crater nearly opposite the parking area at the Halona blowhole.

Ejecta less than 1.5 inches in diameter but greater than 0.25 inch are called *lapilli* (singular: lapillus). The word is Italian and means simply "little stone." No distinction is made on the basis of roundness or angularity of the fragments, which duplicate all of the shapes found in bombs and blocks.

Fragments less than 0.25 inch across are called volcanic *ash* (fig. 13). They may be bits of already solid rock or crystals from solid rock disrupted by explosion, or they may be particles of lava that were thrown up as a liquid spray. In the latter case the bits of liquid freeze in the air to bits of glass. Volcanic ash tends to become cemented together very quickly to form a firm rock called *tuff*. (Volcanic ash, known as pozzulana, was used by the Romans as a natural cement in mortar between building blocks, and it is still used in Italy today.)

Expanding gasses in the lava fountains of Hawaiian-type eruptions tend to tear the liquid into irregular gobs and shreds, which fall back to the ground to build a heap of fragments around the vent. Many of the fragments, being still partly liquid when they strike the ground, flatten out or splash when they hit, forming *spatter*. When these fragments adhere to each other because of the freeezing together of their liquid edges they are said to be *welded*. A mass of welded spatter is sometimes called *agglutinate*. Spatter and agglutinate are distinguished from cinder, in which the fragments are solid, or nearly so, when they strike the ground, and do not stick together.

Some of the bits of liquid thrown up by the lava fountains freeze in the air in typical drop shapes, with one end broadly rounded and the other drawn out into a long thin tapering point. These are called *Pele's tears*, after Pele, the Hawaiian goddess of volcanoes. Because they are frozen very quickly in the air, they consist of glass, usually jet black in color. Many of the drops draw out behind them thin threads of viscous liquid lava that also freezes in the air to form slender glassy filaments called *Pele's hair*. The threads, sometimes several feet long, may drift in the wind for many miles. Many eruptions produce Pele's hair, but commonly that formed by big fountains is coarse and stiff. Large amounts of beautifully fine Pele's hair was formed by the innumerable small fountains on the Halemaumau lava lake (p. 40), and some of it can still be found collected against the lee side of rocks in the Ka'u Desert, downwind from Halemaumau. Although some is dark brown to almost black, much of it is light golden brown.

### Cinder and Spatter Cones

The hill built by fragments falling around the vent commonly has the shape of a cone with the small end cut off, and usually there is a crater at the summit. Hills of this sort are referred to in a general way as volcanic cones.

Figure 9. Spatter-and-cinder cones at the vents of some of the lava flows of 1955 (vents *Y* and *Z,* fig. 68), on the east rift zone of Kilauea.

Those formed of spatter are *spatter cones* (fig. 9), and those of cinder are *cinder cones* (fig. 10). Cow-dung bombs are common in spatter cones, and cinder cones often contain some spherical, ribbon, or spindle bombs. The latter are common in cinder cones near Hale Pohaku, on the south side of Mauna Kea, although otherwise they are quite rare in most of the Mauna Kea cones. Most of the cones on Kilauea and Mauna Loa, and indeed most of those built by Hawaiian-type eruptions, are spatter cones, because of the great fluidity of the erupting magma. A few cinder cones are present, like the Kamakaia Hills on the southwest rift zone of Kilauea, but they are small, and often the cinder is mixed with spatter. A good cross section of a cinder-and-spatter cone can be seen on the Chain of Craters Road in Hawaii Volcanoes National Park, just east of the Devil's Throat.

Because Hawaiian-type eruptions commonly occur along extensive fissures, the heap of ejected material often is very long and narrow, and is called a *spatter rampart.* Rarely in Hawaiian-type eruptions enough pumice is produced to build a pumice cone. The largest of these is the cone of the 1949 eruption at the summit of Mauna Loa (fig. 53), during the early stages of which the magma was so rich in gas that an unusually large amount of pumice was produced.

## Ash and Tuff Cones

Sometimes Hawaiian eruptions near the seashore are moderately explosive because of the abundant steam formed when the hot magma encounters either sea water or water in the pores of the rocks close to the surface. These eruptions are known as *hydromagmatic* because they involve both molten magma and water. (Explosions that result from the heating of ground water to steam are called *phreatic,* after the Greek word for a well; and where magmatic gases also are involved the explosions

18

are called *phreatomagmatic.*) In hydromagmatic explosions the rapidly expanding steam blows much of the magma apart, forming a spray of tiny fragments and droplets which harden in the air to particles of glassy ash, largely of sand or dust size. The ash soon becomes cemented together into tuff. The cone formed of unconsolidated ash is an *ash cone,* and the same cone after the ash has been consolidated into tuff is a *tuff cone* (figs. 11, 12).

The larger fragments of ash tend to fall closer to the vent than do the smaller ones, resulting in a sorting according to size. A single eruption consists of a series of hundreds of more or less separate explosions, and each explosion results in a shower of ash that falls over the surrounding terrain forming layers that mantle the hills and valleys as well as the growing cone (fig. 13). Individual layers arch over the rim of the cone, producing what is known as mantle bedding. Because large fragments fall faster than small ones, each layer tends to grade from coarser at the bottom to finer at the top, although sometimes increasing strength of the explosion results in a reverse grading, and sometimes the material falls so thickly that sorting in the air cannot occur and there is no grading. Often the sorting is far from perfect, and occasional large blocks and bombs are found embedded in fine ash. The large fragments striking the unconsolidated ash may push down the layers beneath them, forming *bomb sags.*

Figure 10. Cinder cones at the summit of Mauna Kea. The astronomical observatory occupies the top of Summit Cone, which is 650 feet high; Goodrich Cone is just in front of it; and in the crater of the older cone in the foreground Lake Waiau lies at the edge of a lava flow. In the right background is Puu Makanaka, 4,000 feet in diameter at its base, with several smaller cinder cones between it and Summit Cone.

Figure 11. Diamond Head, a palagonite tuff cone belonging to the posterosional Honolulu Volcanic Series on Oahu. Note the inward dip (slope) of the beds at the upper right edge of the cone. The beds formerly arched over the rim of the cone, but the outward-dipping beds have been removed by erosion.

Ash and tuff cones usually are much broader in proportion to their height than are cinder or spatter cones, and their craters are very broad and saucer-shaped (fig. 14). Diamond Head (fig. 11) and Punchbowl, on Oahu, are probably the best-known examples. In Hawaii most tuff cones have formed during the very late stage activity of the volcanoes, described on later pages, but some formed during the earlier stages also. Kapoho cone, on Kilauea volcano near the east cape of the island of Hawaii, is a good example; and the nearby cone built during the eruption of 1960 also is very rich in ash, which is not yet cemented into tuff. Other examples include horseshoe-shaped islands such as Molokini, Lehua, and Kaula (fig. 12).

The difference in shape of ash and cinder cones (fig. 14) is the result of the difference in location within the volcanic apparatus of the principal part of the explosion. Cinder cones form from explosions that take place largely down within the feeding conduit, which acts like a gun barrel to shoot the ejecta more or less directly upward. The cinders fall back close around the vent building a narrow cone with a relatively small crater. In the formation of ash cones the explosion occurs mostly at the ground surface, and many of the fragments are thrown out away from the vent at a low angle to pile up at a little distance in a broad cone with a wide shallow crater. If the wind is blowing predominantly from one direction during the eruption, the ash, blown by the wind, piles up more abundantly on the downwind side of the vent, producing a cone which is higher on one side than on the other. This is very conspicuous in Diamond Head (fig. 11), which was built during a time of strong northeast trade wind; as a result more ash was piled up on the southwest side of the crater forming the high crest that towers above Waikiki.

That the eruptions responsible for building the tuff cones were the result of magma coming in contact with water close to the ground surface is shown by the fact that the cones on low ground close to the position of the shoreline at the time of the eruption* are tuff cones, whereas those on higher ground farther from the shore are always normal

*It will be pointed out in a later chapter that the position of the shoreline has shifted greatly many times during the recent history of the Hawaiian Islands.

cinder or spatter cones. As an example, Mt. Tantalus, high on the ridge just west of Manoa Valley in Honolulu, was formed by eruption of the same sort of magma that produced Punchbowl and Diamond Head. Despite the fact that the eruption was moderately explosive and produced a large amount of black glassy sand-size ash, Tantalus is a normal cinder cone.

The very rapid chilling of the droplets of magma during the building of a basaltic ash cone results in the formation of pale brown glass known as *sideromelane,* in contrast to the black glass, called *tachylite,* formed by the slower chilling during ordinary cinder-cone eruptions. The blackness of tachylite is due to the presence in it of innumerable tiny dispersed grains of the black mineral, magnetite, which

Figure 12. Kaula Island, a tuff cone that rises 550 feet above sea level, 22 miles southwest of Niihau. It surmounts a submerged shield volcano. Note the wave-cut bench about 6 feet above present sea level.

Figure 13. Well-bedded ash at the southwest edge of Kilauea caldera. The hammer handle is 15 inches long. The fine ash in the lower part of the picture is glassy, formed by the spray from lava fountains. The blocks near the upper part were thrown out by phreatic explosions.

are absent in sideromelane. All basaltic glass is very unstable, but sideromelane is even more unstable than tachylite, and the fragments are altered by water, descending through the ash, into a brown, waxy- or earthy-appearing substance called *palagonite.* The formation of palagonite is part of the process of cementing

the ash into tuff—referred to as palagonite tuff. Other minerals that help cement the tuff are calcite and zeolite (see chap. 4). Punchbowl and the cones near Koko Head are formed largely of palagonite tuff, which also can be seen readily in the highway cuts on the seaward side of Diamond Head near the Amelia Earhart monument.

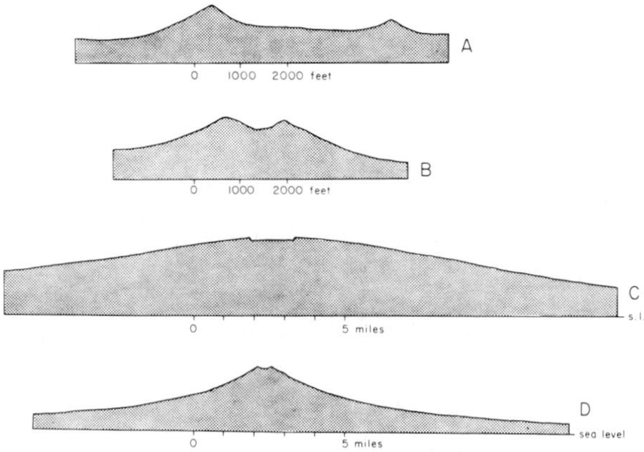

LAVA FLOWS

At the base of the lava fountains of Hawaiian-type eruptions liquid magma pours out to form lava flows. The lava flows of Hawaiian shield volcanoes are of two types. *Pahoehoe* is characterized by smooth, billowy, or ropy surfaces (figs. 15, 16); whereas *aa* has a very rough, spiny, or rubbly surface (figs. 17, 18). The two types intergrade, and occasionally it is difficult to classify a particular flow or part of a flow as definitely one or the other. For the most part, however, they are distinct. Pahoehoe is the more "primitive" of the two types. Most flows

Figure 14. Comparison of the profiles of volcanic cones. *A,* Ash cone or tuff cone; *B,* cinder cone; *C,* shield volcano; *D,* composite volcano.

22

emerge from the vent as pahoehoe, changing to aa as they advance downslope. The reverse change, from aa to pahoehoe, does not occur, although rarely pahoehoe will burrow under an aa flow and emerge at its lower margin giving the false appearance of a flow changing from aa to pahoehoe.

Chemical analyses of congealed fragments of both types of lava show that there is no consistent difference in composition between them. Whether one or the other forms depends on the physical state of the liquid lava and on the amount of stirring it undergoes. The more viscous the lava, the greater is its tendency to change to aa. Likewise, the more stirring it undergoes, the greater is the tendency for this change. The latter is illustrated by the fact that parts of the same pahoehoe flow continuing down a smooth slope in one area and tumbling over a cliff in another remains pahoehoe on the smooth slope but changes immediately to aa where it goes down the cliff. In some instances aa issues directly from the vent, apparently as the result of vigorous stirring of the liquid by unusually violent lava fountaining.

In pahoehoe the gas comes out of solution slowly. Gas bubbles are still expanding when the flow stops moving, and thus they retain fairly regular spheroidal shapes (fig. 7). In aa the gas is rapidly stirred out of solution and the bubbles are no longer expanding, so that continued movement of the flow twists many of the bubbles into very irregular shapes before the flow freezes. Identification of the type of flow from the shape of the vesicles is not entirely infallible, because sometimes some of the bubbles in pahoehoe become deformed, and sometimes some aa bubbles retain their regular shape, but it probably is accurate 7 or 8 times out of 10.

The great fluidity of Hawaiian lava results in rapid movement of the flows. In the main feeding channels, speeds as great as 35 miles an hour have been observed, but the flow as a whole advances much more slowly because the narrow feeding river, seldom more than 50 feet wide, must supply an advancing flow front that may be half a mile to a mile across. The first flows of the 1950 eruption traveled down the steep western slope of Mauna Loa from the vents to the sea at an average speed of 5.6 miles an hour, but most flows move much more slowly. Common rates in Hawaiian eruptions are a few tens of feet to 1,000 feet per hour. In comparison, lava flows in most other regions commonly advance only a few feet or tens of feet per day.

Flows often divide to pass on both sides of a

23

Figure 15. Edge of one pahoehoe flow unit resting on a slightly older pahoehoe surface, on the floor of Kilauea caldera. Note the toes at the edge of the younger unit. The surface is characteristic of pahoehoe.

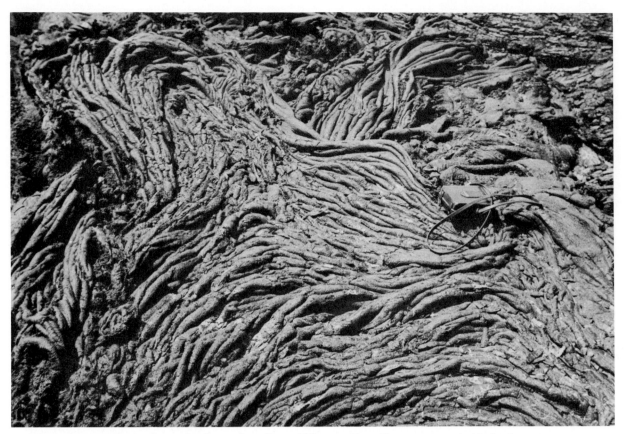

Figure 16. Ropy pahoehoe surface, floor of Kilauea caldera. The camera case is 6 inches long.

Figure 17. Front of an aa lava flow, 10 to 12 feet high, advancing through a field of young sugar cane near Kapoho, March 3, 1955. The light-colored parts are the incandescent, still partly liquid interior portion of the flow.

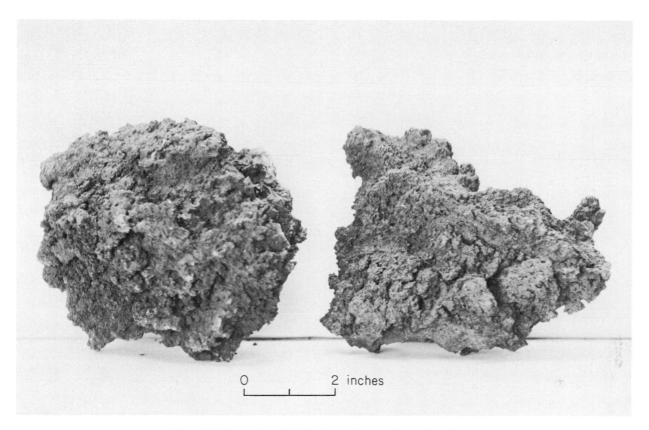

Figure 18. Fragments of clinker from the top of an aa lava flow.

Figure 19. Clinkery surface of a late-prehistoric aa lava flow of Mauna Loa. The hat rests on an accretionary lava ball formed by rolling of a fragment in the viscous, still-molten lava.

25

hill or slight elevation, reuniting on its downhill side but leaving it uncovered. These islands left in flows are known in Hawaii as *kipukas*. Although the original island was a high spot on the pre-flow surface, increase in thickness of the flow may ultimately result in the kipuka being lower than the surrounding lava surface.

## Characteristics of Aa Flows

Aa flows are fed by open rivers of lava located near the center line of the flow. From the river, some lava spreads laterally to moving margins of the flow, but a larger quantity moves downhill to feed the actively advancing flow front. The clinkery surface of an aa flow covers a massive, relatively dense interior. It is this massive central portion that is the really active portion while the flow is still alive. It consists of pasty liquid that actually does the flowing, while the clinkery top is merely carried along. At the front of the flow, clinker fragments from the top tumble down (fig. 17) and are buried by the advance of the lava over them. This results in a layer of clinker at the bottom of the flow also, but the bottom clinker layer is generally thinner and less continuous than the upper one. Locally, clinker may extend all the way through the flow. Many road cuts on all of the islands reveal the massive interior layers of aa flows between the clinkery top and bottom.

The fragments of clinker that cover aa flows are often fantastically jagged and spiny (fig. 18). The spines are so sharp that fragments sometimes cause painful cuts when handled, and the leather boots of a person crossing the flow are soon scarred with a multitude of deep gashes. In mapping the upper parts of Mauna Loa we found that boots would last only a week or two on this cruel terrain.

Roughly spherical balls, ranging in size from a few inches to 10 feet or more, are common on aa flows (fig. 19). Broken open, they often reveal a spiral structure. These *accretionary lava balls* are formed when a fragment of solidified lava, often part of the bank of the lava river, is rolled along and wrapped up in the viscous liquid, growing in size much like a snowball rolling downhill in sticky snow.

At the end of an eruption the fluid lava may partly drain out of the central river of an aa flow, leaving a distinct channel below the adjacent flow surface. Such channels can be seen from the highway on several of the historic lava flows in South Kona and Ka'u, on the island of Hawaii.

## Characteristics of Pahoehoe Flows

The feeding rivers of pahoehoe flows quickly crust over and develop more or less continuous roofs, and thenceforth the lava stream flows within a tunnel of its own making. This tunnel is known as a *lava tube* (fig. 20). Many smaller tubes branch off the main tubes and feed the

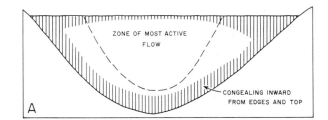

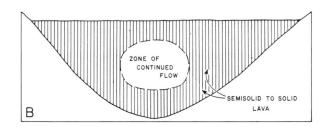

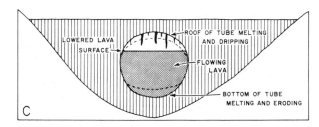

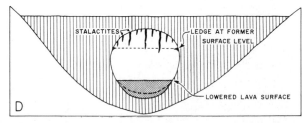

Figure 20. Diagram showing the stages in the formation of a lava tube. *A*, A lava flow (confined in a valley) develops a thin crust and starts to freeze inward from the edges, but the center remains fluid and continues to flow. *B*, The active movement of liquid becomes restricted to a more or less cylindrical, pipelike zone near the axis of the flow. *C*, The supply of liquid lava diminishes and the liquid no longer entirely fills the pipe. Burning gas above the liquid heats the roof of the pipe and causes it to melt and drip. *D*, Further diminution of the supply lowers the level of the surface of the liquid, which eventually congeals to form a flat floor in the tube.

26

front and margins of the flow. Commonly the minor tubes at the flow margin feed many small individual toes that protude one after another (fig. 15), still other toes then filling in between them, so that the flow advances across the ground much in the manner of a giant amoeba.

Natural or artificial cuts often reveal a large number of roughly oval cross sections of pahoehoe toes (fig. 21), fitted into each other at the edges, with a more or less concentric arrangement of rows of vesicles, and in some cases with open centers. These toes resemble superficially what have been called ellipsoidal, or pillow, lavas; but, whereas the latter can form only in water or very wet ground, the pahoehoe toes form either on dry land or in the water, and therefore lack the special environmental significance of pillow lavas.

Most of the toes that project along the edge of pahoehoe flows are solid, or nearly so. However, some are hollow. Since it is apparent that lava could not have drained away from inside to leave the open space, these hollow toes must be balloon-like blisters, inflated by the expansion of gas within them.

The ropy surface of pahoehoe (fig. 16) is the result of dragging and wrinkling of the solidifying but still plastic crust by the moving liquid

beneath. Because the moving of the liquid stream is fastest in the center, the ropy-looking wrinkles are curved, with their convexity pointing in the direction of flow. Often, however, the direction is only that of a local turbulence, and not of the flow as a whole. Geologists sometimes attempt to determine the direction of movement of ancient lava flows by studying the direction of curvature of the wrinkles, but it can only be done statistically, using a large number of separate exposures. Any one exposure may give the wrong answer.

A ropy surface is commonly said to be characteristic of pahoehoe, but actually on many pahoehoe flows it is present only over a small proportion of the total area. Far commoner is a surface that in detail is smooth or covered with tiny sharp protuberances (fig. 15), and in broader view is gently hummocky. Near the vents one often finds on the surface of pahoehoe flows blisters from a few inches to several feet across, which were formed by expanding gas pushing up the overlying plastic skin of the flow. Many of these can be seen on the surface of the 1959 lava in Kilauea Iki Crater, and the 1919 lava flow in Kilauea caldera.

On the surfaces of pahoehoe flows one occasionally finds nearly round to oval, dome-

Figure 21. Cross sections of pahoehoe toes in the sea cliff near Waialua, Molokai. Each toe shows the concentric structure that results from gradual inward consolidation while the lava continues to flow along its axis. The toe in the center is 2 feet across.

27

Figure 22. Tumulus on the surface of a pahoehoe flow on the floor of Kilauea caldera. The step-fault blocks below the Volcano House are visible in the background.

shaped hillocks, commonly 10 to 20 feet high and rarely as much as 40 feet, formed by heaving up of the crust of the flow (fig. 22). These are known as *tumuli* (singular: tumulus), because, viewed from a distance, they somewhat resemble ancient burial mounds. Most of them are on flows that were pooled in a crater, or otherwise confined. The buckled-up crust usually is cracked, and commonly a gaping crevasse crosses the top of the dome parallel to its length. The internal structure of the upper part of the lava flow is clearly revealed in the walls of these cracks. Lava from deeper in the flow often is squeezed out through the cracks and dribbles down the side of the tumulus, and sometimes spattering of the escaping lava builds a steep-sided heap of welded spatter several feet high, known as a *driblet spire* (fig. 23), or a larger cone known as a *hornito* (fig. 24). Sometimes the tumulus is hollow, but usually the molten lava has risen within and filled it.

The heaving up of the tumulus results largely from the moving crust of the flow being pushed against stationary crust farther downstream, much as a tablecloth rises in wrinkles when it is pushed together between one's hands. The rise may be aided, however, by hydrostatic pressure of the underlying liquid. Tumuli are well displayed on the floor of Kilauea caldera.

Tumuli grade into longer and narrower ridges, called *pressure ridges,* that are formed in much the same way. A small pressure ridge on the 1921 lava flow is crossed by the road on the floor of Kilauea caldera a quarter of a mile southwest of the parking area at Halemaumau. Pressure ridges as much as 50 feet high were formed on the 1940 lava flow in the summit caldera of Mauna Loa.

*Lava Tubes*

As the supply of lava diminishes during an eruption, the level of liquid in the feeding tube of a pahoehoe flow often drops, leaving an open space between it and the roof of the tube (fig. 20*D*). Pauses in the lowering of the liquid level may be marked by "shorelines"—nearly horizontal ridges on the sides of the tube formed by remnants of crust adhering to the walls. In the open space between the surface of the flowing stream and the roof of the tube, volcanic gas mixes with air and may burn, producing a temperature actually considerably higher than that of the molten lava. This may cause some remelting of the roof of the tube, which can sometimes be seen, through holes in the crust, dripping like the brick roof of an open-hearth steel furnace. The drips sometimes trickle down, one over another, and freeze to form stalactites hanging from the roof. Some

stalactites are quite regular, slender, tapering cones, like icicles; others are masses of solidified roundish drops resembling bunches of grapes or strings of beads. Still others are very slender, like long thin twigs, as much as 3 feet long; sometimes the lower ends of all of the slender stalactites in a tube are bent sharply over in the same direction—apparently because a strong current of hot gas or air passed through the tube, pushing the still-liquid drops to one side. Falling onto the floor of the tube, the drops solidify to form irregular humps, or stalagmites. One early observer claimed that the lava stalactites and stalagmites were formed by deposition from solution, like the better-

Figure 23. Driblet spire on the surface of the 1919 lava flow in Kilauea caldera. The camera case is 6 inches long.

29

Figure 24. Hornito, built by spattering lava at an opening on the surface of the 1920 lava flow of Kilauea. The vent was rootless, fed from the central portion of the flow, not from depth. Partly ropy pahoehoe surface is visible in the foreground.

known ones in limestone caves in many parts of the world, and he even claimed to have seen them forming in that manner. Indeed, drops of water often form on the tips of lava stalactites and fall onto the floor beneath, but these are only drops of rain water seeping through the roof of the tube. The stalactites have the same texture and structure as the associated lava, even to the presence of gas bubbles, and there is now no question that they formed by the freezing of drops of molten lava.

At the end of an eruption most of the lava may drain out of the main tubes, leaving open tunnels commonly from 10 to 20 feet in diameter, and sometimes as large as 50 feet. Recently, a bulldozer clearing land on the slope of Kilauea volcano broke through the roof of a lava tube and dropped about 30 feet. These open tubes often resemble railroad or highway tunnels, with arched roof and nearly flat floor. The floor is the final congealed surface of the lava stream in the tube. Just before the end of the eruption, the lava sometimes becomes more viscous, and the very last lava flowing through the tube may have changed to aa. We then find the apparently anomalous condition of an aa floor in a pahoehoe tube. Rarely, tubes may form in aa flows, but by far the greater number are in pahoehoe.

A well-known example of a lava tube in a pahoehoe flow is the Thurston Tube in Hawaii Volcanoes National Park. Another is the Kaumana Cave, in the 1881 lava flow on the slope above Hilo. The ancient Hawaiian burial caves in the cliff above Napoopoo Bay, in the Kona district of Hawaii, also are lava tubes.

*Tree Molds*

One of the results of the great fluidity of Hawaiian lavas is the formation of tree molds, found occasionally in aa, but much more commonly in pahoehoe. The fluid lava engulfs the trunk of the tree and freezes against it, often preserving the form of the trunk and branches in considerable detail. The sudden very high temperature often chars the surface of the wood, leaving a network of cracks in the shrinking charcoal; the lava then fills the cracks and preserves a perfect mold of the checked charcoal surface. An extremly fine example of molding of vegetation was found by the late Professor H. S. Palmer, of the University of Hawaii, in Kilauea Iki Crater. There the 1868 lava flow had buried tree ferns, and on the bottom of the flow the forms of the fern fronds were so perfectly preserved that, where the fronds had been upside down, botanists were able to recognize the kind of fern from the pattern of the tiny spores. In some instances, as at the Tree Molds near the north edge of Kilauea caldera, the molds of standing trunks resemble wells dug in the lava. At other places the trees had been knocked over, and the tubular molds are more or less horizontal.

Quite often, as the eruption progresses, the liquid lava surrounding upright trees drains on

downslope, leaving the shells of lava that chilled against the trees standing as pillars above the adjacent lava surface (fig. 25). Excellent examples can be seen in the Lava Trees Park near Pahoa, in the eastern part of the island of Hawaii (fig. 26). Tree molds as much as 20 feet high formed on the surface of the 1962 lava in Aloi Crater, in Hawaii Volcanoes National Park, but were buried in the eruption of December 1965.

### Block Lava Flows

Aa flows grade into another type known as block lava flows, which have much the same structure as that of aa flows, but the upper and lower fragmental parts of the flow, instead of

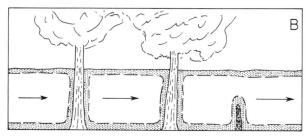
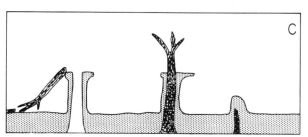

Figure 25. Diagram illustrating the formation of tree molds and "lava trees." A forest (A) is invaded by a lava flow (in B), the lava surrounding the trees and moving in the direction indicated by the arrows. The lava is chilled against the tree trunks and the ground, and at the top by contact with the air, and forms a solid crust on the top and bottom and around the trees. The fluid lava in the center continues to flow and, as the supply diminishes toward the end of the eruption, the remaining liquid portion drains away downslope (or back into the vents), leaving the parts solidified around the trees standing as columns as the surrounding flow surface sinks (C). Some of the tree molds still contain the charred trunks of trees, but in others the wood has burned away and the charcoal has disappeared, leaving the mold hollow.

Figure 26. "Lava tree"—a cylinder of lava formed by freezing of lava around the trunk of a tree—near Lava Trees State Park on the east rift zone of Kilauea. The portion of the lava flow above present ground level later drained away. The light-colored rectangle on the base of the tree is a notebook 8 inches long.

31

Figure 27. Columnar jointing in a late prehistoric lava flow of Mauna Loa that followed the valley of the Wailuku River. The picture was taken at the Boiling Pots, a succession of big potholes along the Wailuku River.

consisting of very irregular jagged pieces of clinker, are made up of blocks with relatively smooth sides. Block lava is formed by more viscous magma than that which forms aa, and consequently the flow tends to be thicker and to move more slowly. There is very little true block lava in Hawaii, but many of the flows erupted in the late stages of activity (see chap. 7) are gradational toward block lava. Excellent exposures are found in the highway cuts through the "hawaiite" (see chap. 4) flows on the flanks of Mauna Kea.

*Joints in Lava Flows*

All types of lava flows are usually broken by cracks into innumerable polygonal blocks. The cracks are known as *joints.* In other parts of the world some of the joints are caused by bending or twisting of the rock during mountain building or similar deforming processes but everywhere most of the joints in lava flows result from the strains that arise in the rock during cooling. As the rock cools, it shrinks. If the shrinkage were completely uniform throughout, and if the tensile strength of the rock were sufficient to overcome the frictional

resistance to movement over the underlying surface, the entire sheet would draw back toward a common center without fracturing. But neither of these conditions is fulfilled. Instead, the rock literally pulls itself apart. The principal cracks develop approximately perpendicular to the cooling surface—in the case of a lava flow, the top and bottom sufaces of the flow. The cracking is analogous to that of mud in a dried-up puddle. In the mud the cracks open at right angles to the drying surface, and the intersection of the cracks ideally forms short columns that tend to be six sided.

In lava flows, also, there is a tendency to form six-sided columns, although the attainment of the regular hexagonal form is rare, and usually five- or seven-sided columns are as common as six-sided ones. In some flows, such as those at the Devil's Postpile in California and the Giant's Causeway in Ireland, and many flows on the Columbia River Plains of Washington and Oregon, columnar jointing of this sort is very well developed. In Hawaii it is rare, but it can be seen at the Boiling Pots on the Wailuku River above Hilo (fig. 27), in the bluff behind the sugar-loading dock at Nawiliwili on

Kauai, in some late lava flows of Haleakala on Maui, and at a few other places. Many dikes show columnar jointing perpendicular to their walls (see chap. 5). In Hawaii the columns in lava flows generally are several feet long and from several inches to a foot or two thick, and are subdivided into short segments by cross joints.

On the island of Hawaii, lava that accumulated in craters to form lava ponds or lakes has developed columnar joints on cooling. However, the columns generally are far less regular than those mentioned in the last paragraph, and of larger size, commonly 10 to 15 feet in diameter. Instead of showing an approach to hexagonal cross sections, they are usually more or less rectangular. The tops of columns of this sort are clearly visible to an observer looking down onto the floor of Keanakakoi Crater at Kilauea, where they are outlined by rows of lighter-colored pumice, and in Kilauea Iki Crater, where they are outlined by white salts and opal deposited by gases rising through the joint cracks.

In most lava flows the joints formed normal (perpendicular) to the surface by cooling are so irregular and so broken by joints resulting from contraction in other directions, and from continued movement of the flow as it consolidated, that they produce little or no columnar appearance. The lava is simply broken into many irregular or rectangular blocks. The complex of irregular joints is important in governing the way in which the rock breaks during quarry operations, and in allowing the passage of ground water through the rock (see chap. 15).

## SHIELD VOLCANOES

The lava flows of Mauna Loa and Kilauea are so fluid that they spread out to great distances from their vents. The flow of 1859, for example, extends for 32 miles from its vent on the upper slope of Mauna Loa into the sea, and no one knows how far beneath sea level. (It poured into the ocean for many weeks.) It covers an area above sea level of 33 square miles. The flows are thin, averaging about 12 to 15 feet thick (fig. 28), and only rarely reaching as much as 50 feet. These far-spreading thin flows build broadly rounded, dome-shaped mountains (fig. 79), known as *shield volcanoes* because of a fancied resemblance in profile (fig. 14C) to that of the round shields of early Germanic warriors. Very little explosive action is involved in their building, and fragmental material resulting from explosion probably forms far less than 1 percent of the part of the shield above sea level. (We shall see in chapter 7 that a considerable amount of explosive material may be present in the part of the mountain built by eruptions a few hundred feet below the sea surface.)

Mauna Loa and Kilauea are typical shield volcanoes. They, and their lava flows, are representative of by far the greatest part—probably more than 95 percent—of all the great Hawaiian mountains. Both pahoehoe and aa flows are present. Pahoehoe is commoner on the upper slopes of the mountain, near the vents from which the flows issued, and aa is commoner on the lower slopes, because as the lava moved downhill it cooled and lost gas, thus becoming more viscous and changing to aa.

### Rift Zones

Eruptions that build shield volcanoes are not confined to the summit of the mountain but occur also along fissures extending across the summit and far down the flanks (fig. 29). The main vent of the 1942 eruption of Mauna Loa was 11 miles from the summit, and the 1955 eruption of Kilauea took place more than 20 miles from the summit. The cracks that provide passageway for the rising magma are not isolated, but are parts of extensive zones of fissures known as *rift zones.* Typically, there are two principal rift zones extending outward from the summit of the mountain, generally not forming a straight line, but lying at an obtuse angle to each other (fig. 30). The main rift zones of Mauna Loa form an angle of 150°. The external angle between the principal rift zones may be bisected, as it is on Haleakala volcano on Maui, by a minor rift zone.

At the surface the rift zones are marked by

Figure 28. Makapuu Head, Oahu, showing the many thin lava flows that built the Koolau shield volcano. The cliff is a sea cliff, as also is the part of the Pali visible in the upper left.

many open cracks, from some of which lava has poured out, by collapse craters, and by spatter cones and rarely cinder cones and long low spatter ramparts. The dominant spatter cones of the shield-building stage are replaced at later stages by cinder cones (fig. 31). Below the surface the rift zone contains hundreds of *dikes,* each formed by lava congealing in one of the fissures that brought it to the surface (see chap. 5).

Most of the ground movement along the rift zones is a simple distension of the surface. The jagged edges of the cracks indicate that they have simply been pulled open by movement in a direction essentially perpendicular to the crack. Occasionally, however, fault movement results in a lateral shifting of one side past the other in a direction parallel to the ground

surface. In 1938, and again in 1963, faults extending across the Chain of Craters Road, in Hawaii Volcanoes National Park, broke the road and shifted the northern side of the fault eastward several feet past the southern side, resulting in a sudden jog of as much as 4 feet on the edge of the pavement.

We really do not know much as yet about the internal structure of Hawaiian volcanoes. Even the deepest exposures in the oldest islands allow us to see only the upper few thousand feet of the shield. It appears most probable, however, that 2 or 3 miles down beneath the active shields, there are broad lens-shaped bodies of magma, extending outward into the rift zones (fig. 6). New magma rising from depth can inflate this reservoir, causing the volcano above it to swell up. This tumescence causes the top of the volcano to rise and the sides to tilt outward at a slightly steeper angle. The rise of the summit can be measured by ordinary surveying methods; the tilting of the flanks is measured by means of instruments called tiltmeters. Inflation of the volcano indicates that it is potentially capable of erupting if fissures break through to the surface. When this happens, eruption may relieve the pressure in the magma reservoir and permit the volcano to shrink.

*Calderas and Pit Craters*

The presence of the body of liquid magma below it seems to create a region of weakness in the summit of the shield. Perhaps the heavier rocks above simply sink into the underlying liquid, or perhaps eruptions on the flank of the volcano drain away part of the liquid leaving a potential void that is filled by sinking of the overlying mountain top. At any rate, we find that usually, if not always, the top of the shield

Figure 29. Vertical air photograph of part of the northeast rift zone of Mauna Loa. Puu Ulaula (Red Hill) is the small, light-gray area just to the left of center. The black lava flows extending northeastward (diagonally upward) and southeastward from the area just to the left of Puu Ulaula are the lava flows of 1880. The black flow in the upper right-hand part of the picture is the flow of 1881 that entered Hilo. Note how the flows radiate from the rift zone.

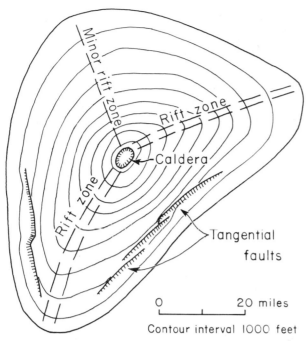

Figure 30. Plan of a typical shield volcano, showing the radiating rift zones, caldera, and tangential faults.

has dropped along a series of steep fractures (faults), producing a sunken-in crater known as a *caldera* (fig. 30). The calderas are more or less oval in outline, are bounded by steep cliffs, and have nearly flat floors formed by lava flows within the depression. Mokuaweoweo, the caldera at the top of Mauna Loa (fig. 32), is 3 miles long, 1.5 miles wide, and 600 feet deep at its western side. Kilauea caldera is 2 by 2.5 miles across. In 1823 it was 900 feet deep, but later activity has partly filled it until its floor is now only 400 feet below the high point on its western rim, at Uwekahuna. The caldera of Kauai volcano was 12 miles wide before it was filled by later flows (see chap. 7).

The caldera walls are fractures along which the rocks beneath the caldera have sunk down in relation to those of the surrounding part of the volcano. In the walls are exposed the broken edges of lava flows that must have been poured out of vents that formerly existed at

Figure 31. Looking seaward along the southwest rift zone of Haleakala. Kahoolawe Island, in the distance, lies directly on the line of the Haleakala rift. Cinder cones on the rift belong to both the Kula and the Hana Volcanic Series. In the foreground are astronomical observatories.

higher levels, where there is now nothing but air. Fractures along which the rocks on one side have moved downwards (or upwards, or sideways) in relation to those on the other side are called *faults,* and the cliffs produced by faulting of the ground surface are *fault scarps.* Caldera walls are fault scarps. At some places a caldera floor has dropped down on a single fault, but at other places there are several parallel faults and the blocks between them have dropped to successively lower levels, like the treads of a flight of stairs. "Step faulting" of this sort can be seen on the western side of Kilauea caldera below the Volcano Observatory (fig. 78), and on the northeastern side below the Volcano House (fig. 22).

Along the rift zones are found smaller craters similar in form and origin to the calderas (fig. 33). These craters, formed by collapse, are called *pit craters.* We believe they are formed as follows: A body of magma works its way upward, eating away the solid rock, until its top is close to the surface. Then in some way, possibly by eruption lower on the flank of the volcano, the magma is partly drained away, removing the support from beneath the overlying rocks and allowing them to sink down. There are several pit craters near Mokuaweoweo caldera, and a whole row of them lies along the east rift zone of Kilauea, easily accessible by means of the Chain of Craters Road (fig. 62). The largest of these, Makaopuhi, is a mile-long, double crater (figs. 34, 35). An older crater, which forms the east

Figure 32. The summit of Mauna Loa, from the southwest. In the foreground are the three pit craters Lua Hou, Lua Hohonu, and South Pit, and behind them is Mokuaweoweo caldera. Mauna Kea is dimly visible in the distance.

37

Figure 33. View westward along the Chain of Craters on the east rift zone of Kilauea volcano. In the foreground are Makaopuhi Crater *(left)* and the small shield volcano Kane Nui o Hamo, with its summit crater. In the middle distance a fume cloud is rising from the vents of the 1969 eruption, and lava flows of that eruption cover the area around the vents and extend to the left across the Chain of Craters Road. The outline of filled Alae Crater is barely visible. Lava flows between Makaopuhi and Alae craters were erupted in February 1969.

end of Makaopuhi, was partly filled by a pond of liquid lava that solidified to form the flat crater floor, much as did the pool of lava that collected in Kilauea Iki Crater during the 1959 eruption. Later, another pit crater collapsed across the western end of the older one, exposing a cross section of the congealed lava pond. This younger pit crater, at the western end of Makaopuhi, was 1,000 feet deep before it was partly filled by the eruption of March 1965.

Three of the pit craters near the summit of Mauna Loa (fig. 32) have formed since the top of the mountain was mapped by Commodore Wilkes, of the U. S. Exploring Expedition, in

1840; and about 1921 a small one (Devil's Throat) appeared on the east rift zone of Kilauea. No one has actually seen one form, although we came very close to seeing the actual collapse of the surface to form a pit crater during the Kilauea eruption of 1955 (see chap. 3).

*Grabens*

Pit craters are not the only collapse features found along a rift zone. Commonly the crest of the broad arch built by eruptions along the rift zone has sunk down between parallel faults, forming a long narrow trench known as a *graben.* The inflation of the rift zone by

38

intrusion of magma at a shallow depth causes stretching of the rocks above it, allowing a wedge-shaped "keystone" at the crest of the arch to sink down a little. The mechanism was illustrated just before the 1955 eruption, which was preceded by a swelling of the east rift zone of Kilauea with uplifting of the ground surface along its crest by several feet, and by sinking of a series of grabens by amounts ranging up to 5 feet. Later measurements indicated that the stretching of the ground surface across the rift zone had been as much as several feet. A graben half a mile wide just north of the later-destroyed village of Kapoho sank more than 5 feet just preceding the 1960 eruption, and the same graben sank as much as 14 feet during the abortive eruption of 1924.

The grabens range from a few feet to 250 feet in depth, from a few tens of feet to half a mile in width, and in length up to 3 miles. In many instances, however, their present depth represents only a part of the total sinking because the graben troughs have been flooded and partly filled by later lava flows.

In some instances sinking of the surface along the rift zone has produced a fault on only one side, the other side merely bending down without breaking. The largest fault of this type is the one that runs nearly north-south just west of South Point, on Hawaii, and has produced the spectacular scarp of the Kahuku Pali, in some places 600 feet high (fig. 36). The scarp is 10 miles long on land, and can be traced southward for 18 miles beneath the ocean.

*Tangential Faults*

Faults of another class lie along, and roughly tangential to, the flanks of the volcanoes (fig. 30). On these faults the lower part of the

Figure 34. Makaopuhi Crater, seen from the west before the recent eruptions. The flat floor formed by the congealed lava lake in the older, eastern part of the crater was truncated by the collapse of the younger, western part of the crater in the foreground. The end of the 1922 lava flow, which came from a vent part way up the western wall of the crater, can be seen at the bottom of the newer part of the crater.

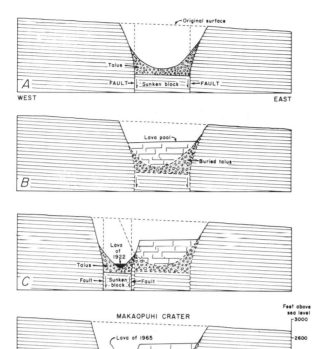

Figure 35. Diagram showing the manner of formation of Makaopuhi, a double pit crater. *A,* A subcircular fault block sinks, leaving a crater at the surface. (The position and attitude of the faults is hypothetical.) The upper walls of the crater collapse to form taluses (piles of rock fragments) that hide the lower walls. *B,* Lava pouring into the crater collects in a deep pool, the surface of which solidifies to form a nearly flat floor. *C,* A second block sinks, making a second crater that cuts across the western edge of the first one. The pool of lava in the bottom of the second crater is from a small eruption in 1922. *D,* A much larger eruption (in 1965) forms a pool 290 feet deep in the second crater. Note the slump scarps at the edge of the new lava floor, formed as lava in the central part of the crater drains back into underlying vents.

mountain slope has been displaced downward in relation to the upper part, forming fault scarps that face toward the ocean. An example is the Kaoiki fault system, just north of the highway between Kilauea caldera and the town of Pahala, that forms a series of steep escarpments at the base of the slope of Mauna Loa. Near the coast south of Kilauea caldera, a similar group of faults forms the Hilina Pali, where vertical displacement has amounted to more than 2,000 feet. The Hilina fault scarps have been largely mantled by later flows of lava pouring down over them. Other faults of the tangential group include the Kaholo Pali near Honaunau in Kona (fig. 43), and the Keala-

kekua fault, which extends inland from the north side of Kealakekua Bay (fig. 194). The Hilina and Kealakekua faults are still active, causing many small earthquakes, and occasional larger ones such as the destructive Kona earthquake of 1951.

The cause of the tangential faults is open to debate. Most geologists have attributed them to a seaward sliding of the lower slope of the mountain under the influence of gravity—a sort of giant landslide. It appears to us, however, that the slope of the mountain (only about 6° both above and below sea level) is too gentle to bring about such instability. Macdonald (1956) has suggested instead that the portion of the volcano inland from the fault has been pushed upward by pressure of underlying magma in the same way that the mountain top is pushed up several feet preceding many eruptions. Earthquakes appear to originate on these faults when the volcano is shrinking as well as when it is swelling, and movements of the inland portion probably are downward as well as upward, although the upward movement has been predominant. A tangential fault may be due simply to the tendency of the lighter magma-filled core of the volcano to rise in relation to the heavier materials around it. No confirmation of actual uplift of the central portion of the mountain on the tangential faults has yet been possible in Hawaii. However, it unquestionably has happened at Mt. Etna, on Sicily. Etna, which in many respects is a twin to Mauna Kea, also lies within a series of faults along which the side toward the mountain has moved relatively upward. But there it can be shown that the movement was an actual absolute elevation of the mountainward side of the fault, not a downward movement of the oceanward side, since marls containing marine fossils of Pleistocene age have been lifted more than 2,000 feet above sea level.

THE HALEMAUMAU LAVA LAKE

Most of the time between 1823 and 1924 Halemaumau, the inner crater in Kilauea caldera, contained a lake of molten lava (fig. 37). The lake was only about 50 feet deep, with a semisolid plastic floor (fig. 38). Liquid lava rose from depth through fissures, moved across the lake, and sank again at other places, thus setting up a convectional circulation in the lake. Lava fountains played above the vents where the lava rose. Black crusts quickly formed on the red incandescent liquid lava. Where the lava sank, the crust was broken up and fragments of it were dragged down into the hotter lava below. There the fragments were partly remelted, and entrapped gas was released to bubble out at the surface with much spattering, producing so-called secondary fountains, which often moved across the lake surface with the circulation of the lava. The plastic floor appeared to have been formed by the accumulation of partly melted, sunken fragments of crust which had lost most of their gas and were being transformed into aa. In

contrast, the lava of the lake surface was pahoehoe.

Portions of the floor of the lake protruded through the liquid as islands (fig. 39). The floor was still mobile and frequently shifted about, though less actively than the more liquid lake magma above it. This shifting caused the islands to change position. Old accounts of "floating islands" refer to these crags of lake-bottom material that were not really floating at all.

Dr. Thomas A. Jaggar, founder and long the Director of the Hawaiian Volcano Observatory, recognized that the lava lake consisted of these two portions. The upper, very fluid part he called *pyromagma.* In it, gases were still actively coming out of solution, bubbles were still expanding, and it solidified to form crusts of pahoehoe. The lower part, though still mobile,

Figure 36. The Kahuku Pali, 2.5 miles northwest of South Point, Hawaii. The pali is a fault scarp in the southwest rift zone of Mauna Loa. The maximum height of the scarp, near the shoreline, is about 600 feet. The cliff extends inland about 10 miles, with gradually decreasing height, and can be traced out to sea for 18 miles. Lava flows of the Kahuku Volcanic Series are exposed in the face of the cliff, and are capped by a layer of yellow ash. The black lava at the base of the cliff is part of the 1868 lava flow of Mauna Loa.

*41*

Figure 37. Lava lake held in a lava ring (a ring-shaped levee) built by spattering of secondary fountains at the edge of the lake and by repeated small overflows such as those visible in the picture; Halemaumau, 1894.

had lost most of its gas and was much more viscous, and many of the projecting crags had the spiny surfaces characteristic of aa. This lower part, obviously closely akin to aa, he called *epimagma.*

At times the level of the lake was very high, and occasionally, as in 1921, it actually overflowed its banks and spilled over onto the surrounding caldera floor. At other times it dropped to levels several hundred feet below the rim, sometimes dropping as rapidly as several hundred feet within a few hours. At such times, the pyromagma drained away first, followed more slowly by the epimagma of the lake bottom. During rapid rises also, the pyromagma flowed in much more rapidly than the epimagma, and for a time the entire lake surface would be unbroken pyromagma. Only later did crags of epimagma make their appearance.

In April 1924, the lava lake suddenly drained away, and though molten lava reappeared briefly in the bottom of the crater in July, the eruption was short. The old "permanent" lava lake was gone. In June 1952, a similar lake developed in Halemaumau, but it lasted only until November of the same year. In November 1967, another eruption in Hale-

42

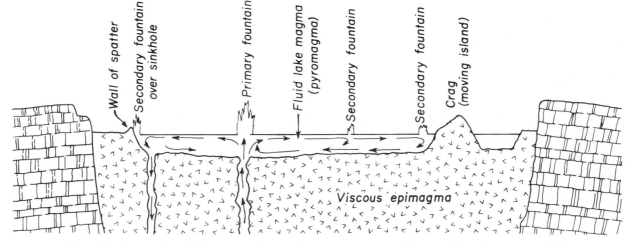

Figure 38. Cross section of Halemaumau Crater, showing the structure of the lava lake as it was about 1919.

maumau established a lava lake which, after a series of 28 short, alternately active and quiescent phases, eventually settled into a long period of continuous activity, ending in July 1968.

A lava lake was present in Kilauea Iki Crater for a month in 1959, and others have existed briefly in Aloi, Alae, and Makaopuhi craters during the last few years.

PHREATIC ERUPTIONS

The draining away of the lava lake in 1924 was followed by one of the rare explosive eruptions of Kilauea (see p. 80, chap. 3). Hundreds of earthquakes along the east rift zone of the volcano, with cracking and faulting of the ground near Kapoho, near the east cape of the island, suggested that the lava had been drained into the rift zone. The whole summit area of the volcano sank several feet, accompanied by wholesale collapse of the walls of Halemaumau that widened that crater from one-quarter to one-half mile in diameter. Much cracking oc-curred in the surrounding rocks. This cracking allowed fresh water (ground water) in the rocks to flow into the hot feeding channels of the volcano, resulting in rapid generation of steam and violent steam explosions. Great clouds of dust rose 4 miles into the air (fig. 40), and fragments of rock weighing as much as 14 tons were thrown out. The debris can be seen scattered over the surface around Halemaumau today. All of the material ejected was simply broken up old rock, even some red-hot fragments which were merely pieces torn from intrusive bodies (sills: see chap. 5) in the rocks of the crater walls, that had not yet cooled. None of the material thrown out was new magma.

Explosions of this sort, caused by ground water coming in contact with magma or with the hot conduits of the volcano, are called *phreatic explosions.* One, even more violent than that of 1924, occurred at Kilauea caldera about 1790. In that eruption pockets of liquid lava appear to have existed in the crater walls, and this magma, draining back into the crater,

Figure 39. Lava lake and islands, Halemaumau, March 30, 1917.

Figure 40. Phreatic explosion cloud of ash-laden steam rising from Halemaumau Crater, seen from the north rim of Kilauea caldera, May 13, 1924.

was thrown out in the explosions along with the much more abundant solid fragments. Some of it formed roundish "bombs," many of which had cores of solid fragments.

Other phreatic explosions occurred in prehistoric time at the summit of Mauna Loa, and at Alae and Puulena craters on the east rift zone of Kilauea. During the 1955 eruption at least one, and probably two, small ones occurred in east Puna.

## LITTORAL EXPLOSIONS

Where aa lava flows enter the sea, water may gain access to the hot central part of the flow, resulting in steam explosions (fig. 41). The out-rushing steam may carry with it a cloud of drops of liquid lava, which on contact with the water or air chill into bits of black or brown glass. Sand-size particles of this sort may be washed alongshore to accumulate at favorable localities to form black sand beaches, such as those of Kalapana (Kaimu) and Punaluu, on the south shore of Hawaii. These littoral explosions, so called because they occur at the beach, usually are produced by aa flows, because the fragmental surface of the flow allows the water to penetrate quite readily to the hottest central portion of the flow. Rarely they may also be produced by pahoehoe flows. Particularly in explosions resulting from pahoehoe, the swarm of liquid droplets in the cloud is accompanied by a varying proportion of fragments of partly or wholly solidified lava derived from the outer portions of the flow.

The littoral ash, together with varying amounts of lapilli and bombs, may pile up on the surface of the lava flow at the side of the main channel to form a hill (fig. 43), known as a *littoral cone.* Often these hills are double, one part forming on each side of the lava river where it enters the sea. Three littoral cones were formed on the 1840 lava flow of Kilauea where it entered the ocean at the site of the village of Nanawale (destroyed by the flow). By the time of the visit of the U. S. Exploring Expedition a few months later, one of the hills had already been almost entirely eroded away by the waves. Puu Hou (New Hill), just west of South Point, Hawaii, is a littoral cone 270 feet high on the 1868 lava flow of Mauna Loa (fig. 42).

## FUMAROLES

Vents from which volcanic gases issue unaccompanied by liquid or solid matter are called *fumaroles,* and fumaroles at which the gases are predominantly of sulfur are *solfataras.* In other parts of the world the gases may be extremely hot (up to 500°C. or more) and commonly include important proportions of hydrofluoric and hydrochloric acid. In contrast, the fumaroles of Hawaii never much exceed the boiling temperature of water at whatever altitude the vent is located, and hydrofluoric and hydrochloric acid are nearly or completely absent (which is true of the eruption gases also). Solfataras often develop on new lava flows, particularly along the edges of the lava rivers, but since the gases are derived only from the lava flow itself, these rootless fumaroles persist only for a few days or weeks. Longer-lived solfataras develop at vents at the end of eruptions, but most of these also die out within

44

a few months. At only a few places on the Hawaiian volcanoes do persistent solfataras exist.

On Kilauea the largest area of solfataras is Sulphur Bank, at the northern edge of the caldera just west of the National Park Headquarters, in a shallow graben just within one of the outer boundary faults of the caldera. The gases issue along the faults at both sides of the graben. As at all Hawaiian fumaroles, the gases are overwhelmingly water—generally more than 95 percent. Other gases include air, sulfur dioxide, hydrogen sulfide, sulfur vapor, carbon dioxide, and just a trace of hydrochloric acid. Although the rotten-egg odor of hydrogen sulfide is conspicuous, this gas is very subordinate in amount to sulfur dioxide. It is probably more abundant at depth, and its oxidation near the surface leads to the deposition of native sulfur, in beautiful clusters of bright yellow, bladed crystals sometimes more than an inch long.

An analysis of the gases at Sulphur Bank by E. T. Allen (1922) showed them to consist of: steam, 96.2 percent; fixed gases, probably largely air and carbon dioxide, 3.7 percent; sulfur dioxide, 0.096 percent; sulfur vapor, 0.004 percent; and a trace of hydrochloric acid. Later periodic collections and analyses of the gases were made by S. S. Ballard and J. H. Payne from 1937 to 1940 (Payne and Ballard, 1940; Ballard and Payne, 1940), and by J. J. Naughton and his associates at the University of Hawaii from 1952 to 1968. The results of the latter studies have not yet been published. The amounts and proportions of the various volcanic gases have been found to vary greatly, as does also the amount of contamination by air.

Changes in composition or temperature of fumarole gases in other parts of the world sometimes herald eruptions of the associated volcanoes. A very small amount of hydrogen sulfide found by Ballard and Payne in the Sulphur Bank gases during March and April 1940, was thought possibly to be related in

Figure 41. The steaming toe of a lava flow in the ocean, April 3, 1955. The flow built a peninsula that extended about 800 feet beyond the former shoreline. In front of the main steam column a black jet of ash-laden steam is being thrown up by a littoral explosion.

45

Figure 42. Puu Hou, a littoral cone formed where the lava flow of 1868 from Mauna Loa entered the ocean. The flow forms the dark surface around the cone. Another segment of the cone is visible on the far left. In the background is the Kahuku Pali, the scarp formed by the Kahuku fault.

some way to the eruption of Mauna Loa on April 7 of that year, although how the activity of one volcano could affect the fumaroles of the other was not clear. The more recent work by Naughton has shown no relation between the changes in composition of the fumarole gases and eruptions of Kilauea volcano. It is unlikely that there is any close connection between the shallow magma chamber of Kilauea and the fumaroles at Sulphur Bank. Most probably the gases of the latter come from a cooling and crystallizing intrusive body at some unknown, but shallow, depth. There is some evidence that the amount of gases has decreased somewhat since the beginning of the century. It is also reported that at one time, late in the 19th century, the Sulphur Bank gases became so hot that the Volcano House steam baths, which then got their steam from Sulphur Bank, could not be used. The increase in temperature probably was caused by a period of drought, when much less surface water entered the fumarole system.

Alteration of the rocks by acid gases at Sulphur Bank is discussed in chapter 4.

Another persistent solfatara is located on one of the boundary faults of Kilauea caldera about a mile east of Halemaumau. One or more solfataras usually exist also in Halemaumau, but their location changes with each eruption. The most persistent one in recent years has been on the 1954 eruption fissure at the northeast wall of the crater. At times, between the 1961 and 1967 eruptions, its output of sulfur dioxide was so voluminous that conditions at the observation stations on the crater rim became unbearable for more than a few minutes. Still another area of persistent solfataric activity since the 1962 eruption has been in and near Aloi Crater.

Many steam vents are located in the Steaming Flats between the inner and outer caldera-boundary faults at the north side of Kilauea caldera, on the caldera floor, and along the east rift zone. In the Steaming Flats, the steam generally is unaccompanied by other gases, although small amounts of carbon dioxide in excess of that normally present in the air are sometimes detected. The steam results from the heating of descending rain water by hot

Figure 43. The southernmost lava flow of 1950 at the Kona coast. The channel of the main feeding river of the flow extends from the sea cliff diagonally upward to the left. Just to the right of it, at the top of the sea cliff, is a small littoral cone. The heavily lava-mantled fault scarp of the Kaholo Pali , one of the tangential faults of Mauna Loa, lies at the inner edge of the coastal flat.

intrusive bodies beneath, probably within a few thousand feet of the surface. The appearance of the steam clouds varies greatly with the weather. On hot dry days they are nearly invisible, but on cold mornings with high humidity the steam condenses in spectacular white plumes, and sometimes in a cloud that rolls over and down the caldera cliff like a great waterfall. The steam cloud can be made more conspicuous also by holding a lighted cigarette or a burning newspaper close to the vent, or racing an automobile engine just to windward. Imaginative tour guides sometimes explain to their charges that the increase in visible steam is the response of Pele, goddess of the volcano, to appropriate incantations which they can recite; but less romantic scientists attribute it to condensation of the steam (which is invisible so long as it remains a gas) on minute smoke particles.

A group of small but persistent solfataras is located near the center of Mokuaweoweo caldera, at the summit of Mauna Loa, in the vicinity of the now-buried cone of the 1914 eruption and the eruptive fissures of the 1940 and 1949 eruptions. During recent years, minor solfataric activity has usually been present also at the cones of the 1949 eruption on the southwestern floor of the caldera. The largest solfataras on Mauna Loa are at Sulphur Cone, at 11,370 feet altitude on the southwest rift zone. There, large amounts of sulfur have been deposited in openings in the rocks and on the surface, but it should be pointed out that the total amount is far too small to constitute a deposit of commercial value. At some time since 1950 the temperature of the gases issuing at Sulphur Cone became high enough to melt some of the sulfur, which trickled down the side of the cone as a miniature "lava flow," a few inches thick and 10 to 15 feet long. Similar sulfur flows are known at a few volcanoes in other parts of the world.

47

## Suggested Additional Reading

Eaton and Murata, 1960; Fisher, 1968; Macdonald, 1953, 1956, 1961, 1967; Moore and Ault, 1965; Peck and Minakami, 1968; Stearns and Macdonald, 1946, pp. 24–33, 96–98, 129–131; Wentworth and Macdonald, 1953

*Kilauea Iki,*
*November 1959*

# CHAPTER THREE

# Historic Eruptions

THE RECORDED HISTORY OF VOLCANIC ACTIVITY IN THE Hawaiian Islands begins with the arrival of the Christian missionaries. In 1823, Rev. William Ellis, an English missionary working with the newly arrived New England Mission, visited Kilauea Crater in the course of a trip around the island of Hawaii. Not only did Ellis leave us the first description of Kilauea Crater, but he also recorded information obtained from Hawaiians about eruptions that had taken place in the recent past. Gas was still rising from the vents of the lava flow that had been erupted from the southwest rift of Kilauea only a few weeks before, and in Puna people told him of two flows that had poured from the east rift zone within the memory of persons then still living. These events, though they had taken place only a relatively short time before Ellis' visit, cannot be dated precisely. Counting back through the years as best they could, Ellis and his informants placed one of the flows about 1790, and the other about 1750.

Of events before that time we have only traditions and legends. Some of the latter obviously are based on actual events, but the dates are vague. Actually, there is surprisingly little in Hawaiian tradition about volcanic activity, and particularly about destruction of property by lava flows. For this reason, it has been suggested that the volcanoes must have been inactive for a long period previous to the early part of the 19th century; but it seems too great a coincidence that activity should happen to resume just at the time of arrival of literate persons capable of keeping a record of events. More probably the Hawaiians, like many primitive peoples, took natural events such as volcanic eruptions very much as a matter of course, the more so as eruptions almost never took human lives and most of the property destroyed was of little permanent value. A grass house was easily replaced, and there was plenty of arable land for the relatively small population. Only rarely did an eruption make a sufficient impression on the people to find its way into tradition.

In 1823, his guides told Ellis that Kilauea Crater, for the reign of "many kings past," had been in much the same condition as it then appeared, and it seems likely that the type of activity of all Hawaiian volcanoes has changed little during at least the past several centuries.

A few of the traditions are worth repeating briefly. We are told, for instance, that in the early days (presumably since A.D. 1000) a chief of Puna, named Keliikuku, went on a visit to the island of Kauai, and there bragged of the superiority of his own magnificent lands. For some reason this annoyed Pele, the goddess of the volcanoes, and she poured lava over broad areas of his district, so that on his return much of Puna was a desolation of bare black rock. It appears likely that this tale refers to the broad expanse of recent pahoehoe erupted from the east rift zone of Kilauea, now crossed by the highway northwest of Pahoa. We are told also that in the days of Liloa, a great king of the island of Hawaii who ruled probably about the year 1500, a lava flow issued from some area east of Kilauea Crater and flowed to the sea at the south coast. This flow cannot be identified with any certainty, but it may be the one that reaches the shore between Kaena and Apua points, and which originated in the general vicinity of Makaopuhi Crater.

Perhaps the most entertaining of the legends is the one that tells of the adventures of Kahawali, also a chief of Puna, who lived about A.D. 1350. It is told that Kahawali was sliding on his holua (sled) on the side of a hill in Puna when an old woman, a stranger, appeared and asked if she could borrow the sled for a while. Kahawali, apparently in an ungenerous mood, refused, telling her to get her own sled if she wanted to slide. She went away but soon returned with a sled and challenged him to a race. He accepted the challenge, and in the ensuing race he beat her. His exultation was short lived, however. She was very angry. She stamped violently on the ground, and immediately cracks opened and red-hot rocks flew out. Too late he realized what perhaps he should have guessed before—this was the goddess Pele!

Terrified, Kahawali ran toward the sea, passing his family en route and pausing briefly to rub noses with his pet pig. All the way Pele chased him, hurling hot rocks after him, until finally he reached the shore, where his brother had just landed with a canoe. He stole his brother's canoe and escaped. The story is fascinating, both for its sidelights on early Hawaiian culture and because it is obviously a legendary account of an actual event. One of

the cinder-and-spatter cones on the east rift zone in Puna bears the name Kaholua o Kahawali (Kahawali's holua slide), and extending from there to the ocean is a line of small spatter cones that marks the path Kahawali took, with Pele's barrage of hot rocks following him.

The following review of the history of Hawaiian volcanic activity is much abbreviated. The history has been recorded in detail, from the time of arrival of the missionaries to the beginning of the 20th century by J. D. Dana, C. H. Hitchcock, and W. T. Brigham (see reference list at the end of the book), and for more recent years by T. A. Jaggar and others in the *Volcano Letter* and the *Bulletin* of the Hawaiian Volcano Observatory, and in Bulletins and Professional Papers of the U.S. Geological Survey. Some of the latter are included in the list of suggested additional reading at the end of the chapter, and many others are listed in a bibliography of the geology and water resources of the island of Hawaii, by Macdonald (1947).

### HALEAKALA

Haleakala, the volcano that makes up the eastern part of the island of Maui, has erupted only once within the memory of man. In about 1790, lava broke from two vents on the southwest rift zone of the volcano and poured into the sea at Makena (figs. 44, 202). The jagged surface of the aa lava is crossed by the old road just west of La Perouse Bay. The first flow issued at a point 1,550 feet above sea level, a short distance downslope from the present road around the island. Shortly afterward, the lower flow broke out from a vent only 575 feet above the sea—its lava rests on top of that from the upper vent. It is common for eruptive vents to open progressively downslope in this manner. The combined lava field from the two vents is 3 miles long above sea level and extends an unknown distance into the sea (fig. 44). Above sea level the total area of lava is about 2.2 square miles, and the volume about 35,000,000 cubic yards.

The approximate date of the eruption was established by the late Lorrin A. Thurston, for many years the editor of the *Honolulu Advert-*

50

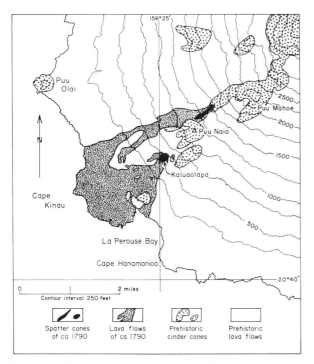

Figure 44. Map of the southwestern part of Haleakala volcano, island of Maui, showing the lava flows of the 1790 eruption and the spatter cones at their vents. (Modified after Stearns and Macdonald, 1942.)

*iser*, and a great volcano enthusiast. When Thurston visited Maui in 1879, Father Bailey, one of the New England missionaries, told him that when he (Bailey) was first stationed on Maui, in 1841, he noticed that one of the lava flows on the southwest slope of Haleakala appeared fresher than the others, much fresher in fact than it did then, 38 years later. Father Bailey questioned people living in the area, who told him that the eruption had been witnessed by their grandparents. In 1906 Thurston was told by a Chinese-Hawaiian cowboy, Charlie Ako, that his father-in-law, who died in 1905 at the age of 92, had told him that his grandfather had seen the eruption when he was a boy big enough to carry two coconuts from the beach to the upper road (4 or 5 miles, and a climb of some 2,000 feet). This was the basis for dating the eruption!

From the two accounts, and assuming the length of time for one generation to be about 33 years, Thurston deduced the date of the eruption to have been about 1750. The well-known Hawaiian ethnologist J. F. G. Stokes suggested 25 years as a more accurate estimate for one Hawaiian generation (Stearns and Macdonald, 1942, p. 107), and, if we calculate

from that assumption, the eruption would have occurred about 1770. A more reliable date has recently been assigned by B. L. Oostdam (1965) on the basis of charts of the coastline of Maui made by members of the exploring expeditions commanded by La Pérouse and Vancouver. La Pérouse's map, made in 1786, shows a broad shallow bay between Puu Olai and Cape Hanamanioa, whereas Vancouver's map, made in 1793, shows a prominent peninsula between those two points, as does the modern map (fig. 44). There can be little question that the peninsula was formed by the lava flow between the visits of the two explorers, and therefore within a year or two of 1790.

A recently published suggestion that the eruption took place at the time of the Maui earthquake, in 1938, is patently ridiculous. Not only would it certainly have been seen by the many people living within sight of that part of the island at the time, but the flow was seen by Father Bailey in 1841, and it is shown on a Hawaiian Government map made about 1880, as well as on the U.S. Geological Survey topographic map made in 1922.

## HUALALAI

Hualalai is the westernmost of the five major volcanoes that make up the island of Hawaii (fig. 45). The two more northerly volcanoes, Kohala Mountain and Mauna Kea, have not erupted in historic time nor for some hundreds of years prior to the beginning of Hawaiian history. The single historic eruption of Hualalai took place in 1800 or 1801. John Young, one of Captain Cook's seamen who had remained in Hawaii and become an important advisor to King Kamehameha I, was living at Kawaihae, 20 miles to the north, at the time of the eruption. Although apparently he did not keep a written record, he was able to recall the approximate date for Rev. William Ellis in 1823.

Like the 1790 eruption of Haleakala, the eruption of Hualalai took place from two main vents and produced two separate big lava flows (fig. 46). The flow from the main vent, between 5,500 and 6,000 feet above sea level, is remarkable in that it contains a tremendous

*51*

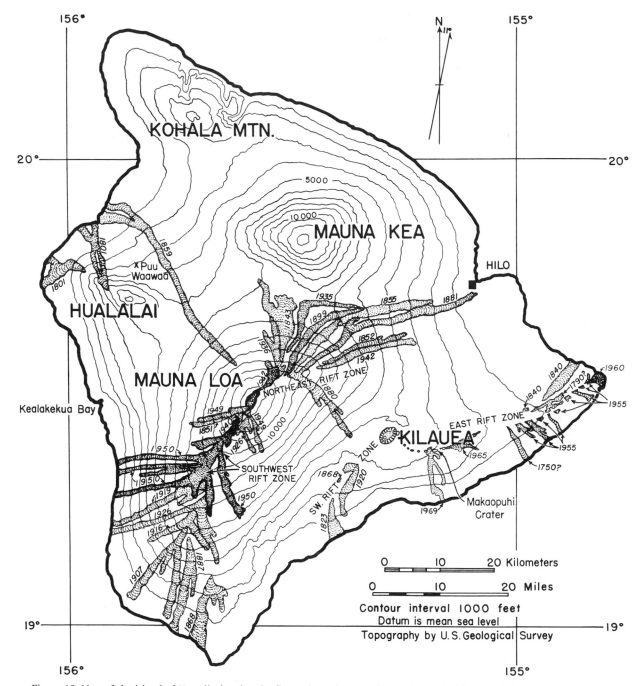

Figure 45. Map of the island of Hawaii, showing the five major volcanoes that make up the island, and the historic lava flows.

number of fragments of rock (dunite) made up largely of the mineral olivine, and other related rock types. The lava, known as the Kaupulehu flow, reached the ocean, destroying a fishpond and villages along the shore.

A second outbreak occurred at a vent below the present highway, 1,600 feet above sea level. The large spatter cone known as Puhiopele, built around the vent, is clearly visible from the highway. The lava flow from this vent also entered the sea, over a front nearly 4 miles wide. Three other, very small lava flows, from vents between the two main vents, were found by H. T. Stearns when he was mapping the geology of the mountain. Their very fresh condition and their composition, which is similar to that of the larger flows, make it almost certain that they were erupted at the same time as the larger ones.

In 1938, Miss Ella Paris told the well-known

volcanologist Dr. T. A. Jaggar that Hawaiians had told her father, a missionary to the Islands, that the upper (Kaupulehu) flow was the earlier, and the lower flow somewhat later. Again we see the downhill progression of vents common in volcanic eruptions.

The length of the Kaupulehu flow above sea level is about 10 miles. The combined flows cover an area of approximately 17.7 square miles, and their total volume is about 410,000,000 cubic yards.

There has been no eruption of Hualalai volcano since 1801. In 1929, however, several thousand earthquakes shook the island of Hawaii, some of them strong enough to do damage in central Kona, and to be felt as far

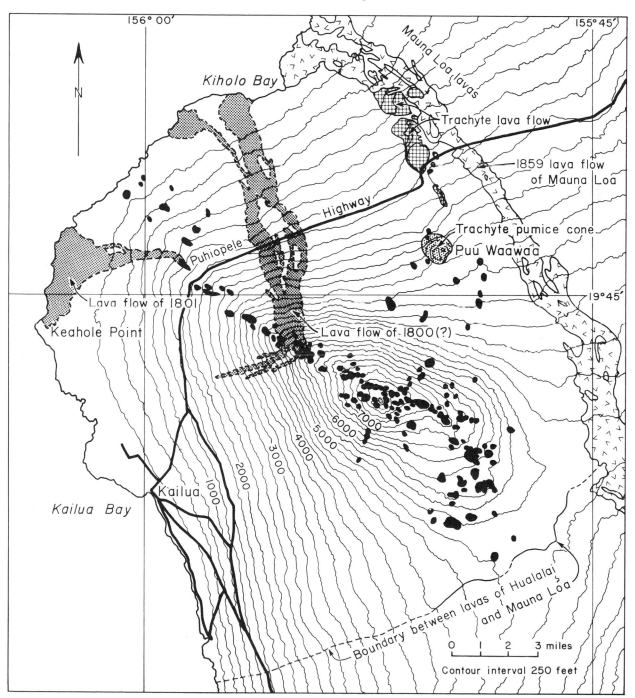

Figure 46. Map of Hualalai volcano, island of Hawaii, showing the lava flows of 1800–1801, the trachyte pumice cone of Puu Waawaa, and the trachyte lava flow from the cone. (After Stearns and Macdonald, 1946.)

53

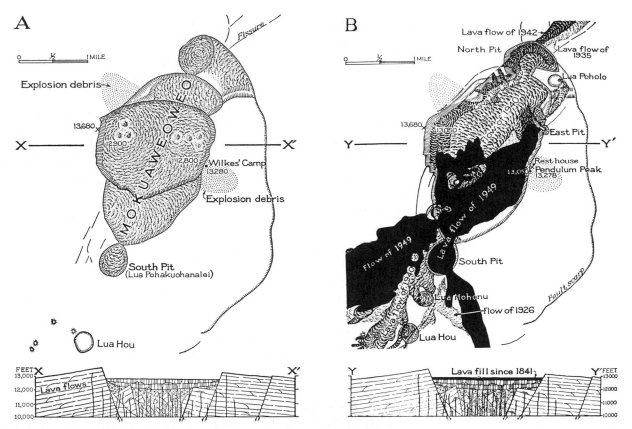

Figure 47. Maps and cross sections of Mokuaweoweo caldera, at the summit of Mauna Loa: *A*, in 1841 (after Wilkes, 1845), and *B*, in 1968. The dotted pattern shows the distribution of prehistoric phreatic explosion debris. The sections (*X–X'* and *Y–Y'*) are along the lines indicated on the maps. (Modified after Stearns and Macdonald, 1946.)

away as Honolulu. These came from a source beneath Hualalai and may have been associated with subsurface movements of magma or with readjustment of the rocks of the mountain as a result of such movement.

## MAUNA LOA

The great shield volcano of Mauna Loa is probably the largest single mountain on earth. It rises 13,677 feet above sea level, and approximately 29,000 feet above its base on the ocean floor. However, Mauna Loa probably is of compound origin, having grown around at least three successive centers. The present active cone has largely buried the earlier ones and integrated the whole into one enormous shield. (See chap. 19.)

The Mauna Loa shield is crossed by two great series of fissures, trending southwest and east-northeast, and intersecting at the summit caldera. These two principal rift zones have been the source of most of the lava flows erupted on the flank of the volcano (fig. 29). A

much less pronounced rift zone, which extends north-northwestward from the summit, gave vent to the great flow of 1859, as well as other, prehistoric flows.

The caldera (fig. 32), which is known as Mokuaweoweo, also has been the site of frequent eruptive activity. It is roughly oval in outline. At its northeastern and southwestern ends it coalesces with nearly circular pit craters named respectively North Bay and South Pit. Southwest of South Pit are two other large pit craters, Lua Hou and Lua Hohonu (fig. 47*B*). At the northeastern edge of the caldera are still two more, smaller, pit craters, named East Pit and Lua Poholo. As mentioned in chapter 2, three of these pit craters (Lua Hohonu, East Bay, and Lua Poholo) have been formed since 1840. It is probable that Lua Poholo was already in existence by 1880, because a flow believed to be of that date cascaded into it. There has been an interesting confusion of names in the craters southwest of the summit. The name "Lua Hou" means "new crater," but from Wilkes' map of 1840 (fig. 47*A*) it is clear that Lua Hohonu, and not Lua Hou, is the new crater.

In 1840 the center of Mokuaweoweo caldera was a deep, nearly circular pit, the bottom of which was nearly 900 feet below the highest point of the western rim. At each end were crescent-shaped benches, called by Jaggar the North and South Lunate Platforms, that stood about 200 feet above the level of the central floor (fig. 47A). Throughout the later part of the century the central basin gradually filled up until finally, in 1914, lava flooded over the northern bench. In 1933 and 1940, lava from vents near the center of the caldera completed the burying of the old northern bench, and in the latter year flooded into North Bay. Not until 1949 did the southern bench finally lose its identity (fig. 53). The 1949 eruption also had the distinction of being the only one in historic time during which lava erupted within the caldera has escaped onto the outer slope of the mountain. At that time lava from a vent at the southwest edge of the caldera poured southwestward through the notch connecting with South Pit, filled the latter to the level of its low south brim, and flowed on 6 miles down the mountainside.

Table 2 summarizes the data on historic eruptions of Mauna Loa. In the period from its first recorded eruption, in 1832, to the end of 1950, Mauna Loa averaged one eruption every 3.6 years, and was in eruption approximately 6.2 percent of the time. Nearly all of Mauna Loa's eruptions take place on one or the other of the rift zones or in the caldera. Depending on whether they center on the slope of the mountain or at the summit, in or near the caldera, the eruptions may be divided into summit and flank eruptions. To some degree there is an alternation between summit and flank activity. In addition, nearly every flank eruption begins with a few hours of summit activity, which is followed within a few more hours by an outbreak lower on the mountain. Repose periods between summit eruptions and the succeeding flank eruptions have averaged only 29 months, as compared with an average of 64 months between flank eruptions and the next summit eruption. Thus the group consisting of a summit eruption followed within 2 or 3 years by a flank eruption may be regarded as

the typical short-term cycle of Mauna Loa. However, variations from the typical cycle are common.

A typical eruption of Mauna Loa consists of the following fairly distinct, though intergrading stages:

*Opening Phase.* The first few hours of the eruption are characterized by a long line of lava fountains, known locally as the "curtain of fire" (fig. 3), extending essentially uninterrupted along a fissure or series of fissures sometimes as much as several miles in length, and by voluminous outwellings of lava along the fissure. Long low spatter ramparts are built, but no true cones. Typically, the opening phase of a flank eruption is directly preceded by a few hours of eruption at the summit having the same characteristics as those of the opening phase lower on the flank.

*Cone-Building Phase.* The second phase of the eruption sees the length of the erupting fissure diminishing, the lava fountains becoming restricted to a short section of it, generally less than a quarter of a mile long. Around themselves the fountains build a cone—or a short chain of cones—of spatter or, less commonly, cinder. The fountains reach their maximum height, sometimes as much as 1,000 feet, or rarely even higher. One or more major lava flows generally issue continuously from gaps in the growing cones, and many minor flows may be formed.

*Declining Phase.* The final stage of the eruption is characterized by a general decrease in fluid pressure at the vents, with a corresponding reduction in the size of the lava fountains accompanying, and probably at least in part caused by, a decrease in the amount of gas in the erupting lava. The outflow of lava may cease with the dying of the fountains, but commonly it continues for days or weeks, and sometimes for months, after the end of fountain activity. Feeble wisps of gas may continue to drift out of the vent for months or even years after the eruption is over.

These three phases are represented to varying degrees in different eruptions. Sometimes, as in the eruption of 1940, the second and

Table 2. Historic eruptions of Mauna Loa

| Year | Month and day | Summit eruption | Flank eruption | Location of principal outflow | Altitude of main vent (feet) | Approximate repose period since last eruption (months) | Area of lava flow (square miles) | Approximate volume of lava (cubic yards) |
|---|---|---|---|---|---|---|---|---|
| 1832 | June 20 | 21 | (?) | Summit | 13,000(?) | — | — | — |
| 1843 | Jan. 9 | 5 | 90 | North flank | 9,800 | 126 | 20.2 | 250,000,000 |
| 1849 | May | 15 | — | Summit | 13,000a | 73 | — | — |
| 1851 | Aug. 8 | 21 | (?) | Summit | 13,300 | 26 | 6.9 | 90,000,000 |
| 1852 | Feb. 17 | 1 | 20 | Northeast rift | 8,400 | 6 | 11.0 | 140,000,000 |
| 1855 | Aug. 11 | — | 450 | Northeast rift | 10,500(?) | 41 | 12.2b | 150,000,000 |
| 1859 | Jan. 23 | <1 | 300 | North flank | 9,200 | 26 | 32.7c | 600,000,000c |
| 1865 | Dec. 30 | 120 | — | Summit | 13,000 | 73 | — | — |
| 1868 | Mar. 27 | 1 | 15d | South rift | 3,300 | 23 | 9.1c | 190,000,000c |
| 1870 | Jan. 1(?) | 14 | — | Summit | 13,000 | 21 | — | — |
| 1871 | Aug. 1(?) | 30 | — | Summit | 13,000 | 18 | — | — |
| 1872 | Aug. 10 | 60e | — | Summit | 13,300 | 11 | — | — |
| 1873 | Jan. 6 | 2(?) | — | Summit | 13,000 | 3 | — | — |
| 1873 | Apr. 20 | 547 | — | Summit | 13,000 | 3 | — | — |
| 1875 | Jan. 10 | 30 | — | Summit | 13,000 | 2 | — | — |
| 1875 | Aug. 11 | 7 | — | Summit | 13,000 | 6 | — | — |
| 1876 | Feb. 13 | Short | — | Summit | 13,000 | 6 | — | — |
| 1877 | Feb. 14 | 10 | 1f | West flank | −180± | 12 | — | — |
| 1880 | May 1 | 6 | — | Summit | 13,000 | 38 | — | — |
| 1880 | Nov. 1 | — | 280 | Northeast rift | 10,400 | 6 | 24.0 | 300,000,000 |
| 1887 | Jan. 16 | — | 10 | Southwest rift | 5,700 | 65 | 11.3c | 300,000,000c |
| 1892 | Nov. 30 | 3 | — | Summit | 13,000 | 68 | — | — |
| 1896 | Apr. 21 | 16 | — | Summit | 13,000 | 41 | — | — |
| 1899 | July 4 | 4 | 19 | Northeast rift | 10,700 | 38 | 16.2 | 200,000,000 |
| 1903 | Oct. 6 | 60 | — | Summit | 13,000 | 50 | — | — |
| 1907 | Jan. 9 | <1 | 15 | Southwest rift | 6,200 | 37 | 8.1 | 100,000,000 |
| 1914 | Nov. 25 | 48 | — | Summit | 13,000 | 94 | — | — |
| 1916 | May 19 | — | 14 | Southwest rift | 7,400 | 16 | 6.6 | 80,000,000 |
| 1919 | Sept. 29 | Short | 42 | Southwest rift | 7,700 | 40 | 9.2c | 350,000,000c |
| 1926 | Apr. 10 | Short | 14 | Southwest rift | 7,600 | 77 | 13.4g | 150,000,000c |
| 1933 | Dec. 2 | 17 | <1 | Summit | 13,000 | 91 | 2.0 | 100,000,000 |

*Source:* Modified after Stearns and Macdonald, 1946.

*Note:* The duration for most of the eruptions previous to 1899 is only approximate. Heavy columns of fume at Mokuaweoweo, apparently representing copious gas release accompanied by little or no lava discharge, were observed in January 1870, December 1887, March 1921, November 1943, and August 1944. They are not indicated in the table.

a. All eruptions in the caldera are listed at 13,000 feet altitude, although many of them were a little lower.

b. Upper end of the flow cannot be identified with certainty.

c. Area above sea level. The volume below sea level is unknown, but estimates give the following orders of magnitude: 1859—300,000,000 cubic yards; 1868—100,000,000 cubic yards; 1887—200,000,000 cubic yards; 1919—200,000,000 cubic yards; 1926—1,500,000 cubic yards. These are included in the volumes given in the table.

d. Flank eruption started April 7.

e. Activity in the summit caldera may have been essentially continuous from August 1872 to February 1877, only the most violent activity being visible from Hilo.

f. Submarine eruption off Kealakekua, on the west coast of Hawaii.

g. 2.5 square miles of this is the area of the thin flow near the summit. An unknown area lies below sea level.

h. About 0.5 square mile of this is covered by the thin flank flow above the main cone and 0.8 square mile is in Mokuaweoweo caldera.

i. 2.8 square miles is in Mokuaweoweo caldera and 1.1 square miles outside the caldera.

j. 2.8 square miles of this is covered by the thin flank flow near the summit, and 0.5 square mile is in the caldera.

k. Amount of lava liberated probably small; eruption was largely a liberation of gas.

Table 2.—*continued*

| Date of outbreak | | Approximate duration (days) | | Location of principal outflow | Altitude of main vent (feet) | Approximate repose period since last eruption (months) | Area of lava flow (square miles) | Approximate volume of lava (cubic yards) |
|---|---|---|---|---|---|---|---|---|
| Year | Month and day | Summit eruption | Flank eruption | | | | | |
| 1935 | Nov. 21 | <1 | 42 | Northeast rift | 12,100 | 23 | 13.8[h] | 160,000,000 |
| 1940 | Apr. 7 | 133 | <1 | Summit | 13,000 | 51 | 3.9[i] | 100,000,000 |
| 1942 | Apr. 26 | 2 | 13 | Northeast rift | 9,200 | 20 | 10.6[j] | 100,000,000 |
| 1943 | Nov. 21 | 3 | — | Summit | 13,000 | 18 | (?) | (?)[k] |
| 1949 | Jan. 6 | 145 | 2 | Summit | 13,000 | 61 | 5.6 | 77,000,000 |
| 1950 | June 1 | <1 | 23 | Southwest rift | 8,000 | 12 | 35.0 | 600,000,000 |
| TOTAL | | 1,328 | 1,352 | | | | 251.8+ | 4,037,000,000+ |

third phases are difficult to separate. In the 1950 eruption the usual restriction of activity was lacking, and until the end activity remained spread out along several miles of erupting fissure. As a result, no major cone was built. The third phase was represented in 1950, however, by a period of greatly diminished fountaining, but of continued, relatively quiet outpouring of lava.

### Eruption of 1940

The typical sequence of summit eruption, followed shortly afterward by flank eruption, is well illustrated by the eruptions of 1940 and 1942.

Following the flank eruption of 1935, Mauna Loa was in repose for more than 4 years. During 1939 and early 1940, however, an increasing number of earthquakes indicated that the quiet would soon come to an end. At 11 P.M. on April 7, 1940, volcanic tremor began recording on the seismographs at the Volcano Observatory, and at 11:30 people in Kona saw the orange glow of eruption at the summit of the mountain.

For the first few hours, a nearly continuous line of lava fountains played along a series of fissures 3 miles long, extending from a point northeast of the center of Mokuaweoweo southwestward across the caldera floor, up over the caldera rim, and nearly 2 miles down the southwest rift zone (fig. 48). At 4:30 the next morning, fountains from a few feet to about 60 feet high were playing along most of the 4-mile

line, with only short dark gaps, but the principal fountain activity was already starting to concentrate near the center of Mokuaweoweo, between spatter cones built during the 1903 and 1914 eruptions. A great flood of very fluid lava was pouring from the vents, and pahoehoe flows had already covered more than two-thirds of the caldera floor and all of North Bay. Vents on the arcuate platform at the southwest side of the caldera spilled lava northward into the caldera to mingle with the flow from the more northerly vents, and

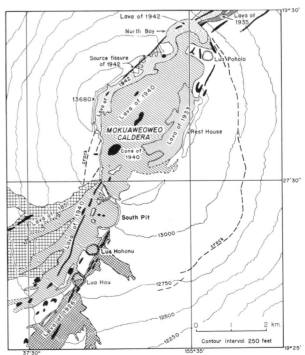

Figure 48. Map of the summit of Mauna Loa, showing the lava flows of 1940 and part of the lava of 1942.

southward into South Pit. From fissures on the outer flank of the mountain, lava poured into all three pit craters (South Pit, Lua Hou, and Lua Hohonu), and formed three principal flows that advanced westward down the mountain-side for a distance of more than a mile.

By mid-morning of April 8 activity had become restricted to a portion of the fissure near the southwestern edge of the main caldera floor. At the same time the remaining fountains increased in height. On the following morning fountains up to 300 feet high were playing along a line a quarter of a mile long. Farther south, on the arcuate platform and the rim of the caldera, formerly active vents were pouring out fume but no lava, and between them and the main vents five other vents contained small fountains less than 50 feet high. By April 11 the latter fountains also had disappeared. On the morning of April 12 the active portion of the fissure was approximately 800 feet long. A line of seven distinct fountains was building a row of coalescing cinder and spatter cones, which were gradually burying the 1933 and 1914 cones. The fountain at the northern end (fountain *1*, fig. 49) was larger

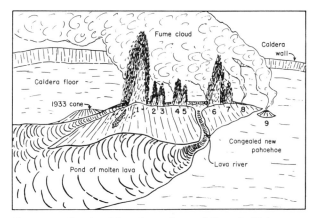

Figure 49. Sketch of lava fountains and flows in Mokuaweo-weo caldera during the eruption of 1940. The view is from the southwestern rim of the caldera on April 12.

and less variable than the others, playing steadily to a height of about 260 feet above the caldera floor, and 200 feet above the top of the cone. Fountain *6* (fig. 49) was the second largest, rising to about 180 feet. The others were smaller, and vents *8* and *9* were giving off only fume, smelling strongly of sulfur dioxide and trioxide.

Two rivers of lava poured down the sides of the growing cone, one toward the west and the

other to the north. The western river, in the foreground in figure 49, originated in a small boiling pool that occupied the top of the cone between fountains *5* and *6*. On leaving the pool, the lava quickly acquired a thin lead-gray crust, which was repeatedly torn open by movement of the flow, exposing the glowing red liquid beneath. Commonly the central part of the river was covered by gray, ropy and twisted crust, but along the stream the moving crust tore away from the banks, leaving two glowing red borders. Where the rivers plunged down the sides of the cone, over a slope of about 20°, the speed of the central part of the stream was about 15 miles an hour.

At the northern end of the cone the two rivers merged to form a broad pool of molten lava north and west of the cone. The rate of movement of the lava surface decreased rapidly away from the base of the cone, until, a quarter of a mile from the cone, no movement was visible. Occasional small secondary lava fountains played briefly on the ponded lava where the crust was fractured, and fragments occasionally sank. Small whirlwinds raised dust clouds on the flanks of the cone and occasionally ripped off sections of the thin crust of the pool, 0.5 to 2 inches thick and as much as 6 feet across, and sent them spinning off through the air. The pool looked in many ways like a lava lake, but there is no indication that convective circulation, characteristic of a true lava lake such as the one that existed for many years in Halemaumau, was ever established. Indeed, no true lava lake appears to have existed in Mokuaweoweo at any time in recorded history.

In daylight the cores of the fountains were a bright orange-red. As the ejecta rose through the air they rapidly darkened, due to cooling, until at the top of their trajectories they appeared nearly or quite black. Falling back onto the outer slopes of the cone, many of them burst in bright red splashes of liquid, showing that their centers were still very hot and unsolidified. Occasional bursts of liquid spatter landed in such abundance on the outer slope of the cone that they coalesced to form thin flows that trickled down the slope for a few minutes before they congealed. (Flows of this sort are called rootless flows because they have no direct connection with the deep-seated source of magma.) When the wind shifted to the southeast, pumice and Pele's hair drifted down on watchers at the observation point at the western rim of the caldera. Some of the pumice fragments were still warm enough to melt little pits, an inch or so deep, in the snow.

Early in the afternoon of April 12 the fountain activity began gradually to increase. By 2:15 the top of fountain *1* reached a height 400 feet above the caldera floor, and by 6 o'clock it was reaching 600 feet. The other fountains showed a corresponding increase. These high fountains continued through the early evening, then gradually died down to heights similar to those of the previous morning.

During the next week there were several periods of high fountains like those of April 12. The cone grew rapidly in size, and by April 15 the remnants of the 1933 and 1914 cones (fig. 49) were almost completely buried. After April 19 the strength of the activity decreased greatly, and by the last week in April the fountains were very weak and no persistent lava flows were visible. By May 1 only a single fountain remained. During the next month the cone grew slowly in height as small amounts of spatter were thrown onto its rim. From time to time small lava flows issued at the base of the cone and spread out to form a broad low shield.

From June 19 to August 18, activity was intermittent, with active periods of a few hours to several days alternating with quiescent periods of 1 to 4 days. The fountain was generally hidden from view within the cone. The last visible glow was seen on the night of August 18, although the cone continued to fume weakly most of the time until the 1949 eruption.

Nearly all of the lava consolidated as pahoehoe. The flows on the outer flank of the mountain, extruded in the first few hours of the eruption, are thin shelly pahoehoe, generally from 3 to about 8 feet thick, and containing numerous open tubes and hollow blisters covered by shells only a few inches thick.

The voluminous outpouring of lava in the caldera during the first few hours covered an area of approximately 2.6 square miles to an average depth of about 20 feet. Adding this to the flows outside the caldera, we calculate that approximately 50,000,000 cubic yards of lava was poured out during the first day of the eruption. The total volume for the entire eruption was about 77,000,000 cubic yards.

The early flood of pahoehoe formed a pool with a nearly level, gently undulating surface. Some of the earliest lava was covered with a skin of pumice, generally about half an inch thick, formed by extreme frothing of the surface of the gas-rich liquid. Later flows advancing over the surface of the congealed early flood developed the toes and ropy or festooned surfaces generally regarded as characteristic of pahoehoe lava, but these features were rare on the early lava.

After its emplacement, the lava which had poured out in the first rapid flood quickly lost volume and shrank back, leaving a low cliff (known as a "slump scarp") around the edges of the caldera where the thin margins of the flow rested against the caldera walls. Similar slump scarps are common around ponded flows in Hawaii. On the morning of April 8 the slump scarp around the northeastern edge of North Bay was 3 to 10 feet high. P. E. Schulz (1943) estimated that the volume decrease indicated by the slumping was about 18 percent. As he pointed out, the loss of volume is much greater than can be explained by cooling of the lava. Other factors that may help to bring it about are (1) draining of liquid lava back into the eruptive fissure (as happened in the 1952 and 1959 eruptions of Kilauea), (2) settling of the lava into incompletely filled depressions beneath it, and (3) loss of gas bubbles from the still-liquid lava beneath the crust. In this case, as in most others, the important factors appear to have been cooling and loss of gas, which together are capable of producing a shrinkage of about 20 percent by volume. Evidence indicates that in Hawaiian lavas most of the gas comes out of solution and escapes within a very short time after the arrival of the lava at the surface; but the early flood of the 1940 lava was so rapid that much of the gas probably was retained until the flow had reached its resting place, only to escape within the next few hours, leaving the slump scarp.

59

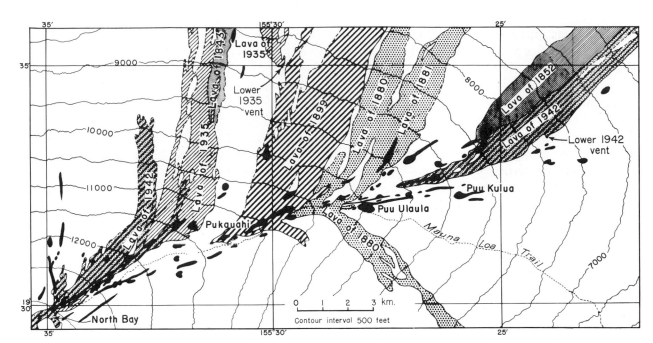

Figure 50. Map of the northeast rift zone of Mauna Loa, showing the upper portions of historic lava flows.

During the middle part of the eruption, pahoehoe flows advancing northward were partly confined by the walls of the caldera. The thrust of additional lava against the obstructed fronts of the flows resulted in much shattering of the flow crust and heaving up of the crust into crude wavelike folds (pressure ridges), resembling asymmetrical anticlines in sedimentary rocks. Some of them are as much as 30 feet high, composed of chunks of lava crust 3 to 10 feet across and up to 3 feet thick.

Not all of the lava added to the fill in Mokuaweoweo was poured out over the surface. Some of it was intruded beneath the surface of the previously solidified lava, and caused an elevation of the surface without overflowing it. This pushing up of the lava surface by new liquid lava beneath resulted in a plateau-like form 1,500 feet across and 15 to 20 feet high in the center of North Bay, and as much as 40 feet high at the northern edge of Mokuaweoweo itself.

The 1940 eruption is one of the few for which it has been possible to estimate the amount of gas contained in the magma as it reached the surface. It was surprisingly small. Even in the unusually gas-rich magma of the early phase of the eruption it comprised only about 1 percent by weight of the total magma, and in the middle phase, only about 0.5 percent.

*Eruption of 1942*

Not only does the 1942 eruption constitute an example of part of a typical summit-flank eruptive sequence of Mauna Loa, but it is noteworthy for two other reasons. It was a "secret" eruption. World War II was in its early stages, and the Pearl Harbor debacle was only 4 months past. Whether to test local security, as claimed by some, or in hopes of delaying the Japanese in learning about the eruption and making use of it as a navigational beacon, publication of any news of the eruption was strictly forbidden. Certainly Pele was a flagrant violator of the blackout! Malihini regulations meant nothing to the goddess of ancient Hawaii. And regardless of regulations, such news could not be long concealed. The eruption was soon general knowledge throughout the islands, although people continued to speak of it surreptitiously, if at all; on the second day Tokyo Rose, speaking over the Japanese radio, congratulated Hawaii on its fine volcanic eruption. The 1942 eruption also has the distinction of having been bombed in an effort to change the course of the lava flow, which was threatening the city of Hilo.

Like most flank outbreaks of Mauna Loa, the 1942 eruption began with a brief period of activity in and near the summit caldera. This summit phase commenced at approximately

60

5:05 P.M. on April 26 with the opening of a fissure part way up the cliff along the western side of Mokuaweoweo, across North Bay, and about 2.4 miles down the northeast rift zone (figs. 48, 50). A voluminous outpouring of very fluid gas-rich pahoehoe formed cascades down the western wall of the caldera (fig. 51), and a ponded flow along the foot of the wall and around the western and northern edges of North Bay. Outside the caldera, along the upper portion of the rift zone, fluid spatter built low spatter ramparts. The flow from the fissure moved northward, dividing into two principal lobes, one of which advanced 4 miles down the mountainside (fig. 50). Near the source the lava was pahoehoe, but it changed to aa as it flowed downslope. By the early morning of April 28 activity at the summit of the mountain had ceased.

At approximately 4:40 A.M. on April 28 lava broke out along a fissure in the rift zone at 9,500 feet altitude, 9 miles northeast of North Bay. At 7:30 that morning a nearly continuous line of lava fountains 200 to 300 feet high was playing along the fissure for a distance of half a mile. During the day the erupting fissure gradually lengthened downslope, until at 3 P.M. its length was about a mile. Fountains were active along both its upper and lower portions, but along its central portion the fountains were largely drowned by a copious flow of very liquid lava that followed the course of the fissure. At 8 P.M. the fountains were playing to a height of more than 500 feet. A spatter rampart 3 to 15 feet high was being built along the fissure, and a rapid flow of lava was moving northeastward toward the city of Hilo (fig. 45).

Within a few hours after the outbreak at 9,500 feet altitude, another occurred 2.9 miles

Figure 51. Part of the northwestern floor of Mokuaweoweo caldera, showing in the foreground the typical undulating surface of ponded pahoehoe lava, formed in 1942. In the middle distance is the "slump scarp," about 15 feet high, formed at the end of the eruption by draining away of the lava and shrinkage due to cooling and loss of gas. Cascades of lava descended the caldera wall from fissures higher upslope during the 1942 eruption.

farther northeast (fig. 50). This lower point of outflow lacked many of the characteristics of most lava vents of Mauna Loa. The lava issued quietly from a series of cracks trending nearly at right angles to the fissures of the rift zone. Much less gas was liberated than at the 9,500-foot vent, and there was very little fountaining. Only a small amount of spatter was formed, and both the spatter and the lava of the flow were denser and much poorer in gas than those at the higher vent. It is believed that lava approaching the surface at the higher vent found its way into an ancient pahoehoe tube and flowed downslope for 3 miles, losing much of its contained gas, before finally issuing at the "rootless" lower vent. A very similar thing happened during the eruption of 1935, when lava plunged into an old pit crater at 11,300 feet altitude and disappeared, to reappear devoid of most of its gas at a vent more than 4 miles downslope (fig. 50).

After attaining its maximum length and strength of eruption on the evening of April 28, the active fissure at the main vent gradually decreased in length, and fountain activity became restricted to its central part as the first phase of the flank eruption came to an end. On the afternoon of May 2 the western 1,500 feet of the fissure was fuming but no longer emitting lava. The eastern part of the fissure likewise was inactive and buried beneath the head of the lava flow. Along the central 1,000 feet of the fissure, at 9,200 feet altitude, lava fountains were playing to a height of 100 to 400 feet, and had built a chain of cinder-and-spatter cones averaging about 75 feet high.

True fountaining was absent at the westernmost vent in the cone chain, which was essentially a gas vent. From it issued dense clouds of fume, and occasional gas explosions within it hurled blocks of pumice high into the air. The more copious ejections of pumice lasted for several minutes and were accompanied by a roaring sound of escaping gas. The blocks of pumice were as much as 10 inches across and were thrown as high as 1,000 feet. At other times, for periods of several minutes' duration, activity at this vent was of Strombolian type (see chap. 6), ejecting ribbon- and spindle-shaped fragments that cooled during flight and struck the ground as solid bombs.

The rest of the cone chain exhibited the usual Hawaiian-type activity. Many of the shreds and blobs of lava struck the ground still in a partly molten condition and accumulated to build steep-sided cones of partly welded cinder and spatter.

A lava river 50 feet wide issued at the eastern end of the cone chain and flowed east-northeastward at a rate estimated in its fastest-moving central part at 15 to 20 miles an hour. Near the cones small fountains burst through the lava river from time to time. By May 4 it was possible to go right to the bank of the river. Its surface undulated and bounded like that of a river of water in flood. Broad standing waves, about 3 feet high, extended entirely across the channel. Where the river left the cone its entire surface consisted of incandescent molten lava, orange-red in daylight, but within 100 feet the river was completely crusted over. Near the cone the crust on the river was frequently broken up and swept away, the fragments sinking into the molten lava.

On May 4 the activity had become largely restricted to a single central fountain which played to an average height of 150 feet above the cone, with occasional bursts reaching 500 feet. The cone around it had grown to a height of about 100 feet.

Throughout this cone-building phase of the eruption, occasional small lava flows escaped from the sides of the cones. Two had already occurred before May 2, spilling over low places in the rim of the cone and flowing only about 100 feet beyond the base. On the afternoon of May 4 the south wall of the western cone partly collapsed, releasing a short, slow-moving flow of pahoehoe. On the evening of the same day a further partial collapse of the cone liberated very fluid lava streams on both sides of the cone chain. That on the northern side soon joined the main flow, but the one on the southern side continued as an independent flow parallel to the main one for a distance of 6 miles. The escape of these lateral flows caused a lowering of the level of the liquid lava within the cones, and simultaneously the fountains increased greatly in height, the highest playing as an essentially constant liquid jet to a height of more than 600 feet for an hour or more afterward. Apparently the lowering of the level of liquid within the cone removed part of the restraint to the free escape of the gas-laden liquid and allowed the fountain to shoot to greater height.

On the morning of May 7 two small flows broke from the western cone. The second of these did not escape over the rim, but forced its way through the cone wall. It appeared first as a slightly glowing bulge on the side of the cone, the bulge then slowly distending and developing into a flow of pahoehoe. The repeated short flows built a flat dome of lava around the central cones, eventually burying their base to a depth of about 30 feet.

The main lava flow advanced east-northeastward, dividing and reuniting like a braided stream, and finally it was joined by the lava from the lower, rootless source to form a single flow. During the first few days the advance of the flow front was rapid. By noon on May 1 it had reached a point 13 miles from the 9,200-foot source, and only 12 miles from the city of Hilo. In the lower portion of its course the flow advanced through wet, swampy jungle. Contact of the lava with the water and wet vegetation resulted in the formation of large volumes of steam and smoke, and, in the breakdown of plant tissues buried by the lava, methane gas was formed (by destructive distillation). The methane, possibly accompanied by other hydrocarbon gases, moved outward away from the flow through tubes in older underlying pahoehoe, occasionally becoming ignited on mixing with air, which caused explosions as far as 300 feet from the edge of the flow. These explosions sent billowing black clouds 500 feet or more into the air, and explosion craters as much as 10 feet across and 2 feet deep were blasted in the old lava. These marginal explosions are among the most dangerous features of lava flows in vegetated regions, because their time and place is wholly unforeseeable.

On May 1 the front of the flow was advancing toward Hilo at a rate of 300 to 500 feet an hour, with a front about half a mile wide. The collapse of the source cone and formation of the new flows on May 4 appear to have depleted the supply of lava to the old flow. Advance of the old flow front continued

until May 5 or 6, but by May 7 it was stagnant, with its terminus about 11 miles from Hilo.

On May 9 activity at the 9,200-foot vents had greatly diminished, and the eruption had entered its final phase. On the morning of May 10 the eastern vent and lava river were dead, and the western vent showed only very weak activity, throwing incandescent fragments 10 to 25 feet above the rim of the cone. By the following morning the cone area was completely dead except for minor amounts of fume.

The lava of the 1942 eruption covered an area of about 10.6 square miles, and had a volume of about 100,000,000 cubic yards. In and near the summit caldera the lava was pahoehoe, but the flows on the flank of the mountain outside the caldera changed to aa near their lower ends. Much of the early lava had a pumiceous top, 0.5 to 1 inch thick. The lava that issued at the 9,200-foot vents also was pahoehoe near its source, changing to aa downstream. The early outpourings formed broad fields of aa along the edges of the flow, but during the cone-building phase of the eruption the actively moving central portion of the flow remained pahoehoe for about 6 miles from the vents, finally changing to aa at an altitude of about 6,500 feet. During later stages of the eruption the point of conversion from pahoehoe to aa retreated toward the vents as the viscosity of the lava increased, and on May 9 it was only about 600 feet from the lower edge of the cone. In the area close to the cone a central stream of pahoehoe flowed through slowly moving marginal fields of aa, and pahoehoe occasionally burrowed under the aa and emerged as small flows from the lower portions of the aa.

The prediction of the 1942 eruption, by R. H. Finch, who in 1940 had succeeded Dr. T. A. Jaggar as Director of the Hawaiian Volcano Observatory, was one of the most successful yet made for any eruption. Unfortunately, Finch was not allowed to publish the prediction because of war-time security restrictions, but the full prediction was submitted to the Superintendent of Hawaii National Park and forwarded by him to Washington.

The prediction was based partly on Finch's knowledge that a flank eruption was likely to follow the 1940 summit eruption within 3 years, but largely on the pattern of earthquakes that developed during the spring of 1942. The pattern started with two small quakes originating 25 to 30 miles beneath the northeast rift zone of Mauna Loa some 8 or 10 miles southwest of Hilo. Then, through February and early March, came many quakes, most of them too small to be located with accuracy. Some, however, could be located. The situation for determining the horizontal position of the quakes was especially favorable, since the Hilo and Kona seismograph stations lay almost directly in line with the rift zone, one in each direction from the points of origin of the shocks. (The placement of the stations was not accidental; their value in just such a situation had been foreseen by Jaggar when they were established.)

On February 21 and 22 a swarm of 10 earthquakes came from shallow foci on the northeast rift in the area between 9,000 and 10,000 feet altitude (approximately where the main vent of the eruption later developed). The epicenters (the epicenter is the point on the ground surface above the point of origin of the earthquake) of succeeding quakes continued to migrate southwestward, reaching Mokuaweoweo on March 7, and moving on down the southwest rift zone, reaching a point 5 miles beyond Mokuaweoweo by March 21. On March 28 another small earthquake centered under the upper end of the northeast rift. Normally during February and March the ground surface at the old seismograph station beneath the Volcano House, at the northeast edge of Kilauea Crater, tilts westward at a fairly regular rate. In 1942 the westward tilting was appreciably less than usual, suggesting that the seasonal tilt was being partly offset by a tendency to eastward tilting due to the swelling of Mauna Loa.

On the basis of the earthquakes and the abnormal ground tilting, in addition to the usual periodicity of the volcano, Finch issued a written statement early in April that Mauna Loa probably would erupt within the next few months, with a flank outbreak on the northeast rift; orally he suggested that the point of outbreak probably would be at about 9,000 to 10,000 feet altitude. The outbreak came at 9,200 feet altitude on April 28, fulfilling his expectations in a very gratifying manner.

*Bombing of the Lava Flow.* During the first four days after the outbreak at 9,200 feet, the lava flow made rapid progress toward the city of Hilo (fig. 45). On May 1 there appeared to be imminent danger of the flow successively cutting the flume that supplied water to the town of Mountain View, blocking the highway around the island, which was vital to military operations, and destroying part of Hilo and possibly its harbor. It was decided to try to divert the flow by aerial bombing.

The idea of diverting a lava flow by blasting away the banks of the lava river was first suggested by L. A. Thurston (the same who investigated the date of the eruption of Haleakala), and the idea was later developed by T. A. Jaggar and R. H. Finch. It should be emphasized that the intention was not to stop the eruption, as is sometimes claimed, but simply to change the course of the flow to one where it would do less damage. Ideally, the lava would be made to spread out in the wastelands on the upper slopes of the mountain, sparing productive areas lower down. There appear to be three principal ways in which bombing can be used to divert a part or all of a flow: (1) bombing of the main feeding tubes of a mature pahoehoe flow, (2) bombing of the natural levees along the main feeding river of an aa flow or a young pahoehoe flow in which tubes have not yet formed, and (3) bombing of the walls of the cone at the source of the flow.

Bombing a mature pahoehoe flow aims at breaking in the roof of a main feeding tube and partly or entirely blocking it with fragments of the shattered roof or with viscous aa lava formed from the fluid pahoehoe as a result of violent stirring caused by the explosion. The fluid pahoehoe then escapes through the shattered roof of the tube at the site of the bombing—high on the mountainside—to spread there harmlessly or to form a new flow. Even if the new flow advances downslope parallel to the former one, a distinct advantage is gained because it probably will require days, or even weeks, to reach the point previously reached by the earlier flow; if the flow can be delayed long enough in this battle against time, the eruption will end, halting the flow, before an important area is reached. If necessary, the bombing might be repeated.

This method of bombing a pahoehoe flow was tried during the 1935 eruption of Mauna Loa, when Keystone bombers of the old U.S. Army Air Corps, under the direction of Dr. Jaggar, dropped bombs on the flow at an altitude of 8,800 feet on the north slope of the mountain. The success achieved was sufficient to demonstrate the feasibility of the method if circumstances are favorable. The main tube was broken open, lava overflowed on the slope around the break, and the advance of the flow front toward Hilo stopped two days later. Stirring of the lava in the tube by the explosion may have increased its viscosity sufficiently to cause stagnation of the flow. It is not certain, however, that other, natural causes were not responsible for stopping the flow.

In 1942 the lava flow endangering Hilo was aa throughout its lower course, and the bombing method proposed was that of breaking down the banks of the open aa river to allow the liquid of the river to escape laterally and thus reduce or eliminate the supply of lava to the main flow front. Repeated small overflows of an aa river build embankments along its edges similar to the natural levees that develop along the banks of rivers such as the Mississippi, and between these natural levees the lava river may be flowing at a level several feet above the ground surface on each side. Breaking down of the levee will allow the liquid to escape. If the topography is favorable, a new flow may form and take a very different course from that of the former flow; but even if the new flow parallels the earlier one, the advance of the main flow front is delayed. R. H. Finch made a reconnaissance flight over the flow and selected the most favorable sites for bombing, but before the planes could reach the area the best sites were hidden by clouds and smoke, and the bombs had to be dropped at less favorable places. Nevertheless, the levee was broken and a small lava flow escaped to one side of the main one. The new flow moved along the edge of the older one and rejoined it only a few hundred yards below the point of diversion. Though no great success in delaying the flow could be claimed, the general possibili-

ties of the method were demonstrated. In this case, further bombing was made unnecessary by the stagnation of the flow through natural causes.

The third method, that of bombing the cone at the source of the flow, has not yet been tried. There seems little doubt, however, that it would be successful under favorable circumstances. Many of the cones of Mauna Loa and other Hawaiian volcanoes are elongate, with relatively thin side walls that probably could be broken down by heavy bombs. The consequent release of the contained pool of lava, that commonly stands at high level in these cones, would divert much of the supply of liquid from the earlier main flow. That the method might work has been demonstrated in nature. On May 4, 1942, the walls of the cone at the 9,200-foot vent partly collapsed, liberating floods of lava around the cone and thereby depleting the supply to the main flow. A day later the advance of the main flow had nearly ceased.

Diversion of lava flows by bombing is still in an early experimental stage. It bears great promise of usefulness, and certainly should be tried, if conditions are favorable, when a lava flow threatens some area of sufficient importance. On the other hand, there are certainly many flows not susceptible to diversion by bombing. For those it may be possible to use other methods of diversion, such as the barriers mentioned on a later page. It must be pointed out, however, that the success of any method depends on topographic and other conditions, and any attempt to divert a flow when conditions do not warrant is a waste of effort and money. There is no panacea for lava flows.

### Mauna Loa Since 1942

During late 1942 and early 1943, Mauna Loa was inactive except for quiet steaming of vents on the rift zones and weak fume liberation from the 1940 and 1942 cones. On several occasions in October and November 1943, and again in August 1944, conspicuous clouds of fume were seen over Mokuaweoweo. Almost surely magma was moving in the conduits beneath the mountain, but none reached the surface.

The 1949 and 1950 eruptions again constitute a typical summit-flank sequence. The

1949 eruption began in the late afternoon of January 6. The mountain top was hidden in clouds, and the first indication of eruption was a deep rumbling sound, clearly audible at the Volcano Observatory at Kilauea, 20 miles air line from the summit of Mauna Loa. And the seismographs were recording the harmonic tremor (the trembling of the ground that accompanies movement of molten lava in the feeding conduits of the volcano), which is characteristic of eruption.

The outbreak took place along a series of fissures that opened south-southwestward across the floor of Mokuaweoweo, through the caldera wall, and about 1.6 miles down the southwestern flank of the mountain (fig. 52).

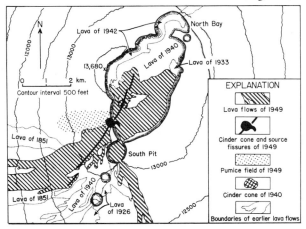

Figure 52. Map of the summit of Mauna Loa, showing the location of the eruptive fissures and cone of 1949, and the upper portions of the lava flows.

The total length of the active fissure zone was nearly 3 miles. During the first few hours of the eruption, lava fountains, from 10 to 100 feet high, formed an almost continuous "curtain of fire" along the fissures. The lava was unusually rich in gas, and large amounts of very light pumice (thread-lace scoria, or reticulite) were formed. Fragments of pumice half an inch across were found 7 miles east of Mokuaweoweo, and Pele's hair drifted down as far away as 20 miles.

The eruptive fissure passed directly through the 1940 cone. Early in the eruption lava rose in the crater of the old cone to the level of a low notch on its northern rim, and sheets of gas-rich pumiceous pahoehoe issued from the fissure on both the northern and southern slopes of the cone. Spattering in the crater covered the upper flanks of the cone with liquid bombs that accumulated so rapidly that they ran together and formed three small rootless flows that trickled down the north side of the cone. The lava extruded along the fissures in the southwest rift zone, outside the caldera, formed a thin flow that advanced rapidly westward along the north side of the lava flow of 1851 (fig. 52). By January 8 the flow had stopped, 7 miles from its source.

As usual, the opening phase of the eruption was of short duration. Within 72 hours activity was entirely confined to a quarter-mile length of fissure at the southwest edge of the caldera. During the first 24 hours the volume of lava poured out was about 50,000,000 cubic yards —approximately two-thirds of the amount extruded during the entire eruption.

On January 9 two lava fountains still played at the southwest edge of the caldera, one of them directly against the caldera wall. They continued active, and in mid-January began to increase in height until on January 23 the larger of the two was reaching heights of at least 800 feet, and probably more than 1,000 feet. Because of the large amount of gas in the lava, vast quantities of pumice were formed. It piled up around the vent to form a pumice cone 1,500 feet across built partly against the cliff at the edge of the caldera and partly on the caldera rim (fig. 53), and, blowing westward from the cone, it formed a pumice blanket several feet thick and a mile long. Lava from the fountains flooded most of the floor of Mokuaweoweo and poured southward through the connecting gap into South Pit. On January 25 it reached the level of the southern rim of South Pit and spilled over, forming a flow that had extended 5.5 miles down the mountainside by January 31, when it came to a stop. Early in February fountain activity began to weaken, and on February 5 it stopped. For a time it seemed that the eruption was over.

About the first of April, however, activity resumed. Repeated small overflows of gas-poor pahoehoe built a cone of lava just east of the earlier cones (fig. 53). On April 7 the lava cone was 80 feet high, and a crater in its top contained a seething pool of liquid lava. Lava leaving the vent through two tubes beneath the surface fed two sluggish flows of aa, one of which spread over the caldera floor a mile northeast of the vent, and the other tumbled through the gap into South Pit.

Activity at the vent continued for several weeks with little change, but by early May the flows to the east had ceased. A mile northeast of the cone, lava issued from a tube to form a small active flow of pahoehoe. From time to time the pool at the top of the lava cone rose to the level of the brim and overflowed, sending small tongues of fluid lava down the cone flanks, thus demonstrating the manner in which the cone had been built. Similar feeble activity continued until about the end of May, when the eruption finally came to an end.

The 1950 eruption (figs. 54, 55) has already been described (pp.9-11). It was one of the two greatest historic eruptions of Mauna Loa. The eruption of 1859, which lasted for nearly 10 months, equalled it in the volume of lava extruded, but in 1950 the huge volume of lava was poured out in only 3 weeks, and between one-third and one-half of it during the first 36 hours. The extent of the flows and the course of the eruptive fissure in the southwest rift zone are shown in figure 56.

Since 1950 Mauna Loa has not erupted. To the date of writing, the volcano has been quiet more than 19 years—by far the longest period of quiescence in its recorded history. When activity returns at last, what can we expect? Will it be more or less like the eruption we have seen before? Or will the gradual accumulation of gas in the upper part of the magma body during the long period of quiet result in unusual violence? Only time will tell.

## KILAUEA

Kilauea is a broad shield volcano built against the southeastern slope of the larger Mauna Loa (fig. 45). At its summit is a caldera 2.5 miles long, 2 miles wide, and 400 feet deep at its western side. Toward the south the height of the caldera walls gradually decreases to zero.

Figure 53. The cone area of the 1949 eruption, at the southwestern edge of Mokuaweoweo caldera. The main cone, of cinder and pumice, is on the skyline. Just below and to the left of it is a small fuming spatter cone, and to the left of that is the lava cone, built by repeated small overflows of pahoehoe. The lavas are mantling the edge of the old South Lunate Platform that formerly existed as a "mezzanine" level in the end of the caldera.

67

Figure 54. Clouds of steam rising where lava flows of Mauna Loa enter the ocean, June 2, 1950.

Within the caldera, southwest of the center, lies the crater of Halemaumau (fig. 78), which for the past century has been the principal site of activity of the volcano. Extending southwestward and eastward from the caldera are two rift zones, along which most of the flank eruptions have taken place. The east rift zone is unusual in that it bends sharply. For 5 miles the rift zone trends southeastward from the caldera, but then it bends and extends east-northeastward to Cape Kumukahi and onward along the ocean floor. In the 10 miles of its course nearest the caldera, the east rift zone is marked by 13 pit craters (fig. 33).

Along the southern edge of the Kilauea shield the Hilina fault system consists of a series of high-angle faults nearly parallel to the rift zones. Another series of faults, the Kaoiki fault system, lies northwest and west of the caldera near the junction of the slopes of Kilauea and Mauna Loa. Evidence suggests that this fault system acts as an adjustment joint on which slippage occurs during swelling or shrinking of either volcano.

As on Mauna Loa, eruptions of Kilauea commonly occur in pairs, a summit eruption being followed by one on the flank. The 1954–55 and 1959–60 eruptions, described on later pages, are typical pairs of this sort. More commonly than on Mauna Loa, however, groups of several eruptions have occurred in the same general area, as along the upper part of the east rift zone since 1961.

The many thin flows of which the shield volcano of Kilauea is built are clearly exposed to view in the walls of the caldera. Pahoehoe flows predominate in this region because they are close to the vents from which they issued, but, with increasing distance from the caldera and from the rift zones, aa becomes more abundant, and near the coast aa commonly predominates. In the immediate vicinity of the caldera the lava flows are covered by several feet of volcanic ash (fig. 13), at the base of which there is generally a layer of very light basaltic pumice (thread-lace scoria or reticulite), from a few inches to 18 inches thick. The ash is of two sorts: glassy (vitric) ash derived from the spray of lava fountains, and stony (lithic) ash formed by the pulverization of older solidified lava during phreatic eruptions. In general the lower part of the ash deposit consists mostly of greenish-brown glassy ash, and the upper part of stony ash mixed with angular blocks of lava. This is readily seen in the road cuts around the southwest side of the caldera. In places, as near Halemaumau and around the southeast side of the caldera near Keanakakoi Crater, there are many large angular blocks, the largest weighing about 8 tons, which were thrown out by the phreatic explosions of 1790 and 1924.

## 1790 to 1923

When William Ellis stood on the brink of Kilauea Crater in August 1823, he was awe-struck by its immensity and by the spectacular volcanic activity within. His camp for the first night was on the northern edge of the crater, near where Kilauea Military Camp now stands, probably at the foot of the low cliff nearly opposite the east edge of the camp where there was some protection from the mist-laden trade winds. The view was indeed awe-inspiring. Most of the area of the caldera was occupied by an inner pit, the floor of which was about 900 feet below him. Around the inner pit was a narrow "black ledge," about 600 feet below the crater rim (and about 200 feet lower than the present crater floor). On the floor of the inner pit molten lava "rolled to and fro its fiery surge." Apparently the activity had at least

Figure 55. Lava flow entering the ocean on the west side of Mauna Loa on June 2, 1950. Note the steam cloud rising from the ocean and the boiling water at its base.

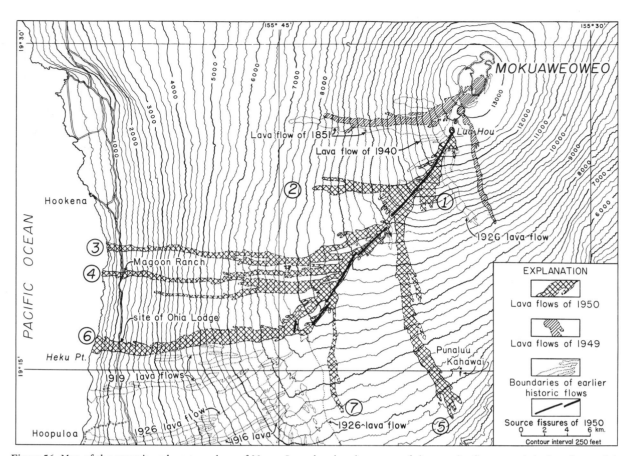

Figure 56. Map of the summit and western slope of Mauna Loa, showing the course of the eruptive fissures and the lava flows of the 1950 eruption.

some of the features of a lava lake. As Ellis left the caldera the next day on his way toward Kalapana, he passed on the eastern rim the remains of the important heiau, Oalalauo, the site of which has now been lost. He does not mention the heiau at Uwekahuna, which of course was already abandoned, where tradition tells us priests kept up a constant vigil and chanting to appease the goddess Pele. (The name, Uwekahuna, means "weeping of the kahunas.")

The caldera appeared much the same during the visit of Lord George Byron, in 1825. Byron, a cousin of the poet, was in command of the British warship *Blonde,* which returned the bodies of King Liholiho and Queen Kamamalu from London, where they had died of measles. Byron's lieutenant, James Malden, made a map of Kilauea Crater—the first ever made (fig. 57*A*). The deep inner pit and its encircling "black ledge" were still there.

In the spring of 1823, only shortly before Ellis' visit, an outflow of lava had taken place in the southwest rift (fig. 45). By draining

away supporting magma from beneath the caldera, this outflow had probably caused the central part of the caldera to sink, leaving the black ledge. Before the sinking the whole caldera floor had been at the level of the black ledge.

A similar sinking of the central part of the caldera probably occurred at the time of the eruption on the east rift zone about 1790. At that time a great phreatic explosion took place. Judging from the debris it left around the caldera, it must have been considerably more violent than that of 1924. Blocks of rock as much as 18 inches in diameter, thrown out by the explosions, are common along the eastern edge of the caldera. Some of them are wrapped in a shell of lava to form "cored bombs." The lava is denser than that generally thrown out from primary volcanic vents, and probably represents liquid that drained from unsolidified but largely degassed intrusive bodies back into the collapsing inner pit.

The 1790(?) explosions were witnessed by members of the army of Keoua, chief of Puna, southern Hawaii, who was then on his way toward Ka'u after battles with Kamehameha. Stopping near Keanakakoi Crater, on the edge of Kilauea caldera, to make appropriate offerings to Pele, they were thrown into consternation by the beginning of explosions (a very rare phenomenon at Kilauea). Some accounts say that some of the soldiers were killed by falling blocks. Further offerings failed to stop the explosions; indeed their violence increased. Apparently deciding that further propitiation was useless, and that the neighborhood had become decidedly inhospitable, Keoua ordered his army to proceed as three separate groups. As was usual in those days, the army consisted not only of warriors, but also of women and children and domestic animals. The first division proceeded down the Ka'u trail, apparently without mishap, followed at 2-hour intervals by the other two divisions. The third division, traveling southwestward about 6 miles from the caldera, looked ahead and saw the second division, apparently resting. On reaching them they found that every single person was dead, though some pigs were unharmed. Footprints that can still be seen in ash along the route of the trail are said to have been left by this

stricken army. There is no real confirmation of this, of course, and since footprints can be found in several different layers in the ash they certainly cannot all have been made at that one time. The route was a common one, followed frequently by travelers between Ka'u and Puna.

Just what happened to the second army division we do not know. The trail lay directly along the southwest rift zone of Kilauea volcano, and there appears to be no question they were killed by volcanic action. But the bodies were not burned, or apparently otherwise injured. They must have been caught in a cloud of poisonous gas liberated from vents along the rift zone. Perhaps the pigs had greater resistance to the poison, or perhaps they simply breathed air that was closer to the ground. Caught in a fume cloud, one can find pockets of fresh air in hollows in the ground surface, and make one's way out of the cloud by moving from hollow to hollow.

The sinking of the caldera floor in 1823 does not appear to have been accompanied by explosions. The 1823 lava flow issued from the Great Crack—a crack that extends uninterrupted for 14 miles along the southwest rift zone near the sea (fig. 58). In places the crack

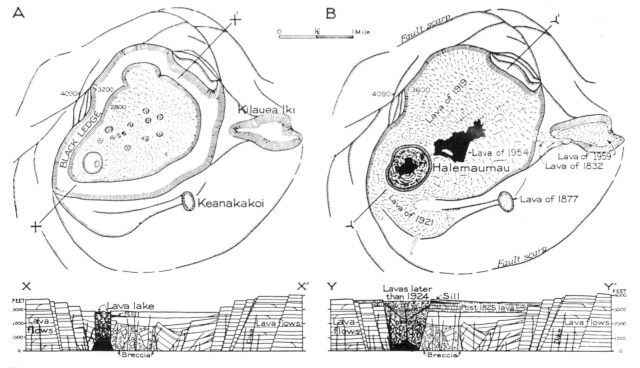

Figure 57. Maps and cross sections of Kilauea caldera: *A*, in 1825 (after Malden), and *B*, in 1960. The large central pit that existed in 1825 had been entirely filled with lava before 1900. The structure beneath the caldera is hypothetical. (Modified after Stearns and Macdonald, 1946.)

71

is as much as 50 feet wide, but it narrows rapidly downward, and nowhere is it possible to climb down into it more than 60 or 70 feet. Lava rose in the crack to the surface and welled out quietly, with almost no explosion. Even spattering was very slight. The flow is remarkable, however, for the large number of lava balls, a few inches to several feet in diameter, formed of blocks of older lava wrapped up in the new liquid lava. The balls are well exposed along the edges of the crack. The lava was very fluid and flowed rapidly, sweeping high up onto the slopes of a group of older spatter cones (now called the Lava-Plastered Cones) before rushing on to the ocean. It is said that the flow came so fast that the people of a village in its path could not launch their canoes in time to save them, and that a few old people and small children could not run fast enough to escape.

Additional data on the eruption of 1823, and later eruptions, are given in table 3.

During the years immediately following 1823, eruptions within the central basin of the caldera gradually filled it, and by 1832 the new floor stretched entirely across the caldera at a level about 45 feet above the former level of the black ledge. In 1832 a small eruption occurred on Byron Ledge (the flat area around the eastern side of the caldera that had been the camping place of Lord Byron in 1825) between the main caldera and the adjacent pit crater, Kilauea Iki. This event no doubt marked the opening of the east rift zone allowing magma to drain into it from beneath the caldera because, although there does not appear to have been any outbreak lower on the rift, the caldera floor once again collapsed. In 1834 the caldera was much as it had been in 1823, with an inner basin 400 feet deep surrounded by a narrow black ledge.

No account of Kilauea in the early 19th century is complete without mention of one of the most courageous acts in Hawaiian history. Since time immemorial the Hawaiian people had lived in dread of the volcano goddess, Pele. Even today, long after most of the old religion has disappeared and been forgotten, during eruptions one commonly finds offerings to Pele on the lava flows or around the edges of the crater, left by people who still firmly believe in her. Yet in 1824, in the face of this deeply ingrained belief, to prove once and for all that the old gods were false and the God of the newly arrived Christian missionaries was all-powerful, the high Chiefess Kapiolani descended with her followers into Kilauea caldera, the very home of Pele, and deliberately violated her tabus: they ate of the sacred ohelo berries without first offering some to Pele, threw stones instead of offerings into the crater, and in every possible way defied her. Kapiolani had been a convert to the new religion for only a year. It is difficult today to appreciate fully the strength of conviction and the tremendous courage represented by her deed.

Again in the years following 1832 eruptions filled the inner basin and overflowed the black ledge with new lava. By 1840 a broad dome 100 feet high occupied the caldera floor, with its summit in the southwestern part of the caldera. A crater at the apex of the dome held an active lava lake. The site of this lake, first referred to as Halemaumau by Count Strzelecki in 1838, has ever since been the principal focus of activity of Kilauea volcano.

During late May 1840, activity in the Halemaumau lava lake was unusually strong. On May 30 a small outbreak of lava took place on the east rift zone 6 miles east-southeast of the caldera, followed by several others within the next 6 miles farther east. Two days later outbreaks 18 to 25 miles east of the caldera sent a lava flow into the ocean (fig. 45), destroying the coastal village of Nanawale. This flow is crossed today by the road that runs northward from Kapoho to Honolulu Landing.

The flank flow of 1840 was accompanied by another collapse of the central portion of the caldera, but the area of the central basin was smaller than in 1823 and 1832. The caldera was mapped by the U.S. Exploring Expedition, under Commodore Charles Wilkes, in late 1840. The inner pit was about 340 feet deep and irregularly oval in outline, with a long diameter of approximately 1.75 miles and an average shorter diameter of a little less than 1

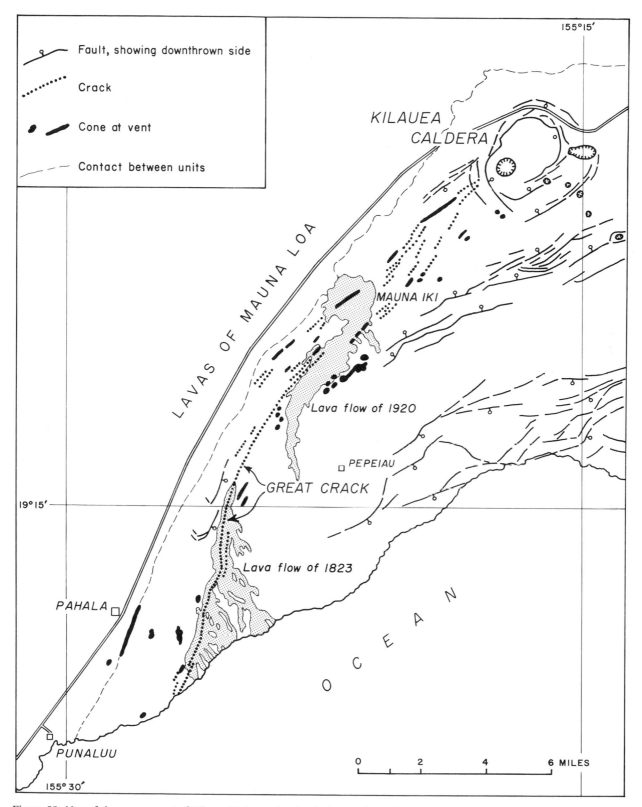

Figure 58. Map of the western part of Kilauea Volcano showing faults, cracks, and cones along the southwest rift zone, and the lava flows of 1823 and 1920.

Table 3.  Historic eruptions of Kilauea

| Year | Date of outbreak | Duration (days) | Altitude (feet) | Location | Approximate repose period since last eruption (months)[a] | Area (square miles) | Volume (cubic yards) |
|---|---|---|---|---|---|---|---|
| 1750 (?) | — | — | 1,700 | East rift | — | 1.57 | 19,500,000 |
| 1790 (?) | — | — | 1,100-750 | East rift | — | 3.04 | 37,670,000 |
| 1790b | November (?) | — | — | Caldera | — | No lava flow | No lava flow |
| 1823 | Feb.-July | Short | 1,700-250 | Southwest rift | — | 3.86c | 15,000,000c |
| 1832 | Jan. 14 | Short | 3,650 | East rim of caldera | — | (?) | (?) |
| 1840 | May 30 | 26 | 3,100-750 | East rift | — | 6.60c | 281,000,000c |
| 1868 | April 2 | Short | 3,350 | Kilauea Iki | — | 0.07 | (?) |
| 1868 | April 2 (?) | Short | 2,550 | Southwest rift | — | 0.04 | 250,000 |
| 1877 | May 4 | 1 (?) | 3,500 (?) | Caldera wall | — | (?) | (?) |
| 1877 | May 21 (?) | — | 3,450 (?) | Keanakakoi | — | 0.04 | (?) |
| 1884 | Jan. 22d | 1 | −60 (?) | East rift | — | (?) | (?) |
| 1885 | March | 80 (?) | 3,640 (?) | Caldera | 14 | (?) | (?) |
| 1894 | Mar. 21 | 6+ | 3,690 | Caldera | 108 | (?) | (?) |
| 1894 | July 7 | 4 (?) | 3,690 | Caldera | 3.5 | (?) | (?) |
| 1918 | Feb. 23 | 14 | 3,700 | Caldera | 283 | 0.04 | 250,000 |
| 1919 | Feb. 7 | 294e | 3,700 | Caldera | 11 | 1.60 | 34,500,000 (?) |
| 1919 | Dec. 21 | 221 | 3,000 | Southwest rift | 1 | 5.00 | 62,000,000 |
| 1921 | Mar. 18 | 7 | 3,700 | Caldera | 7.5 | 0.77 | 8,800,000 |
| 1922 | May 28 | 2 | 2,650-2,400 | Makaopuhi and Napau | 14 | 0.04 | (?) |
| 1923 | Aug. 25 (?) | 1 | 3,000 | East rift | 15 | 0.20 | 100,000 |
| 1924f | May 10 | 17 | — | Caldera | 8 | No lava | No lava |
| 1924 | July 19 | 11 | 2,365 | Halemaumau | 2.5 | 0.02 | 320,000 |
| 1927 | July 7 | 13 | 2,400 | Halemaumau | 35 | 0.04 | 3,160,000f |
| 1929 | Feb. 20 | 2 | 2,500 | Halemaumau | 19 | 0.06 | 1,920,000 |
| 1929 | July 25 | 4 | 2,560 | Halemaumau | 5 | 0.08 | 3,600,000 |
| 1930 | Nov. 19 | 19 | 2,600 | Halemaumau | 15.5 | 0.09 | 8,480,000 |
| 1931 | Dec. 23 | 14 | 2,700 | Halemaumau | 12.5 | 0.12 | 9,640,000 |
| 1934 | Sept. 6 | 33 | 2,800 | Halemaumau | 44 | 0.16 | 9,500,000 |
| 1952 | June 27 | 136 | 2,870 | Halemaumau | 212.5 | 0.23 | 64,000,000 |
| 1954 | May 31 | 3 | 3,180 | Halemaumau and caldera | 18.5 | 0.44 | 8,500,000 |

| 1955 | Feb. 28 | 88 | East rift | 150-1,310 | 8.9 | 6.10 | 120,000,000 |
| 1959 | Nov. 14 | 36 | Kilauea Iki | 3,500 | 53.5 | 0.24 | 51,000,000 |
| 1960 | Jan. 13 | 36 | East rift | 100 | 0.8 | 4.1 | 155,000,000 |
| 1961 | Feb. 24 | 1 | Halemaumau | 3,150 | 12.2 | 0.02 | 30,000g |
| 1961 | Mar. 3 | 22 | Halemaumau | 3,150 | 0.2 | 0.1 | 350,000 |
| 1961 | July 10 | 7 | Halemaumau | 3,150 | 3.5 | 0.4 | 17,300,000 |
| 1961 | Sept. 22 | 3 | East rifth | 2,600-1,300 | 2.2 | 0.3 | 3,000,000 |
| 1962 | Dec. 7 | 2 | East rifti | 3,250-3,100 | 14.4 | 0.02 | 430,000 |
| 1963 | Aug. 21 | 2 | East riftj | 3,150-2,700 | 8.4 | 0.06 | 1,100,000 |
| 1963 | Oct. 5 | 1 | East riftk | 2,750-2,300 | 1.4 | 1.3 | 9,000,000 |
| 1965 | Mar. 5 | 10 | East riftl | 3,000-2,300 | 17.0 | 3.0 | 23,000,000 |
| 1965 | Dec. 24 | <1 | East riftm | 3,150-3,000 | 9.5 | 0.23 | 1,160,000 |
| 1967 | Nov. 5 | 251 | Halemaumau | 3,150 | 23.3 | 0.25 | 110,000,000 |
| 1968 | Aug. 22 | 5 | East riftn | 2,900-1,900 | 1.3 | 0.01 | 50,000o |
| 1968 | Oct. 7 | 15 | East riftp | 3,000-2,400 | 1.3 | 0.8 | 9,000,000 |
| 1969 | Feb. 22 | 6 | East riftq | 3,100-2,900 | 4.0 | 2.3 | 22,000,000 |
| 1969 | May 24 | – | East riftr | 3,150 | 2.0 | 4.8s | 71,000,000s |

*Source:* Modified after Stearns and Macdonald, 1946. Data for eruptions since 1960 from Hawaiian Volcano Observatory reports.

*Note:* Many eruptions have occurred on the floor of the caldera, but only a few of the later ones are listed here, data being inadequate or totally lacking for the earlier ones. On January 11, 1928, a small amount of lava was extruded on the floor of Halemaumau, but this is believed to have been squeezed out by the weight of a heavy landslide on the crust of the 1927 lava which was still fluid beneath (Jaggar, 1932).

a. During the early historic period Kilauea caldera was observed only occasionally, and no definite record exists of the many caldera flows which are known to have occurred.
b. Violently explosive.
c. Area above sea level. The volume below sea level is unknown, but estimates give the following orders of magnitude: 1823–3,000,000 cubic yards; 1840–200,000,000 cubic yards. These are included in the volumes given in the table.
d. *Pacific Commercial Advertiser,* Feb. 2, 1884. "A column of water, like a dome, shot several hundred feet up into the air, accompanied with clouds of smoke and steam." No further eruption was observed next day.
e. Several separate flows, with short intervals without extrusion.
f. Violent phreatic explosions, possibly accompanied by a submarine lava flow on the east rift.
g. About 320,000 cubic yards of lava poured into Halemaumau, but most of it drained back into the vents.
h. 14 outbreaks along a 13-mile stretch east of Napau Crater.
i. 5 outbreaks from Aloi Crater to Kane Nui o Hamo.
j. In and near Alae Crater.
k. In and near Napau Crater.
l. Makaopuhi Crater to Kalalua Crater.
m. In and east of Aloi Crater.
n. In Heake Crater and at scattered points for 13 miles eastward.
o. About 4,000,000 cubic yards poured into Heake Crater, but most of it drained back into the feeding fissure at the end of the eruption.
p. From the east flank of Kane Nui o Hamo for about 2 miles eastward.
q. Between Alae and Napau craters.
r. Between Aloi and Alae craters. Eruption still in progress on November 20, 1969.
s. Based on preliminary mapping and volume estimate by Hawaiian Volcano Observatory in late October 1969.

Figure 59. Sketch of Kilauea caldera by J. Drayton, in 1840. (From Wilkes, 1845.)

mile (fig. 59). The black ledge around it averaged about 1,600 feet in width, and stood at an average level about 600 feet below the western rim of the caldera.

By 1846 the inner basin was again filled to and above the level of the black ledge. The filling was accomplished partly by flows over the floor of the basin, but also partly by a bodily elevation of the floor, which rose like the top of a piston being pushed upward from beneath, carrying not only the relatively flat lava floor of the basin, but also the banks of rock fragments (taluses) that had accumulated at the foot of the cliffs enclosing the basin. The result was a narrow curved ridge of rock fragments a mile long that stood 50 to 100 feet above the adjacent caldera floor and enclosed the area of the former pit (fig. 60). During 1848 a dome 3,000 feet across and 200 to 300 feet high was built, with its apex at Halemaumau.

Early in April 1868, a small flank eruption took place on the southwest rift zone, accompanied by still another collapse in the caldera. At the same time lava from a fissure along the crater wall flooded the floor of the adjacent Kilauea Iki Crater. These eruptions occurred nearly simultaneously with one from the southwest rift of Mauna Loa—one of the rare examples of simultaneous eruptions of the two volcanoes. Within Kilauea caldera an area ap-

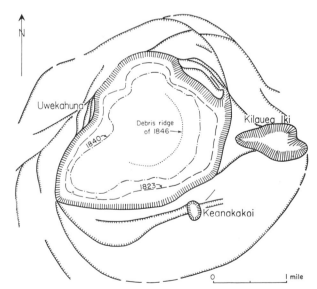

Figure 60. Map of Kilauea caldera, showing the approximate margin of the inner depression of 1823 (adapted from the map by Malden in 1825), and 1840 (after Wilkes, 1845), and the approximate position of the ridge of elevated talus in 1846 (after Lyman, 1851).

proximately the same as that of the inner basin of 1840 sagged downward about 300 feet, and in the southwestern portion of this sunken area an inner conical pit 3,000 feet across and 500 feet deep contained Halemaumau. The volume of collapse was somewhat less than in 1840.

During the following months the central depression was again gradually filled. By 1874 the dome-shaped cone around Halemaumau had reached a height nearly equal to that of the southern wall of the caldera, and lava streams from it were pouring northward into the central depression. The lava disappeared briefly from Halemaumau in April 1879, but there was no general sinking of the caldera floor, and the lake of molten lava reappeared within about a month.

In 1886 a subsidence of the area immediately surrounding the location of the present Halemaumau produced a roughly triangular pit about 3,000 feet across and 350 feet deep, with a small pit 280 feet deep in its floor. This depression was soon refilled, but 5 years later, in 1891, another similar collapse occurred, this one also restricted to the immediate neighborhood of Halemaumau. Refilling of the basin started almost immediately, and by July 1892, the lake was again overflowing. It is noteworthy that neither of these subsidences (1886 and 1891), nor another in 1894, was accompanied by any known flank eruptions. Possibly lava simply drained into opening fissures in the rift zones, or eruptions may have taken place beneath the ocean in water so deep that no sign of activity reached the surface. A submarine eruption in shallow water just east of Cape Kumukahi (East Cape) in 1884 may well have been followed by one in deeper water in 1886.

From 1823 to 1894 activity at Kilauea was essentially continuous. There were many brief intervals, especially just after the great subsidences, when liquid lava disappeared from the crater, but active lava always returned within a few days or weeks.

During July 1894 a spectacular minor subsidence occurred at Halemaumau. Within a few hours the level of the lava fell 250 feet below the rim of the pit. The lava level then continued to sink, though at a slower rate, but activity continued in Halemaumau at depths of 300 to 600 feet until December 1894, when the molten lava disappeared altogether. The diameter of the pit at that time was approximately 1,000 feet. From then until 1907 only occasional brief spells of activity occurred deep in the pit, although some fume was always visible. This 13-year interval of relative quiescence was the first real interruption in the continuity of activity at Kilauea since before 1823.

From 1907 until 1924 the volcano was again almost continuously active (fig. 61). Spectacular but minor collapses occurred at Halemaumau in 1916, 1919, 1922, and 1923. The collapse in 1919 was associated with drainage of lava from the Halemaumau lake into fissures in the southwest rift zone, and an eruption on that rift zone 5.5 miles from the caldera built the small shield volcano of Mauna Iki and sent a lava flow 6 miles downslope toward the sea. In 1922 the collapse at Halemaumau accompanied a small eruption on the east rift zone 6.6 miles from the caldera, in Makaopuhi and Napau craters; and still another small eruption occurred in August 1923 between Makaopuhi and Alae craters (fig. 62).

In 1919 Halemaumau was filled nearly to the brim with molten lava, which drained northwestward through the Postal Rift (so-called because visitors used to scorch souvenir postcards by lowering them into the hot crack) and spread northward to cover the whole northern part of the caldera floor (fig. 57B). The present trail from the Volcano House to Halemaumau crosses this flow. In 1921 the molten lava in Halemaumau rose even higher, overflowed the edge of the crater, and sent flows both northeastward and southwestward. The northeastern flow was short lived, but the one toward the southwest continued spreading, covered the southern part of the caldera floor, and sent a tiny tongue through a low gap in the wall to become the only flow that has spilled out of the caldera onto the outer flank of the mountain in historic time.

## Steam Explosions of 1924

At the beginning of 1924 the lava lake in Halemaumau stood at a high level. In late January its surface was only 105 feet below the southeast edge of the crater, and the lake was vigorously active. In February, however, the

Figure 61. Small circular lava lake in Halemaumau Crater, September 7, 1920. The main lake is in the left background, behind the row of crags.

level dropped to 370 feet below the rim, where it remained through March. In April a great swarm of earthquakes commenced on the east rift zone of Kilauea. At first the quakes came from centers near the caldera, but the points of origin shifted farther and farther east, indicating a progressive opening of the rift zone. At Kapoho, 28 miles east of the caldera, about 200 earthquakes were felt on April 22 and 23, and there must have been many hundreds more, too small to be felt. Some of the quakes were quite violent. They were accompanied by cracking open of the ground and the sinking of a graben along the crest of the rift zone. It was reported that some cracks opened to a width of several feet, and then closed again. Such reports are commonly discounted by scientists as gross exaggeration, but in this case the

stories were true. At least one of the cracks opened wide enough for a cow to fall in, then closed, crushing the unfortunate animal. The event is documented by a photograph showing the cow's leg protruding from the crack.

The fault forming the southern edge of the graben ran nearly eastward from Kapoho, reaching the coast just north of Cape Kumukahi; half a mile to the north the nearly parallel fault that formed the north edge of the graben crossed the road just south of Koae village. Between the two faults the block dropped irregularly. About half a mile inland from Koae, near the old quarry which supplied rock for the Hilo breakwater, the railroad tracks were bent downward 8 feet. Along the coast just north of Cape Kumukahi the ground sank 14 feet, allowing the ocean to flood inland

78

nearly half a mile. The waves soon built a bar of gravel across the mouth of the flooded area, leaving a brackish pond, fed by ground water, that received the formal name, Ipoho Lagoon, though it was generally known locally as Higashi's Fishpond. Until 1959 the stumps of coconut trees could still be seen standing in the water around the head of the pond.

No visible eruption took place, though lava may have poured out on the ocean bottom in deep water somewhere east of Cape Kumukahi. At any rate it is quite certain that magma did move out from beneath the caldera into the east rift zone. Seismographs recorded the characteristic tremor that accompanies subterranean movement of magma, and on April 29 rapid subsidence resumed at Halemaumau. By May 6 the lake surface was more than 600 feet below the crater rim.

On May 11 small explosions threw rock

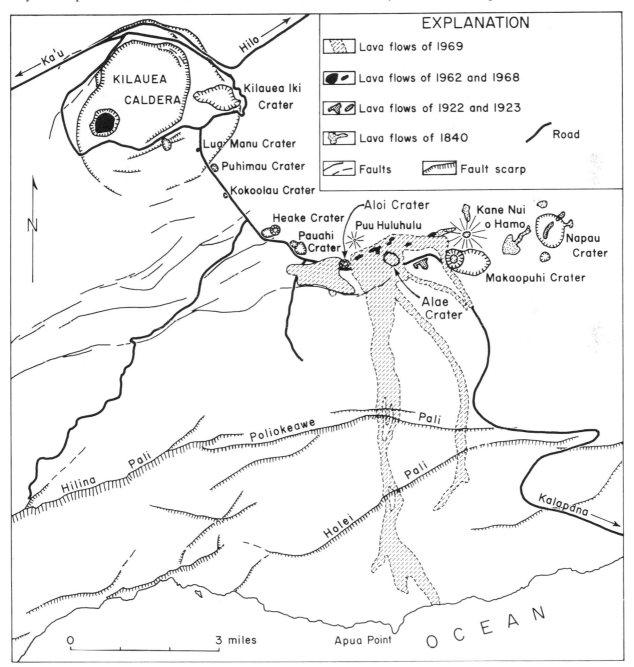

Figure 62. Map showing Kilauea caldera, the Chain of Craters along the east rift zone of Kilauea, and historic lava flows. The road extending southeastward from the caldera toward Kalapana is the Chain of Craters Road.

79

fragments out of the still-deepening pit. Similar small explosions had accompanied the subsidences of 1922 and 1923. But on May 13 five stronger explosions threw rocks half a mile into the air (fig. 40), and thereafter the explosions increased rapidly in violence. They reached a maximum on May 18. Great boiling cauliflower-like clouds of dust rose 4 miles above the crater. Spectacular lightning bolts flashed through the cloud, caused by static electricity generated by friction between the dust particles in the cloud. Condensing steam caused heavy rains of mud, and some of the mud gathered together to form little mudballs, known as pisolites, or accretionary lapilli. Blocks of rock weighing several tons were thrown as far as half a mile from the crater. One block weighing about 8 tons can still be seen about a quarter of a mile east of the parking area at Halemaumau.

The explosions resulted in the only death caused by Hawaiian volcanoes during historic times. One man, bent on photographing the interior of the crater between explosions, ventured too close despite a warning from the volcanologist that another explosion was due. He ran too late and was overtaken by the flying debris. Burned by the hot ash and with one leg severed by a falling block of rock, he died before he reached the hospital.

As the explosions continued, the walls of Halemaumau gradually slid in, and the width of the crater increased from about 1,400 to 3,000 feet. After May 18 the explosions decreased in strength and the dark gray cloud of dust gradually gave place to a white cloud of pure steam. The last violent explosions were on May 24. When the bottom finally could be seen again, the crater was found to be about 1,300 feet deep. The entire bottom was formed of slides of rock fragments from the crater walls; no molten lava was visible. How much deeper the crater would have been without the accumulation of rock fragments in its bottom we do not know, but probably several hundred feet.

The explosions threw out a total volume of about one million cubic yards of broken rock and dust. However, the material thrown out represents less than 1 percent of the great increase in size of the crater. The enlargement was due chiefly to collapse—an engulfment of the top of the volcano when support was removed from beneath it as magma moved out into the rift zone. Not only the area close to Halemaumau was involved in the sinking. Level lines run from Hilo in 1921 and again in 1926 showed that during the intervening time the whole mountaintop sank, and tiltmeters showed that the sinking occurred at the time of the collapse of Halemaumau. Beginning near Mountain View, 15 miles from the caldera, the sinking gradually increased in amount as the summit was approached. The benchmark near the Volcano House, on the northeast rim of the caldera, sank about 4 feet, and the ground near the edge of Halemaumau, 11 feet.

Many of the rock fragments thrown out by the explosions were red hot, but none of them were new volcanic material. All were derived from the old solid walls of the crater. Most of the red-hot fragments came from intrusive bodies, some of which could still be seen glowing in the walls of the crater after the explosions were over. The gas that caused the explosions was entirely steam; no gases characteristic of magma could be detected. This fact, and the absence of new magmatic liquid among the ejecta, indicate that the explosions could not have been of magmatic origin. They were *phreatic,* caused by suddenly heated ground water. It is believed that the sinking of the mountaintop caused the rocks to crack, which in turn allowed water in the surrounding rocks to flow into the hot feeding channels of the volcano. Almost instantaneous boiling of this water caused rapid generation of steam, and the violently escaping steam carried with it clouds of rock fragments and dust. The water must have entered the conduit at relatively high levels, because no trace of chlorine from seawater could be found.

The outrush of steam was not continuous, but rhythmic. It has been suggested that collapse of the walls of the crater resulted in accumulation of a mass of debris that clogged the vent until the pressure of steam trapped beneath it became great enough to blow out the obstruction. It has also been suggested that the rhythmic surging of liquid lava observed in

80

the lake before the collapse continued afterward at depth, and that each surge resulted in rapid formation of steam and an explosion. Still another mechanism may have been involved. With each sudden generation of steam resulting from water entering the hot conduit, the temporarily high steam pressure would force back the water in the surrounding rocks, perhaps for many tens of feet, and then, with the release of pressure in the explosion, the water would again flow toward and into the conduit. The periodicity of the explosions would thus result from the length of time required for a rush of water to return to the conduit and be heated into steam—a mechanism similar to that which governs the rhythmic action of some geysers, such as Old Faithful in Yellowstone National Park. It is unlikely that a mass of loose fallen rock fragments could clog the vent effectively enough to allow the high steam pressure evident in the explosions to accumulate beneath it, but either or both of the other two mechanisms may have been operative.

### 1924 to 1953

Molten lava soon returned to Halemaumau after the great collapse of 1924, but the new activity was of a different sort. The old familiar lava lake was gone. Starting in July 1924, a series of seven short eruptions poured lava into the bottom of the crater and gradually filled it, reducing the depth of the crater from 1,300 to 770 feet (measured from the visitors' observation area at its southeast edge). The eruptions are listed in table 3. In the autumn of 1934 began the longest quiet period in Kilauea's recent history. For 18 years there was no eruption; visitors to the caldera found only a bleak expanse of black lava, and the yawning empty hole of Halemaumau. People began to say, "The volcano is dead!" and some of them believed it. But volcanologists, who could "feel" the volcano repeatedly stirring in its sleep, knew that it was only a question of time until it would awaken.

In November 1944 earthquakes and swelling of the volcano indicated a rise of magma beneath the caldera region. It appeared for a time that an eruption was likely, but in

December a reversal of the direction of ground tilting showed that the volcano was shrinking again. The magma had been withdrawn, but had it moved into the rift zones, or back into the depths where it was formed? We do not know.

Again in early 1950 tilting of the ground surface indicated a swelling of the volcano. The swelling was interrupted briefly by the eruption of Mauna Loa in June, but it resumed immediately afterward and continued through the autumn. Once more eruption seemed imminent; but again, in December, the magmatic pressure was suddenly relieved. Many earthquakes and cracking of the ground in the area south and southwest of the caldera suggested that the magma may have moved into the southwest rift zone, though this is far from certain.

On April 22, 1951, a powerful earthquake (magnitude 6.5: see chap. 16) originated at a depth of 20 or more miles beneath the summit region of Kilauea southeast of the caldera. It was felt as far as Oahu, 200 miles away. This marked the beginning of a period of very numerous earthquakes that culminated with the return of molten lava to Halemaumau in June 1952. In the meantime, however, a brief interlude indicated that the island of Hawaii is probably still growing southward beneath the ocean. South of the island inadequate soundings delineate a broad bulge that is probably another shield volcano built against the flank of Kilauea just as the latter is built against Mauna Loa. In late March and April 1952, more than 4,000 earthquakes came from sources south of the island, probably from a fault or zone of faults parallel to the Hilina fault system (see chap. 19). No evidence of eruption was seen, but eruption may have occurred beneath the cover of more than a mile of ocean without giving any sign at the surface of the water.

Simultaneously with the submarine activity other earthquakes originated along the east rift zone of Kilauea, and still others along the Hilina fault system near the south shore of the island. Movement along the latter faults probably resulted from uplift of the summit region of Kilauea caused by inflation of the magma

*81*

reservoir beneath it. Tilting of the ground surface indicated that the volcano was swelling. Finally, just before midnight on June 27, came the long-awaited event. Halemaumau was again in eruption.

Spouting lava formed a "curtain of fire" along a newly opened fissure that extended southwestward across the floor of Halemaumau and part way up the walls. Along most of the length of the fissure, the lava fountains were between 50 and 150 feet tall, but at the southwestern end a fountain more than 800 feet high shot well above the rim of the crater. Pumice fragments, borne upward by the rising heated air and volcanic gas, fell as a thick blanket over the caldera floor southwest of Halemaumau. A car, driven the short distance from the Volcano Observatory to Halemaumau, through the blinding, choking cloud of fume and falling pumice, was so thoroughly sand-blasted that it had to be entirely repainted and have its windshield replaced.

The rate of lava outpouring at the beginning of the eruption was tremendous—probably close to 8,000,000 cubic yards per hour. Within half an hour the whole crater floor was covered to a depth of more than 30 feet. The lava appeared extremely fluid. Waves swept outward from the fountains across the surface of the pool and washed up and down 10 to 15 feet on the crater walls, recalling vividly Ellis' description in 1823 of the lava lake's "fiery surge." A dark semisolid flexible crust quickly formed on the lava, but was repeatedly torn apart to reveal the glowing, golden-yellow liquid beneath it. Convectional circulation quickly developed in the growing pool, and a true lava lake was established. Lava rising along the feeding fissure moved outward in all directions, but in well-marked currents, and sank again along the edge of the lake. Where the lava sank, secondary fountains a few feet high were formed as a result of the release of gas that had been entrapped in the fragments of crust that were dragged down with the fluid lava and at least partly remelted. Most of these secondary fountains lasted only a few seconds, then disappeared, to be replaced by others at different places; but where the principal currents plunged downward, more or less continuous secondary fountains were formed.

The rate of lava outpouring soon started to decline and the fountains to lose height. At 4:00 A.M. on June 28 the southwestern part of the fissure had become completely inactive, and it appeared that the eruption might already be nearly over. But 20 minutes later, fountains reappeared along the southwestern end of the fissure and lava extrusion again gradually increased. By daylight the lake was again in full activity. This was the beginning of a period of lava lake activity that was to continue until November. As time progressed, the supply of lava decreased, and the lake occupied an ever smaller proportion of the crater floor. The conditions on July 1 and July 25 are shown in figure 63. Around the lake was a black ledge or bench; and the lake stood 10 to 20 feet above the level of the bench, held in by a natural dike or ring-wall built by spatter from the secondary fountains at its edge (fig. 64). Occasional streams of lava overflowed the wall and spread over the surface of the bench. Spatter cones built by primary lava fountains over the vents rose as islands in the lake, and from time to time islands consisting of accumulations of crust or segments of the spatter cones shifted slowly and majestically across the lake, borne by currents in the slow-moving pasty epimagma that lay beneath the fluid lava of the lake. The whole floor of the crater slowly rose, partly as a result of overflows from the lake, but partly by bodily elevation as it was pushed up by new lava being intruded beneath it. When the eruption finally ended, in early November, the thickness of the new fill of lava in the crater was a little more than 400 feet.

It has been calculated that the energy released during the 1952 eruption was about $1.8 \times 10^{24}$ ergs—roughly the equivalent of the total amount of electrical energy consumed in the entire state of Hawaii during 11 months of 1967. In terms of energy release, the eruption is classified as of only moderate size.

The 1952 eruption produced only a small shrinking of Kilauea volcano (fig. 65), and swelling had already resumed by the end of the year. Throughout 1953 and early 1954 the volcano continued to swell intermittently. Between December 1952 and May 1954 the caldera floor east-northeast of Halemaumau rose approximately 1 foot in relation to the

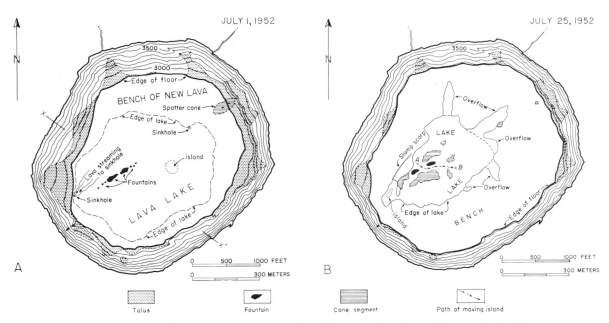

Figure 63. Map of Halemaumau Crater on July 1 (*A*) and July 25 (*B*), 1952. (After Macdonald, 1955.)

station at the Volcano House, at the northeast rim of the caldera. The total rise of the top of the mountain was not measured, but it probably was more than 2 feet. Many earthquakes were recorded. Some came from the Kaoiki fault zone, which separates Kilauea from Mauna Loa, and probably were caused by the rocks of the swelling Kilauea shield slipping past the stationary Mauna Loa. Others stemmed from beneath Kilauea caldera, and still others, from the east rift zone, were harbingers of disaster soon to come in that region. This seismic prelude culminated in the 1954 and 1955 eruptions—a summit-flank pair much like those already described as typical of Mauna Loa.

Figure 64. Lava rivers pouring through wide gaps in the central cinder-and-spatter cone to feed the surrounding lava lake; Halemaumau, July 29, 1952. The lava ring along the southeastern edge of the lake, in the foreground, is holding the surface of the lake at a level about 20 feet above the adjacent crater floor. Fumaroles are active along the lava ring.

*83*

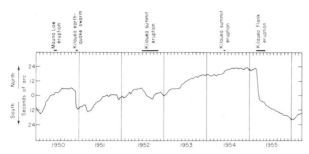

Figure 65. Graph of ground-tilting at the northeastern edge of Kilauea caldera during the years 1950 to 1956. (After Macdonald, 1959.)

## 1954 and 1955 Eruptions

At 3:42 A.M. on May 31, 1954, a strong local earthquake was felt at Kilauea caldera. Immediately afterward I (Macdonald) became aware of a peculiar low-pitched humming or roaring sound, somewhat resembling that of a heavily laden truck laboring up a long grade at a considerable distance, and never getting closer. Several old residents of the Volcano region had mentioned hearing a noise of this sort before eruptions (they referred to it as "hearing Pele"). It produced no recognizable record on the seismographs, but these are very insensitive to vibrations of such short period as sound waves. Despite the fact that it was a very definite phenomenon, the nature and origin of this "Pele noise" remains obscure. Several other strong earthquakes occurred during the

next 15 minutes, and at 3:54 one dismantled the writing mechanism on the seismograph beneath the Volcano House. (The old mechanical seismographs were made in such a way that the writing mechanism was automatically disconnected by a very strong earthquake, to prevent damage to the mechanism.) The instrument was quickly restored to operation, and at 4:09 it started to record volcanic tremor, indicating that the volcano was in eruption. The sky above Halemaumau quickly became suffused with a rapidly expanding orange-red glow as molten lava gushed into the crater.

The flood of lava into Halemaumau was fed from a series of fissures in almost the same position as those of 1952. At about 4:30 the fissure started to extend up over the wall of the crater and across the caldera floor to the northeast. A pool of lava issued from the fissure (fig. 66) and spread out rapidly to become the only lava flow on the floor of the caldera since 1921. Lava fountains along the fissure built a rampart of coalescing spatter cones. Both the flow and the spatter rampart are crossed by the trail from the Volcano House to Halemaumau. The activity on the caldera floor lasted only a few hours.

Within Halemaumau the scene closely resembled that of the opening hours of the 1952 eruption, but in addition to the surging pool in the bottom of the crater a tremendous cascade

84

Figure 66. Lava fountains, and flow spreading on the floor of Kilauea caldera, May 31, 1954. On the right a cloud of fume rises from Halemaumau.

of yellow-orange lava poured from the crack in the northeastern crater wall and tumbled 300 feet down the cliff to join the pool at its base. More than 12,000,000 cubic yards of lava were poured into Halemaumau in the first 2.5 hours of the eruption.

By mid-morning, however, activity had greatly decreased, and, although weak spattering and lava outflow, and occasional small showers of cinder continued until June 3, the eruption was essentially over. Not even all the lava poured out in the first few hours remained in the crater. During the later stages of the eruption part of the fluid bottom of the pool drained back into the conduits beneath it, and this, together with shrinkage of the lava because of cooling and loss of gas, reduced the mass to less than half its original volume and produced a slump scarp 30 feet high around the edge of the floor. The average final thickness of the new layer of lava was only about 20 feet (fig. 67A).

The brief outbreak of 1954 scarcely interrupted the swelling of the volcano (fig. 65). Throughout the fall of 1954 the top of the mountain continued to rise. Earthquakes at depths of 25 to 30 miles probably marked the zone of generation of new magma, which rose into the shallow reservoir only 3 miles or so beneath the surface. Shallow earthquakes reflected the readjustment of the rocks to the swelling of the mountain. The stage was set for further eruption.

On April 1, 1954, a new low-sensitivity seismograph was put into operation at the town of Pahoa, 20 miles east-northeast of Kilauea caldera. The station is 2.5 miles north of the crest of the ridge along the east rift zone of Kilauea, and its main purpose was to aid in location of earthquakes along the rift zone. For the first several months of its operation the instrument recorded an average of about 25 quakes a month; but in November 1954, the frequency of quakes increased to an average of two a day, in December to three a day, and in January to six a day. Most of them originated along the east rift zone 3 to 6 miles from Pahoa, and indicated the shifting of rocks along that part of the rift zone. Between February 1 and 23, 1955, the number of quakes recorded at Pahoa increased still more, to about 15 a

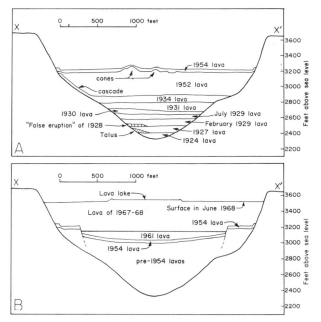

Figure 67. Cross sections of Halemaumau Crater: *A* in June 1954, and *B*, in April 1968. The sections are along the line *X–X'* in figure 63A.

day, and at the same time the tiltmeters indicated a northward tilting of the ground at the Pahoa station, resulting from swelling of the rift zone. During this interval the ground surface along the crest of the rift zone south of Pahoa rose about a foot. On February 24 approximately 100 earthquakes were recorded, on February 25 about 300, on February 26 about 600, and on February 27 about 700. Many were accompanied by dull rumblings and explosion-like noises from the ground. One did not need to be a volcanologist to realize that something out of the ordinary was going on in east Puna.

At the Nanawale Ranch, 5 miles east-southeast of Pahoa, through all of the afternoon of February 26 the ground was in almost constant vibration, like that caused by the passing of a heavy truck. The ranch lies within the area of the rift zone, and very close to the lower vents of the 1840 eruption. Dogs on the ranch were much disturbed, running around excitedly, digging holes in the ground, and snuffling in the holes as though in pursuit of burrowing animals. So far as could be ascertained, however, there were no animals for the dogs to chase, nor could any odor of sulfur gas be detected in the holes they dug. The area was almost directly in line with eruptive fissures that developed later, less than a quarter of a

*85*

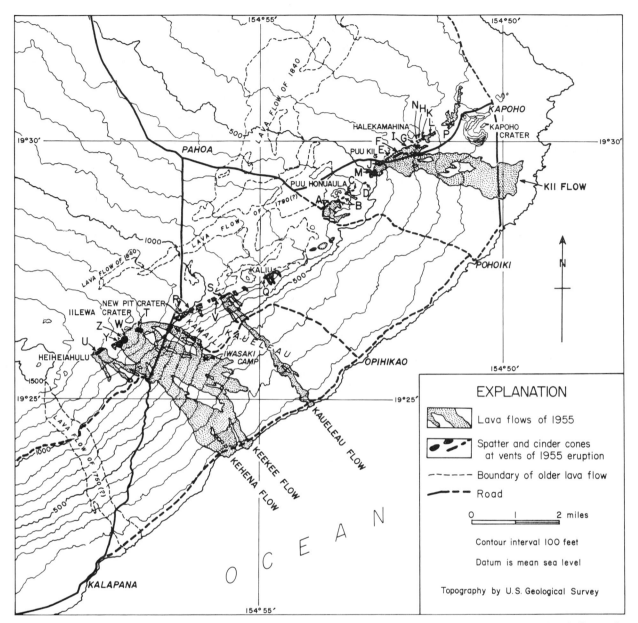

Figure 68. Map of east Puna (the east rift zone of Kilauea), showing the lava flows of 1955 in relation to older historic flows. The 1955 vents, identified by letters, are discussed in the text. (After Macdonald and Eaton, 1964.)

mile away. Whether gas was already rising in amounts detectable to the dogs but not to humans, or whether their uneasiness was wholly due to the ground vibration and earthquakes, we do not know, but dogs do not usually behave in that manner during earthquake swarms.

All things taken together, by the evening of February 27 it seemed quite certain that the volcano was about to erupt somewhere in the vicinity of the Nanawale Ranch. The eruption came even sooner than expected. At approximately 8:00 A.M. on February 28, lava fountains broke out along a fissure that crossed the Pahoa-Pohoiki road just northeast of the Nanawale Ranch, 3.6 miles from Pahoa (fig. 68).

Despite the numerous earthquakes that preceded it, the eruption came as a surprise to most people living in east Puna. It should not have. To volcanologists it was obvious that the east rift zone of Kilauea was a very active region in which there were apt to be frequent eruptions: more than 60 small cones, each marking the place of a former eruption, can be counted between Kilauea caldera and Cape Kumukahi, the easternmost point of the island of Hawaii; and these were only the last of thousands of eruptions along the rift zone that have built the ridge that makes up most of Puna. Early in his career as Hawaii's volcanologist, Dr. T. A. Jaggar had warned that more eruptions would come in Puna. But the last eruption had been in 1840, and 115 years of volcanic quiet had lulled the people of Puna into a false sense of security. Even the tremendous number of earthquakes failed to alarm them, because they remembered that even more numerous and more intense quakes 30 years earlier (1924) in the same area had brought no visible eruption.

During the first few hours of the eruption four short rows of lava fountains (areas *A* and *B,* fig. 68) played along fissures that ran in an east-northeastward direction between the road and the south base of Honuaula hill (a spatter and lava cone built by a prehistoric eruption). The fountains were less than 75 feet high, and a sluggish lava flow spread out from them. The two groups of fountains farthest northeast *(B)* diminished greatly during the late morning, and by early afternoon only the more southwesterly fountains *(A)* were still active. About 2:00 P.M., however, a new group of fountains formed still farther northeast (*C,* fig. 68), and another lava flow started to develop. The eruptive fissures were opening directly toward the village of Kapoho, and Civil Defense authorities were advised to move people out of the village. They acted promptly and effectively, and by nightfall evacuation of people and their belongings from the village was already well along.

Only a brief summary of the events of the 1955 eruption can be given here. Detailed accounts have been published elsewhere (Macdonald, 1959, pp. 41–63; Macdonald and Eaton, 1964).

By mid-morning of March 1, activity in area *C* (fig. 68) had ceased. In areas *A* and *B* lava fountains 50 to 75 feet tall had built spatter cones as high as 50 feet. Just after noon the fountains suddenly began to shrink, and within 5 minutes had completely disappeared. Immediately, clouds of steam began to pour from the vents in areas *B* and *C.* Most of the steam escaped gently, in great rolling white clouds, but at some places it rushed forth violently, tearing loose fragments from the walls of the vents, and carrying up clouds of black sand-

sized ash. There was very little odor of sulfur in the steam clouds, and no taste of salt. The steam was derived from the body of fresh ground water (the Ghyben-Herzberg lens: see chap. 15) that underlies that part of the island. Withdrawal of liquid lava from the shallow parts of the eruptive fissures probably was accompanied by cracking of the rocks, which allowed ground water to flow into the hot fissures and quickly become heated to steam. The resulting phreatic explosions continued for about 2 hours.

At about 4:00 P.M. a small outbreak of lava occurred at *D* (fig. 68), but it lasted for only a few minutes and produced a flow less than 50 feet long.

On the morning of March 2 all appeared quiet except for a small amount of steam drifting gently out of the vents at *B* and *C.* In the field it appeared that the eruption might be over, but the Pahoa seismograph indicated otherwise. After the outbreak on February 28, earthquakes had nearly ceased, but on the night of March 1 they began again, and on the following morning abundant quakes indicated the opening of new fissures, this time farther to the northeast. Early in the morning a crack opened across the Pahoa-Kapoho road (*E,* fig. 68). By 11:00 it was nearly 2 feet wide, and the ground surface southeast of it had dropped about 18 inches. This was one side of a long narrow sunken area (graben) that formed along the rift zone. About 1:00 P.M. another group of cracks started to open farther northeast (*N).*

Two hours later, lava broke out in area *E,* just southeast of the old spatter-and-cinder cone known as Puu Kii. The first sign of the actual outbreak was the appearance of wisps of white, sulfurous fume along one of the fissures. These quickly became larger and more numerous. Then, less than 5 minutes after the first appearance of fume, small shreds of red-hot lava began to be blown from the fissure, and were quickly followed by flowing liquid lava. Within 5 minutes the line of lava fountains was 1,000 feet long, and the largest fountains were shooting up 45 to 50 feet. Very fluid flows of pahoehoe lava spread outward from the fissures at a rate of about 40 feet a minute.

The remainder of that day and the following

87

day saw the opening of a series of other fissure vents farther northeastward (*F* to *L,* fig. 68) and one *(M)* just to the southwest, and the gradual increase in the size of the fountains at vent *E.* At 7:00 P.M. on March 3, new fissures appeared at the western edge of Kapoho village *(P)* and gradually spread on through the center of the village. Only a police detail and a few persons still loading heavy appliances and store stocks into trucks still remained in the village, and they were quickly moved away. At 9:30 lava fountains broke out within the western edge of the village. The village seemed doomed. But by great good fortune the lava-spewing vents did not spread into the center of the village, and the flow of lava was deflected from the main part of the village by a low spatter rampart built by an eruption in prehistoric time. About 15 houses in the outskirts of the village were destroyed by lava, and several others were made uninhabitable when fissures opened in the ground directly beneath them. But the main part of the village was spared—only to be annihilated by another eruption 5 years later.

The lava fountains at vents *E* and *M* continued to grow, and by the evening of March 4 the largest of them was more than 750 feet tall. A flood of lava poured from the vents at a rate probably exceeding 500,000 cubic yards per hour. Thereafter activity decreased, however, and by daylight on March 7 all activity had ceased. The lava flow came to a halt just short of the ocean.

Once again it looked in the field as though the eruption might be ended—but once again the seismograph told a different story. On March 5 a new series of earthquakes had begun, this time predominantly from centers along the rift zone southwest of the point where eruption had started on February 28. For several days quakes were recorded at Pahoa at a rate averaging between one and two a minute. A new outbreak in the vicinity of the Pahoa-Kalapana road was expected. Kalapana and Opihikao villages were evacuated, not because of any direct threat from the eruption, but because it was anticipated that lava flows would block the roads and isolate the residents from supplies and medical care. Finally, on the morning of March 12, cracks appeared in the

road at *R* (fig. 68). At 5:05 that afternoon lava broke out at *Q*, and 2 hours later at *S.* Both outbreaks lasted only a few hours.

At 7:50 A.M. on March 13 a new outbreak occurred just east of the road at *R*, followed throughout the day by a series of small outbreaks in the same area. These were in cleared land (fig. 69), and even in the paved road (figs. 70, 71) affording an opportunity for close-range observation such as no volcanologist had ever had before. The following account of the development of one of the new vents is quoted from a report (Macdonald and Eaton, 1955) published shortly afterward:

First, hairline cracks opened in the ground, gradually widening to 2 or 3 inches. Then from the crack there poured out a cloud of white choking sulfur dioxide fume. This was followed a few minutes later by the ejection of scattered tiny fragments of red hot lava, and then the appearance at the surface of a small bulb of viscous molten lava. The bulb gradually swelled to a diameter of 1 to 1.5 feet, and started to spread laterally to form a lava flow. From the top of the bulb there developed a fountain of molten lava which gradually built around itself a cone of solidified spatter.

The outbreak in the road was like the others except that instead of the lava emerging first as a small nearly hemispherical bulge, it pushed up through the pavement as a ridge 6 or 8 feet long and several inches thick, gradually increasing in length and thickness, and growing in height to 12 to 18 inches (fig. 70). We were viewing the top of a growing dike. Gas bursting out of the top of the dike then started to form a small lava fountain, which by the next morning had built around itself a spatter cone 20 feet high (fig. 71).

The eruption of lava at vent area *R* was short lived, but a week later the same area provided us with another unique experience. As noted on an earlier page, pit craters are numerous on Mauna Loa and Kilauea volcanoes, but, although a few are known to have been formed in historic time, never before had the formation of one actually been observed. At 4:03 P.M. on March 20, a sharp explosion in vent area *R* threw a billowing black cloud to a height of 500 feet, and several other similar but smaller explosions took place during the next hour. The closest ground observers were half a

Figure 69. Small lava fountains starting to build spatter cones along a newly opened fissure on the east rift zone of Kilauea, March 13, 1955. The highest fountains are about 5 feet high.

89

Figure 70. Ridge of molten lava 6 to 10 inches high pushing up through a fissure in the pavement of the Pahoa-Kalapana road, on the east rift zone of Kilauea, March 13, 1955. The ridge is the top of a growing dike, and the beginning of a new volcanic vent.

Figure 71. Spatter cone forming around a small lava fountain on the Pahoa-Kalapana road, east rift zone of Kilauea, March 14, 1955. The cone, built in about 18 hours at the new vent shown in figure 70, is about 20 feet high. A small flow of pahoehoe has advanced along the road toward the camera.

mile away, but within a few minutes of the first explosion an observer in an airplane found a new hole about 25 feet across in the ground surface at the site of the explosion. The interior of the hole was so brightly incandescent that it was difficult to see inside, but it appeared to be between 50 and 100 feet deep, with the walls overhanging so that it grew larger in diameter downward. The surface of the ground around the crater was covered with a thin layer of black glassy ash and fine cinder. Undoubtedly, this black ash was the cause of the dark color of the explosion cloud. The volume of ash thrown out was only a very small fraction of the volume of the crater, and the ash was wholly glassy—no stony material derived from old rocks was present. Obviously, therefore, the old rocks formerly occupying the site of the crater had not simply been blown out by the explosion, but must instead have dropped in. This illustrated nicely the manner of formation of pit craters and calderas

which previously had been only deduced from geological studies.

On March 14 activity had resumed at vent *S*, and an aa flow advanced southeastward from it, entering the ocean on the morning of March 16. Some Hawaiians said that once Pele got her feet wet she would go home, and the eruption would end; but events proved otherwise.

The rest of the eruption consisted of a series of outbreaks along the rift zone southwest of the Kalapana-Pahoa road (vents *T* to *Z*, fig. 68). Two more lava flows poured into the ocean (fig. 73), accompanied by small steam explosions that threw up showers of black glassy sand-size ash, some of which washed up on shore to form the beginning of black sand beaches. For two weeks, from April 7 to 24, the only surface activity was the quiet emission of gas; but the eruption then resumed, with fountaining and copious outpouring of lava (fig. 72), and continued until late May. On May 23 the lava fountain at vent *Z* was 250 to 300 feet tall, and a magnificent cascade of lava, 12 to 15 feet wide, was rushing down a 30° slope north of the cone at a speed of about 30 miles per hour. Similar activity was still in progress on the morning of May 26. Then, at approximately 11:15 A.M., activity started to die down. A professional photographer had been taking moving pictures of the fountain at intervals all morning. Grown blasé through three months of watching the eruption, he sat down with his back toward the fountain to smoke a cigarette. When he turned again to look, after three or four minutes, the whole grand display had come suddenly to an end. The eruption was over.

We cannot leave the 1955 eruption without some mention of events in the caldera region of Kilauea while lava was pouring out in Puna, 20 to 30 miles to the east. For several days before the beginning of the eruption there had been a slow sinking of the summit of the volcano, shown by slow southward tilting of the ground at the north edge of the caldera. This continued through the first week of the eruption, but on March 7 the rate increased greatly, and this more rapid sinking continued until mid-April. After that, the rate again decreased, but slow sinking continued until the end of the eruption, and indeed on until about the end of the year (fig. 65). The calculated volume of sinking of the summit of the mountain is about the same as that of the erupted lava (141,000,000 cubic yards), and it is tempting to think that the lava simply

Figure 72. Feeding river of an aa flow cascading down a steep embankment near the Pahoa-Kalapana road, May 17, 1955. The river is about 25 feet wide at the top of the embankment, and 50 feet wide at the widest point. A large accretionary lava ball is being rolled along about half-way down the cascade.

Figure 73. Lava flow pouring over a 50-foot sea cliff into the ocean, March 28, 1955. A steam cloud rises from the boiling water.

drained out from beneath the summit to produce the flows 20 miles to the east. However, the relationships cannot be that simple. The lava erupted in east Puna was quite different in composition from that of the 1954 eruption in Kilauea caldera, which almost surely was representative of the magma body remaining beneath the summit region. Probably magma moved out from beneath the top of the volcano into the rift zone to replace other magma which was removed from temporary storage in the fluid core of the rift zone to form the surface flows.

### 1959 and 1960 Eruptions

The eruptions of 1959 and 1960 form another typical summit-flank pair. After the end of the 1955 eruption there was no eruptive activity at either Kilauea or Mauna Loa until the outbreak of lava in Kilauea Iki Crater in November 1959. Kilauea Iki is a pit crater approximately a mile

in length, situated immediately adjacent to the eastern edge of Kilauea caldera and separated from it only by a low ridge known as Byron Ledge. The last activity in it had occurred in 1868, when a fissure part way up the southwestern wall poured lava onto the crater floor.

From late September until mid-November 1959, swelling of the top of Kilauea volcano was rapid, and in early October the frequency of earthquakes increased greatly. From October 7 to 20 there was an average of about 400 earthquakes a day (recorded on seismographs that were far more sensitive than those in Puna in 1955, and which consequently picked up many much smaller earthquakes than the older seismographs could detect). The Hawaiian Volcano Observatory reported that most of the quakes came from very shallow points of origin, many of them less than half a mile deep, close to Halemaumau Crater in Kilauea caldera.

Also, during October the amount of fume issuing from the 1954 eruption fissure in Halemaumau increased notably.

In the second week of November earthquakes became still more frequent, as many as 1,500 being recorded in a single day. On November 12 the rate of swelling of the mountaintop also increased and continued to be rapid through November 13 and 14. On the latter day about 2,200 quakes were recorded. An eruption of the volcano was expected momentarily, but the most probable site for the outbreak appeared to be Halemaumau. Just before 8:00 P.M. on November 14, harmonic tremor began recording on the seismographs, and at 8:08 the eruption commenced.

The outbreak was on the southwest wall of Kilauea Iki Crater (fig. 57B), roughly half way between the floor and the rim, and very near the line of outbreak of the 1868 eruption. The fissures quickly lengthened both eastward and westward, and there was soon a line of lava fountains 45 to 50 feet high nearly half a mile long. From the eastern fountains golden cascades of lava plunged 300 feet to the main crater floor; but from the western fountains the lava poured down onto a platform that stood 180 feet above the main crater floor. From this "mezzanine" floor the lava cascaded eastward down a steep slope to the main floor, where it joined the stream from the eastern fountains to form a rapidly spreading pool of liquid. However, most of the fountains lasted for only a few hours. By the morning of November 15, only two small fountains were still active, on the end of the fissure above the mezzanine, and the volume of lava being poured out was small.

Activity continued at a low level through November 16, but on November 17 it began to increase again. By the afternoon of the 18th a single fountain was playing to a height of 750 feet, and on the 19th it occasionally reached 1,000 feet. Lava poured out at the base of the fountain at a rate of about 1,000,000 cubic yards per hour, spread over the mezzanine, and cascaded onto the main floor. (See photograph on page 48.) Behind the fountain a semicircular cone of cinder and pumice was accumulating on the rim of the crater.

On November 21 the height of the fountain varied greatly, at times being as low as 150 feet, but at others reaching as high as 1,250 feet. Hot pumice fell in showers on the road just south of the crater, and large crystals of olivine, blown clear of the lava, trailed behind them long strands of Pele's hair. During the morning the surface of the deepening lava pool reached the level of the vent, until, in the afternoon, the fountain threw up both bright orange-red lava that was rising from depth, and cooler deep red lava that had flooded into it from the surrounding pond. Perhaps the effort of bursting its way through the pool was too great, or perhaps partly cooled lava trickling back down the conduit tended to clog it. At any rate, early that evening the fountain suddenly disappeared. At 10:00 P.M. the eruption appeared dead.

For the next four days there was no surface activity, but seismographs recorded continuing harmonic tremor, indicating that magma was still moving within the volcano. Just after midnight on the night of November 25–26, the lava fountain resumed activity as suddenly as it had stopped. This was the beginning of a series of brief eruptive episodes, ranging from 2 to 32 hours in length, separated by short periods of quiet. At the end of each episode lava drained back down the feeding conduit as it had at the end of the first episode. This had occurred also during the 1954 eruption in Halemaumau and at times during the existence of the lava lake previous to 1924. The lava fountain reached a height of at least 1,500 feet, and perhaps more than 1,700 feet—a height unprecedented in the records of Hawaiian volcanoes. The great height probably resulted from an unusually great abundance of gas in the erupting lava. At times the temperature of the lava in the core of the fountain was in the neighborhood of 1200° C., which is the highest temperature yet measured in erupting Hawaiian lava. Cinder and spatter from the fountain piled up on the crater rim, building a conical hill 150 feet high, and pumice formed a blanket that was 5 feet thick at a distance of half a mile from the vent. The effects of the falling pumice and bombs are spectacularly displayed along the Devastation Trail southwest of the cone. The final depth of new lava in the crater was 380 feet, and the final volume of lava remaining in the

93

crater after the last drainage back into the vent was about 80,000,000 cubic yards.

The 1960 eruption is noteworthy for its unusually great explosiveness, and because walls were constructed in an attempt to reduce the spreading of the lava flows. It also completed the destruction of Kapoho village.

Soon after the end of the Kilauea Iki eruption, in December 1959, very numerous earthquakes commenced on the east rift zone of Kilauea volcano in east Puna. Some of them were felt by persons living in the area, but most were so small that they were detected only by seismographs. About January 10, 1960, the earthquakes became centered in the vicinity of Kapoho (fig. 68). Many of the quakes were felt in the village and persons who were working in the area north of the village on January 12 and 13 reported that some were so strong that it was difficult to remain standing. East-northeast-trending cracks opened in the ground in and near the village. A crack across the paved road within the village was about an inch wide on the morning of January 13, but had widened to 3 feet by early afternoon, and the ground surface on the north had dropped nearly a foot. Along another crack just north of the road, the ground to the north had dropped 2.5 feet. These displacements of the ground were along faults that define the south edge of a long narrow graben. The north edge of the graben consisted of a line of cliffs as much as 50 feet high—fault scarps formed by sinking of the ground to the south. On January 13 renewed movement on the faults along the north edge of the graben dropped the ground surface to the south an additional 4.5 feet.

The village of Kapoho lay almost entirely within the graben. (The name "Kapoho" means "the sunken place.") Much of the sinking of the graben had taken place in prehistoric times, and still more had occurred in 1924. At the north base of the wooded prehistoric cinder cone, Puu Kukae, a fault crack along the south border of the graben exposed the surface of the body of fresh water (basal ground water) that floats on salt water within the rocks of the island (see chap. 15). The water was slightly warm, heated by intrusive bodies in the rift zone of the volcano, and the locality was known as Warm Spring. Beautifully landscaped, it was a favorite resort for both islanders and visitors.

During the morning of January 13, Kapoho was declared a disaster area by the Civil Defense Agency, and evacuation of the village was begun. At 12:30 the seismograph operator at the Pahoa station reported the beginning of harmonic tremor, indicating that magma was moving in the volcanic conduits; and at 7:30 P.M. molten lava broke out in a sugarcane field half a mile northwest of the center of Kapoho village.

The erupting fissure opened gradually eastward, and a line of lava fountains along it gradually grew until it was more than half a mile long. By midnight the three westernmost fountains had ceased activity, but farther east was a row of 15 to 20 fountains, ranging in height from a few feet to more than 300. Steam blasts accompanied the fountains, showering the surrounding area and plastering the vegetation with wet black ash containing a large amount of salt. Similar steam explosions continued sporadically, along with the lava fountains, through much of the eruption. The surface of the ground-water body (the same one that was exposed at Warm Spring) was only about 80 feet below the ground surface in the eruption area, and the hot lava rising through the water generated steam, which rushed out carrying with it a mixture of frozen glassy spray from the molten lava, pulverized old solid rock, and salt from the ocean water beneath the fresh-water layer. As compared with eruptions of many other volcanoes, the explosions were very mild indeed, but they were much more violent than those of most Hawaiian-type eruptions.

By the morning of January 14 the number of lava fountains was reduced to seven, and through most of the eruption there were only one to three fountains. The activity centered approximately a third of a mile northwest of Kapoho village, and at times the main fountain reached a height of nearly 1,000 feet, looming as a great ominous red column over the village. Ash clouds shot upward 2,000 feet. During most of the eruption spectators were allowed to come as far as the edge of the village, but from time to time, when the wind veered, they had to be moved quickly away to avoid being showered with cinder and pumice. A cone of cinder and spatter gradually grew around the vents, and by the end of the eruption what had been a broad shallow valley was transformed into a hill 350 feet high.

Lava flows poured out at the base of the fountains. Occasional small flows spread westward, and some circled the base of the cone toward the village, but most of the lava flowed eastward along the course of the graben valley. On the morning of January 15 the flow entered the ocean, with a spectacular burst of rolling clouds of white steam. Although the first lava reached the ocean at the north edge of the valley, the slope of the underlying surface was so gentle (only 38 feet per mile) that the tendency of the lava to congeal because of chilling where it met the water caused a clogging of the channel, and as a result the flow turned southward along the beach. By the afternoon of January 18, lava had buried the Warm Spring area. On that afternoon the

volume of lava being extruded was estimated to be approximately 6,600,000 cubic yards a day.

Gradually the fiery flood engulfed nearly the whole village of Kapoho. At first the lava was confined to the graben valley, but the drainage of lava into the sea was very slow, and it soon became apparent that, if the eruption continued, the valley would be filled and lava would spill out over the surrounding countryside. There was no way to prevent it from burying most of Kapoho or from spreading northward toward the village of Koae, but, if the height of the ridge along the south side of the valley could be increased, it was thought that it might be possible to prevent overflow in that direction and save Kapoho School, Kumukahi Lighthouse and the lighthouse keepers' residences, and the 20 or so houses along the shoreline between Waiakaea Bay and Kapoho Point (fig. 74). For that purpose a series of walls was built along the ridge south of the valley.

It should be noted that the walls built at

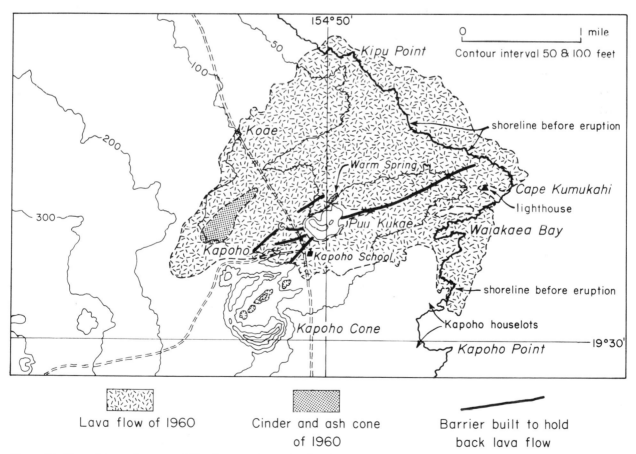

Lava flow of 1960     Cinder and ash cone of 1960     Barrier built to hold back lava flow

Figure 74. Map of the region around Kapoho, east Puna, showing the cone and lava flow of 1960 and the position of barriers built in an attempt to hold back the flow. (After Macdonald, 1962.)

95

Kapoho were not intended as diversion barriers such as those described at the end of this chapter. They were simply dams raising the height of the ridge, and it was recognized that, as with any dam, if the liquid behind them became too deep it would overflow. To a considerable extent, it was a battle against time. If the eruption ended soon enough, the walls would not be overflowed.

The walls, about 20 to 30 feet high, consisted of loose rock fragments and soil pushed up by bulldozers. Some, near Kapoho village, were intended largely to delay the encroachment of the lava until the main wall could be built along the crest of the ridge. Altogether, about 3 miles of walls were built.

The degree to which the walls succeeded in their purpose is a matter of controversy. Some people, including some of our professional colleagues, claim that they failed. Unfortunately, this can be only a matter of opinion. On January 28 the ground surface just north of the wall near Kapoho School was heaved upward, apparently by an intrusion of magma at very shallow depth, a fault developed along the wall, and lava, issuing from a vent on the fault on the south side of the wall, destroyed the school. This cannot be attributed to mechanical failure of the wall; it was just bad luck. Later, lava overflowed the long wall farther east and destroyed the residences of the lighthouse keepers. But though lava crawled to its very base, the lighthouse itself remained unscathed. Lava filled Waiakaea Bay, and advanced into the edge of the settlement along the beach, destroying several houses; but most of the beach houses remained intact. Although there is no way to prove it, it is our opinion that the walls prevented additional southward flooding of lava which would have destroyed not only the lighthouse, but most or all of the remaining beach houses as well. Thus we believe that, although to some extent they did fail, the walls served a real purpose. And no one can deny what is perhaps the greatest dividend of all: a great deal of information was gained on how such walls behave, and how they should be built when they are again needed.

By the time the eruption had ended, on February 19, a total volume of about 150,000,000 cubic yards of lava had been poured out, and the shoreline had been pushed seaward in places nearly half a mile. Approximately half a square mile of new land had been added to the island of Hawaii.

As in 1955, the eruption in east Puna was accompanied by a general sinking of the whole top of the volcano, the volume of the subsidence being approximately equal to that of the extruded lava. On January 25 the Hawaiian Volcano Observatory staff estimated that the amount of sinking to that date was equal to about half the amount of swelling that had preceded the eruption, and on that basis it was believed that the eruption would continue for some time. As we have seen, the expectation proved correct. The eruption continued another three weeks.

On January 28 an ash-laden steam cloud rose from the 1959 vent in Kilauea Iki Crater, probably as a result of subsurface cracking that admitted ground water to the still-hot conduit that had fed the 1959 eruption.

The effects of the subsidence of the volcano were most conspicuous in Halemaumau. On February 7 the whole floor of the crater suddenly sank, leaving a broad basin about 130 feet deep. Within this was formed a steep-walled inner pit 300 feet across and 150 feet deep. Liquid lava, part of the still-molten interior of the lava lake of 1952, drained from beneath the surface of the former crater floor into the central pit, filling it to a depth of about 75 feet. (The interior of a thick pool of molten lava cools and solidifies very slowly. Three years after the end of the 1959 eruption the solidified crust on the pool of lava in Kilauea Iki was only 50 feet thick.) On February 9 a second collapse formed another pit on the floor of Halemaumau southwest of the first, and on March 11 a third pit formed at the northeast edge of the crater floor (fig. 75). The three pits lay approximately on the line of the eruptive fissures of 1952 and 1954.

The center of sinking of the mountaintop, however, was somewhat northeast of Halemaumau, as it had been also in 1955 and 1950. The general subsidence of the mountaintop was probably caused by withdrawal of magma from a reservoir 2 or 3 miles beneath the surface, whereas the local collapse within Halemaumau

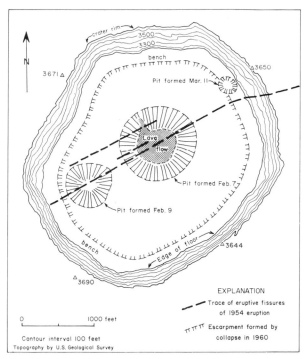

Figure 75. Map of Halemaumau Crater, showing the pits formed during the collapse of 1960. (After Macdonald, 1962.)

resulted from drainage of magma from the still-fluid lower portion of the lava lake of the 1952 eruption and from small feeders of recent eruptions, probably of dike form, that were still liquid to a level very close to the surface.

## Activity Since 1960

After the shrinking of Kilauea and the collapses in Halemaumau that accompanied the 1960 eruption, swelling of the volcano quickly resumed. In February and March 1961, two small eruptions partly filled the pits formed in Halemaumau by the 1960 collapses, and in July the filling was completed by a larger eruption that once again established a flat floor in the crater (fig. 67B). In September still another outbreak occurred—the fourth in a single year, constituting a record for frequency of eruption of a Hawaiian volcano during historic time. The September eruption consisted of outbreaks and small lava flows at 13 places in a 12-mile stretch along the east rift zone beyond Napau Crater (figs. 62, 76). The volume of lava poured out on the surface was less than 2,000,000 cubic yards, but measurements by the Volcano Observatory staff indicate that the volume of the accompanying subsidence of the mountaintop was about

70,000,000 cubic yards. What had happened to the 68,000,000 cubic yards of magma drained out of the underlying reservoir, but not accounted for in the lava flows? A small part of it was seen to run back down from the surface into open fissures. Some of the rest may have receded into the depths from which it originally came, but most of it probably was injected into rift-zone fissures below the surface. In this way are formed many of the vast number of dikes seen in the exposed parts of the rift zones of the older volcanoes.

In December 1962 a small eruption produced six small lava flows along the east rift zone between Aloi Crater and the prehistoric lava shield called Kane Nui o Hamo, that lies just north of Makaopuhi Crater (fig. 62). Lava pouring into Aloi Crater quickly filled it to a depth of about 40 feet, but most of this lava drained back into the feeding fissures, leaving a permanent fill only about 6 feet deep.

The 1962 eruption was accompanied by movements along the rift zone causing many cracks across the Chain of Craters Road. Displacement of the center line on the pavement indicated that the ground south of these faults had moved westward in relation to that north of them. The summit area of the volcano sank, while the eruption area rose about 2.5 feet.

Again in July 1963, great numbers of earthquakes accompanied further movement in the same general area on the east rift zone. Faults crossing the road displaced the pavement as much as 4 feet horizontally and 3 feet vertically. The direction of horizontal movement

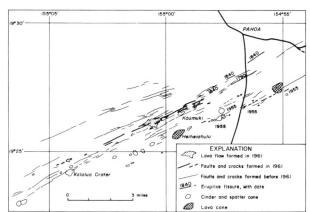

Figure 76. Map of part of the east rift zone of Kilauea, showing faults, cracks, and lava flows formed in 1961. (Modified after Richter and others, 1964.)

97

was the same as in the previous year. Soon afterward, in August, a small eruption occurred in Alae Crater, a mile east of Aloi Crater. Then in October still another took place in Napau Crater, 3 miles farther east, followed shortly afterward by outbreaks at two other places 2 miles east-northeast of Napau Crater. At the latter place the lava formed two flows, one of which reached a length of nearly 8 miles. However, all the activity was over in about 30 hours, and the total volume of lava extruded in the two eruptions of 1963 was only about 9.5 million cubic yards. It seemed almost as though the rift zone had become so full of lava that a little simply spilled out at points of weakness. The eruptions caused very little sinking of the mountaintop, and the east rift zone and summit area of the volcano remained highly inflated.

At 8:00 A.M. on March 5, 1965, the seismographs began recording volcanic tremor and the summit of the volcano started to sink, indicating that magma was again moving into the east rift zone. At 9:43 A.M., eruption commenced on a series of fissures that extended across the western end of Makaopuhi Crater and eastward through Napau Crater for a total distance of 8 miles (fig. 77). During the

first 8.5 hours more than 20,000,000 cubic yards of lava poured out, forming a pool 260 feet deep in the deeper, western part of Makaopuhi and another 40 feet deep in Napau Crater. In the early afternoon the rate of this outflow decreased and lava began to drain back into the feeding fissures. During the next few hours the drainback reduced the thickness of the fill in Napau Crater to about 20 feet. By 4:30 P.M. surface activity was over, but, after about 16 hours of quiet, weak fountaining resumed and continued intermittently for 4 days, pouring out an additional 2 or 3 million cubic yards of lava at several vents scattered along the rift zone for 6 miles eastward from Makaopuhi.

About 1:30 P.M. on March 10, the weak activity at the more easterly vents stopped, and simultaneously the strength of fountain activity and the volume of lava pouring out in Makaopuhi increased. The temperature and the fluidity of the lava also increased, suggesting that a new batch of magma had arrived from a deep-seated source. The rate of lava outflow continued to increase until, on March 15, it reached about 200,000 cubic yards an hour; but on that day the level of the lava pool reached the higher of the two active vents on

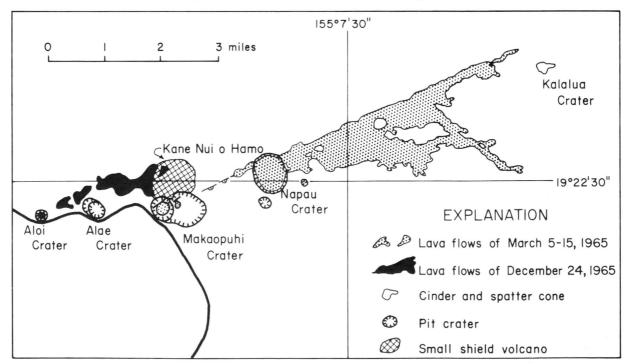

Figure 77. Map of the eastern part of the Chain of Craters, Kilauea, showing the lava flows of March and December 1965. (After J. G. Moore and D. L. Peck, Hawaiian Volcano Observatory Summary 37, 1966.)

the crater wall, and fountaining ceased. Outflow of lava into the crater also stopped at about the same time. By then the pool of new lava was 340 feet deep and its volume was about 9,000,000 cubic yards; but during the next 4 days about 2,500,000 cubic yards drained back into the vents, decreasing the thickness of the permanent lava fill to about 290 feet.

On Christmas Eve 1965, still another small eruption took place. For 6 hours a few small fountains played along a series of fissures extending for 2 miles eastward from Aloi Crater (fig. 77), and a cascade of lava streamed down the crater wall, forming a pool about 50 feet deep. During the next few hours nearly all of this lava drained back into the vents, leaving a layer only 5 feet thick over the 1962 lava on the crater floor.

The year 1965 ended with sinking of the summit region of Kilauea and inflation of the nearby portion of the east rift zone. However, tumescence of the summit region soon began again, and continued through 1966 and most of 1967. For the first time, the Volcano Observatory tiltmeters were now sufficiently sensitive and adequately distributed to reveal some of the details of the tumescence. At first the apex of the bulge was northeast of Halemaumau, approximately at the place where the maximum sinking had occurred during the periods of deflation in 1950 and 1955. Over several months, however, the center of swelling slowly shifted southward to a point southeast of Halemaumau. During the single month of October 1967, the ground surface a mile southeast of Halemaumau rose about 0.4 foot. Obviously, the volcano was in a condition of potential eruption.

At approximately 1:30 A.M. on November 5, 1967, volcanic tremor started to record on the seismographs. At 2:05 a sharp earthquake was felt, and at about 2:30 lava fountains broke out along a line extending nearly north-south across the floor of Halemaumau. Lava poured into the crater at a rate of more than 1.5 million cubic yards an hour. At first the fountains were only 50 or 60 feet high, but they gradually increased in height, and by mid-afternoon occasionally were reaching 200 feet. By midnight the pool of new lava was 100

feet deep, but the strength of the activity was decreasing. An hour later it stopped, and the molten rock started to drain back into the vents from which it had come. By mid-afternoon the level of the lava surface in the center of the crater had sunk 45 feet.

To an observer at the edge of the crater it looked as though the eruption was over, but the seismographs continued to record weak tremor, showing that magma was still moving at depth. On the morning of November 9 eruption resumed, at first very weakly, but gradually increasing in strength over the next several days, with occasional jets of lava being thrown as high as 50 to 100 feet. The tremor increased in strength as greater volumes of magma moved upward. A lava lake was formed (fig. 78), covering about 40 percent of the crater floor, and around its edges the spattering of secondary fountains built a levee that confined the lake surface at a level 20 to 30 feet above the floor outside the lake. Occasional overflows sent tongues of lava down the outside of the levee and across the surrounding floor. By November 19 the lake surface was 110 feet above the former bottom of the crater, but at 7:45 P.M. on that day extrusion of lava again ceased, and by noon of the next day drainage of the liquid lava back into the conduits had lowered the lake surface 20 feet. Extrusion began again on November 21, and lava from fountains 50 to 100 feet high quickly refilled the lake basin and restored the condition of November 19.

This alternation of periods of fountaining and lava-lake activity with periods of drainback and inactivity continued for three months, through a total of 28 active periods, with a continual net rise of the level of the lake surface and surrounding crater floor.

Phase 29, which began on February 27, 1968, brought a change in the pattern of the eruption. The periods of inactivity ceased, and lava lake activity became continuous. Overflows from the lake onto the surrounding floor became much less frequent, but the entire floor continued to rise as a unit, pushed upward from beneath and carrying the lake upward on its back. As the floor moved upward between the upward-flaring crater walls, a space was left between it and the walls through which molten

Figure 78. Lava lake and fountains in Halemaumau Crater, seen from the air on November 15, 1967. The step-fault blocks at the edge of Kilauea caldera at Uwekahuna are visible in the background.

lava oozed up from time to time and made small flows at the foot of the walls. The main fountain activity was near the northeast edge of the lake, and there a group of spatter cones was built up, the highest of them about 50 feet high, steeply conical and symmetrical. The lake gradually decreased in area, until by June it was only about 800 feet across. The average rate of rise of the crater floor was more than a foot a day, and by early July the lake surface was only about 100 feet below the crater rim. The depth of the fill in the crater since the eruption began was about 440 feet (figs. 79 and 67B). Activity ceased on July 8, however, and draining back of the lava lowered the lake surface about 40 feet. The rest of the floor of Halemaumau remained stationary. On July 13 lava trickled quietly into Halemaumau for a few hours, forming a small pool in the bottom of the basin left by the sinking of the lake.

Early on August 22, 1968, another outbreak occurred on the east rift zone along a fissure that crossed Heake Crater (fig. 62). Rapid outpouring of lava soon formed a molten pool 300 feet across and 100 feet deep in the bottom of the crater. At about the same time another small flow issued from another fissure a mile to the east. About three hours after its beginning, the activity in Heake Crater started to diminish and the lava began to drain back into the vent. By noon the activity in Heake Crater had ceased and drainback had removed most of the lava, leaving only a thin veneer over the crater bottom. During the next four days cracks opened progressively eastward for more than 13 miles. Along the more westerly cracks a violent release of hot gas occurred, but no lava reached the surface. Along the more easterly cracks small amounts of spatter and very short lava flows were ejected. The last activity occurred on August 27.

During the first 36 hours of the August eruption the summit of Kilauea sank a little more than 6 inches, but deflation of the volcano then ceased, and after another 36 hours the summit region began to inflate again.

Inflation of the volcano continued through September.

On October 7, 1968, another eruption began on the east rift zone, along a series of fissures from the eastern flank of Kane Nui o Hamo (fig. 77) to 2 miles east of Napau Crater. Activity continued, with varying strength and occasional periods of quiet, until October 21. The source fissures lay along, or a little to the north of, the 1965 fissures, but considerably south of the August 1968 fissures. Lava covered a large part of the flows of March 1965, and again poured into Napau Crater. It is interesting to note that the lava differs considerably from that erupted on the same rift zone only two months earlier. It contains only a few very small crystals of olivine, whereas that of August 1968 contains many fairly large ones.

Still again, on February 22, 1969, eruption occurred in the same general area. A line of lava fountains formed between Alae and Napau craters, lava cascaded into Alae and Makaopuhi craters, and flows buried 3 miles of the Chain of Craters Road. Activity came to an end on the night of February 27. The top of Kilauea deflated slightly during the eruption, but tumescence resumed soon afterward as once more the magma reservoir was refilled in preparation for the next eruption.

It came on May 24, with lava fountains at three places along the rift zone from just southwest of Aloi Crater to Kane Nui o Hamo. Lava poured into Alae Crater and again covered part of the Chain of Craters Road. Activity ceased on the evening of May 25, but recommenced on May 27, to continue until May 29.

Figure 79. Halemaumau Crater, in Kilauea caldera, after the 1967–68 eruption. Comparison with figure 78 shows the amount of filling of the crater during the eruption. The wall of the caldera is in the middle distance, and in the background is the broad rounded shield of Mauna Loa.

Next came 2 weeks of quiet, in which new magma moved into the reservoir and the volcano regained the swelling it had lost during the brief eruption. On June 12 lava fountains as much as 500 feet high again shot from the fissure between Aloi and Alae craters, and a voluminous rapid flow of lava poured southward across the road and down over the cliffs of the Hilina fault system to form a broad pool on the flat land a mile from the coast. The activity stopped abruptly on June 13. Again, on June 25, activity resumed in the same area, with fountains 400 feet high and voluminous outpouring of lava, and again a flow cascaded down the Hilina fault scarps (fig. 198), sending a narrow tongue into the ocean just east of Apua Point (fig. 62). The activity stopped early on June 26, and the volume of lava that went into the sea was very small. Still another outbreak occurred at the same place on July 15, and so on through the summer and autumn.

At the time of writing, the eruption is still in progress after 11 eruptive phases since its beginning in May, separated by periods of relative quiet. Most of the eruptive phases have lasted about 10 hours, with interphase periods from 2 days to 4 weeks. During quiet periods the summit of Kilauea has swelled up, only to sink back rapidly during the eruptive phases. The fountains lie along an open fissure about 300 feet long, and during the interphase periods molten lava has usually been visible from 20 to 75 feet below the surface in the fissure. At times it has risen to the surface and overflowed slightly. During one of the eruptive phases, in early September, the fountains reached a height of 1,500 feet. A cone of welded spatter, more than 100 feet high, has been built on the leeward side of the vents, and beyond it stretches a broad blanket of spatter and pumice. Lava flows have covered an area of about 5 square miles. Flows spilled into Alae Crater, filling it to overflowing, so that only a low crescentic scarp remains to mark the former highest side of the crater (fig. 33).

DIVERSION BARRIERS TO PROTECT HILO

The walls built near Kapoho bring to mind those that have been suggested to protect Hilo, the chief port of the island of Hawaii, but the principle behind the Hilo walls is quite different. The walls at Kapoho were simply dams, built in the hope of preventing the spread of lava toward the south. The Hilo walls, on the other hand, would be intended not to act as dams, but to change the course of the lava flow sufficiently to miss the city and harbor. Other land would be destroyed, of course, but it would be land of less value.

Hilo Bay lies at the intersection of the slope of Mauna Loa with that of Mauna Kea to the north (fig. 80). The valley of the Wailuku River is formed by the intersecting slopes of the two mountains, and the river itself follows approximately the edge of the lava flows of Mauna Loa where they have overlapped onto the slope of the older Mauna Kea. The part of the city of Hilo south of the Wailuku River is built almost wholly on very recent lava flows of Mauna Loa, many of them probably less than 2,000 years old. During the interval between 1830 and 1950 Mauna Loa erupted on an average of once every 3.6 years, and most of those eruptions took place along the rift zones of the volcano or in the summit caldera. One of the rift zones extends from the summit east-northeastward directly toward Hilo (fig. 45). The three small cinder cones known as the Halai Hills, within the city of Hilo, are old cones on the northeast rift zone, but fortunately, in recent millennia, very little activity has taken place along the part of the rift zone below 6,000 feet altitude. Only one vent, marking an eruption which occurred between 2,000 and 4,000 years ago, has been found on the lower extension of the rift.

The northeast rift zone is approximately a mile wide, and it lies along the crest of a broad ridge built by its own eruptions. Outbreaks near the southern edge of the rift zone send flows down the south slope of the ridge, but those near the northern edge send flows northward or northeastward into the Wailuku River valley, where they are directed toward Hilo. Along the part of the rift zone below about 11,500 feet altitude, all the lava flows from the northern part of the rift zone advance toward Hilo. Whether or not they reach Hilo depends largely on the volume of lava. Between 1850 and 1950, five major flows were directed

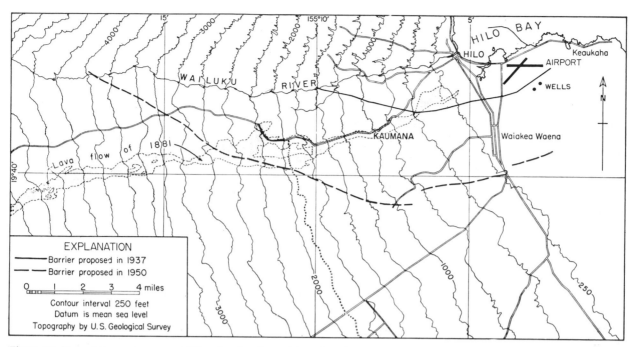

Figure 80. Map of the region around Hilo, showing the position of barriers suggested to divert lava flows from the city and harbor. (From Macdonald, 1958.)

toward Hilo. In 1881 a flow actually entered what is now Hilo, and only the end of the eruption prevented it from entering, and eventually filling, Hilo harbor. The volume of the 1881 lava flow was about 250,000,000 cubic yards. If it had had a volume equal to that of the 1859 flow on the northwest slope of Mauna Loa—more than 600,000,000 cubic yards—there is little question that the harbor and the town would have been destroyed.

Fortunately, the belt through which lava flows may advance toward Hilo is only about 6 miles wide. It is this restriction of possible lava paths that makes feasible any serious consideration of the possiblity of protecting Hilo by building diversion barriers. The barriers were first suggested by Dr. T. A. Jaggar. The scheme consists of a series of walls, averaging about 30 feet high, built in much the same manner as those near Kapoho, extending across the slope of Mauna Loa above Hilo. Details of the suggested manner of construction are given in papers by Jaggar and by Macdonald (see list of

suggested reading at the end of this chapter). The position of the barriers suggested by Jaggar, and also those suggested later by Finch and by Macdonald (1958, p. 272), are shown in figure 80. The change was made only because of the growth of the city during the interim. From the standpoints of shorter length, and hence lower cost of construction, and of the dimensions of the channel behind the barrier, the position suggested by Jaggar is still the better. All the barriers are intended to divert flows away from the city and harbor, and direct them southward into lands of lesser value to the community as a whole.

The scheme suggested by Dr. Jaggar was studied in detail by the Corps of Engineers of the U.S. Army and considered feasible. The whole idea of diversion barriers has been attacked by other geologists (Wentworth, Powers, and Eaton, 1961). The question remains controversial, but the possiblity that barriers would be successful appears great enough that the idea should not be abandoned.

## Suggested Additional Reading

Brigham, 1909; Dana, 1890; Finch, 1940, 1941; Finch and Macdonald, 1953; Hitchcock, 1911; Jaggar, 1936, 1945, 1947; Jaggar and Finch, 1924; Kinoshita and others, 1969; Macdonald, 1943, 1947, 1955a, 1958, 1962; Macdonald and Eaton, 1957, 1964; Macdonald and Orr, 1950; Moore and Koyanagi, 1969; Murata and Richter, 1966; Richter and others, 1964; Schulz, 1943; Stearns, 1926; Stearns and Macdonald, 1942, pp. 102–107; 1946, pp. 82–96, 112–128, 146–148; Wright, Kinoshita, and Peck, 1968.

*Cross sections of aa lava flows, Halawa Quarry, Honolulu*

# CHAPTER FOUR

# Hawaiian Minerals and Igneous Rocks

BEFORE CONTINUING OUR ACCOUNT OF THE GROWTH OF Hawaii's volcanic mountains it is necessary to consider briefly the materials of which they are made. These materials are *rocks,* which in turn are made up of *minerals* and mineral-like substances called *mineraloids.* Full considerations of rocks and minerals constitute respectively the sciences of petrology and mineralogy. Here we can take space only for a very elementary treatment of the two subjects, carrying the discussion only as far as is necessary to understand the essential features of Hawaiian rocks.

Rocks in general are divided into three great classes. *Igneous rocks* are those which have solidified from a once-molten condition. *Sedimentary rocks* have been formed by the transportation and deposition of rock material by one of the geologic agents of transportation: water, wind, ice, and gravity. *Metamorphic rocks* are former sedimentary or igneous rocks which have been changed from their original condition (metamorphosed) to bring them into equilibrium with a new physical or chemical environment—that is, to adjust them to a higher or lower temperature or pressure than that in which they were originally formed. The change is aided at times by water or steam containing other substances in solution. The igneous rocks, which make up by far the greatest bulk of the islands, and the comparatively minor (in Hawaii) metamorphic rocks, are treated in this chapter. Sedimentary rocks are discussed in chapter 12, but for convenience the minerals found in them are treated here, along with the igneous minerals.

## MINERALS

A mineral may be defined as a natural solid compound having a definite and invariable internal structure and a narrowly delimited chemical composition. Minerals consist of combinations of atoms of various chemical elements arranged in certain patterns in relation to each other. To take a simple example, the mineral halite or rock salt ($NaCl$) consists of atoms of sodium (Na) and chlorine (Cl) arranged in such a way that they alternate in all directions in three planes at right angles to each other (fig. 81$A$). The resulting mineral grains have the shape of cubes. Similarly, other minerals have

105

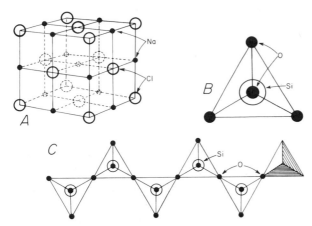

Figure 81. Diagrams showing the atomic structure of a halite crystal (A), a silica tetrahedron (B), and the chain of silica tetrahedra that forms the basic structural element of a pyroxene crystal (C).

characteristic arrangements of their constituent atoms. The mineral quartz ($SiO_2$), for example, consists of atoms of silicon (Si) and oxygen (O) arranged in such a way that one silicon is surrounded by four oxygen atoms at the four corners of a tetrahedron (fig. 81B). Adjacent tetrahedrons share the oxygen atoms at their corners, and in that way the pattern is extended throughout the grain of the mineral, and the final proportion of oxygen atoms to silicon atoms is only two to one. Other minerals consist of the same basic groups of silica tetrahedra, arranged in chains or in sheets (fig. 81C). These chains or sheets are tied together by atoms of other chemical elements, or in some instances simply enclose atoms of other elements. The nature of the atoms, their arrangement, and the strength of the bonds between them determine the physical properties that give each mineral the individual characteristics by which we recognize it.

Nearly all of the known chemical elements are present to some degree in minerals and rocks, but most of them form only a very small proportion of the earth as a whole. Eight elements make up 98.5 percent of the earth's crust. The average chemical composition of the crust is as follows (expressed as percent by weight):

| | | | |
|---|---|---|---|
| Oxygen (O) | 46.6 | Sodium (Na) | 2.8 |
| Silicon (Si) | 27.7 | Potassium (K) | 2.6 |
| Aluminum (Al) | 8.1 | Magnesium (Mg) | 2.1 |
| Iron (Fe) | 5.0 | All others | 1.5 |
| Calcium (Ca) | 3.6 | TOTAL | 100.0 |

Thus nearly one-half the earth's crust is oxygen. It is perhaps surprising to realize that rocks contain a larger proportion of oxygen than does the air we breathe (16 percent).

Several thousand different minerals have been recognized, but again, only a very few make up most of the earth's crust. Most of these are silicates—combinations of silicon and oxygen with the other common chemical elements. The common minerals that make up the rocks of Hawaii are listed in table 4.

The division into light- and dark-colored minerals is an easy and fundamental one. Note that the dark minerals all contain magnesium (Mg) or iron (Fe) or both: they are often referred to as ferromagnesian or femic minerals. Parentheses enclosing two chemical symbols in a formula indicate that the proportion of the two elements in that mineral is variable. In olivine, for instance, the proportion of magnesium and iron can vary all the way from 100 percent Mg to 100 percent Fe. To a lesser extent, other elements also may substitute for those listed in the formulae, but, with these relatively minor exceptions, the chemical composition of any mineral is always about the same.

Although each mineral has a nearly constant chemical composition, the same chemical elements may combine in different patterns to form different minerals. Thus the compound $CaCO_3$ (calcium carbonate) may form either of two minerals—calcite or aragonite—found in Hawaiian limestones. Therefore, even if it were practicable to make a chemical analysis of every mineral we want to identify, which would be both time-consuming and expensive, the analysis still might not give us a definite identification of the mineral. For these reasons, in identifying minerals, we rely heavily on their physical properties. Table 5 lists the characteristic physical properties of the common Hawaiian minerals.

When it is present, crystal form (fig. 82) is characteristic of a mineral species. However, well-formed crystals are generally found only where the mineral was free to grow with little or no interference from surrounding mineral grains. This condition prevails during the early stages of cooling of some igneous-rock melts,

## Table 4. Minerals in Hawaiian rocks

| Light-colored minerals | | Dark-colored minerals | |
|---|---|---|---|
| Quartz | $SiO_2$ | Biotite (black mica) | $H_4K_2(Mg,Fe)_5Al_4Si_5O_{24}$ |
| Orthoclase feldspar | $KAlSi_3O_8$ | Olivine | $(Mg,Fe)_2SiO_4$ |
| Anorthoclase feldspar | $(K,Na)AlSi_3O_8$ | Pyroxene { Hypersthene | $(Mg,Fe)SiO_3$ |
| Plagioclase feldspar { Albite | $NaAlSi_3O_8$ | Pigeonite | $(Mg,Fe)SiO_3$ |
| Anorthite | $CaAl_2Si_2O_8$ | Augite | $Ca(Mg,Fe)Si_2O_6 \cdot (Al,Fe)_2O_3$ |
| Nepheline | $NaAlSiO_4$ | Magnetite | $Fe_3O_4$ |
| Calcite | $CaCO_3$ | Hematite | $Fe_2O_3$ |
| Gypsum | $CaSO_4 \cdot 2H_2O$ | Ilmenite | $FeO \cdot TiO_2$ |
| Kaolinite (clay) | $Al_2Si_2O_5(OH)_4$ | Melilite | $Ca(Mg,Fe)_2Si_2O_7$ |
| Gibbsite | $Al(OH)_3$ | Chlorite | $(Mg,Fe)_5(Al,Fe)_2Si_3O_{10}(OH)_8$ |
| Montmorillonite | $(Al,Mg)_8(Si_4O_{10})_3(OH)_{10} \cdot 12H_2O$ | Limonite | $Fe_2O_3 \cdot nH_2O$ |

## Table 5. Physical properties of common Hawaiian minerals

| Name of mineral | Color | Streak | Luster | Cleavage | Specific gravity | Hardness |
|---|---|---|---|---|---|---|
| *LIGHT-COLORED MINERALS* | | | | | | |
| Kaolinite } Clay Montmorillonite | White, yellow, brown | White | Earthy, dull | None | 2.6 | 1.5–2.5 |
| Gypsum | White, gray, yellow | White | Glassy, silky | Good | 2.3 | 1.5–2 |
| Calcite | White, yellow | White | Glassy | Good | 2.7 | 3 |
| Gibbsite | White to brown | White | Pearly to dull | Good | 2.4 | 2.5–3.5 |
| Opal | White, brown, gray | White | Glassy to waxy | None | 2.2 | 5.5–6.5 |
| Nepheline | White, gray | White | Glassy, greasy | Poor | 2.6 | 5.5–6 |
| Orthoclase } Feld- Anorthoclase } spar Plagioclase | White, gray | White | Glassy | Good | 2.4–2.6 | 6–6.5 |
| Quartz (and chalcedony) | White, gray | White | Glassy | None | 2.6 | 7 |
| *DARK-COLORED MINERALS* | | | | | | |
| Limonite | Brown, yellow | Brown, yellow | Earthy, dull | None | 3.6–4.0 | 1–4 |
| Chlorite | Grayish green | Grayish green | Glassy, pearly | Good | 2.8 | 1.5–2.5 |
| Biotite | Brown | White | Glassy, pearly | Good | 2.7–3.1 | 2.5–3 |
| Hematite | Brownish red | Red | Earthy, dull | None | 5 | 1–4 |
| Ilmenite | Black | Brownish black | Metallic | None | 4.5–5.0 | 5.5–6 |
| Magnetite | Black | Black | Metallic | None | 5.2 | 5.5–6.5 |
| Hypersthene } Pyro- Pigeonite } xene Augite | Black | White | Glassy | Good | 3.2–3.5 | 5–6 |
| Olivine | Green, brownish green | White | Glassy | Poor | 3.3–3.4 | 6.5–7 |

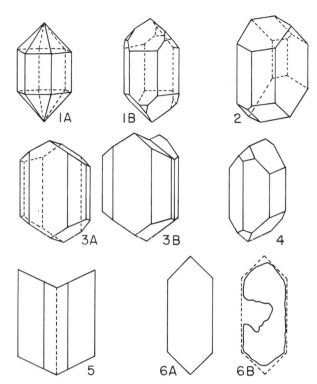

Figure 82. Drawings showing typical crystal outlines of quartz (*1A* and *1B*), plagioclase feldspar *(2)*, a simple augite crystal *(3A)*, a twinned augite crystal *(3B)*, and olivine *(4)*; typical cross sections of a twinned augite crystal showing the "arrowhead" form *(5)*; typical cross section of an olivine crystal *(6A)*, and the rounded and embayed outline resulting from partial resorption *(6B)*. In *1* to *5* the dashed lines represents the backs of the crystals; in *6B* the dashed line represents the outline of the crystal before resorption.

when crystals of early-formed minerals may form at scattered points in the liquid magma. Such minerals—particularly olivine, augite, and plagioclase feldspar—may grow into well-formed crystals. Likewise, grains of calcite or quartz growing outward from the walls into water-filled openings may take on good crystal forms. In the later stages of cooling of igneous rocks, however, many mineral grains form simultaneously and close together at many places in the magma, and the adjacent grains interfere with each other, so that they are not free to develop well-formed crystals. In the same way calcite crystals growing within the body of a limestone interfere with each other and cannot develop their characteristic shape. Technically, even though they do not have well-developed external crystal forms, the grains are crystals because they do have the definite internal structure mentioned above, and their other physical properties are the same

whether the external crystal form is developed or not.

Perhaps the most conspicuous property of most minerals is their color, although unfortunately the color of a mineral often varies widely. For example, potassium feldspar may be colorless (like clear window glass), white, pink, green, or gray. Nevertheless, with minor exceptions, the color of a mineral remains always in the same general color group—either light or dark.

More characteristic than the color of the whole mineral grain is the color of its powder. This is obtained by rubbing the grain across a plate of white unglazed porcelain, leaving a streak of powdered mineral on the plate. The streak of feldspar is always white, no matter what the color of the uncrushed grain. (It should be noted, however, that where the mineral is harder than the plate the streak is powdered white porcelain. The hardness of a porcelain streak plate is approximately 6, and so the streak of all minerals harder than 6 is white.)

Hardness of minerals is generally measured by a scratch test. If one mineral will scratch another, it is the harder of the two. A scale of hardness known as the Mohs scale, is in very general use. Each of the steps in the scale is represented by a specific mineral, from soft to hard, as follows: 1, talc; 2, gypsum; 3, calcite; 4, fluorite; 5, apatite; 6, feldspar; 7, quartz; 8, topaz; 9, corundum; 10, diamond. The steps are approximately equally spaced except for the last: diamond is several times as hard as corundum. These standard minerals can be used to test the relative hardness of any other mineral. If a mineral grain will not scratch calcite, but will scratch gypsum and is scratched by fluorite, it has a hardness of 3; if another mineral grain scratches feldspar but is scratched by quartz its hardness is about 6.5; and so forth. In the field, however, where a set of the standard minerals is seldom available, other common objects can be used: one's fingernail usually has a hardness of about 2.5, a U. S. penny about 3.5, an ordinary knife blade about 5, and window glass about 5.5.

Cleavage is the property of many minerals to split along smooth plane surfaces. Some minerals cleave in only one direction, others in two or even three directions. Feldspar, for instance, cleaves in two directions at about 90° to each other, yielding blocks with square angles and four smooth sides. The other two sides of the blocks are rough and sometimes very irregular —a type of break called fracture. Calcite cleaves in three directions, yielding smooth-sided blocks shaped like rhombohedrons. Mica has one very good, close-spaced cleavage direction, yielding many thin plates. Other minerals, such as quartz, do not cleave in any direction; all the breaks are irregular fractures. The lack of cleavage is as characteristic as its presence.

Luster is the general appearance of the mineral surface in reflected light: metallic (like the shiny surface of an ordinary metal), glassy, pearly, waxy, or dull and earthy. Density, or specific gravity, also is characteristic. Some minerals, like olivine, are conspicuously heavy; others are notably light.

The most abundant single mineral in Hawaiian rocks is feldspar. It is white or gray, with glassy or pearly luster, and the cleavages generally can be seen on broken surfaces. The large white crystals in Hawaiian lavas are nearly always plagioclase feldspar. Only in the rare trachytes (see next section) are large crystals of alkalic feldspar (anorthoclase) found. The plagioclase feldspar is a mixture of two chemical compounds, albite ($NaAlSi_3O_8$) and anorthite ($CaAl_2Si_2O_8$), usually with more anorthite than albite. In large grains it can be distinguished from orthoclase or anorthoclase by the presence on one cleavage surface of fine parallel striations, visible with a magnifying glass. Small light-colored grains of feldspar are abundant in the fine-grained matrix of the lavas also, but individual grains generally are difficult to make out without a microscope. Nepheline also is light colored, but it lacks good cleavage and has a greasy or waxy luster; furthermore, it is never found in large grains in Hawaiian rocks. Quartz is rare in Hawaii but is found in a few places. It can be recognized by its glassy appearance and lack of cleavage, and sometimes by its crystal form—a 6-sided prism terminated by a 6-sided pyramid.

Among the dark minerals, the most conspicuous are olivine and augite. Olivine is easily recognized by its green or brownish green color, glassy luster, and lack of good cleavage. It is common as big grains in Hawaiian lavas. Augite is black to very dark green, and shows good cleavage in two directions at nearly right angles to each other. It is much less widespread than olivine but is common on some of the volcanoes, such as Haleakala and Mauna Kea, and at Koko Head on Oahu. Hypersthene is not readily distinguishable from augite except by use of a microscope, but it rarely occurs as large grains in Hawaiian rocks. Pigeonite, which is identical in appearance with augite, is found only in small grains that can be identified only by use of a microscope. Tiny brown to black flakes of biotite can be found in some Hawaiian rocks (particulary the hawaiites, mugearites, and trachytes described in the following section). Magnetite and ilmenite form tiny black metallic grains sometimes recognizable with a magnifying glass. Melilite, present in some of the nepheline-bearing lavas, cannot be identified without a microscope.

Calcite is by far the most abundant mineral of Hawaiian limestones. It is generally white to yellowish brown in color, and is easily recognized by its rhombohedral cleavage and by the fact that it can be scratched by a penny. Gypsum is white to gray, and so soft that it can be scratched with one's fingernail.

Worthy of special mention because of their interest to "rock hounds" are two of the silica minerals, chalcedony (pronounced *kălsĕd'o nĭ*) and opal. Chalcedony is simply very fine-grained quartz. Opal is uncrystallized—a mineraloid composed of $SiO_2$ with a variable amount of water. Both sometimes form masses showing more or less regular bands of different colors, known as jasper or agate.

Another mineraloid is volcanic glass—once-molten rock that solidified so quickly that minerals did not have time to form in it. Very thin skins of glass occur on many Hawaiian lava flows, but large chunks of glass have been found only in the trachyte cone of Puu Waawaa, on the west side of the island of Hawaii. Volcanic glass in general is called obsidian, and the small amounts of obsidian on basalt lava flows receive the special designation tachylite.

Various localities of interest to mineral collectors are listed in appendix A.

## IGNEOUS ROCKS

Rocks are aggregates of grains of minerals and mineraloids. Some rocks consist largely, or even entirely, of a single species of mineral. Examples in Hawaii are limestone, made up of calcite, and the rock known as dunite, which is almost wholly olivine. Most rocks, however, consist of mixtures of several different minerals. Rocks are more variable in composition than minerals, but each type of rock contains certain specific minerals in more or less definite proportions. Thus granite always consists mostly of feldspar and quartz, with smaller amounts of other minerals. (The name granite is used by geologists in two different senses: as a general term for any coarse-grained igneous rock consisting predominantly of quartz and feldspar, and in a specific sense for a rock of this type in which the feldspar is potassium feldspar.) Basalt always consists largely of anorthite-rich plagioclase feldspar and pyroxene, although other minerals, such as olivine, may also be present. The lesser minerals, whose presence do not affect the classification of the rock species, often are used to indicate a particular variety of the rock. Thus, basalt containing olivine is called olivine basalt.

It has been stated earlier that *igneous* rocks are those that have solidified from molten magma. The consolidation may take place on the earth's surface after the magma has been poured or blown out by volcanic activity, forming what are known as *extrusive* rocks. Sometimes, however, rising magma fails to reach the surface and consolidates within the earth's crust as *intrusive* rocks. The magma may halt its rise and solidify at a depth of a mile or more beneath the surface, in which case the resulting rocks are referred to as *plutonic* (formed in the realm of Pluto, the ancient god of the underworld). It may, on the other hand, rise nearly to the surface. Shallow-seated intrusive rocks are included with those extruded onto the earth's surface in the class of *volcanic* rocks. All of the igneous rocks of Hawaii are volcanic except for a few fragments brought up from great depths as inclusions in the rising magma.

Extrusive rocks cool quickly. The minerals forming in the magma have only a short time in which to grow, and as a result the rock is fine grained. Small masses of intrusive rock, such as narrow dikes (see chap. 5), may also cool quickly and be fine grained. Larger intrusive bodies cool more slowly. The greater supply of heat brought in with the large intrusive mass quickly warms the surrounding rocks without appreciably lowering the temperature of the intrusive magma itself, and, since the rocks are poor conductors of heat, they act thereafter as insulators, retaining the heat in the intrusive body. Consequently the atoms have more time in which to move through the magma to a smaller number of centers of crystallization, and as a result the rock becomes coarse grained. Plutonic rocks in particular are characterized by coarseness of grain, but the rock in some of the large volcanic intrusive bodies also is coarse. Examples are the masses of gabbro (the coarse-grained equivalent of basalt) found in West Maui and in a few other places.

Volcanic magmas often are halted temporarily within the earth's crust before their final rise to the surface. Cooling within the crust may bring the temperature of the magma down to the point at which some minerals reach the saturation level and begin to crystallize. Which mineral starts to crystallize first depends on the composition of the magma. In many Hawaiian magmas the first to crystallize is olivine, in others it is feldspar. As the temperature continues to fall, these first-crystallizing minerals may be joined by others. For instance, olivine may be joined later by feldspar and augite. Under intrusive conditions the magma cools slowly, and the mineral grains may grow to large size. If the magma containing these large crystals then rises to the earth's surface, the remaining liquid part cools quickly and develops the fine granularity characteristic of extrusive rocks. We then have a rock made up of large crystals scattered through a fine-grained matrix. The large crystals are called *phenocrysts*, and the rocks containing them are *porphyritic*. The porphyritic texture is exceedingly common in all volcanic rocks.

Igneous rocks can be classified in various ways. The simplest and most satisfactory for ordinary purposes depends on grain size and the relative proportions of different minerals in the rock. A primary separation can be made on the basis of whether light or dark minerals are predominant. Several hundred different names have been given to special types of igneous rocks, but the simplified classification given in table 6 is adequate for our present purposes.

Note that along with the increase in proportion of dark minerals toward the right-hand side of the table goes a general decrease in the amount of silica in the chemical composition of the rock. Simultaneously there is an increase in the proportion of iron, magnesium, and calcium. The light-colored rocks are also light in weight, and the dark-colored rocks are relatively heavy.

The types of rocks found in the Hawaiian Islands are further defined as follows:

*Andesite*—a rock composed largely of sodic plagioclase feldspar and pyroxene, fine grained and generally medium to light gray, though sometimes dark. Olivine phenocrysts may be present, and less commonly phenocrysts of feldspar or augite. Hawaiian andesites are now usually referred to by the special names hawaiite and mugearite (defined below), to emphasize the fact that they are not the same as the common andesites of continental regions.

*Ankaramite*—a rock containing abundant phenocrysts of black augite and green olivine in a fine-grained dark matrix. Feldspar, which makes up less than one-third of the total rock, is mostly in the matrix. With a decrease in the abundance of phenocrysts the rock grades into the alkalic olivine basalts (defined below). Together with oceanite, ankaramite is also referred to by the more general name *picrite-basalt.*

*Basalt*—a rock composed almost wholly of calcic plagioclase feldspar and pyroxene, fine grained and usually dark colored. Two general varieties are recognized in Hawaii: *tholeiite,* or *tholeiitic basalt,* which is relatively rich in silica and poor in the alkalies (sodium and potassium), and in which the pyroxene is partly or largely pigeonite containing little calcium or aluminum; and *alkalic basalt,* which is comparatively poor in silica and rich in alkalies, and in which the pyroxene is generally augite containing calcium and aluminum. Both tholeiitic and alkalic basalts commonly contain olivine, often as phenocrysts, and they may also contain phenocrysts of plagioclase feldspar or pyroxene.

*Dunite*—a coarse-grained rock composed largely of olivine. A few dark green grains of pyroxene and tiny black grains of magnetite are often present.

*Gabbro*—the coarse-grained equivalent of basalt, like the latter composed largely of calcic plagioclase feldspar and pyroxene, often with olivine.

*Hawaiite*—an andesitic rock composed largely of andesine (plagioclase feldspar containing between 50 and 70 percent albite) and pyroxene, generally with some olivine. Hawaiite grades on the one hand into alkalic olivine basalt and on the other into mugearite.

*Mugearite*—an andesitic rock resembling hawaiite but in which the feldspar is oligoclase (plagioclase containing 70 to 90 percent albite). Mugearite grades on one side into hawaiite and on the other into trachyte.

*Nephelinite*—a fine-grained, dark-colored rock resembling basalt, but containing nepheline in place of feldspar. Olivine is generally present. Nephelinites are also called nepheline basalts. A rock containing both nepheline and plagioclase feldspar, and thus transitional between nephelinite and ordinary alkalic basalt, is known as *basanite.*

*Obsidian*—a general name for volcanic glass. It may have the same chemical composition as any volcanic rock, but is most abundant in those rich in silicon and poor in iron and magnesium. The only even moderately large masses found in Hawaii have the composition of trachyte or rhyodacite (defined below). Obsidian is generally black or dark gray no matter what its chemical composition, though it may also be red or brown, or more rarely green. The small amounts of black obsidian found on some basaltic lava flows receive the special name *tachylite.*

*Oceanite*—a basaltic rock containing very abundant (more than 35 percent) olivine phenocrysts. A variety of picrite-basalt, it grades into the olivine basalts.

*Olivine basalt*—a basalt containing a moderate amount (5 percent or more) of olivine. Olivine basalts may be either tholeiitic or alkalic.

*Peridotite*—a coarse-grained rock composed of olivine and some other dark minerals—in Hawaii either augite or hypersthene. Dunite is a special variety of peridotite.

111

## Table 6. Simplified classification of igneous rocks

| | Light-colored minerals dominant | | | | | | Dark-colored minerals dominant | |
| | Potassium feldspar dominant | | | Plagioclase feldspar dominant | | | Feldspar essentially absent | |
| | Quartz present | Quartz and nepheline absent | Nepheline present | Quartz present | Quartz and nepheline absent | | Nepheline present | Quartz and nepheline absent |
|---|---|---|---|---|---|---|---|---|
| Fine grained or porphyritic with fine matrix | Rhyolite | *Trachyte* | Phonolite | *Rhyodacite* Dacite | *Andesite (hawaiite, mugearite)* | *Basalt Olivine basalt Oceanite Ankaramite Basanite* | *Nephelinite ("nepheline basalt")* | |
| Coarse grained | Granite (in the strict sense) | Syenite | Nepheline syenite | Granodiorite Quartz diorite | *Diorite* | *Gabbro Olivine gabbro* | | *Peridotite Dunite* |
| Silica ($SiO_2$) content | 75% ◄————————————————————————————————————————► 40% | | | | | | | |

NOTE: Names of rocks found in Hawaii are shown in italics.

*Rhyodacite*—a fine-grained rock consisting predominantly of a mixture of potassium feldspar and plagioclase feldspar, with minor amounts of dark minerals. Excess silica is present either as grains of quartz or dissolved in glass. Alkalies are less abundant than in trachyte, and silica relatively more abundant. Only one body of rhyodacite is known in Hawaii, at Mauna Kuwale, between Lualualei and Waianae valleys in the Waianae Range of Oahu.

*Trachyte*—a fine-grained, generally light-colored rock composed predominantly of potassium feldspar, with minor amounts of dark minerals. Trachyte consisting mostly of glass may be black, as at Puu Waawaa on the island of Hawaii.

Granite and rhyolite, dacite, granodiorite, and quartz diorite are not found in Hawaii, though they are among the commonest igneous rocks of continental regions. Most of the andesites of continental volcanoes are quite different in appearance and in mineral and chemical composition from the andesitic rocks (hawaiite and mugearite) of the Hawaiian Islands. Phonolites and nepheline syenites also seem to be absent in Hawaii, though they are found in some other central Pacific islands, such as Tahiti. For the most part, however, the same types of rocks that make up the Hawaiian Islands are found also in the other central Pacific islands, and indeed in most other oceanic islands throughout the world. The reasons why the mid-oceanic rocks differ from those of the continents are of course related to the processes by which they were formed—processes which are still far from being completely understood.

Hawaiian rock types can be arranged in two general series (or suites), the tholeiitic and the alkalic. The tholeiitic series comprises the following rocks, arranged in order from most basic (magnesium- and iron-rich) to most acid (silica-rich): oceanite, tholeiitic olivine basalt, tholeiitic basalt, tholeiitic quartz basalt, rhyodacite. The alkalic series similarly consists of: ankaramite, alkalic olivine basalt, alkalic basalt, hawaiite, mugearite, trachyte. The nephelinites and basanites are even more strongly alkalic than the rocks of the main alkalic series, and form a group more or less by themselves. Within each of the principal series, the variations among the rock types can be accounted for largely by fractional crystallization—a process of separation of crystals formed in early stages of consolidation of the magma that leaves a liquid residue with a composition different from that of the original magma.

It is far more difficult to explain the origin of the nephelinites and of the original ("parent") magma of the alkalic series. There appears to be little question that the parent tholeiitic magma is formed by melting of part of the earth's mantle, not far beneath the crust (see chap. 2). Most petrologists believe that the parent alkalic magma is formed in the same way, but at a level deeper in the mantle. The question is too theoretical and involved to pursue further here.

The average chemical compositions of the principal types of Hawaiian igneous rocks are given in table 7.

## XENOLITHS AND MAGMATIC REACTIONS

Fragments of older solid rock enclosed in the magma, and eventually left frozen into the solidified igneous rock, are known as *xenoliths* ("foreign rocks"), or simply *inclusions.* Some of them may be picked up from the ground surface over which lava is flowing, but most are acquired as the magma rises toward the surface. Some are torn from the walls of the conduit; others are loosened from the roof of the magma chamber and sink into the magma. (The latter is the process of "stoping" described in chapter 5.) Xenoliths may be fragments of older lava similar to that in which they are found, or they may be pieces of coarser-grained intrusive rock. In other parts of the world they may also be pieces of sedimentary rock, but very few sedimentary xenoliths have been found in Hawaii.

Some xenoliths remain intact and little altered; others gradually disappear in the magma. To some extent the disappearance may be brought about by a mechanical fraying, grains and fragments of the xenolith being broken off and strewed through the adjacent magma. To some extent also the xenolith may be melted, the molten material becoming a part of the magma. However, it appears that magmas rarely have enough extra heat, beyond that

necessary to keep themselves molten, to melt much foreign material. Commonly the magma is on the verge of crystallizing, or actually contains phenocrysts already being formed, indicating that it is no longer hot enough to remain completely molten. The amount of heat liberated into the surrounding magma during the formation of these phenocrysts (latent heat of crystallization) has been shown to be too small to have any important effect in heating the magma. However, not all the minerals start to form in the magma at the same time. In Hawaiian tholeiitic basalts, for example, olivine starts to form first, then on further cooling feldspar and pyroxene make their appearance. It is entirely possible that, even had olivine started to crystallize, the magma might still have been capable of melting feldspar and pyroxene in a relatively small volume of xenoliths, the loss of heat used up in remelting these minerals resulting in a more rapid crystallization of olivine. Melting of some of the grains in the xenolith aids of course in the mechanical disintegration of the remaining solid part.

In lava flows, and in places where intrusive rocks have been exposed at the surface by erosion, it is possible to study xenoliths in all stages of disappearance, and it seems that in most instances the process is largely one of reaction between the xenolith and the enclosing magma. This process is known as *assimilation.* It is the result of instability of the minerals in the xenolith when they are brought into this new magmatic environment. The minerals that compose rocks are generally stable in the physical and chemical environment in which they are formed. Some minerals,

Table 7. Average chemical compositions of Hawaiian rock types (in weight percent)

| | Melilite and nephelinite | Nephelinite | Ankaramite | Nepheline basanite | Oceanite | Alkalic basalt | Hawaiite | Tholeiitic basalt | Mugearite | Trachyte | Rhyodacite |
|---|---|---|---|---|---|---|---|---|---|---|---|
| $SiO_2$ | 36.7 | 38.7 | 44.1 | 44.3 | 46.4 | 46.5 | 48.6 | 49.4 | 51.9 | 61.7 | 66.0 |
| $TiO_2$ | 2.8 | 2.7 | 2.7 | 2.6 | 2.0 | 3.0 | 3.2 | 2.5 | 2.6 | 0.5 | 0.7 |
| $Al_2O_3$ | 10.8 | 10.8 | 11.2 | 12.8 | 8.5 | 14.6 | 16.5 | 13.9 | 16.6 | 18.0 | 15.5 |
| $Fe_2O_3$ | 5.6 | 5.8 | 2.8 | 3.4 | 2.5 | 3.3 | 4.2 | 3.0 | 4.2 | 3.3 | 1.4 |
| $FeO$ | 8.8 | 7.8 | 9.9 | 9.2 | 9.8 | 9.1 | 7.4 | 8.5 | 6.2 | 1.5 | 1.9 |
| $MnO$ | 0.1 | 0.1 | 0.2 | 0.2 | 0.2 | 0.1 | 0.2 | 0.2 | 0.2 | 0.2 | 0.1 |
| $MgO$ | 12.7 | 13.6 | 15.1 | 11.0 | 20.8 | 8.2 | 4.7 | 8.4 | 3.6 | 0.4 | 1.5 |
| $CaO$ | 13.7 | 13.0 | 10.7 | 10.5 | 7.4 | 10.3 | 7.8 | 10.3 | 6.3 | 1.2 | 2.8 |
| $Na_2O$ | 3.9 | 4.0 | 1.7 | 3.6 | 1.6 | 2.9 | 4.4 | 2.1 | 5.2 | 7.4 | 4.4 |
| $K_2O$ | 0.9 | 1.1 | 0.5 | 1.0 | 0.3 | 0.8 | 1.6 | 0.4 | 2.0 | 4.2 | 3.3 |
| $P_2O_5$ | 1.1 | 1.0 | 0.3 | 0.4 | 0.2 | 0.4 | 0.7 | 0.3 | 0.9 | 0.2 | 0.5 |

such as quartz, are stable through a wide range of physical conditions, but the stability range of many other minerals is very limited. Thus a mineral formed under low pressure is apt to be unstable under high pressure, and it tends to change to some other mineral that is stable under the new high-pressure condition. Similarly, minerals formed at low temperature commonly are unstable at high temperature, and vice versa. Stability is also affected by the chemical environment, and minerals formed by crystallization in one magma commonly are not stable in another magma of different chemical composition even though the temperature and pressure may be the same. Reaction with the magma results in the removal of some chemical elements from the mineral and the addition of others, generally with simultaneous recrystallization, producing a wholly new mineral that is stable under the new conditions. Thus the minerals in xenoliths tend to be changed to new minerals—often the same ones that are forming in the surrounding magma—and the xenolith gradually loses its identity.

The results of disintegration of xenoliths are clearly visible in some Hawaiian rocks, both intrusive and extrusive. Fragments of older lava rock are common in many lava flows and occasionally are found in intrusives. Not infrequently it can be seen that they were in process of mechanical breaking up at the time the surrounding magma solidified. Angular fragments of the xenolith may be scattered around it or may be strung out by flowage of the enclosing liquid. The originally angular blocks may show all degrees of progressive rounding as their corners are gradually removed. Some of the rounding is purely mechanical, simply by breaking off of bits from the corners, but in other cases there is no evidence of mechanical fraying, and the rounding appears to be largely the result of melting away of the corners. The latter is particularly common in some of the tholeiitic lava flows of Kauai, but it has been observed at many other places as well.

The 1801 lava flow of Hualalai volcano contains a very large number of xenoliths of dunite and other types of peridotite and of gabbro. Many of the dunite xenoliths obviously were undergoing mechanical disintegration at the time the lava consolidated. Grains of olivine separated from them are scattered through the surrounding lava.

One type of reaction with the magma, involving phenocrysts rather than xenoliths, is very common in Hawaiian rocks. As tholeiitic basalt magma cools, normally the first mineral to start to crystallize from it is olivine, which forms phenocrysts suspended in the magma. As they originally form, these phenocrysts are sharply angular. Those in Hawaiian tholeiitic basalts, however, commonly are much rounded and embayed (fig. 82, 6B), showing that they were being redissolved by the magma at the time it suddenly cooled and solidified. This is because in tholeiitic magmas olivine that formed at high temperature becomes unstable at lower temperature. It then tends to be taken back into solution in the magma, its constituents combining with additional silica to form pyroxene, which is stable at the lower temperature. In contrast, in alkalic olivine basalt magma, olivine remains stable down to the temperature of complete consolidation of the magma, and here the olivine phenocrysts seldom are reabsorbed into the magma. Since the phenocrysts of olivine in Hawaiian tholeiitic basalts are nearly always rounded, the presence of olivine phenocrysts with well preserved crystal faces is almost certain indication that the rock is alkalic.

Somewhat different is the magmatic reaction that often produces a thin shell of the brick-red or brown mineral iddingsite on the olivine phenocrysts. Iddingsite contains essentially the same materials as olivine, with water added and the iron oxidized. Probably at certain times, perhaps usually just before eruption, gases are concentrated in the upper part of the magma body, causing the olivine phenocrysts in that part of the magma to become unstable and to change on the outside to iddingsite. On eruption of the magma onto the surface, much of the gas escapes and olivine may again become stable; crystallization may then resume, forming a thin outer shell around the shell of iddingsite.

Peridotite inclusions, like those mentioned above in the 1801 lava flow, are common in alkalic basalts and nephelinites of the Hawaiian

Islands, and quite common also in the hawai-ites and mugearites. It has been claimed that they are absent from the tholeiitic basalts, but actually they are present, though very rare, in them also. Some of the inclusions are dunite, composed almost wholly of pale green olivine, but generally with a few larger grains of darker green olivine and tiny grains of magnetite. Others contain either hypersthene or augite, or both. Still others contain in addition small amounts of feldspar, and, as the abundance of feldspar increases, we find a complete grada-tion into olivine gabbro, and even into a rock called anorthosite that consists almost wholly of plagioclase feldspar. In some, the mineral grains have a tendency to be aligned in parallel position, and alternate layers are of slightly different mineral composition and color.

Usually there are only a few peridotite inclusions scattered through the rock, but occasionally they are exceedingly abundant. Nephelinite in the headwaters of Hanalei Val-ley on Kauai and ankaramite near Paia on Maui contain angular inclusions of dunite that lo-cally are more abundant than the enclosing rock matrix. In cuts along the highway just north of the Huehue Ranch headquarters, many inclusions of dunite and gabbro can be seen in the 1801 lava flow of Hualalai, but higher on the mountainside (close to the telephone relay station) they are almost unbe-lievably abundant. There the slope of the surface beneath the lava became gentler and the speed of the flow decreased, allowing the heavy inclusions to settle to the bottom. Most of the liquid then drained away from between them, leaving each one coated with a thin crust of black or brown lava. Individual inclusions range from less than an inch to about a foot across. Seen from a little distance, where they are exposed on the side of the channel of the flow, they resemble a huge heap of potatoes; broken open and viewed at close range they look more like big bonbons, with a chocolate shell enclosing bright green or gray centers.

Recent work (White, 1966; Jackson, 1968) has shown that the peridotite inclusions belong to two separate groups, of different origin. Those of one group, which includes the banded peridotites and the associated gabbros and anorthosites, closely resemble in composition and texture rocks found in big gabbro intrusive bodies in other parts of the world (such as the Stillwater intrusion in Montana). These rocks have formed by settling of crystals of olivine and pyroxene to the bottom of the cooling magma body. Inclusions of this type in Hawai-ian lavas are believed to have been torn by the rising magma from solidified intrusive bodies somewhere in the substructure of the volca-noes. They are abundant in rocks of the alkalic and nephelinic suites, but rare in those of the tholeiitic suite.

The second group includes the peridotites that contain both hypersthene and augite (the lherzolites), and the dunites with large, dark green olivines. The olivines contain innumera-ble tiny bubbles of carbon dioxide under high pressure, indicating that the crystals containing them were formed at depths of 6 to 9 miles (Roedder, 1965)—the lower part of the earth's crust or the upper part of the mantle. The large olivines are surrounded by smaller grains of lighter-colored olivine which appear to have formed by crushing and recrystallization of the edges of the larger grains. The lherzolites likewise show clear indications of shearing and crushing. On the basis of both their mineral composition and the signs of distortion, it is believed that the inclusions of this group have come from the upper mantle (Jackson, 1968).

Still more clearly of mantle origin are garnet-bearing peridotite inclusions found at Salt Lake and nearby posterosional craters on Oahu. They consist of olivine and pyroxene, with pink or brown garnet. The pyroxene contains sodium, and more aluminum than ordinary augite. Laboratory experiments have shown that when olivine basalt or gabbro is subjected to very high pressure at a high temperature it recrystallizes into garnet peri-dotite ("eclogite"). Because this combination of pyroxene and garnet forms only at pressures equivalent to depths of 35 or more miles—well below the crust in the Hawaiian region—it is virtually certain that the garnet peridotite (and garnet pyroxenite) inclusions at Salt Lake were brought up from the mantle. There are very few places in the world (the diamond pipes of South Africa are another) where we can see rocks that are so certainly pieces of the mantle, and consequently the Salt Lake inclusions have attracted the attention of petrologists from all over the world.

The garnet-bearing inclusions are found only in rocks of the nephelinic suite.

METAMORPHIC ROCKS

Metamorphic rocks are those which have undergone a change in mineral composition to bring them into equilibrium with a new environment. Commonly there is little change in chemical composition during the metamorphism of one rock to another, other than the loss or gain of a small amount of water. The change of olivine gabbro to garnet peridotite mentioned above is of this type. At other times, however, considerable amounts of new material may be brought in by steam and other gases, and the new rock is quite different in chemical composition from the original.

Alterations of the sort ordinarily included by geologists in the category of metamorphism are of very minor importance in Hawaii. In ordinary volcanic rocks the minerals are formed at high temperature, and as the rock cools they are no longer in equilibrium and tend to change to new forms stable at lower temperatures. At ordinary air temperature the rate of reaction is so slow that the changes are imperceptible, but when the rock is kept at a temperature of a few hundred degrees for a long period, reaction is rapid enough to produce recognizable changes. Thus olivine basalts in the walls of the volcanic conduits at Halemaumau, in Kilauea caldera, have been kept warm long enough for the olivine and pyroxene to readjust to the lower temperature and the greater abundance of oxygen. Tiny grains of magnetite have separated from the olivine, and the olivine phenocrysts have become iridescent and rather metallic in appearance, or sometimes quite black. Pigeonite, which is stable in the tholeiitic basalts at the high temperature of lava consolidation, is no longer stable and separates into a mixture of augite and hypersthene. Fragments of these altered basalts, torn loose from the walls of the conduits and hurled out by the explosions of 1924, are abundant in the vicinity of Halemaumau.

Around a few of the larger intrusive bodies, also, the temperature remained high long enough to bring about some readjustment in the rocks. For example, around part of the edge of the laccolith (see chap. 5) exposed in the boundary cliff of Kilauea caldera at Uwekahuna the adjacent rocks have been recrystallized through a thickness of a few inches into a tough metamorphic rock known as hornfels.

Metamorphism, in a broad sense, includes two other types of alteration. Both are brought about by volcanic gases.

The rocks that fill the former calderas of some of the volcanoes have been much changed by volcanic gases rising through them for long periods of time, probably many thousands of years. The rocks were originally tholeiitic basalts composed primarily of pyroxene and plagioclase feldspar, with or without olivine. In the calderas of the East Molokai volcano and the Koolau volcano on Oahu, the rocks have been extensively altered, the original pyroxene and olivine being almost wholly destroyed. The olivine has been changed to serpentine or talc—a change that chemically involves little but the addition of water. (It is noteworthy that much talc has been found in the muds of Kaneohe Bay that were derived by erosion of the rocks of the Koolau caldera.) The pyroxene has been changed largely to chlorite, and this mineral imparts a prevailing greenish tinge to the altered caldera rocks, as can be well seen along the part of the Pali Highway just above the hairpin turn. The change to chlorite also involves the addition of water. However, chlorite contains less silica than does pyroxene, and after the change some free silica remains. This silica has moved in solution through the rock and has been deposited as quartz, chalcedony, or opal in any available open spaces. In the vicinity of Olomana Peak, in the Koolau caldera area, these silica minerals are found in thin sheetlike veins formed by filling of cracks in the rock, and in irregular masses up to several inches across. Both have been prized by local "rock hounds" for cutting and polishing. More commonly, however, the silica minerals were deposited in the former vesicles of the lavas, forming spheroidal or more or less almond-shaped masses up to about an inch across, known as *amygdules*. The lavas contain-

116

ing them are said to be *amygdaloidal.* Some of the amygdules are hollow, with tiny quartz crystals projecting into the central cavity. In addition to the silica minerals, some of the amygdules contain calcite and minerals of a group known as *zeolite,* which in composition resemble feldspar with the addition of water.

So characteristic are these alterations that green amygdaloidal lavas can almost be taken as the identifying stamp of caldera-filling rocks in the Koolau and East Molokai volcanoes. The alteration extends at most only a few hundred feet into the adjacent lavas outside the caldera, and for the most part it does not extend beyond the caldera boundary. That similar alteration is probably going on at depth in Kilauea caldera is indicated by the deposition of tiny mounds of silica (generally opal) around steam vents on the floor of the caldera.

The other type of alteration by gases occurs in the immediate vicinity of fumaroles—places where volcanic gases are reaching the surface (see chap. 2). The most accessible example is at Sulphur Bank, near the Volcano House on the north rim of Kilauea caldera. Others are found along the southeast boundary cliff of Kilauea caldera and the southwest rift zone of Mauna Loa in the vicinity of Sulphur Cone. The gases issuing at Sulphur Bank are mostly steam, but contain also some sulfur gases and carbon gases. The presence of sulfur is made conspicuous by the strong "rotten egg" odor of hydrogen sulfide. As the gases approach the ground surface, part of the hydrogen sulfide is oxidized to sulfur dioxide and trioxide, which combine with water to form sulfurous and sulfuric acids. The acid attacks the rock and breaks down the original minerals. Some of the products of the breakdown are carried away in solution, but others are left behind. What is left depends on the strength of the acid. The action of moderately strong acid (pH 4.5–5.5) leaves a white to pale gray rock that is made up almost wholly of opal and a lesser amount of clay mineral (kaolinite). The structure of the original rock generally is almost perfectly preserved: vesicles are easily recognized with the naked eye, and under the microscope the outlines and even the cleavages of the original minerals (now wholly transformed to opal and kaolin) are clearly discernible. Because the gases have risen along fractures in the rocks, alteration has proceeded from the outside of joint blocks inward. Blocks that have been broken open often reveal a dark gray center of fresh original rock, passing outward through progressively lighter shades of gray, partly altered rock to the nearly white, wholly altered rock at the outside.

Where the acid was less concentrated (pH 6–6.5) the products of alteration are very different. Silica is removed in solution, along with most of the other constituents of the original rock, and the residual product consists largely of clay minerals and the hydrated iron oxides limonite and goethite. The result is a red to reddish brown clay. The general process of alteration and the resulting product closely resemble the lateritic weathering described in chapter 8.

*Suggested Additional Reading*

Dunham, 1933, 1935; Eakle, 1931; Kuno, and others, 1957; Macdonald, 1949*a*, 1968; Macdonald and Katsura, 1962, 1964; Powers, 1955; Tilley and Scoon, 1961; Yoder and Tilley, 1962.

*Uwekahuna laccolith, Kilauea caldera*

# CHAPTER FIVE

# Intrusive Bodies

THUS FAR WE HAVE BEEN CONCERNED MOSTLY WITH extrusive rocks—lava flows and tephra—formed by eruption of magma onto the surface of the earth. In this chapter we will deal very briefly with *intrusive rocks,* formed by consolidation of magma within the crust of the earth. Masses of such rock are called *intrusive bodies,* or often simply *intrusives.*

The forms of intrusive bodies vary greatly, depending largely on how the space they occupy was produced. The rising magma may simply fill cracks opened in the overlying rocks. In other instances the rising magma forces the surrounding rocks apart, or lifts the overlying rocks, to make room for itself. In still other instances the space may be acquired by removal of the overlying and surrounding rocks, fragments of which sink into the magma—a process known by the miners' term, *stoping.* The entire mass of overlying rock, bounded by more or less circular fractures, may sink into the magma essentially as a single block (en masse stoping); or many small fragments of the roof may be detached and sink (piecemeal stoping). The two mechanisms are illustrated in figure 83 *A, B.* In either way the chamber occupied by the magma is enlarged upward. En masse sinking of the roof block above the magma chamber closely resembles the sinking that has been envisaged as the mechanism by which Hawaiian calderas are produced (fig. 83C), except that the break does not extend all the way to the surface.

Other means by which the chamber roof may be eaten away are assimilation of the roof rocks in place (see section on xenoliths and magmatic reaction, chap. 4), and melting of the roof rocks. Melting may result from extra heat in the magma itself, or from heat generated by reactions between gases accumulating at the top of the chamber. It seems doubtful that the latter process can have much, if any, importance more than a few thousand feet below the earth's surface, because gases probably are not released from solution in the magma under the high pressures prevailing at greater depths; but close to the surface heat generated by gases may indeed be the direct cause of opening some volcanic conduits through to the surface.

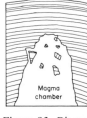
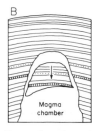
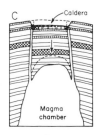

Figure 83. Diagrams illustrating the enlargement of a magma chamber by *(A)* piecemeal stoping, and *(B) en masse* stoping. The latter is probably related to the mechanism of sinking that forms Hawaiian calderas, illustrated in *(C)*.

## GRANULARITY OF INTRUSIVE ROCKS

Intrusive magmas consolidate within the earth, surrounded by solid rocks. The latter, being poor conductors of heat, act as insulators and retard the loss of heat from the magma. Near the surface the surrounding rocks may be cold, chilling the edge of the magma body where it comes against them. Once the initial chilling is accomplished, however, the wall rocks slow up the cooling of the magma body. At greater depths the wall rocks are warm, chilling of the magma against them is less, and cooling is even slower.

Slower cooling permits material to travel to centers of crystallization through greater distances in the magma, and thus build larger crystals. Diffusion, and other movements of material in the magma are also aided by lower viscosity. The viscosity of magma is reduced by the presence of dissolved volatile substances, such as water, and since intrusive magmas tend to lose their dissolved gases less readily than extrusive ones, the cooling intrusive magma remains somewhat less viscous. Thus, both because of slower cooling and lower viscosity, conditions for the formation of large crystals are more favorable in intrusive magmas than in those poured out onto the earth's surface. Where it is possible, large crystals will form instead of an equal mass of small crystals because the large crystals have a smaller surface area, and therefore there is less tendency for their constituents to escape from the crystal structure and return into the magma. As a result, intrusive rocks tend to be coarser grained than extrusive ones. Thus typical granites or gabbros, formed by crystallization at considerable depth in the earth's crust, are coarse grained, with crystals easily visible to the unaided eye, whereas lava rocks are usually very fine grained, requiring the use of a microscope to see most of their constituent crystals.

However, since the place of intrusion may vary from almost at the surface of the earth to a depth of many miles, resulting in a great range of temperature in the wall rocks, and since large intrusives cool more slowly than small ones, the grain size of intrusive rocks varies greatly. The enormous bodies of deep-seated igneous rocks (intrusive batholiths) found in parts of the continents typically are moderately to very coarse grained. In contrast, thin bodies intruded close to the surface, such as the abundant dikes in the Hawaiian volcanoes, typically are fine grained like extrusives.

The importance of volatiles in developing coarse granularity is shown in continental areas by the extremely large crystals commonly found in volatile-rich intrusives known as *pegmatites*. There are no true pegmatites in Hawaii, but some thick lava flows are cut by thin, veinlike bodies of material formed from the portion of the magma that still remained liquid when most of the flow had solidified. This last-remaining liquid, squeezed into cracks in the already solid portion of the flow, contained volatile materials concentrated in it during the solidification of the rest of the rock. These veins, known as *pegmatoids* ("pegmatite-like"), are much coarser than the surrounding rock. In the old Moiliili Quarry, at the lower end of the University of Hawaii's Honolulu campus, veins of pegmatoid with an average grain size as great as 1.5 mm cut melilite-nepheline basalt with a grain size of about 0.3 mm. (The Moiliili lava flow itself is a very thick one which cooled unusually slowly, so that its average grain size is much greater than that in most Hawaiian lava flows—generally less than 0.1 mm.) In the Hawaii Rock Company quarry (Dillingham Quarry) at Mokuleia, Oahu, even coarser veins of pegmatoid cut hawaiite in which the grains are too small to be seen without a microscope. There the pegmatoid veins contain the same minerals as the surrounding rock, and on the basis of texture and mineral content are classified as diorite (table 6).

Some small bodies intruded at shallow

depths are cooled so quickly at the periphery by contact with cold wall rocks that little or no crystallization can take place in their marginal zones. Instead, thin glassy borders may form on the intrusive body. These glassy selvedges, usually less than half an inch thick, are common on dikes and other small intrusives in Hawaii. They grade inward, through a zone of very fine crystallization in which some glass is often still present, to rock of ordinary fine granularity.

Thus, the granularity in Hawaiian intrusive rocks ranges from coarse to very fine, or even glassy. All the intrusive rocks visible in the Hawaiian Islands were formed at very shallow depths, in no instance exceeding 4,000 feet (the depth to which the deepest canyons have been cut below the original surface of the volcano), and probably seldom exceeding 1,000 to 2,000 feet. All are integral parts of the volcanic edifices, and all may be grouped with the extrusive rocks under the general term *volcanic.* The only igneous rocks in Hawaii that can be regarded as nonvolcanic are the fragments of eclogite at Salt Lake Crater on Oahu, and possibly the inclusions of peridotite and dunite that are widespread in the lava flows of several of the volcanoes (see chap. 6).

DIKES

An intrusive body is classified according to its general form and its relationships to the adjacent rocks and rock structures. Tabular, sheet-like bodies (fig. 84) intruded into nonbedded rocks or across the bedding in bedded rocks are *dikes* (fig. 90). They may have any attitude, but most of them have a high angle to the

Figure 84. Dike complex exposed in the highway cut on the Pali Highway, Oahu, just north of Castle Junction. The dikes are cutting caldera-filling lavas of the Koolau volcano. The prominent dike near the center of the picture is about 2 feet thick.

horizontal and many are essentially vertical.

In Hawaii, dikes range from less than a foot to as much as 50 feet in thickness. However, it is rare to find one more than 10 feet thick, and the great majority are between 1 and 3 feet. The unusually thick dikes generally are hawaiite, mugearite, or trachyte; those formed by the more fluid basaltic magmas are thin. As in other parts of the world, most of the dikes are nearly vertical. For the most part they follow pre-existing breaks in the wall rocks—joints that tend to stand approximately at right angles to the surfaces of the lava flows. Thus, locally the inclination of a dike tends to be at right angles to the slope of the flow it is crossing, and, since the latter is generally between 0° and 10°, the dikes stand at an angle of 80° to 90°. The upward course of a dike tends to be zigzag, crossing each lava flow at approximately right angles, but offsetting occasionally between flows in such a way as to keep the overall path nearly vertical. Low-angle dikes, some of them dipping (sloping) as little as 20°, are known, but, except in two localities to be mentioned later, they are rare.

Commonly the dikes have glassy or very fine-grained selvedges, caused by chilling of the magma against cold wall rocks. Typically, a glassy selvedge 0.25 to 0.5 inch thick is succeeded inward by a zone of stony appearance 0.5 to 2 inches thick in which crystallization is very fine and moderately abundant glass is still present between the crystals. This grades into the central part of the dike, which is largely crystalline with an average grain size usually around 0.05 mm (0.002 inch). The dikes may be either porphyritic or nonporphyritic, like the lava flows. All the compositional types found among the lava flows are present also in dikes.

Characteristically, the dikes show moderately- to well-developed columnar jointing approximately normal to their margins. In the central parts of the dikes the columns range from a few inches to a foot or a little more across. In any one dike they are fairly uniform in size, and there is a crude correlation between the diameter of the columns and the thickness of the dike, thicker dikes having larger columns. The size of the columns unquestionably is related to the rate of cooling of the dike.

There is a tendency for the columns to be 5- or 6-sided in cross section, but variation in shape is common, and 4- or 3-sided columns are abundant. The major columnar joints generally extend all the way to the edge of the dike, but in the fine-grained marginal zone additional joints appear, making the mesh of the joint pattern much finer. A still further increase occurs in the glassy selvedge, resulting in an irregular checking with the fractures usually spaced 0.5 to 0.25 inch apart (fig. 85).

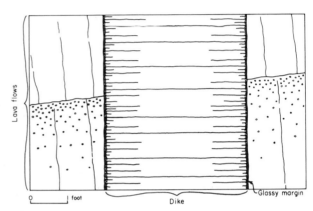

Figure 85. Diagram of columnar jointing in a dike.

The rock in most of the dikes is very dense, with almost no vesicles. Only in the zone that must have been within a few hundred feet of the surface at the time of intrusion does a significant number of vesicles appear. It seems probable that below that level the pressure due to the weight of overlying magma in the dike fissure was too great to permit rapid release of gas from solution to form bubbles in the magma. The conclusion that vesiculation occurs only close to the surface may apply only to the very abundant tholeiitic dikes of the gas-poor, shield-building stage of Hawaiian volcanism. For dikes of later stages, in which the magma appears to have been richer in gas, data are too sparse to warrant conclusions on depth of vesiculation; and it is certainly not warranted to extend the conclusion in a general way to other parts of the world.

In the vesicular portions of dikes the vesicles commonly are arranged in bands parallel to the dike walls. Individual vesicles generally are of pahoehoe type, and often are elongated in the direction of flow (parallel to the wall of the dike). Aa-type vesicles are found occasionally, but it is very rare to find a dike containing aa clinker, like the one on Lanai described by Stearns (1940c, pp. 39–40).

For the most part there is no indication that the opening of the fissures was caused by forceful intrusion of the magma of the dikes that occupy them. There seems to have been little tendency of the magma to be squeezed out into the wall rocks, even where the dike crosses rather open-textured aa clinker, and there is no sign of disturbance of any structures in the adjoining rocks. Only minor amounts of stoping of the walls by the rising magma took place: generally projections of one wall would fit well into reentrants in the opposite wall. The magma seems to have simply risen in the opening fissure—the opening probably caused by doming (tumescence) of the crust above an expanding magma body. However, this should not be interpreted to mean that the magma could not have been under appreciably more pressure than was necessary to lift it to the surface. It may mean only that the direction of easiest movement was upward, and the fluid magma followed the easiest path without intruding laterally or disturbing adjacent rock structures.

Evidence of the direction of movement of magma in the dikes is almost lacking. In a general way, of course, the movement must have been upward, because the dikes served as channels through which magma came from depth to erupt at the surface. In a few places direct, confirming evidence has been seen in the orientation of elongated xenoliths and the stretching of vesicles by flowage in a vertical direction. There is no question, however, that lateral movement also can occur. At the beginning of the 1919–20 eruption of Kilauea, at Mauna Iki, Jaggar (1947, pp. 137–139) observed magma moving almost horizontally at several places along the fissure that was opening from Halemaumau Crater toward the point of eruption.

It was mentioned earlier that low-angle dikes are common in two areas in Hawaii. These are near the edge of the Koolau caldera in the vicinity of Castle Junction, on Oahu, and around the periphery of the principal Kauai shield volcano, especially in the Kalepa-Nonou ridge and along the Napali Coast. In both, the low-angle dikes dip toward the center of the

caldera. In other parts of the world also, similar low-angle dikes dip toward the center of circular structures which often are associated with central sunken areas that may have underlain calderas at the surface. These inward-dipping dikes, known as cone sheets, are believed to result from the opening of conical fractures by the upward thrust of an underlying magma body (fig. 86). Possibly the low-angle dikes in

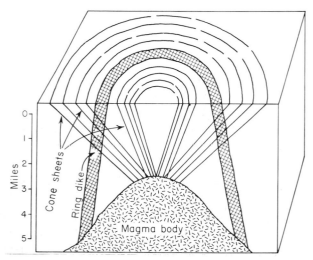

Figure 86. Diagram of cone sheets and a ring dike. (Modified after Richey and others, 1930.)

Hawaii likewise are the result of thrust by a magma body beneath lifting the central part of the overlying volcanic structure.

Cone sheets commonly are curved in ground plan, following part of the arc of a circle. Arcuate intrusives that are vertical or slope away from the center of the circle, instead of toward it, are generally believed to result from sinking of the central block, as in en masse stoping or caldera subsidence (fig. 86). They are called ring dikes. None has been identified in Hawaii, although, in view of the common in-sinking of the top of the shields to form calderas, it would be logical to expect them. Arcuate intrusions of some sort, either ring dikes or cone sheets, probably served as feeders for some of the late eruptions of Mauna Kea, thus accounting for the curved alignment of cinder cones of certain types of rock on that mountain (fig. 87).

SILLS

Sills are sheetlike intrusive bodies that lie parallel to the stratification in bedded rocks

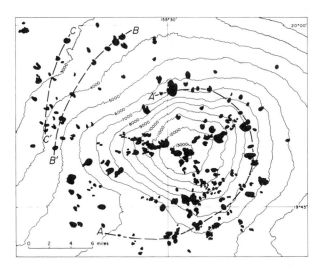

Figure 87. Map showing curved alignment (heavy dashed lines A–A', B–B', and C–C') of cinder cones on Mauna Kea, Hawaii. The curvature is probably the result of feeding of the surface eruptions by arcuate intrusions. (After Macdonald, 1945.)

(fig. 90). Some geologists add the restriction that the beds must have been more or less horizontal at the time of intrusion, and the restriction is justified from the standpoint of the dynamics of intrusion. Where tabular bodies are intruded parallel to beds standing at a high angle, the mechanism that provides the space probably is essentially the same as in the case of dikes, whereas typical sills, intruded into nearly horizontal beds, must lift the overlying beds to make room for themselves. In the latter case, considerably more magmatic pressure may be necessary than in the case of dikes, that simply rise passively into fissures opened by other forces. In Hawaii, however, the restriction is of only academic interest; the bedded rocks all have low angles of dip, hence intrusive sheets that parallel them are sills by either definition.

Sills are common in Hawaii (fig. 127), though far less numerous than dikes. Their range of thickness is about the same as that of dikes, but their average thickness is a little greater—probably about 10 feet. Their textures and jointing are closely similar to those of dikes of comparable thickness. Some sills follow a single bedding plane. Commonly, however, they follow one bedding plane for only a few tens of feet, then break across the bedding and follow another plane for a similar distance. Several such steps may be recognizable. Sills

with this pattern are well displayed in the highway cuts along the Pali just west of Makapuu Point, on Oahu.

All sills, of course, are lenses, pinching out laterally within relatively short distances. In other parts of the world some can be traced continuously for many tens of miles. In Hawaii, however, they rarely can be followed for more than two or three hundred feet, though Wentworth (Wentworth and Macdonald, 1953, p. 88) succeeded in following some in the Koolau Range behind Honolulu for more than 1,000 feet. Short thick sills grade into laccoliths.

Though the great majority of sills probably were fed by dikes, it is only rarely that a dike is seen actually passing into a sill. A particularly good example can be seen in the west wall of Mokuaweoweo caldera, at the summit of Mauna Loa, where a dike cuts up across bedded lavas to connect with a short lenticular sill, 26 feet thick, that spread about 250 feet in both directions along a bedding plane.

If the rule deduced for dikes is applied, most sills recognized in Hawaii were formed at depths of several hundred feet below the contemporaneous surface, since they are essentially devoid of vesicles. This means that, in making room for themselves, they must have lifted a load of several hundred feet of overlying rock. The magmatic pressure must have been considerably greater than would have been necessary to lift the molten rock (appreciably lighter than solid rock) to the surface. Considering the highly jointed nature of the overlying lava flows, it is not entirely clear why the magma intruded as a sill rather than forcing its way up through joints to the surface.

LACCOLITHS

Laccoliths are lenticular bodies that have been intruded parallel to the bedding in enclosing stratified rocks and have made room for themselves by sharply arching up the overlying beds. They are numerous in some plateau areas of the world where nearly horizontal bedded rocks have been intruded by viscous magma. Well-developed ones are rare in Hawaii, prob-

124

ably because most Hawaiian magmas are very fluid and spread far out laterally to form sills instead of causing a conspicuous bulging of the overlying beds. It would not be surprising to find laccoliths of the viscous trachyte magma that formed domes at the surface, but none have been found.

The only well-formed laccolith known in Hawaii is seen in the lower part of the wall of Kilauea caldera at Uwekahuna (see photograph, p. 118). It is a two-humped lens, about 940 feet long and 90 feet high, composed of porphyritic olivine-rich gabbro (picrite-gabbro). Along part of the boundary, the overlying lavas are arched up over the intrusive, but at other places the intrusive seems to have broken sharply across the bedding. (This latter feature is shared by laccoliths in other parts of the world, for example some of those in the San Francisco Mountains of Arizona.) The Uwekahuna laccolith probably was fed laterally from a conduit somewhere within the area of the present floor of Kilauea caldera, much as some of the famous laccoliths of the Henry Mountains in Utah were fed laterally from parent stocks.

Small laccolith-like structures formed nearly on the surface have been observed at several localities. On the side of the Mauna Iki shield, built during the 1920 eruption of Kilauea, a shallow intrusion heaved up the surface lava bed into an asymmetrical dome about 25 feet high before lava burst through the side of the dome as a flow of aa. An identical structure can be seen on the floor of Kilauea caldera about a quarter of a mile east of Halemaumau. In the same way a shallow intrusion heaved up part of the edge of a cinder cone along a fissure vent during the 1960 eruption of Kilauea.

## PLUGS AND NECKS

The congealed fillings of more or less cylindrical volcanic conduits are called *plugs*. Left standing in high relief by the erosion of less-resistant surrounding rocks, they are known as volcanic *necks*. Neither is well developed in Hawaii, partly because the feeders below Hawaiian vents generally are dikelike rather than cylindrical, and partly because

there is less contrast in the erodability of conduit fillings and surrounding lava flows than there would be between the same fillings and surrounding tephra.

At a few places dikes have been found that swell out at the surface into thicker masses, more or less ovoid in ground plan. Several of these are known in the Koolau Range on Oahu, one of them the intrusive that supplied the rock at the now-abandoned Palolo Quarry. In a broad sense these are plugs, but they are not typical of volcanic plugs elsewhere in the world. The name "bud" has been suggested for them (Wentworth and Jones, 1940, p. 988). Similar plugs formed by swelling out of dikes are exposed near the west end of Haleakala Crater south of Kalahaku Pali.

Another, better exposed, plug can be seen on the northwest wall of Ukumehame Canyon, West Maui. The feeding dike of trachyte can be traced for more than a mile along the canyon wall. Beneath the vent, about 500 feet below the contemporaneous land surface, the dike started to swell out into an ovoid mass that apparently occupied a funnel-like vent formed partly by explosive blasting out of the older tholeiitic basalt lavas and partly by thrusting the lavas aside. At the surface the viscous mass protruded as a dome about 200 feet high. This mass, Puu Koai, thus is part dome and part plug.

Haupu, southwest of Lihue on Kauai, is a massive mountain formed of the thick, resistant lava flows that accumulated in a large pit crater on the flank of the Kauai shield volcano. Erosion has removed the less-resistant lavas that once surrounded it, leaving the crater fill standing in magnificent relief. Though different from most, it can be regarded as a volcanic neck of a special type. Imposing mountain massifs in the Marquesas and Society islands probably have a similar origin.

## BATHOLITHS, STOCKS, AND BOSSES

The largest intrusive bodies known are *batholiths*. Tremendous batholiths, with outcrop areas of many thousands of square miles, occupy the cores of folded mountain ranges on the continents. Some are now known to have been formed by replacement processes ("granitization") in which invading volatiles bearing alkalies and other materials have brought about the transformation of older rocks of igneous, sedimentary, and metamorphic origin into such rocks as granites, granodiorites, and quartz diorites. Others, though perhaps formed by magma that in turn was created by extreme granitization at greater depths, are clearly intrusive at the presently visible level. These huge intrusions grade into smaller ones. Arbitrarily, at a limit of 40 square miles of outcrop area, we stop calling them batholiths. Smaller, but otherwise similar, bodies are known as *stocks*, and stocks that are essentially circular in ground plan are called *bosses*. Stocks, like batholiths, have walls that are vertical or diverge downward, and their bottoms are at levels so deep that they have never yet been revealed by erosion.

There are no batholiths in Hawaii. They seem to be restricted to the areas in which former great down-warps of the earth's crust (geosynclines), with their accumulations of thirty or forty thousand feet of sedimentary and volcanic rocks, have been deformed into folded mountains, and to certain "shield" areas that formed the original cores of the continents. But on several islands small stocks and bosses are known. Some are exposed in the big valleys of East Molokai, and others in West Maui (fig. 88). The largest of those on Maui, in

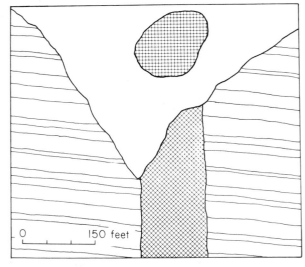

Figure 88. Cross section and plan of a small boss in Black Gorge, a tributary of Iao Valley, West Maui. (After Daly, 1911.)

the head of Ukumehame Canyon, is oval in ground plan, half a mile long and averaging about 0.1 mile wide. Another, in the canyon of Kahoma Stream, is 0.3 mile long and nearly 0.2 mile wide. These masses consist of gabbro, with average grain size from about 1 to 2 mm. Some are dense, resembling many gabbros from continental areas, but others consist of a meshwork of crystals separated by open spaces. These rocks must have formed at very shallow levels. A mass of similar gabbro that intruded posterosional lavas is exposed just north of Kalaheo, on Kauai. Since it seems impossible that erosion at that locality could have removed more than a very few hundred feet of overlying rock, the coarse granularity of the gabbro formed at this very shallow level must have resulted from the presence of volatiles.

If the cover of solid rock over the Kalaheo stock was indeed only one or two hundred feet, it is difficult to see why it remained intact instead of breaking up and sinking into the magma. Actually, that may have happened; but if it did the surficial cap of finer-grained rock that would have formed by chilling of the top of the gabbro has been removed by later erosion. Caving in of the roof over stocks of this sort is very probably the mechanism by which were formed pit craters such as those along the Chain of Craters Road in Hawaii Volcanoes National Park.

There is no evidence of thrusting aside or of other disturbance of the lava rocks in the walls of the stocks. The chambers occupied by the stocks must have been excavated in the older enclosing rocks by stoping, by assimilation, or by melting.

## DIKE COMPLEXES

Where erosion has cut deeply into the older Hawaiian volcanoes, the downward extension of the surficial rift zones is seen to contain a very great number of dikes (fig. 84). These represent the fissure conduits through which magma rose toward the surface. Some did not reach the surface, but most did, and fed volcanic eruptions. These zones of abundant dikes are referred to as *dike swarms*, or in Hawaii more commonly as *dike complexes*.

In Hawaiian volcanoes the rift zones and

126

dike complexes range in width from about 1.5 to 3 miles and average about 2 miles. They extend radially outward from the summit of the volcano, and in the case of elongate shields such as that of the Koolau Range on Oahu they trend parallel to the long dimension of the shield. Within it the trend of most of the dikes is approximately parallel to that of the complex, but minor divergences are common and dikes frequently cross each other.

Sills also are quite common in the dike complexes, but are far less abundant than dikes. A few stocks also have been found, and stocks probably underlie the pit craters we see at the surface.

The dike complex is sharply truncated at the edge of the caldera. Some dikes are present in the caldera-filling lavas, but they are not nearly as abundant as they are in the adjacent dike complex.

The abundance of dikes in a dike complex varies greatly, both from one complex to another and from place to place within the same complex. It is generally greatest in the central part, where the average number is about 100 to 200 per mile of horizontal distance. The dikes generally are separated by segments of the pile of lava flows of which the volcanic mountain is built. Where dikes are sparse the segments of flow rock greatly predominate in volume, but as dikes become more abundant the proportion of intervening flow rock decreases. In some areas the number of dikes approaches 1,000 per mile, and when it is remembered that the average thickness of dikes in Hawaii is between 2 and 3 feet it becomes apparent that they may constitute half or more

of the total rock volume. In some places they are so numerous that one dike lies directly against another, without any intervening flow rock.

In the outer part of the complex the number of dikes decreases to about 10 to 100 per mile, and at the edge the number drops off very abruptly. Relatively few are found outside the dike complex. Downward, the number of dikes in the complex must increase, because dikes that fed the surface eruptions during all stages of the volcano's growth must pass through the lower part of the structure. It has been calculated that at the level of the ocean floor the number of dikes must be 650 to 700 per mile of width of the complex.

Dike swarms similar to the Hawaiian dike complexes are known in other parts of the world. One of the best examples is the swarm of northwest-trending dikes in western Scotland, generally believed to represent the feeders for the flood eruptions that built the great Thulean basalt plateau that once occupied much of the northeastern Atlantic Ocean. It seems clear that dike swarms such as these can only have resulted from a local stretching of the earth's crust, the rocks being pulled apart to provide the open fissures into which the magma rose. The necessary stretching is quite considerable. Calculations indicate that the amount represented by each of the dike complexes of the two major volcanoes on Oahu at the level of the sea floor is about 0.45 mile, making a total crustal stretching during the formation of these two parallel dike complexes of about 0.9 mile. The cause of this stretching can still only be conjectured.

*Suggested Additional Reading*

Daly, 1933, pp. 74–110; Frankel, 1967; Wentworth, 1951, pp. 10–12; Wentworth and Jones, 1940; Wentworth and Macdonald, 1953, pp. 86–91

*Cinder cones on
Mauna Kea*

# CHAPTER SIX

# Classification of Volcanic Eruptions

VOLCANOES OF OTHER PARTS OF THE WORLD DIFFER SIG-nificantly from those of Hawaii. The differences, however, appear to be more of degree than of kind. That is, the same general principles and processes are operative everywhere, but certain factors are more important in some regions and less so in others. Therefore, the concentrated study of Hawaiian volcanoes over the past half-century has importance far beyond the boundaries of Hawaii. Here the eruptions are relatively gentle, and we can work at close range making observations and physical measurements on the erupting lava, whereas elsewhere we are generally kept at a distance by the violence of the eruption. Of course other factors also are important, among them the frequency of eruptions in Hawaii, their reasonably easy accessibility, and their closeness to a good base of supply of both materials and technical skills. We are confident that the principles we develop working here in Hawaii can be applied elsewhere.

We are far from alone in our work. Very important studies are being made of active volcanoes in Japan, Kamchatka, New Britain and New Guinea, New Zealand, Iceland, and Italy. Also, all over the world geologists are studying ancient volcanoes, trying to deduce from postmortem studies the processes that went on when the volcanoes were alive. All these are important, but the particular importance of Hawaii is shown by the frequency with which visitors from other lands come here to study our volcanoes.

Even within the confines of the Hawaiian Islands the volcanic activity has not always been the same as that of recent eruptions. In particular, the late stage eruptions differ significantly from the earlier ones that built by far the largest part of the great Hawaiian shield volcanoes, and they bear more resemblance to eruptions in other parts of the world. Therefore, before we proceed to a description of the late stages of Hawaiian volcanic activity, we should consider some of the other types of eruptions. This is best done in a very brief manner by outlining a classification of volcanic eruptions. The most commonly used classification is based on the strength of the explosions and the nature of the erupted material; and several eruption types are named for volcanoes that characteristically exhibit that type of activity. It should be emphasized, however, that nearly

129

all volcanoes vary somewhat in type of activity from one eruption to another, or even during the course of a single eruption. Thus, although the usual activity of Stromboli is of the Strombolian type in the following classification, Stromboli also sometimes has Hawaiian- or Vulcanian-type eruptions. Arranged in the order of increasing strength of explosion, the usual classification is as follows:

FLOOD ERUPTIONS

We have already seen that Hawaiian basaltic lava flows are very fluid and voluminous, but in some regions basalt flows even more fluid and of far greater volume have been erupted. These have received the name *flood basalts,* and the eruptions that produce them are called *flood eruptions.* They are also sometimes called "plateau basalts," because they build broad, nearly level plains, some of which are at high altitude.

Flood basalt magma reaches the earth's surface through fissures, as does the Hawaiian basalt. However, the fissures are scattered over a wide area instead of being concentrated in narrow "rift zones." As a result, successive eruptions do not build a mountain, but instead leave a broad, nearly level surface. The lavas are so fluid that they spread out to great distances. Actually they do build very broad low mounds or shields around the principal vents, but these are so flat that they are easily overlooked. Explosive activity is almost wholly absent. Only diminutive spatter cones are formed along the fissure vents. Commonly, the flows are poured over an older eroded land surface, sometimes of considerable vertical relief. Gradually the lava flows bury the hills and valleys of the older surface, transforming it into a relatively featureless plain.

Individual flood basalt flows tend to be much thicker than those of Hawaii. Often they are 50 to 100 feet thick, and some are several times that. Many are very extensive. One flow near the Grand Coulee, in Washington, has been traced over an area of about 20,000 square miles and has a volume of about 74 cubic miles (as compared to about 0.11 cubic mile for the largest historic flows in Hawaii). The only flood eruption witnessed by modern man took place in southern Iceland in 1783, when the 10-mile-long Laki fissure gave vent to about 3 cubic miles of lava.

Several areas of flood basalts are known. The Columbia Plateau, in northwestern United States (fig. 89B), is the surface of a mass of flood basalts with a volume of more than 100,000 cubic miles, erupted in fairly recent geologic time (about 60 million years ago). Just to the east, the more recent Snake River Plain is a similar mass of flood basalt. Some of the eruptions on the Snake River Plain were more explosive than the typical flood eruptions, and built moderately big cinder cones such as those at Craters of the Moon National Monument. Other areas of flood basalts are found in the Deccan region of northwestern India (fig. 89A), in southeastern Brazil, Patagonia, and South Africa. A great mass of flood basalt formerly occupied the northeastern part of the Atlantic Ocean, but it has broken up. Only small remnants are now visible in Ireland and western Scotland, the Faroes, Iceland, and southeastern Greenland. Far back in geologic time, more than 500,000,000 years ago, a broad area of flood basalt was formed in what is now north-central United States. Its lavas contain the famous Michigan copper deposits.

HAWAIIAN ERUPTIONS

Eruptions of Hawaiian type have already been discussed in chapter 2. They are characterized by the extrusion of fluid basaltic lavas, typically with little or no explosion, and the building of broadly rounded shield volcanoes. Small spatter or spatter-and-cinder cones are built along the fissure vents. The ejecta commonly are still fluid when they fall back to the ground, and flatten out on impact, giving rise to cow-dung bombs.

STROMBOLIAN ERUPTIONS

Stromboli volcano, on an island off the coast of Italy, gives its name to a type of eruption characterized by less fluid lavas, typically erupted with considerable explosive activity. Either lava flows or tephra may predominate, but both generally are present. Tephra is usually much more abundant than in Hawaiian eruptions, and builds large cinder cones. Many of the bombs take on spindle shapes in the air, and are hard enough to retain those shapes on striking the ground. Ash often blankets the surrounding country. The lavas of Strombolian eruptions may have any composition from picrite-basalt to trachyte, but basalt and andesite are most common.

Single Strombolian eruptions build cinder cones accompanied by lava flows. An example is the eruption of Paricutin volcano, in Michoacan, Mexico. The eruption began in 1943 with the opening of a fissure in a field that was being plowed for the spring planting. At first gas oozed from the fissure, and a few hours later red-hot cinders began to fly out. Within 5 days a cinder cone had grown to a height of 300 feet, and in less than a year it was 1,200 feet high. The eruption continued for 9 years, and, although the cone remained at about the same height, a broad field of lava flows was formed around it, burying two villages but taking no lives.

Repeated Strombolian eruptions from the same vent may build high mountains composed of interbedded tephra and lava flows (fig. 90). These are called *composite volcanoes*. (Most composite volcanoes are built partly by eruptions of the Vulcanian type, described in the next paragraph.) They include some of the most beautiful mountains in the world—the ones ordinarily considered typical volcanoes— Mt. Shasta in California, Shishaldin in the Aleutian Islands, Fuji-san in Japan, Vesuvius in Italy, and the most perfectly formed of them all, Mayon in the Philippines. These very symmetrical steep-sided cones depend on tephra falling back around a small pipelike central vent to give them their regular conical shape, but if the cone is to reach large size and remain symmetrical it must be strengthened by ribs of interbedded lava. Without them, the loose cinder and ash is unstable and the cone tends to slump down, losing its graceful shape.

VULCANIAN ERUPTIONS

Vulcanian eruptions (named after Vulcano, another island volcano off the west coast of Italy) are characterized by more viscous lavas, forming only small stubby flows, and a great preponderance of tephra. Viscous lava in the

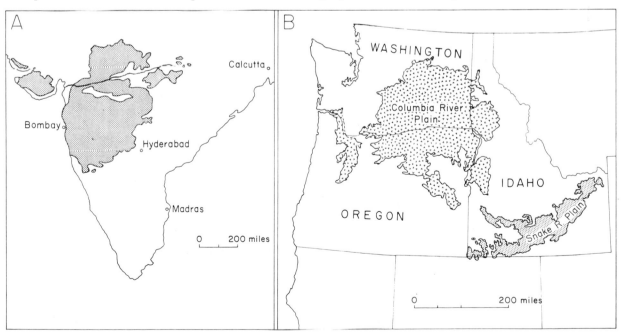

Figure 89. Maps showing the flood basalt areas of *(A)* the Deccan, India; and *(B)* the Columbia and Snake River plains of northwestern United States.

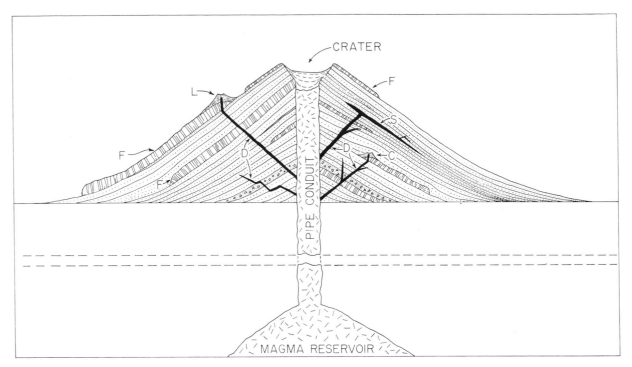

Figure 90. Diagrammatic cross section of a composite volcano, showing interbedded lava flows and beds of pyroclastic material, and the typical cone shape. C, Cinder cone; D, dike; F, lava flow; L, lateral cone; S, sill. (After Macdonald, in preparation.)

crater is torn apart by violent explosions and hurled into the air as angular fragments that remain angular during flight. Thus, the large ejecta are blocks rather than bombs. Vulcanian eruptions also may build cinder cones, and play a part in the building of composite volcanoes. Often the most conspicuous feature of a Vulcanian eruption is the tremendous dark cloud of ash-laden gas that rises several miles into the air and spreads ash over the country-side for many miles to leeward of the volcano. The lava of Vulcanian eruptions may be basalt, but more commonly it is andesite or dacite.

## PELÉEAN ERUPTIONS

Peléean eruptions are named after Mt. Pelée ("Bald Mountain"), on the island of Martinique in the West Indies. The lava is always very viscous, but the eruptions may have very different aspects, depending largely on the abundance of gas and its condition in the magma. If the magma is relatively poor in gas it may push up with little disturbance, forming over the vent a rounded hill known as a *dome.* Often, however, the uppermost part of the

magma body approaching the earth's surface contains a large amount of gas, which has risen in the liquid leaving little gas in the lower part. This results in an explosive outbreak, throwing the magma into the air as a shower of froth, which hardens and falls back around the vent to build a cone of pumice or mixed pumice and cinder. Then, as the lower, gas-poor magma rises to the surface, a dome is built in the crater of the cone. A beautiful example of this sequence is Panum Crater, a rhyolitic pumice cone and dome at the south shore of Mono Lake, in California. At other times the early lava is relatively gas-poor, and a dome is built as the first phase of the eruption, followed later by explosive eruption of more highly gas-charged lava. Thus domes had already grown in the crater of Hibok-Hibok in 1950, and of Mt. Pelée in 1902 and 1929, before explosive eruptions generated the deadly hot avalanches to be described later.

Volcanic domes grow in part by overflows of lava onto their flanks, but largely by expansion, the whole structure being inflated as viscous lava is forced from below up into its interior. Sometimes the outer shell of the dome remains plastic enough to stretch as the dome is inflated, but often it cools and becomes brittle. Then as it is stretched it breaks up into vast numbers of loose blocks that tumble down the flanks of the dome and form a bank of angular fragments (breccia) at its base (fig. 91). These "crumble breccias" may cover a large part of the dome, leaving only its very top exposed and giving the hill the general shape and appearance of a cinder cone. Very commonly, viscous magma is squeezed up through fractures in the shell of the dome, much as molten metal is squeezed out through a die in making wire. This results in protrusions on the dome surface known as *spines.* Many domes rather resemble sea urchins, bristling with small spines. Occasionally they attain huge size. In 1902 a spine on the summit of a dome on Mt. Pelée grew to a height of 1,200 feet, but, like most spines, it was short lived, and over a few months it collapsed, to add to the volume of crumble breccia around the dome.

Peléean eruptions may also give rise to one of the most terrifying and destructive phenomena of volcanism—the *glowing avalanche* or *nuée ardente* ("glowing cloud"), which is, in a sense, an explosion of gas-charged magma, but an explosion of a very special type. The initial blast may be directed either upward or laterally. In either case the explosion cloud is so heavily laden with fragments that it falls immediately to the ground and flows down the mountainside as an incandescent avalanche,

above which rises the billowing cloud of ash and dust that gives the phenomenon the name "glowing cloud." Each glowing fragment in the avalanche continues to give off gas, and in addition the air trapped in the avalanche is being heated up and consequently expanding. In this way each fragment in the avalanche is surrounded by a shell of expanding gas that serves as a cushion to hold it away from the surrounding fragments and from the ground. The mass is really an emulsion of solid fragments in gas. As a result it moves with very little friction, and the pull of gravity added to the impetus of the initial blast may give it tremendous speed down steep mountain slopes. At Mt. Pelée, in 1902, a glowing cloud rushed down the mountainside with a speed of about 100 miles an hour, and in a matter of moments destroyed the entire city of St. Pierre with its 30,000 inhabitants.

## PLINIAN ERUPTIONS

The most violent of known eruptions are named after Pliny the Elder, the great Roman naturalist, who was killed while investigating the eruption of Vesuvius and trying to rescue friends, in A.D. 79. These eruptions are cataclysmic in power, and frequently very destructive. They often accompany the collapse of the top of the cone of a composite volcano to form a caldera. Crater Lake, in Oregon, was formed in that way during a Plinian eruption some

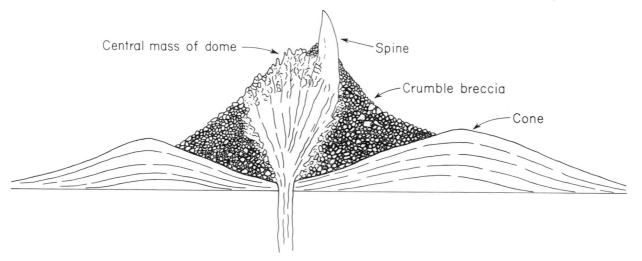

Central mass of dome — Spine

Crumble breccia

Cone

133

Figure 91. Diagrammatic cross section of a volcanic dome with banks of crumble breccia and a spine.

6,000 years ago. It is commonly said that the ancient mountain, called Mt. Mazama, blew its head off, leaving the great hole now occupied by Crater Lake. However, Howel Williams, of the University of California, has shown that the material blown out during the tremendous eruption consisted, not of fragments of the old mountain, but of a froth of new magma (pumice) brought up from below. In Williams' words, the mountain did not blow its head off; it eviscerated itself and its head fell in! Part of the outrushing pumice was hurled high into the air as towering eruption clouds, but most of it swept at tremendous speeds down the mountainsides as glowing avalanches, with their attendant glowing clouds of ash. The hole formed in the mountaintop by collapse slowly filled with water from rains and melting snow, forming the lake, and a new series of small eruptions within the caldera formed Wizard Island.

The greatest volcanic eruption of modern times was the Plinian eruption that destroyed the whole top of Tamboro volcano in Indonesia, in 1813. Unfortunately, we know little about it. In 1883 the famous eruption of Krakatoa, in the strait between Java and Sumatra, destroyed a large part of the old volcanic cone and formed a caldera on the sea floor, causing "tidal waves" that killed many thousands of persons on nearby shores, and throwing up a cloud of dust that rose high into the stratosphere and drifted around the earth three times, causing brilliantly colored sunsets even in Europe and the United States. The noise of the explosion was heard 1,500 miles away, in Australia.

The collapses that have formed the calderas of many composite volcanoes appear to have resulted from the draining away of magma from the underlying magma reservoir by great Plinian eruptions. Even greater collapses of the earth's crust, forming broad down-faulted basins (volcano-tectonic depressions), have occurred in several parts of the world, for instance in New Zealand and Sumatra, accompanying the eruption of enormous amounts of rhyolite pumice. Much of the pumice is blown apart by expanding gas to form tiny fragments

of ash; and the masses of ash, with their enclosed lumps of pumice, flow in much the same way as do glowing avalanches. These are known as *ash flows*. Like the flood eruptions of basalt, these ash flows erupt from fissures and have enormous volume, the flows spreading out to great distances to form broad flat "plateau" surfaces. The ash fragments commonly are so hot that when the flow comes to rest the fragments become partly fused, forming so-called welded tuffs, or *ignimbrites*. The "ignimbrite plateau" in the center of the North Island of New Zealand was formed in this way.

Neither Plinian eruptions nor the disastrous glowing avalanches of Peléean eruptions are at all likely to occur in Hawaii. They seem to belong wholly to the volcanoes of mountain-building belts on the continents.

Plinian eruptions often, if not always, terminate a long period of volcanic repose. In 1951, some 2,000 persons were killed by an eruption of Mt. Lamington, in New Guinea—an ancient volcano that had not been active in the memory of the New Guinea natives. Before A.D. 79 Vesuvius had not erupted in all of earlier Roman, and probably Etruscan, history. Unfortunately, signs of the coming great eruption were not recognized. When it came, the eruption destroyed the entire top of the mountain, leaving a great hole surrounded by a crescentic ridge—the present Mte. Somma. The new cone of Vesuvius has grown up within the crescent of Mte. Somma since that time.

VOLCANIC MUDFLOWS

Seldom is it realized that flows of mud are among the most destructive of volcanic phenomena. In the eruption of Vesuvius which destroyed three Roman cities (Pompeii, Stabia, and Herculaneum), Pompeii was buried by thick beds of pumice and ash, but Herculaneum was inundated by mud.

Of course, not all mudflows are of volcanic origin. Some are simply the result of the soil on hillsides becoming saturated with water and flowing as mud down into the valleys, generally following destruction of vegetation by fire or by artificial clearing (see chap. 10).

134

Volcanic mudflows form in several ways. Eruptions beneath glaciers, melting the ice, or the breaking down of the walls of crater lakes, can release great floods of water that rush down the mountainside carrying with them masses of loose rock debris. Glowing avalanches may sweep into lakes or rivers and generate mudflows. But most mudflows are simply the result of heavy rains. At the end of a Strombolian or Vulcanian eruption the upper slopes of the mountain commonly are covered with an unstable layer of loose debris. Heavy rains may saturate this material, turn it into a slurry, and start the entire mass moving downslope. As it moves, its velocity increases and may reach as much as 60 miles an hour. The onrushing muck sweeps along with it everything movable in its path on the steeper slopes of the mountain, only to deposit the whole mass on the gentler slopes around the mountain foot. Whole villages have been buried in that way.

The destructive mudflow of 1868 in Wood Valley on the island of Hawaii was generated in a somewhat different way. It is described in chapter 10.

*Suggested Additional Reading*

Bullard, 1962; Macdonald, in preparation; Rittmann, 1962

*Haleakala Crater*

# CHAPTER SEVEN

# Life Stages of Hawaiian Volcanoes

VOLCANOES, LIKE PEOPLE, PASS THROUGH A SUCCESSION of stages in their development. Volcanoes are born, and eventually they die, though it is often hard to be sure they are really dead—that no hidden spark of life remains to bring the somnolent giant back for a last violent fling of destruction. We can think of a volcano as young, or middle aged, or old, or when new life has entered into the old mechanism, as rejuvenated.

Geologic mapping of the Hawaiian volcanoes has led to the recognition of four stages in their activity. These, in turn, can be assigned to two general periods, separated by a very long time of volcanic quiet. During the principal period of volcanism, comprising three successive stages, the main volcanic mountain is built. Following this period, erosion by streams and waves makes great inroads on the mountain, cutting sea cliffs hundreds or thousands of feet high and valleys thousands of feet deep. As yet we cannot say just how long this period continued, but certainly it must have lasted for many tens of thousands of years—about 2 million years at the Koolau volcano of Oahu. Judging these deeply eroded volcanoes by any criteria we yet possess, we would be justified in calling them dead. But some of them have come back to life and have produced cones and lava flows that rest on, and bury, the old eroded topography. This time of renewed volcanic activity is called the rejuvenated period or, rather prosaically, the posterosional period.

Following is a list of the life stages of Hawaiian volcanoes, arranged as is traditional in geology, with the first one at the bottom and the last at the top, just as in nature the older rocks are buried by successively younger ones. The wavy line indicates an *unconformity*—a buried erosional surface that marks a period during which no new rocks were being deposited and the older rocks were being eroded. Minor erosional unconformities are found in the rocks of the principal period of volcanism, particularly in those of the last stage, but the great erosional unconformity separates the principal period from the rejuvenated period.

Rejuvenated period

~~~~~~~~~~~~~~~~ Great erosional unconformity

Old age, or
postcaldera, stage

Principal period | Mature, or
caldera, stage

Youthful, or
shield-building, stage

Not all Hawaiian volcanoes have necessarily gone through all these stages. Kilauea and Mauna Loa are still in the mature stage. Other volcanoes have died before reaching the later stages, or have simply skipped some stages. Thus, there is no evidence that Hualalai ever possessed a caldera—the principal feature of the mature stage—though its rocks and the character of its eruptions indicate that it is approaching old age. On Mauna Kea also, there is no direct evidence of a former caldera, though indirect evidence suggests that there may have been one which has been wholly hidden by later lavas. West Molokai volcano seems to have stopped erupting before a caldera was formed, although a few small late lava flows resting on soil are of rock types that belong to the old age stage. Lanai volcano went out of action before its caldera had quite become filled—that is, while it was still in the mature stage.

The various stages in the development of the Hawaiian Islands, and many other islands of the central Pacific, are illustrated in figure 92.

## YOUTHFUL STAGE

By far the largest part of the Hawaiian mountains has been built by frequent eruptions of very fluid lava. The characteristics of the resulting shield volcano have been described in chapter 2. The lavas are tholeiitic basalts, ranging from basalt wholly devoid of olivine, through olivine basalt to oceanite, more than half of which may consist of large green phenocrysts of olivine.

The character of the eruptions must change greatly as the huge shield volcanoes are built. Building starts on the ocean floor, where the

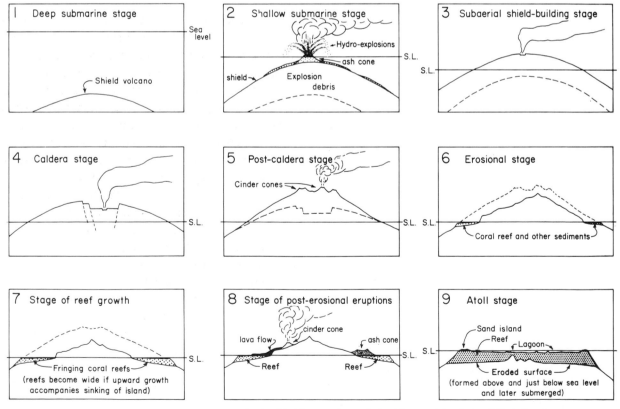

Figure 92. Diagrams illustrating the typical stages in the life history of a typical mid-ocean island. (Modified after Stearns, 1946.)

138

weight of three or more miles of ocean water above restrains any tendency to explosive release of gas, and even prevents the gas from separating as bubbles and forming the vesicular lava familiar to us above sea level. Therefore, on the deep ocean floor the lavas are dense and nonvesicular. Until recently this last statement would have been based wholly on theory. Recently, however, geologists of the U.S. Geological Survey have obtained photographs and samples of the rocks along the east rift zone of Kilauea beneath the ocean east of the island of Hawaii. In shallow water the rocks have the familiar vesicular character, but as the depth of water increases the size of the vesicles decreases until at a depth of about a mile they are only tiny holes the size of the head of a pin and smaller, and below about a mile and a half of water they disappear altogether. The surfaces of the flows show the familiar characteristics of pahoehoe, but in addition there are rounded, bulbous "pillows," characteristic of eruption in water (Moore, 1965).

In deep water, then, the shield is built by repeated outpourings of dense lava. But as the mound builds close to the surface of the ocean the restraining pressure of the water decreases to the point where explosion is possible. The amount of gas in Hawaiian lavas is so small that even mild magmatic explosions probably do not take place in water more than a few hundred feet deep, but the hot lava coming in contact with water may, under some circumstances, form large volumes of steam, with resulting violent and spectacular steam explosions. Steam explosions in shallow water throw up huge amounts of pulverized rock material and an "atomized" spray of liquid lava—the materials which built the tuff cones of the Hawaiian Islands. We have never seen an eruption of this sort in Hawaii, but eruptions just off the shore of Fayal Island in the Azores (in 1957), and at Surtsey, south of Iceland (in 1963) showed what they must be like. Explosion after explosion shoots up out of the water long black jets of ash-laden steam; if the water is shallow enough there appears around the eruption point a ring-shaped or horseshoe-shaped island—the top of an ash cone. Molokini, between Maui and Kahoolawe, is just such

an islet. Probably most islands of this sort are short lived, and are soon eroded away by the ocean waves. Examples of such "disappearing" islands are numerous. One is Falcon Island, in the Tonga group, a cinder-and-ash cone that every now and then is built above sea level by eruption, only to be attacked immediately by the waves and in a few years eaten away. The same sort of thing must have happened repeatedly as the Hawaiian shield volcanoes were built through shallow water; but the ash was minor in amount as compared with the volume of lava, and the mountain continued to consist largely of flows and its shape continued to be that of a shield. To some extent the ash eroded from the cones by the waves probably was dropped on the summit of the submarine mountain and remained there interbedded with the flows. (We hope some day to check on this by means of deep drill holes.) For the most part, however, it probably was carried out into the deep water surrounding the growing mountain.

Once the top of the shield emerged from the water and further eruptions were on land, the character of the activity again changed. Except for occasional off-shore outbreaks, the eruptions were of the sort witnessed in recent years on Kilauea and Mauna Loa—the Hawaiian-type eruptions described in chapters 2 and 6.

MATURE STAGE

Toward the end of their growth the tops of most of the Hawaiian shield volcanoes have collapsed to form calderas. Volcanism is still very active at that stage, however, and the caldera is no sooner formed than lavas start to refill it. The period of collapse and refilling of the caldera is known as the mature stage.

Caldera formation is not a one-step process. Evidence points to a gradual enlargement of the calderas by the collapse of additional areas around their edges. Thus, even in historic time, the caldera of Mauna Loa has grown by the addition of the small coalescing pit crater known as the East Bay. The rather lobate outlines of several calderas indicate that they have grown in a similar manner. Also, caldera formation does not take place in a single period

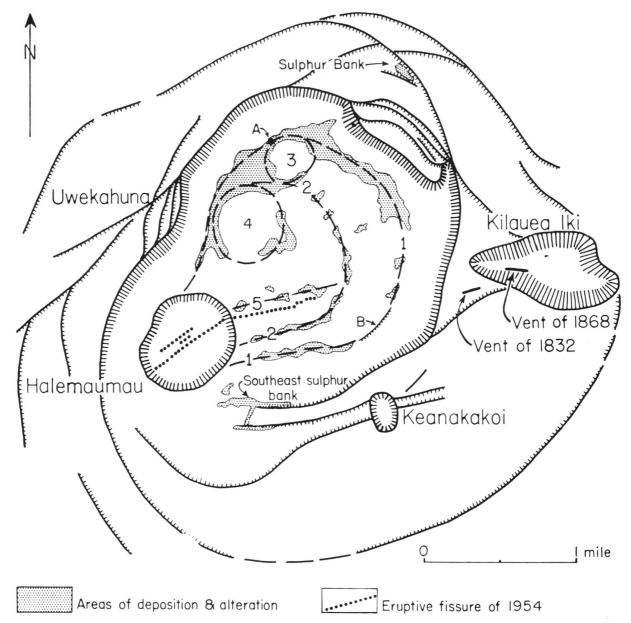

Figure 93. Map of Kilauea caldera, showing outlines of several former pits (*dashed lines 1, 2, 3, 4*), and areas of rock alteration and deposition of opal controlled by the fractures bounding the former (now buried) pits. *A*, A very persistent steam vent; *B*, a scarp formed by uplift along the border of the former pit. (From Macdonald, 1955a.)

of collapse. Instead, there are repeated collapses and refillings. The history of Kilauea caldera has already been outlined in chapter 3, and it will be recalled that during the 19th century several collapses of the central part of the caldera occurred, each being followed by refilling to or above the level of the floor before the collapse. Furthermore, a study of the old records of Kilauea and the pattern of rock alteration on the present caldera floor—a pattern caused by gases rising through cracks that mark the position of buried pits that formerly existed within the caldera (fig. 93)—

shows that the caldera contains many separate collapses within the larger collapse area. Evidence of a similar condition is found in other Hawaiian calderas. For example, the rocks in the hills just inland from Lanikai and Kailua, on the windward side of Oahu, accumulated in a crater within the caldera of the Koolau volcano. The downbowing of the lava beds within this crater (fig. 94), caused by further subsidence of the crater floor, can be clearly seen from points along the lower part of the Pali Highway where it descends the Pali.

The sinking of the top of the volcano is

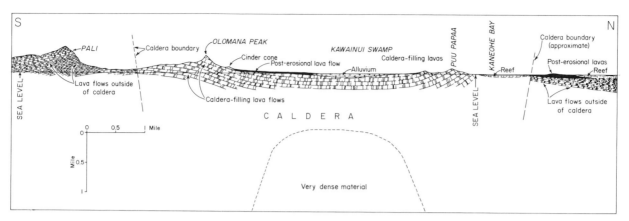

Figure 94. Geologic section across the caldera of the Koolau volcano, Oahu.

usually attributed to removal of support from beneath the summit area of the mountain by withdrawal of magma from the underlying shallow magma chamber during flank eruptions. We have already seen that the mountain top does sink, by an amount easily measurable by tiltmeters and ordinary leveling, during a big flank eruption. However, magma withdrawal is probably not the only mechanism which causes the sinking. The rocks of the mountain top are denser than the liquid magma beneath and would tend to sink into it, at the same time displacing the magma upward. In this way a single mechanism would bring about both the sinking and the refilling of the caldera. But the fact that an actual depression results indicates that material must actually be removed, and lateral drainage of the magma, either as surface eruptions or as intrusions into the rift zones, probably is the dominant mechanism.

We do not actually know how early in the history of the volcano the caldera starts to form, but it is probably a late development. Analyses of the tilt measurements that indicate swelling and shrinking of the volcano show that the magma chamber that makes possible the collapse of the caldera lies at a depth of only 2 or 3 kilometers (1.5 or 2 miles)—above the level of the surrounding ocean floor—and thus actually within the body of the volcano. Hence it must have formed by melting of the once-solid lavas in the heart of the mountain. As the shield grew, by addition of one lava flow after another on its outside, lava passing upward through the feeding fissures contributed heat to the surrounding rocks. At first the heat escaped as rapidly as it was supplied, but

eventually the insulating effect of the thickening blanket of overlying lavas became great enough so that heat was added faster than it could escape, and eventually the melting temperature was reached. From that time onward the body of molten rock in the core of the mountain gradually grew in size, enlarging outward along the rift zones as well as on the central axis of the volcano, and eventually it became large enough to allow the overlying mountain top to sink in, forming the caldera at the summit and, commonly, grabens along the rift zones.

Volcanism remains vigorous, and eruptions frequent, through most of the life span of the caldera. Perhaps the end of the period of collapse and the eventual filling of the caldera are related to the freezing of the shallow magma body as the activity of the volcano decreases and eruptions become less frequent, allowing heat to escape faster than it is added. Certainly, whatever the reason, the caldera finally becomes filled, and in the last stage of activity a caldera no longer exists. In this, the Hawaiian volcanoes differ markedly from volcanoes of the mountain-building belts of the continents, since in those the formation of a caldera commonly is nearly the last event.

OLD AGE STAGE

The last stage of the principal period of volcanism consists in the building of a cap of lava and pyroclastic material over the top of the shield, completely hiding the caldera. The cap is commonly steeper and less extensive than the shield, and perches on it like a limpet

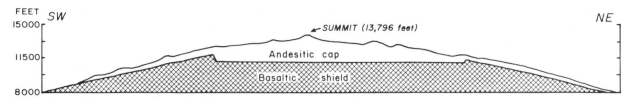

Figure 95. Cross section of Mauna Kea, showing the hypothetical caldera buried beneath the cap of andesitic lavas of the old age stage. (After Stearns and Macdonald, 1946.)

on a larger shell. The clearest illustration of this is Mauna Kea, as viewed from Hilo, the steep slopes of the cap contrasting clearly with the gentle (but ash covered) slope of the shield that projects from beneath it and forms the agricultural lands of Hamakua (fig. 95). The remnants of the steep cap on East Molokai also are clearly discernible on a clear day from Oahu and West Maui.

The steeper cap is a reflection of a change in composition of the lavas erupted in the late stage as compared with those of the earlier stages, and consequently in the character of the eruptions. In several of the volcanoes it has been possible to show that the change takes place late in the mature stage, when the caldera is almost filled; but it occurs also in volcanoes, such as Hualalai, that may never have had a caldera. Because the general character of these late-stage eruptions is the same both at the end of the caldera-filling, or mature, stage and in the old age stage, and because it probably is wholly unrelated to whether or not a caldera ever was present, it seems best to consider it independently of these stages. This is done in the following section.

END OF THE PRINCIPAL PERIOD OF VOLCANISM

The change in composition of the lava erupted in the late mature stage of Hawaiian volcanoes is at first hardly detectable except by chemical analysis. The predominant lavas are olivine basalt, in appearance very much like those of the earlier parts of the period, but containing a little less silicon and a little more sodium and potassium than the earlier lavas. These are called alkalic basalts, in contrast with the so-called tholeiitic basalts of the earlier stages.

The alkalic basalts are still very fluid, but when they reach the surface they contain

somewhat more gas than is typical of the tholeiitic basalts, and consequently the eruptions are a little more explosive. The spatter cones that are built at the vents tend to be somewhat larger than those of the earlier stages. Hualalai volcano has entered this phase, and comparison of its profile with that of Mauna Loa, which is still erupting tholeiitic basalts, shows it to be much bumpier—the cones scattered along the rift zones are larger.

Along with these alkalic basalts there appear also a few black phenocrysts of augite, and sometimes numerous white phenocrysts of feldspar. Temperatures in the magma chamber beneath the volcano have become lower, and more minerals are crystallizing before the magma reaches the surface. Lavas containing very abundant feldspar phenocrysts can be seen along the trail into Waipio Valley in the uppermost part of the basaltic shield of the ancient Kohala volcano. Similar lavas in the Koolau Range on Oahu contain some feldspar crystals as much as 1.5 inches long; and both on Lanai and at Pohakea Pass on the Waianae Range, Oahu, some of the large feldspar crystals are very clear and free enough of fractures that they can be cut as gem stones.

The next change is much greater, and the rocks produced are radically different in appearance from those of the primitive tholeiitic phase, and so different in composition that they are given wholly different names. The eruptions are of sufficiently different character that they can no longer be considered as typically Hawaiian. The change occurs in two distinct ways, at different volcanoes. At some, such as Mauna Kea and Haleakala, there is a gradual appearance of the new types of rocks, interbedded with alkalic basalts, without any prolonged interruption of volcanic activity. At others, such as Kohala and West Maui, volcanic activity comes to a halt for many thousands of years, and then resumes with the eruption of wholly new rock types without any admixture of basalts.

The new rocks are of four kinds: ankaramite, hawaiite, mugearite, and trachyte (see chap. 4). They appear to derive from alkalic basalt magma largely as a result of crystallization in the magma chamber beneath the volcano. As the magma in the chamber cools, more and more crystals of olivine, calcium-rich plagioclase feldspar, and augite form in it. Being heavier than the surrounding liquid, the crystals sink. This leaves, at the top of the magma body, liquid magma that contains less of the components of the sunken crystals—particularly magnesium, iron, and calcium—than the original magma. Erupted to the surface, this magma makes the hawaiites, mugearites, and trachytes. The ankaramite, on the other hand, appears to be formed largely by the addition of sunken crystals to alkalic basalt magma in the lower part of the magma body. The magma in the reservoir thus comes to differ in composition from one level to another, being rich in magnesium and iron at the bottom, poor in those elements but rich in sodium and potassium at the top. Fissures opening in the rock above it may tap the magma body at any level, thus bringing to the surface ankaramite, olivine basalt, andesite, or trachyte (fig. 96). Indeed, these rocks are commonly interbedded on such volcanoes as Mauna Kea, showing that the different magmas were erupted alternately and must have existed simultaneously in the reservoir.

This process of change in chemical composition of the magma is known as magmatic *differentiation.* It begins early in the history of the volcano, as soon as olivine crystals start to form in the subterranean magma. In tholeiitic basalt the sinking crystals enrich the lower part of the magma body in olivine, and eruption of magma from this part of the reservoir may yield a rock containing more than 50 percent olivine phenocrysts—the picrite-basalt of oceanite type. However, throughout most of the life of the volcano the temperature of the magma remains too high for crystallization of much

else besides olivine, and furthermore, frequent eruptions keep the magma in the reservoir well stirred and probably frequently replenished with fresh additions of new magma from below. As a result relatively little differentiation occurs. Only in late stages does the temperature drop low enough to permit extensive crystallization, and the magma body remain quiescent long enough to allow much differentiation. Thus the strongly differentiated types—ankaramite, hawaiite, mugearite, and trachyte—mark a stage approaching the end of the volcano's activity. These rocks form only a relatively very thin shell over the surface of the basaltic shield volcano. At the most, they constitute only a few percent of the bulk of the mountain (table 8).

On volcanoes of the Mauna Kea or Haleakala type the late-stage rocks include alkalic basalt, ankaramite, and the hawaiite type of andesite. On those of the Kohala type they include only trachyte and the mugearite type of andesite. The rhyodacite of Mauna Kuwale, in the Waianae Valley on Oahu, appears to be an unusual, extreme type derived by differentiation of the tholeiitic basalt magma in the same way that trachyte is derived from alkalic basalt.

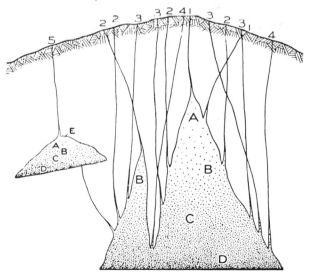

Figure 96. Diagram showing the eruption of magmas of different composition from different levels in the same magma body. *A,* Hawaiite; *B,* basaltic hawaiite; *C,* basalt; *D,* picritic basalt; *E,* trachyte. *1,* Dike erupting hawaiite; *2,* dike erupting basaltic hawaiite; *3,* dike erupting basalt; *4,* dike erupting picritic basalt; *5,* dike erupting trachyte. The various types of lava may be erupted alternately, and may be interbedded in the volcano. (From Stearns and Macdonald, 1942.)

143

Table 8. Estimated proportions of rock types in the late stage cap of Hawaiian volcanic mountains

| Mountain | Proportion of cap to whole mountain, % | Proportion of rock type in cap, % | | | | | |
| --- | --- | --- | --- | --- | --- | --- | --- |
| | | Alkalic basalt | Ankara-mite | Hawaiite | Mugearite | Trachyte | Rhyodacite |
| Kohala | 0.1 | 40 | — | 1 | 58.5 | 0.5 | — |
| West Maui | 3 | 30 | — | — | 67 | 3 | — |
| Mauna Kea | 0.7 | 25 | 5 | 70 | — | — | — |
| Haleakala | 0.8 | 30 | 5 | 65 | 1 | — | — |
| Waianae | 0.3 | 30 | — | 70 | — | — | 0.07 |
| Hualalai | — | 90 | — | 7 | — | 3 | — |
| Overall average | 1.0 | 41 | 1.7 | 56.5 | | 1.1 | |

*Source:* After Macdonald, 1963*b*

The trachyte flow and pumice cone of Puu Waawaa and Puu Anahulu, on the north flank of Hualalai, appear inconsistent with the relatively small amount of differentiation (indicating a still fairly early stage in the evolution of the underlying magma body) of the surrounding lavas of Hualalai. The trachyte may have been formed in a magma chamber separate from that of the main magmatic reservoir of Hualalai, or the magma may have migrated laterally from the reservoir of Kohala volcano.

Most of the late-stage eruptions are decidedly more explosive than the typical Hawaiian eruptions that build the major part of the basaltic shield volcanoes. In this they resemble the common types of eruptions of continental volcanoes. The greater explosiveness appears to result partly from the fact that the erupting andesitic or trachytic magma is more viscous than basalt magma. The gas does not escape as easily, and it accumulates under greater and greater pressure until it finally blows the enclosing sticky magma apart and escapes. These explosive eruptions consist essentially of the bursting of a succession of big bubbles, each burst throwing into the air a shower of molten rock fragments. The general process is well illustrated by a pot of oatmeal mush boiling on the stove. With a moderate amount of heat beneath the pot, the thin fluid at first boils gently. But as the mush cooks the mixture becomes more and more viscous and spatters more and more as the bubbles burst.

The greater viscosity of the erupting andesite and trachyte magmas is shown by the fact that the flows are thicker and shorter than those of basalt. They spread out from their vents less readily. To cite an extreme example, as compared with an average thickness of about 15 feet for Hawaiian basalt flows, the trachyte flow from Puu Waawaa is more than 900 feet thick. Some magma, particularly trachyte, is so viscous that it hardly flows away from the vent at all, but piles up over it as a rounded hill that we call a volcanic dome. There are several small domes of trachyte and mugearite on Kohala Mountain, but the best examples are on West Maui. One of the latter is Puu Mahanalua Nui, formerly called Launiupoko Hill, just south of Lahaina. At Kahakuloa an explosive eruption built a broad cone of pumice and cinder; then extrusion of viscous trachyte magma formed a dome (Puu Koae) in the crater of the cone (fig. 97).

The flows of hawaiite, mugearite, and trachyte are all of aa type, but the fragments are somewhat more regular, and less spinose, than those of typical basaltic aa flows, and tend toward block lava. Rarely, fairly typical block lava is found. Pahoehoe is absent.

The explosive character of the late-stage eruptions is not wholly due to the greater viscosity of the magma. Most of the flows of ankaramite and alkalic basalt, and some of those of hawaiite and mugearite, are thin and widespread, and can have been not much if any more viscous than those of tholeiitic basalt. For example, many of the flows exposed in highway cuts near Kailua, on the western slope of Hualalai, must have been very fluid. Yet the eruptions were distinctly more explosive. In these the greater explosiveness must have resulted from a greater abundance of gas in the magma reaching the surface.

The explosions throw high into the air great showers and clouds of rock fragments. Most of the fragments are ejected in a molten condition, but cool and solidify in the air. The fine material is blown away by the wind to form extensive deposits of volcanic ash. Big dunes of black glassy volcanic ash are found on the upper slopes of Mauna Kea, drifted by the wind from the late-stage cones on the upper part of the mountain. They are well exposed in road cuts near Waikii. The coarse material falls close to the vent, piling up around it to form cinder cones, some of which are as much as a mile across and several hundred feet high. The large size of the cinder cones makes the profile of the late-stage volcanic mountains, such as Mauna Kea or Kohala, very irregular and strikingly different from the smooth profile of the shield volcano, Mauna Loa, or even the slightly bumpy profile of Hualalai.

The cones are built largely of scoriaceous, sponge-like cinder. For the most part the cinder is loose, and not welded together like much of the spatter in the cones of the earlier stages. Ribbon and spindle-shaped bombs and lapilli are common in the cinder, which indicates that the eruptions were mostly of the Strombolian type.

145

Figure 97. Puu Koae, at the east edge of Kahakuloa Bay, West Maui. The precipitous hill is a dome of trachyte which has been cut in two by wave erosion. The dome was built in the crater of a cone of cinder and pumice that has since been mostly eroded away. The cone rests on thin-bedded lava flows of the Wailuku Volcanic Series, exposed in the sea cliff below the dome. Another smaller dome lies to the right.

Thus by far the greatest part of the building of the Hawaiian volcanic mountains is by tholeiitic basalts liberated in eruptions of Hawaiian type. This is followed by a late stage in which eruptions, largely of Strombolian type, form a relatively thin cap of differentiated lavas belonging to the alkalic suite (alkalic olivine basalts, ankaramites, hawaiites, mugearites, and trachytes) that tend to obliterate the caldera and considerably alter the form of the mountain. This sequence is not complete at all the volcanoes. It may be interrupted at any stage.

## REJUVENATED PERIOD

The cessation of the principal period of volcanism, at whatever stage it occurs, may mark the permanent end of activity of a volcano. At other volcanoes, however, a very long period of quiescence is followed by a resumption of volcanic activity.

The lavas of this posterosional period include more alkalic olivine basalt, but also other rock types not previously erupted. These are the nephelinites and basanites. Some of the nephelinites contain less silicon than any other lavas.

The eruptions are rather explosive, like those of the late stage of the principal period, but the flows appear to be quite fluid and the explosiveness must be caused by abundant gas in the erupting lava rather than by high viscosity. Phenocrysts, even of olivine, are few and small, suggesting that temperatures in the magma reservoir were high—probably higher than at the end of the principal period of volcanism.

Explosive eruptions in the Tantalus-Sugarloaf area above Honolulu not only built large cinder cones, but scattered black glassy ash over a large part of the present area of the city. Nearly all building excavations in downtown Honolulu reveal this sandy black Tantalus ash.

Many of the lava flows were poured into valleys and accumulated there to thicknesses of several tens of feet. These pools of lava remained liquid long enough for most of the gas to bubble out of them, and the resulting rocks often are very dense. An unusually dense, thick flow of nephelinite poured down Manoa Valley from Tantalus, forming the flat valley floor on which the Manoa campus of the University of Hawaii is situated, and providing the site for the former Moiliili Quarry. That quarry, now occupied by the University's athletic field and gymnasium, and by apartment buildings, was the best quarry thus far developed in the islands.

Many posterosional eruptions took place on the eastern end of the Koolau Range on Oahu, in the Honolulu and Kailua areas, and many also on the islands of Kauai and Niihau. On all these islands, but particularly on Kauai, the series of posterosional eruptions extended over a long period of time, probably at least several tens of thousands of years. Activity was not continuous. There were long periods of quiet between eruptions, some of them long enough for soil layers to form on the lavas of earlier posterosional eruptions, and for valleys several hundred feet deep to be cut into them by streams. Some of these quiet periods must have lasted several thousand years—probably somewhat longer than the time between the last eruptions on Kauai and Oahu and the present. Thus there is a distinct possibility that more eruptions may take place on those islands, or, indeed, on any of the other islands.

A few small posterosional eruptions took place near Lahaina on West Maui, on Kahoolawe, and in the Kolekole Pass area on the Waianae Range of Oahu. The peninsula of Kalaupapa, at the foot of the great sea cliff of East Molokai, is a broad low volcano built by an eruption that took place after the cliff had been cut by the waves, and the little island of Mokuhooniki, near the east end of Molokai, also was built by a posterosional eruption.

On Oahu several of the late eruptions were accompanied by violent phreatomagmatic explosions, caused by hot magma rising into shallow ocean water or into water-saturated lava or reef rock. The explosions built broad saucer-shaped tuff cones—Diamond Head, Punchbowl, Ulupau Head, Manana (Rabbit) Island, Koko Head and Koko Crater, and the Salt Lake and Aliamanu cones near Pearl Harbor. Kilauea cone, on the

146

north side of Kauai, also is a tuff cone of the posterosional series.

Eventually, erosion by streams and waves will remove all of the islands above sea level. Streams alone can never cut an island entirely away. Waves, on the other hand, in the long sweep of geologic time, can complete the task and cut the entire top off an island, leaving a broad, very gently sloping, shallow wave-cut platform. Many volcanic mountains of the central Pacific have had their heads cut off, and now lie thousands of feet below sea level. Other similar sunken platforms are crowned with coral reefs, forming at the sea surface the ring-shaped islands called atolls—the final evolutionary stage of a mid-ocean island. The erosional and reef-building processes are discussed in the succeeding chapters.

*Suggested Additional Reading*

Macdonald, 1949*a;* Macdonald and Katsura, 1962; Stearns H. T., 1946, 1966*a*

*Spheroidal weathering
in basalt, Oahu*

CHAPTER EIGHT

# Rock Weathering and Soils

ROCKS AT AND NEAR THE SURFACE OF THE EARTH ARE attacked by air and water and other surficial agents and broken down into new materials. This process of surface change is known as *weathering.*

Two different sorts of changes are involved. In one the rock is merely broken into smaller and smaller pieces without change in its chemical or mineral composition. This *physical disintegration* of the rock is called *mechanical weathering.* The other sort of change involves an alteration in chemical composition and the formation of new minerals. This *decomposition* of the original rock minerals is called *chemical weathering.* The two types of weathering are often difficult to separate in nature, as they commonly go on together. In certain limited regions mechanical weathering may take place with little or no accompanying chemical change, but chemical weathering probably never takes place to any important degree without simultaneous mechanical weathering. The process of chemical decomposition of the minerals in a rock brings about a change of volume, and even if no other process is operating this generally results in mechanical disintegration.

## MECHANICAL WEATHERING

The physical disintegration of rock goes on everywhere in the Hawaiian Islands, as it does throughout the world, but in most parts of the islands its effects are masked by those of chemical decomposition. Only on the cold dry tops of the high mountains, where chemical weathering is partly or largely inhibited, are the effects of mechanical weathering conspicuous.

The upper parts of Mauna Loa and Mauna Kea, on the island of Hawaii, are partly covered by a mantle of broken rock debris that is very little decomposed chemically. On these high mountain tops the temperature drops below freezing nearly every night of the year, and rises far above the melting point of ice during the daytime. Freezing and thawing is a daily event. During the daytime, water penetrates joint cracks and other openings in the rocks, and at night freezes there. As the temperature drops a few degrees below freezing the ice expands and exerts a tremendous pressure (theoretically as much as 30,000 pounds per square inch) on the enclosing walls, splitting the rocks apart and gradually producing smaller and smaller fragments. In spite of the relatively low rainfall and consequently small amount of water available, the rocks are thoroughly frost riven. The result is a carpet of angular rock fragments, ranging in size from sand to a few inches, or rarely a foot or two, across, lying between projecting outcrops of bedrock. The fragments commonly are arranged in stripes of varying coarseness, or in polygonal patterns ("Polygonboden"), that also are the result of frost action. Patterned ground, stripes, and polygons of this sort have been described by many authors in northern latitudes, and are usually attributed to alternate freezing and thawing. The patterns on Mauna Kea have been studied and described by Gregory and Wentworth (1937).

Similar effects are shown, though less clearly, on Haleakala (East Maui), but are not encountered on the lower mountains where freezing does not occur. Near the summit of Haleakala, at about 10,000 feet elevation, large boulders resting on a surface mantled by black pumice and cinders mark out definite geometric patterns, the most common of which is a straight line. The effect is particularly accentuated near the southwestern corner of Haleakala Crater because the boulders are a light gray andesite from White Hill, and the alignment of the boulders on the black cinder is boldly displayed.

Also in the same area on Haleakala are seen fine-textured patterns in the black cinders which appear as though some fastidious custodian of the mountain had neatly swept the surface with a broom, being careful to make scrolls and waves and curves in the lee of the boulders and projecting ledges. This "Japanese-rock-garden-like" appearance is probably due to action of frost and also of the wind that sometimes shrieks across the barren upper slopes of the mountain.

Frost action is certainly the principal, and perhaps the only, important agency of mechanical weathering on the high mountain tops. In other parts of the world alternate heating and cooling of the rock may be important in causing rock disintegration, but on the Hawaiian mountains its importance seems to be minimized by local conditions. Thus, heating by forest or grass fires is ruled out on the high mountains because these areas are essentially bare of vegetation, and in lower areas the vegetation is commonly too wet to burn well. Possibly, rare fires have caused the breaking of surface rocks, but if so the evidence has been hidden by chemical weathering.

In other parts of the world, where coarse-grained rocks such as granite are exposed at the surface, the change of temperature from day to night may cause the rock to disintegrate into a mass of loose grains. As they are heated by the sun all the mineral grains in the rock expand, but grains of some minerals expand more than those of others. This sets up strains along the boundaries of the grains, and eventually may cause them to break apart. The effectiveness of this process is a matter of debate among geologists, because hundreds of repeated heatings and coolings in the laboratory do not cause rocks to break apart in this manner; but field evidence seems to indicate that in some desert regions rocks have disintegrated by this process. The question is of no practical importance here, because in fine-grained lava rocks such as those of Hawaii the grains are so interlocked that the strains are inadequate to break them apart.

At lower altitudes in Hawaii the wedging action of growing plant roots and stems undoubtedly is of great and widespread importance in breaking up the rocks, but it is only rarely observable because its results usually are hidden by the predominant chemical weathering.

Also very important is breaking of the rocks as a result of volume changes brought about by chemical weathering. The ordinary processes of chemical weathering (discussed later) produce an increase of volume of the weathered product as compared to the original volume of the unweathered rock. This sets up strains between the outer more weathered and the inner less weathered portions, which may cause the rock to break apart. Disintegration of this sort may leave a mass of irregularly arranged chips of weathered rock; but often it produces *spheroidal weathering,* in which a core of less weathered rock is surrounded by separate concentric shells of progressively more and more weathered material (fig. 98). Often, successive shells can be peeled off much like the layers in an onion. Because the water enters from two or more directions at the corners of a block of rock such as a joint block, it penetrates deeper there and the detached shell is thicker, resulting in rounding of the corners of the residual central part of the block. As weathering continues the residual core becomes more and more spherical. Weathering spheroids, ranging from an inch or two to several feet across, can be seen in many roadcuts and natural outcrops throughout the Islands.

A type of mechanical weathering of little volumetric importance, but of considerable local interest because it is seen in few other areas, is the flaking off of the outermost surface of many pahoehoe flows. On these flows an outer skin, commonly only a millimeter or so thick, is separated from the rest of the flow by a nearly continuous layer of gas bubbles. After consolidation of the flow this thin skin soon disintegrates into flakes that are washed or blown away. Disintegration is greatly speeded by animals or men walking over the flow, and in areas where trails cross recent flows, both ancient and modern trails commonly are very distinct because of this destruction of the fragile flow surface.

Breaking up of the rock by mechanical weathering, or by any other means, greatly aids chemical weathering because it increases the area of rock surface exposed to chemical action. For example, if a 1-foot cube is broken

151

Figure 98. Rounded boulder formed by spheroidal weathering of basalt, Koolau Range, Oahu. A smaller weathering spheroid is visible just below the hammer head.

$$2NaAlSi_3O_8 \quad + \quad H_2CO_3 \quad + \quad H_2O \quad \longrightarrow \quad Al_2Si_2O_5(OH)_4 \quad + \quad Na_2CO_3 \quad + \quad SiO_2$$

albite $\quad + \quad$ carbonic acid $\quad +$ water $\quad$ yields $\quad$ kaolinite $\quad +$ sodium carbonate $\quad +$ silica

$$CaAl_2Si_2O_8 \quad + \quad H_2CO_3 \quad + \quad H_2O \quad \longrightarrow \quad Al_2Si_2O_5(OH)_4 \quad + \quad CaCO_3$$

anorthite $\quad +$ carbonic acid $\quad +$ water $\quad$ yields $\quad$ kaolinite $\quad +$ calcium carbonate

into 3-inch cubes, the area of exposed surface is increased from 6 to 24 square feet.

## CHEMICAL WEATHERING

In Hawaii, as in other parts of the tropics and subtropics, conditions for chemical decomposition of rocks are nearly optimum. In general, the effects of chemical weathering predominate greatly over those of mechanical weathering. Climatic conditions in Hawaii are abetted by the mineralogical composition of the rocks, in which all of the abundant minerals are readily decomposed. Quartz ($SiO_2$), the principal resistant rock-forming mineral in rocks of continental areas, is essentially absent in those of Hawaii.

All of the principal processes of chemical weathering—oxidation, hydration, hydrolysis, carbonation, cation exchange, and solution— are active in the decomposition of Hawaiian rocks. In the drier areas chemical weathering is less rapid than in the wet areas, but the processes involved are the same. As everywhere, oxygen from the air and water from rainfall are of primary importance. The abundant vegetation releases large amounts of carbon dioxide in the soil. This, in addition to a lesser amount derived from the atmosphere, combines with water to produce carbonic acid ($H_2O + CO_2 \rightleftharpoons H_2CO_3$), which in turn dissolves calcium carbonate, both in limestone and in the decomposition products of calcium-bearing minerals in the lavas. Humic acids, resulting from the decay of vegetation, also bring about decomposition of the rock-forming minerals. Hydrogen ions in the soil water around plant roots are exchanged for cations ("bases") such as sodium and potassium in the rock-forming minerals, thus both speeding the decomposition of the latter and making available dissolved bases for plant use.

As an example of the chemical weathering of the rock-forming silicate minerals, let us consider the decomposition of plagioclase feldspar, one of the two most abundant minerals in Hawaiian lavas. It will be recalled that plagioclase consists of a mixture of two compounds: albite ($NaAlSi_3O_8$) and anorthite ($CaAl_2Si_2O_8$). The ordinary reactions involved in the weathering of these two compounds may be written as shown in the box.

Both reactions yield as one of their products kaolinite, one of a group of hydrous aluminum silicates known as clay minerals. The kaolinite is quite insoluble, and most of it remains in the weathered rock. The sodium carbonate derived from albite is readily soluble and is carried away in solution in the water seeping downward through the weathered material to join the ground water body, and thence by way of springs and streams to the ocean. In the presence of carbonic acid, calcium carbonate derived from the anorthite is transformed into calcium bicarbonate

$$CaCO_3 + H_2CO_3 \rightleftharpoons Ca(HCO_3)_2$$

and in that form also is soluble and is carried away in solution.

Potassium feldspar ($KAlSi_3O_8$) behaves much like albite, but more of the potassium is left behind, adsorbed on clay particles. Potassium feldspar is present in very small amounts in most Hawaiian rocks, however, and consequently Hawaiian soils tend to be deficient in potassium.

The silica released by feldspar decomposition is in part dissolved and carried away, and in part left behind, largely as amorphous material, in the weathered mantle. In time it may crystallize, yielding some of the tiny grains of quartz commonly found in Hawaiian soils. Other evidence suggests, however, that many of the quartz grains in Hawaiian soils have been carried by the wind at high levels in the atmosphere all the way across the ocean from the continent of Asia. Present with the quartz grains in the soils are tiny flakes of mica that are entirely unlike the minerals in Hawaiian rocks, but are like mica in granitic rocks of the continents.

Other minerals in Hawaiian lavas decompose in a similar manner, yielding carbonates or bicarbonates that are carried away in solution, and relatively insoluble clay minerals, oxides, and hydroxides that are mostly left behind. Table 9 lists the principal products of decomposition of the minerals in Hawaiian volcanic rocks, arranged in the order of rapidity of alteration under ordinary conditions of weathering. In a partly weathered rock the minerals at the top of the list may be entirely decomposed, whereas those lower in the list remain unaltered. Other clay minerals than kaolinite are present in some Hawaiian weathered materials. These include halloysite (amorphous hydrated aluminum silicate) and allophane (amorphous hydrated aluminum oxide), and in some areas montmorillonite (hydrated aluminum silicate containing iron and magnesium). Montmorillonite is particularly abundant in some rather dry regions and in some marshy areas with very poor drainage such as taro patches. However, in most areas kaolinite appears to be by far the most abundant clay mineral. Some of the iron oxide ($Fe_2O_3$) is present as hematite, but recent investigations have shown that much of it is in the form of the magnetic

ferric oxide, maghemite. Iron hydroxide is partly in the form of amorphous limonite ($Fe_2O_3 \cdot nH_2O$), and partly as the crystalline mineral goethite ($Fe_2O_3 \cdot H_2O$). The common red, brown, or yellow color of weathered rock and soil is due to these compounds of iron, red colors resulting from hematite, and brown to yellow colors from goethite and limonite.

Under the conditions prevailing in the temperate zones the principal residual products of chemical weathering are clay minerals and hydrated iron oxides. However, under more tropical conditions such as exist in much of Hawaii, the silica may be still more completely removed, giving rise to a further alteration of the clay minerals into hydrated aluminum oxides, known under the group name of *bauxite*. The principal minerals in the bauxite group are gibbsite ($Al_2O_3 \cdot 3H_2O$), diaspore ($Al_2O_3 \cdot H_2O$), and allophane ($Al_2O_3 \cdot nH_2O$). The first two are crystalline, and the last is amorphous. All of these have been recognized in Hawaiian weathering products, but gibbsite is much the most abundant. Sometimes the rock minerals (especially the feldspars and nepheline) may pass directly into bauxite (Abbott, 1958) without pausing for any appreciable interval of time in the kaolinite stage. It also appears that in Hawaii under some conditions part of the silica released by weathering may recombine with bauxite to form a secondary generation of kaolinite.

Solution is important in the weathering and

Table 9. Common weathering products of the principal minerals in Hawaiian lava rocks

| Original mineral | Ease of alteration | Decomposition products | |
|---|---|---|---|
| | | Residual, or only partly removed in solution | Soluble and carried away in solution |
| Olivine | Greatest | Limonite, hematite, silica | $MgCO_3$, $FeCO_3$, some silica |
| Magnetite | | Limonite, hematite | — |
| Anorthite | | Clay minerals, bauxite, silica | $CaCO_3$, some silica |
| Nepheline | | Clay minerals, bauxite | $Na_2CO_3$ |
| Pyroxene | | | |
| Pigeonite or hypersthene | | Limonite, hematite, silica | $MgCO_3$, $FeCO_3$, some silica |
| Augite | | Clay minerals, bauxite, limonite, hematite, silica | $CaCO_3$, $MgCO_3$, $FeCO_3$, some silica |
| Ilmenite | | Limonite, hematite, titanium oxide | — |
| Albite | Least | Clay minerals, bauxite, silica | $Na_2CO_3$, some silica |

*153*

erosion of limestone the world over, and Hawaii is no exception. The raised coral reefs around the edges of the islands show abundant evidence of solution. Rainwater containing dissolved carbon dioxide accumulates in pools on the surface of the limestone, and enlarges the hollows by dissolving the rock around their edges. This may result in broad shallow depressions known as pans. Water spilling out of the pans and trickling down the inclined surface of the rock dissolves out groove-shaped channelways. Commonly also, the surface of the limestone as a whole undergoes a general attack, resulting in a miniature "karst" topography consisting of very irregular spines and intervening hollows with a vertical relief of several inches.

Solution is not usually considered an important agent in the weathering of igneous rocks, except for the removal of the soluble products of breakdown of the rock-forming silicates (such as $Na_2CO_3$ from weathering of albite). In Hawaii, however, we have many examples of the dissolving of lava rock, although this action probably plays a minor role in the weathering process as a whole. At many places rainwater trickling over the surface has dissolved the rock and produced a series of smooth-sided grooves, known as lapiés, from less than an inch to 6 or 8 inches across and about equally deep. Where the grooves are formed on the side of a boulder, the fragment may later roll over and a new set of grooves be formed crossing the first set at an angle. As many as three successive series of grooves have been observed. One of the best-known examples of such grooved boulders is the "altar stone" in Kolekole Pass in the Waianae Range, Oahu. Very well developed solution grooves have been found on both nepheline basalt and olivine basalt, and less perfect ones on mugearite.

One special aspect of chemical weathering in Hawaii is the alteration of glassy basaltic ash to the material known as *palagonite*. Much of the tuff in the tuff cones of Oahu (Diamond Head, Punchbowl, Koko Crater, and others) and the Kilauea tuff cone on Kauai, shows this type of alteration. The basaltic ash of such widespread formations as the Pahala ash on Hawaii Island commonly show alteration that is similar in appearance, and whereas in some areas the resulting material consists of mixtures of clay minerals, limonite, and goethite, in others it seems to be true palagonite. Rocks that have suffered palagonitic alteration are generally olive brown, buff, yellowish brown, or reddish brown, and commonly they have a slightly waxy appearance on fresh breaks. Palagonite itself takes two different forms: gel palagonite which is isotropic under the microscope and appears to be a mineraloid, like opal, rather than a true mineral; and fibrous palagonite, a truly crystalline mineral. It appears that gel palagonite forms first but may later recrystallize to fibrous palagonite. The alteration of basaltic glass to palagonite involves the loss of some of the silicon, aluminum, magnesium, calcium, sodium, and potassium, the oxidation of most of the iron, and the addition of a large amount of water. The fact that these substances are removed from the palagonite does not mean, however, that they are removed from the rock. Most of them are deposited between the grains of palagonite as calcite, zeolite, and opal, and it is to these cements that the rock owes its moderately hard character.

Formerly it was believed that the alteration of basaltic glass to palagonite in the tuff cones took place very rapidly after the eruption that built the cone, as a result of high temperature and saturation of the cone with steam. However, the ash of cones formed during the last decades, such as Surtsey volcano in Iceland and Capelinhos volcano in the Azores, shows little sign of alteration even several years after the cone was built. Recent studies of the tuff cones of Oahu by Hay and Iijima (1968) have shown that the palagonitization of the tuff of Koko Crater almost certainly is the result of weathering long after the cone was built. The weathering is unusual in that the uppermost tuff is less altered than that beneath it. In an upper zone 10 to 50 feet thick the basaltic glass fragments of the tuff are relatively little altered, whereas in the zone just below, the glass is largely changed to palagonite. The passage from one zone to the other is quite sharp, usually within a thickness of a few inches, and the contact cuts across the bedding of the tuff. The cone has been gullied by erosion, and the upper surface of the palagonite zone is approximately parallel to the erosional surface, not to the original surface of the cone, which shows that the palagonitization took place after the cone was eroded. The relationship of the top of the palagonite zone to the eroded topography is clearly visible in highway cuts toward Hanauma Bay from the Halona Blowhole (fig. 99). There

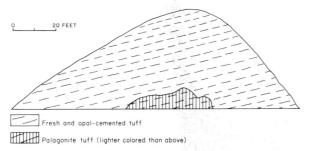

Figure 99. Cross section through a spur exposed in a highway cut on the south side of Koko Crater, Oahu, showing the boundary between the upper zone of largely unaltered or opal-cemented tuff and the lower zone of lighter-colored palagonitized tuff. The dashed lines show the bedding of the tuff cone, cutting sharply across the boundary between palagonitized and unpalagonitized tuff. (After Hay and Iijima, 1968.)

it can be seen that the cores of the ridges consist of buff to reddish brown palagonite tuff, overlain by darker, relatively unaltered tuff, with the contact roughly parallel to the surface of the ridge.

Although in the upper zone the glass fragments remain mostly fresh, in the lower part of that zone the tuff commonly is well cemented by opal, deposited between the original grains. This cementation makes the tuff much more resistant to erosion, and wherever erosion has broken through this resistant shell there is a tendency for it to cut away the less resistant tuff beneath more rapidly than the resistant rock, undercutting the opalized zone and forming shallow caves along the sides of the gullies. The overhanging portion often breaks off, leaving the opalized tuff standing in a steep cliff. The prominent cliff near the top of the Koko Crater cone (fig. 256), easily seen from the Halona Blowhole, was formed in that way.

Hay and Iijima (1968) suggest that the alteration pattern in Koko Crater is the result of the following process: Rainwater percolating downward dissolves a little of the glass of the tuff. As the solution moves downward it becomes more concentrated, and by hydrolysis its alkalinity increases. At a critical level some of the dissolved silica is deposited as opal cement; and with a small further increase of alkalinity, to a pH of about 9, the solution begins to react rapidly with the glass, altering it to palagonite. Although the resulting palagonite is quite different in composition from the original glass, much of the calcium, alkalies, aluminum, and silica removed from the glass during its alteration to palagonite is redeposited between the palagonite grains as a cement of calcite and zeolites, so that there is little change in the composition of the total rock. More studies are needed on other Hawaiian tuff cones to confirm this process.

## SOILS

Soil, the surficial material that supports plant life, is a mixture of mechanically and chemically weathered rock fragments and humus, the latter largely the product of plant decay. In temperate regions, where soils have been studied most thoroughly, it is generally possible to divide a soil into three zones, or *horizons,* one above the other.

Soils that have been formed by breakdown of the rocks beneath them are known as *residual soils.* Other soils, however, have been formed from rock debris brought in from other areas by natural agents of transportation. The latter, known as *transported soils,* may have no genetic relationship to the rocks beneath them. Most Hawaiian soils are residual, and grade downward into their parent rocks; but some, such as those on flat alluviated valley floors or those derived from air-laid volcanic ash, are transported soils and have no genetic connection with the underlying rocks.

Where soil zonation is well developed, the lowest, or C, horizon is a subsoil grading into the underlying rocks. Above the C horizon is the B horizon, in which chemical weathering is more advanced, and commonly only the most resistant of the original rock-forming minerals remain. At the top is the A horizon, from which nearly all of the most soluble products and some of the less soluble ones have been

*155*

leached out and carried downward by percolating water. The dissolved material includes much of the silica and iron oxide, and even part of the clay minerals. Much of the iron oxide and clay are redeposited in the B horizon. The A horizon usually contains an abundant admixture of humus.

In moderately to very wet parts of the tropics and subtropics, such as Hawaii, where drainage is good there is a general leaching and removal of sodium, calcium, and magnesium from the upper part of the weathered rock mantle by downward-moving water. The alumina, titania, and particularly the iron oxide, are largely left behind by the percolating water, and as a result they undergo a concentration relative to their abundance in the original rocks. Because of the concentration of iron oxide, the soil-forming process is commonly referred to as laterization, and although few of the resulting soils can be considered true laterites, the soils in general have received the name *latosols* (Cline and others, 1955).

Like those of other tropical and subtropical regions, Hawaiian soils generally show a much less marked division into horizons of differing chemical and physical properties than do those of temperate regions. Nevertheless, a recognizable zonation is present. Cline and his associates (1955) have classified Hawaiian soils into three orders: zonal, intrazonal, and azonal soils. The zonal soils, which include the typical latosols, are those that show a recognizable division into horizontal zones. They form where drainage is good, where the water table lies at a level below the base of the soil, and where weathering has been in progress for a long period. As elsewhere, the zones consist of an A horizon in which leaching is the dominant process, and beneath it a B horizon in which there is commonly some redeposition of material dissolved in the A horizon. The A and B horizons together constitute the solum, or true soil. Beneath them, in residual soils the C horizon comprises a gradation through partly weathered rock into the underlying fresh bed rock. Instead of being strongly concentrated in the A horizon, as it commonly is in temperate-zone soils, humus generally is well distributed throughout the solum.

156

Azonal soils show no appreciable horizontal zonation. They include transported alluvial soils and the stony soils of high altitudes where mechanical weathering predominates. At lower altitudes they are generally young soils in which soil-forming processes have as yet made little headway. Intrazonal soils are those that are transitional between azonal and zonal soils, or have developed in areas of poor drainage or where other special conditions have prevented the formation of zonal soils. The zonal soils are by far the most important, areally, of Hawaiian soils, and the following discussion is largely restricted to them.

The classification of latosols given in table 10 is largely that used by Cline and his

Table 10.  Latosols of the Hawaiian Islands

| Great soil group | Conditions during formation | |
| --- | --- | --- |
| | Annual rainfall (inches) | General environment |
| Latosolic brown forest soils | 40–150 | Moderately wet |
| Low humic latosols | 10–80 | Dry part of year |
| Humic latosols | 40–150 | Constantly wet |
| Hydrol humic latosols | 100–500 | Constantly wet |
| Humic ferruginous latosols | 25–150 | Constantly wet |
| Aluminous ferruginous latosols | 40–250 | Constantly wet |

associates in the soil survey of Hawaii (1955). The final great soil group, aluminous ferruginous latosol, has been proposed more recently by Sherman (1958). Among the zonal soils, the low humic latosol forms in regions of relatively low rainfall, where the soil is dry during part of the year. The other zonal soils form in regions of higher rainfall where the soil is constantly wet, the hydrol humic latosols being characteristic of the rainiest regions. The latosolic brown forest soils are strictly classified as intrazonal, but they bear strong resemblances to the zonal soils, and probably are younger soils representing a stage in the development of humic and low humic latosols. The humic ferruginous latosols form under special conditions that result in a completer-than-usual removal of other substances and reprecipitation of some dissolved iron, resulting in a marked concentration of iron oxide in the B horizon. The aluminous ferruginous latosols form in the same general environment as the humic and hydrol humic latosols, but under special conditions that result in the bauxite deposits discussed on a later page.

In their normal condition, the zonal soils are quite plastic. Low humic and humic latosols which have been dried out will regain their plasticity on rewetting, but the hydrol humic latosols, once dried out, will not. In their normal state the hydrol humic latosols may contain exceedingly large amounts of water. The water content of the wet soil commonly is more than 300 percent—that is, the weight of the contained water is more than three times that of the other material present—and soils containing more than 600 percent water have been reported, for instance on the Hamakua Coast of Hawaii. This material commonly exhibits thixotropic properties—that is, it behaves as a solid so long as it is not disturbed, but if disturbed, and particularly if agitated, it behaves as a liquid. A carefully cut out block of the material will remain on a table top and hold its shape indefinitely so long as it is not disturbed, but if it is shaken it assumes the characteristics of a liquid and flows out into a flattened, puddle-like mass. Heavy construction equipment, such as bulldozers, left on such soil will sometimes sink several feet into it overnight.

The average chemical compositions of the types of zonal soils are shown in table 11, which also gives compositions of some of the common parent rocks for comparison. The iron oxide is the least soluble product of weathering, and becomes concentrated in the soil by the removal of the more soluble materials. As rainfall increases there is a progressively more complete removal of silica. Under ordinary circumstances titania is concentrated along with iron oxide.

In some drier areas calcium carbonate has been deposited in the lower part of the soil, as in the B horizon of certain types of soils (pedocals) of temperate regions. The calcium carbonate forms irregular sheets, concretions around plant roots and other vegetable mate-

Table 11. Average composition of Hawaiian soil types and some parent rocks

| | Soil type[a] | | | | | | | | | |
|---|---|---|---|---|---|---|---|---|---|---|
| | 1 | 2 | 3 | 4 | 5 | 6 | 7 | 8 | 9 | 10 |
| $SiO_2$ | 36.7 | 46.4 | 50.9 | 30.0 | 22.7 | 11.2 | 7.4 | 10.0 | 4.0 | 33.9 |
| $Al_2O_3$ | 10.8 | 14.3 | 13.2 | 29.5 | 21.1 | 25.0 | 12.3 | 30.0 | 48.5 | 25.7 |
| $Fe_2O_3$[b] | 15.2 | 10.2 | 12.1 | 21.8 | 30.3 | 28.2 | 52.5 | 35.0 | 26.0 | 8.6 |
| $MgO$ | 12.7 | 8.7 | 8.0 | — | — | — | — | — | — | 0.9 |
| $CaO$ | 13.7 | 10.6 | 10.6 | — | — | — | — | — | — | 0.1 |
| $Na_2O$ | 3.9 | 3.0 | 2.2 | — | — | — | — | — | — | 0.2 |
| $K_2O$ | 0.9 | 0.8 | 0.4 | — | — | — | — | — | — | 2.8 |
| $TiO_2$ | 2.8 | 3.0 | 2.8 | 4.4 | 4.6 | 6.1 | 15.4 | 5.0 | 3.1 | 15.3 |
| $H_2O$ | — | — | 0.3 | 14.3 | 21.3 | 29.5 | 12.4 | 20.0 | 17.3 | 8.2 |

a. The columns are numbered as follows:
    1. Melilite-nepheline basalt, average of 6 analyses. Macdonald, 1949a, p. 1571.
    2. Alkali olivine basalt, average of 20 analyses.
    3. Tholeiitic basalt, average of 32 analyses, Kuno and others, 1957, p. 213.
    4. Low humic latosol. After Cathcart, 1958.
    5. Humic latosol. From Cathcart, 1958.
    6. Hydrol humic latosol. From Cathcart, 1958.
    7. Ferruginous humic latosol. From Cathcart, 1958.
    8. Bauxitic soils. From Cathcart, 1958.
    9. Aluminous ferruginous latosol. Sherman, 1958, p. 13.
    10. High-titania clay. Wentworth, Wells, and Allen, 1940, p. 24.
b. Total iron as $Fe_2O_3$.

rial, or casts deposited in the holes left by decayed roots and stems. Among other places, these are well shown in the South Point area and near Waikii, both on the island of Hawaii. In some areas magnesium has been deposited with the calcium, forming dolomite, $(Ca, Mg)CO_3$.

In a few swamp areas, where drainage is poor and acidity high, very pale gray, cream colored, or white clays, high in titanium, have been found (Wentworth, Wells, and Allen, 1940). Under these conditions iron has been largely removed, alumina has remained in clay minerals instead of altering to bauxite, and titania has been greatly concentrated (see column 10 of table 11). Potassium also has been concentrated, partly in a mineral known as illite that can be regarded as intermediate between a clay mineral and a mica, but probably also in part by adsorption on the kaolinite. The clays are plastic, slip well, and have been used in the amateur manufacture of pottery.

Space does not permit a complete or detailed discussion of Hawaiian soils in this book. For additional information the reader is referred to the list of references at the end of this chapter.

BAUXITE DEPOSITS

In some of the humic and hydrol humic latosols, alumina has to some extent been taken into solution in the A horizon and carried downward, to be partly redeposited in the B horizon. Thus there results an enrichment of the lower part of the solum in alumina, both by removal of bases and silica in solution, and by deposition of additional alumina carried downward from above.

The causes of the redeposition of the alumina are not wholly understood. If the alumina is carried in true solution, the redeposition is probably at least in part brought about by the very rapid change in solubility of alumina with change of acidity in weakly acid solutions. (With a change of acidity from about pH 3.8 to pH 4 the solubility of alumina changes from more than 10 to about 1 millimole per liter; between pH 4 and 5 the change

is much less, from about 1 to 0.2 millimoles per liter—see Keller, 1957, p. 26). Soil solutions more acid than pH 4.5 are rare in Hawaii, but acidities of pH 4.8 to 4.6 are recorded frequently. It is possible that alumina is dissolved in such acid solutions, carried downward as the solution descends toward the water table, and redeposited as the acidity is lowered to pH 5 or less by reaction of the solutions with various soil constituents. No doubt the solutions are always very dilute, but over many decades and centuries they have accomplished a considerable concentration of alumina in the lower part of the soil.

It is quite probable, however, that much, if not most, of the alumina is carried downward as a colloid, rather than in true solution (T. F. Bates, personal communication), and if so, G. D. Sherman has suggested that its deposition may be largely the result of abstraction of water from the colloid by plant roots. The greater abundance of gibbsite on and close to the face of some roadcuts as compared to that in the soil back away from the cut is probably due to evaporation of water from the face of the cut.

The concentrations of alumina are creamy white to pale reddish brown in color, and they occur in the form of irregular spiny nodules and sheets of bauxite which take the shape of the cavities in which they were deposited—ellipsoidal amygdules occupying the former vesicles of lava flows, and pseudomorphs of the original angular feldspar grains of the parent lava rock (Abbott, 1958). (A pseudomorph is a grain of a mineral that possesses the external form, and sometimes the structure, of another mineral, usually one being replaced by it.)

G. Donald Sherman, of the University of Hawaii, was the first to recognize that some of these alumina-enriched deposits have potential value as ores of aluminum (Sherman, 1954). Some hand-picked samples of the material contain more than 50 percent aluminum oxide (as compared with 66 percent in pure gibbsite). However, the greater part of the richest deposits generally contains 30 to 40 percent (Cathcart, 1958).

Several areas contain sufficiently large concentrations of alumina of sufficiently high grade to be of possible commerical interest. These are shown in figure 100. The deposits on

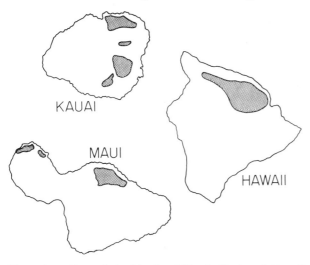

KAUAI

MAUI

HAWAII

Figure 100. Maps of the islands of Kauai, Maui, and Hawaii showing the distribution of bauxitic soils. (After Patterson, 1962.)

the island of Hawaii have resulted from the alteration of volcanic ash of originally basaltic to andesitic (hawaiitic) composition. Those of East Maui have been formed from olivine basalt and hawaiite lava flows, those of West Maui from mugearite and trachyte lavas, those of Oahu from tholeiitic basalts, and those of Kauai from nepheline basalts and alkali olivine basalts. Thus, weathering of nearly the entire range of volcanic rocks of Hawaii has yielded end products of closely similar composition.

The largest deposit is that on the northeast (Hamakua) coast of the island of Hawaii, which is estimated by Cathcart (1958) to contain about 112 million short tons of bauxite. The highest grade material, in large masses, has been found on West Maui, but there a thick overburden would have to be removed in mining the ore. The deposits that thus far have attracted the most attention commercially are those on Kauai, estimated by Cathcart to contain about 58 million tons of bauxite. Experiments by Sherman and his associates appear to indicate that the Kauai deposits can be mined by stripping methods without seriously detrimental effects on the growing of sugar cane, and probably even with some gain in field fertility, because the less-leached portion of the soil below the level of alumina concentration contains a greater amount of mineral nutrient for plants than does the top soil above the bauxite.

*Suggested Additional Reading*

Abbott, 1958; Cline, 1955; Jackson and Sherman, 1953; Keller, 1957; Sherman, 1954, 1955, 1958, 1962; Sherman and Uehara, 1956; Sherman and others, 1967; Sherman and Ikawa, 1968; Swindale and Sherman, 1964; Tamura and others, 1953; Walker, 1964

*Honopu Valley,*
*Napali Coast, Kauai*

# CHAPTER NINE

# Stream Erosion

THE HILLS THAT MAY APPEAR SO EVERLASTING ARE IN reality quite evanescent. No sooner does a land area appear than various forces set to work to destroy it. Weathering prepares the way, by softening the rocks and breaking them down into smaller fragments. Rainwater and streams running over the surface of the ground wash loose material downslope, moving it ever closer to the ocean; gravity moves soil and rock fragments downhill; wind blows fine material away; ice scrapes off the surface material and carries it away; the ocean waves gnaw at the edges of the land, and ocean currents sweep away the broken debris. These processes of destruction of the land are known as *erosion.* All of them modify the shape of the land surface, in the course of destroying it. In Hawaii the most important are stream and wave erosion and the downslope movements that, for want of a better name, we call mass transfer.

### PRINCIPLES OF STREAM EROSION AND DEPOSITION

During rains some of the water runs off as a sheet over the land surface, gradually gathering into rills and runlets, and these in turn uniting to form streams. Unless it is diverted by some obstacle, the water of each stream runs down the steepest available slope toward the sea, drawn by the force of gravity. The speed of flow is governed largely by the steepness of the slope—what we call the stream gradient. On gentle slopes the water may scarcely move, but on steep slopes it rushes down. The steeper the gradient, the more rapid the flow.

As the water flows, it tends to move any loose rock debris along with it. The size of the fragments it can move depends largely on the speed of flow. Slowly moving water can transport only very fine material—clay and silt. More rapidly flowing water can move coarser debris—sand or gravel, or even large boulders. Some of the debris is carried in suspension in the water, but much of it is pushed or rolled along the bed of the stream, or propelled by a series of jumps (the latter is called saltation).

The velocity of a stream depends not only on the steepness of its gradient, but also on its volume. We have all observed that a stream which in ordinary weather flows rather slowly becomes a rushing torrent in times of heavy rains. The water must flow faster for the stream to carry the increased volume.

The size of the individual fragments and the total amount of material that a stream can move along depend on the velocity and the volume of the stream. We refer to the material being moved as the *load* of the stream, and the amount of load the stream can transport depends directly on the volume of the stream, but varies as the square of any change in velocity. Thus, if the velocity of a stream is doubled, the amount of load it can carry increases not twice but as the square of 2, or 4; if the velocity is tripled, the load-carrying power increases 9 times. We must distinguish clearly between the load (the total amount of material) that can be carried and the size of the largest fragments that can be moved along by the stream. As the velocity of the flowing water increases, the size of the fragments that can be moved increases far more than might be expected. Thus, doubling the velocity increases the size of the moving fragments not by two, but by a factor approaching the sixth power of the change in velocity—in this case the sixth power of 2, or 64.

Changes in a stream's velocity greatly affect the power of the stream to erode. Increase in velocity enables a stream to transport more and larger material and to erode more rapidly, but decrease in velocity causes a fully loaded stream to deposit some of its load.

Since both the increased volume and the increased velocity of a stream in flood greatly increase the load it can carry, in one great flood it may accomplish more work than it does over many years of more normal flow. Thus the great flood in Waimea Canyon on Kauai, in 1951, produced changes in the floor of the valley far greater than any that could be recalled by old residents of the area for many years before, or any that have happened since.

The load of the stream consists largely of rock material that has been broken down and loosened by weathering. Removal of this material constitutes one phase of erosion; but as it rolls, slides, and bumps along, this moving material itself erodes the stream bed by abrasion and impact. As the fragments wear away the rock of the stream bed they themselves are worn in turn; corners are ground and knocked off, and the fragments are reduced in size and become rounded. Other fragments along the stream bed are pushed loose by water pressure or undermined by removal of finer material around and under them, and once loosened they become part of the stream load and are transported away. Thus the stream gradually saws its way downward, cutting first a gully, and eventually a valley. As the valley deepens, it may intersect the water table (see chap. 15), and springs may be formed, contributing more water and thus adding to the stream's volume and still further increasing its eroding power.

The world over, cobbles and pebbles found along stream beds are typically rounded as a result of abrasion during transportation. In Hawaii, however, many of the fragments picked up by the streams are already moderately to well rounded cores of weathering spheroids (see page 151). Thus, to geologists from other areas the degree of rounding of cobbles along Hawaiian streams may suggest a longer distance of transportation than actually has taken place.

Not all of a stream's load consists of solid particles. You will recall that some of the products of weathering are dissolved and carried away in solution. Additional material is dissolved by the stream from the rocks it flows over. In temperate regions only such rocks as limestone and rock salt are directly dissolved by running water to any important extent, but we have already seen (chap. 8) that in Hawaii grooves (or lapiés) are commonly formed by solution of silicate rocks. Similarly, there undoubtedly is a considerable amount of solution of rock in the beds of streams, though its effects are hidden by those of mechanical erosion. Even in Hawaii, however, direct solution of silicate rocks is a very minor factor in erosion.

Streams cannot cut downward indefinitely. Temporary checks to downcutting may result from unusually hard rock ledges along the stream course. If the hard ledge is cut down less rapidly than the bed upstream from it the stream gradient is flattened, the velocity is reduced, and the stream's transporting and cutting power decreases. These local controls are numerous, and the longitudinal profile of most streams consists not of a smooth slope to the sea, but of a series of steps, each held up by a more-than-ordinarily resistant ledge of rock. This is particularly conspicuous in Hawaii, where the rocks being cut by the streams consist of occasional very resistant layers formed by the central massive portions of aa lava flows, alternating with less resistant layers of aa clinker and pahoehoe.

Where the stream passes beyond the resistant ledge onto less resistant rock, it erodes more rapidly, thus locally increasing its gradient, and this in turn still further increases its cutting power. Its cutting power is also often augmented because some of the load has been dropped in the flattened part of its course upstream from the ledge, the less loaded stream being able to pick up more material just downstream from the ledge than it could otherwise have done. As a result, immediately downstream from the ledge, cutting often is very rapid and a steep cliff is formed, over which the stream plunges in a waterfall. Erosion at the base of the waterfall undercuts the resistant ledge, blocks of which gradually fall away causing the waterfall to migrate slowly upstream.

The ultimate limit to downcutting by streams is sea level. Where a stream enters the ocean its velocity is checked, and the stream itself gradually loses identity. Only a few very large and underloaded rivers continue to cut appreciably after they enter the ocean. Usually a stream quickly stops eroding and drops its load, which may accumulate to build a delta. The level that limits the depth to which a stream can cut is known as *base level.* The hard ledges along the course of streams constitute local base levels, as also do any lakes that may be present. In terms of geologic time, however, both are very short lived. The ledges are quickly worn away, and lake basins are filled with deposited rock fragments (sediment) and

the lakes eliminated. The final and general base level is sea level. The mouth of the stream cannot be cut below base level to any important amount, and as erosion continues the upper course of the stream is gradually lowered in relation to the mouth, so that the entire longitudinal profile of the stream bed becomes less and less steep and the velocity and transporting power of the stream decrease.

Streams erode sidewise, against their banks, as well as downward. However, as long as the gradient remains steep the erosional attack is predominantly vertical and downcutting overshadows lateral cutting. The stream tends to cut a vertical-walled notch. Some Hawaiian streams have actually done this. Much more commonly, however, other processes widen out the upper part of the notch and transform it into a V. This widening is accomplished by the processes of mass transfer discussed in chapter 10.

As the valley grows older and the stream gradient less steep, vertical cutting is slowed and lateral cutting becomes relatively more important. The valley floor becomes widened as the stream starts to swing from side to side across it attacking the walls of the channel more vigorously. In time the stream flows in great meandering loops along a nearly flat broad valley floor. There are no good examples of meandering streams in Hawaii—the valleys are not old enough. The slight tendency to meander shown by a few streams, such as Hanalei River on Kauai, is the result, not of the age of the topography, but of alluviation of the valley floor due to causes discussed later.

## THE GEOMORPHIC CYCLE

A landscape being sculptured into hills and valleys by stream erosion goes through a series of more or less definite stages, differing somewhat from one type of region to another, but, except in very arid regions, having overall similarity. This "geomorphic cycle" is commonly divided into stages known as youth, maturity, and old age (fig. 101).

The rather regular constructional surfaces of the Hawaiian shield volcanoes provide ideal

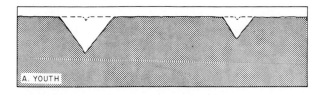

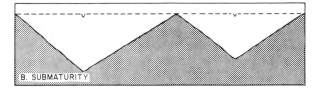

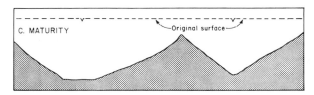

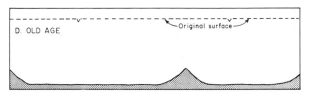

Figure 101. Diagrammatic cross sections of ridges and valleys, showing the profiles characteristic of the successive stages in the ideal geomorphic cycle of humid regions.

starting surfaces to demonstrate the progress of geomorphic evolution. The initial gullies give way in the *youthful stage* to V-shaped valleys between which broad areas of the original surface remain almost unchanged. The northeastern slopes of Mauna Kea, on the island of Hawaii, are in this stage (figs. 102–104). As the valleys are deepened and widened the areas of original surface between them grow smaller, until they finally disappear (fig. 105). The stage in which the valleys are still V-shaped, but the original surface of the shield has been destroyed, is called *submature*. The ridge tops still mark the approximate level of the original land surface. Most of the Koolau and Waianae ranges on Oahu, and the West Maui Mountains arc in this stage, but the windward side of Kohala Mountain, on Hawaii, has not quite reached it.

Following the submature stage the valley floor is slowly widened and the cross profile of the valley changes from a sharp V to a broad V with a flat floor. Steep mountains still stand, and their tops are not far below the original land level. This is the *mature* stage. Parts of northeastern Kauai and windward Oahu are in this stage (fig. 106). Hawaiian valleys in other areas show the same sort of cross profile, with a broad flat floor, but this configuration is due to a different cause discussed on a following page. None of the major Hawaiian islands has reached a mature stage throughout. None is old enough.

The *old age* stage is marked by a reduction of the mountains to rolling hills with broad flat-floored valleys between them. The entire surface has been lowered well below the original level. The ultimate end of the process of stream erosion is a broad, almost-plain surface with very low relief, known as a *peneplain*. Near the ocean a peneplain is cut down almost to sea level. Inland, however, the peneplain surface can never be reduced by stream erosion all the way to sea level, because without some gradient, streams cannot flow. Therefore, although the Hawaiian Islands will ultimately be eroded to and below sea level, it will not be accomplished entirely by streams. The final triumph over the volcano's constructive forces will belong to the ocean.

INITIAL STAGE OF EROSION BY STREAMS

Through most of the period of growth of the Hawaiian volcanoes, erosion of the surface by streams is almost nonexistent. This is true not only because successive lava flows follow each other at such short intervals that running water has little opportunity to affect one surface appreciably before it is buried and a new surface created, but also because the rocks for the most part are so permeable that all of the water that falls on the land surface sinks into it before flowing very far. On the island of Hawaii not a single stream flows into the ocean along the entire coast from Hilo to South Point and up the west coast to the southern boundary of Kohala Mountain. The rocks of Mauna Loa, Kilauea, and Hualalai absorb the rainwater nearly as fast as it can fall. Occasionally a very heavy rain may cause streams to flow locally for a short distance, but a stream flow of as much as 10 or 15 million gallons a day at one point commonly has completely disappeared a

Figure 102. Youthful valleys of consequent streams on the Hamakua Coast, Hawaii. The narrow V-shaped valleys are separated by broad areas, covered with a blanket of ash, that are almost untouched by erosion. The ghosts of an earlier lava-flow topography can be seen through the ash.

Figure 103. Laupahoehoe gulch and peninsula on the Hamakua Coast of Hawaii. The peninsula was formed by a late lava flow that descended the valley. The topography is very youthful. Note the "slip over" of captured consequent streams as they enter the main valley on the left side. At Laupahoehoe Point, in the foreground, 26 persons were killed during the tsunami of April 1, 1946.

165

Figure 104. Sacred Falls, Oahu. The canyon of Kaluanui Stream is typical of the very deep, V-shaped, youthful canyons that incise the slopes of the Koolau shield toward its northern end.

mile or two downslope; and even in the area of heaviest rain the streams dry up within a few hours after the cloudburst.

It is not until the surficial rocks become partially sealed by chemical weathering or by a cover of fine volcanic ash that they are able to sustain prolonged surface runoff. This condition generally is reached in the late stage of volcanism, while the volcano is still active but flows have become less frequent. In this stage it

Figure 105. East end of Molokai. The gently sloping shield surface of this section of Molokai is disappearing under the attack of stream erosion, but a few patches of little-dissected original slope remain. The topography is transitional from youthful to submature.

Figure 106. Mature stage of topography in Kailua Hills, Oahu. Prolonged weathering and erosion have produced a subdued, gentle landscape. Ulupau Head is visible in the distance. The Pali Golf Course lies in the foreground.

is common to find small, or occasionally even large, valleys that have been excavated by stream erosion and sometimes partly filled with alluvium, and then buried by lava. Many lava-filled valleys of this sort occur, several, for example, in the Nahiku area on Maui.

The initial streamways make use of any existing depressions on the constructional surface of the shield. The commonest are the channels of former lava rivers (fig. 190) and depressions following the boundaries of adjacent lava flows. These are especially favorable paths because they extend great distances up and down the slope. On the middle slopes of Mauna Kea, where erosional topography is still in an initial stage of development, nearly every one of the shallow stream channels can be recognized as one or the other of these two types of depression. Occasionally a stream of water following a former lava river channel disappears underground into a lava tube, sometimes reappearing on the surface farther downslope.

## STREAM PATTERNS

The initial stream pattern is determined almost wholly by simple runoff down the steepest slopes of the circular to elliptical volcanic shields. The streams are *consequent* (that is, they follow the original steepest slope of the land surface), and their courses are radial with respect to the shield summit or the crest line of the more elongated shields (fig. 107*A*). Because

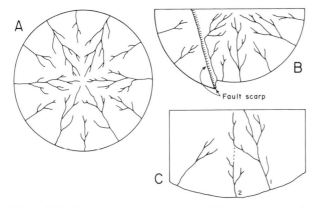

Figure 107. Diagrams showing (*A*) the radial arrangement of dendritic consequent streams on a circular volcanic mountain; (*B*) deviation from the radial pattern caused by faulting; and (*C*) stream piracy, in which stream *1* has captured the headwaters that formerly belonged to stream *2*.

the materials on which they work are fairly uniform in resistance to erosion, the young stream patterns are generally dendritic. The courses of a few streams are determined by fault zones that cut sharply across the direction of the dominant drainage (fig. 107*B*). Thus the lower part of Waimea Canyon, on Kauai, cuts diagonally across the trend of consequent streams flowing southwestward from the central Waialeale plateau (figs. 112, 260). Most of these fault-controlled streams were guided by the original fault topography, and are therefore also consequent streams. A few, however, may be *subsequent* streams, in that their courses were determined by greater ease of erosion in the somewhat broken and crushed rocks along the fault zone.

The general courses of both slope-controlled and fault-controlled initial streams persist through all the later stages of erosion.

Sometimes one stream cuts into the valley of another, and when this happens the stream which is cutting faster steals the headwaters of the other (fig. 107*C*). This is known as *stream capture,* or *stream piracy.* The faster downcutting by one stream may be due to steeper gradient, greater water supply, smaller load, or weaker rocks along its course. Once the headwaters of the other stream have been captured, the larger water supply causes even faster cutting by the capturing, or "master," stream, and the loss of its headwaters results in slower cutting along the lower course of the other stream. Capture of one consequent stream by another is common in Hawaii. It takes place by lateral excavation and side-wall retreat in the valleys. Numerous examples of such "lateral capture" can be pointed out on topographic maps or, even better, viewed from the air (figs. 103, 185). Excellent examples of near capture can also be found, for example on the northeastern slope of Mauna Kea, on Hawaii, where they can be seen during the flight to and from Hilo.

Conditions that appear to favor particularly this type of piracy in Hawaii are: (1) All streams flowing off the volcanic shield tend to converge in the direction of their headwaters, so that even without lateral cutting the valleys may be very close together. (2) The heaviest rainfall is in the upper and middle sections of the streams; the lower sections are generally drier and sometimes arid. (3) In the initial stages of stream formation much, if not most, of the water is traveling underground and is subject, therefore, to control by underground structures such as dikes or other permeability barriers which may channel the subsurface water to a particular youthful consequent gully, thereby providing it with more water and consequently more erosive power than its neighbors. The gully so endowed will enlarge faster and may eventually intercept the channels of the nearby streams, which in turn will provide it with an even greater capacity to enlarge. As a final result of such a process one side of the volcanic shield may be cut by only two or three huge master valleys where originally there were many small ones. (4) None of the above conditions would be as effective in promoting lateral capture were it not for the layers of strongly resistant lava flows interbedded with very weakly resistant layers of clinker, ash, and cinder that are readily eroded. Removal of these weak beds leaves the resistant layers unsupported, and they collapse to form a steepened slope. Eventually the side walls of the valleys become very steep and maintain that attitude as they continue to spread, capturing the smaller streams as they go.

It is of interest here to note that some of the giant grooves or "vertical valleys" that corrugate the sides of the larger valleys, such as Manoa, Nuuanu, and Kalihi on Oahu, Waipio and Waimanu on Hawaii, and many others, are apparently the result of "slip over" of the captured streams into the valley of the master stream (figs. 103, 109). In Honokohau Canyon on West Maui several streams in varying stages of "slip over" may be observed.

EVOLUTION OF STREAM VALLEYS

As soon as streams start to flow more or less continuously, they quickly cut narrow V-shaped notches into the land surface (figs. 102, 103). Because wave attack is not dependent on previous weathering of the land surface, but starts immediately upon the entrance of a lava flow into the ocean, it gains a head start on stream erosion, and on the more exposed coasts the initial streams drop over wave-cut cliffs to enter the sea. Throughout the youthful stage many of the streams, such as most of those along the Hamakua Coast of Mauna Kea, continue to plunge down the sea cliff in waterfalls. Even when the master streams have reached maturity and enter the ocean at grade, the minor stream valleys between them may remain hanging far above sea level (fig. 183), because wave erosion is far more effective than the downcutting by these minor streams. This is well shown on the north coast of Molokai, where the minor streams on the shield segments between the major valleys drop into the ocean in a series of spectacular waterfalls as much as 2,000 feet high.

The violent impact and turbulence of the water at the foot of a waterfall causes much more rapid erosion there than elsewhere along the stream. Undercutting in the plunge pool at the base results in repeated collapses of the face of the fall, and an upstream recession of the fall (fig. 108). In this way the stream cuts a gorge that gradually extends headward as the waterfall retreats. Commonly the waterfall that starts at the coast is only one of a series. Others develop on resistant rock ledges, as already mentioned. Typical youthful Hawaiian streams have a stairlike longitudinal profile, steps of relatively gentle gradient being separated by risers traversed by waterfalls or cascades (fig. 109).

In the youthful stage of stream erosion downcutting is greatly predominant over lateral cutting, and the stream tends to cut a vertical-sided notch. Generally, mass transfer widens the upper part of the notch concurrently with its deepening by the stream, and the valley becomes V-shaped. Occasionally, however, downward cutting is much more rapid than widening, and the stream occupies the bottom of a very steep-walled trench several hundred feet deep. The very narrow canyon of Kaluanui Stream at Sacred Falls, Oahu (fig. 104), is an example. Even more extreme is the gorge of Koula Stream, a branch of Hanapepe River on Kauai, which is so narrow and steep walled that sunshine is said to strike the canyon floor for only about an hour a day.

*169*

Figure 108. Akaka Falls, 420 feet high, is one of many waterfalls along the Hamakua Coast of Hawaii. The stream is cutting through a blanket-like mantle of Pahala ash into a resistant Mauna Kea flow which forms the lip of the falls. The deep gorge in the foreground has been left by the retreat of the fall. Note the right-angle diversion of a tributary stream by an ancient lava flow on the right side of the canyon.

As the streams approach grade, lateral cutting becomes more and more important, and eventually a broad, gently sloping valley floor is produced. This is essentially a cut floor, although for several reasons in many valleys it is largely buried by alluvium. The valley floor becomes wider as the valley walls are cut back. However, unlike the usual situation in humid temperate climates, the slope of the walls does not decrease appreciably (figs. 110, 111). Even as old age is approached, the side slopes of the valley remain steep. Apparently this is because, of all the common methods of mass wasting, only soil avalanching is active enough in the wet parts of Hawaii to have any significant effect, and this can take place only on slopes of at least a minimum steepness—probably about 45°. If in any way the slope is reduced to a lower angle, slope retreat simply ceases until lateral widening of the valley floor has pushed back the base of the slope sufficiently to reestablish the critical steepness.

Because of the radial pattern of drainage, as the canyons of the more actively eroding streams are extended headward and broadened, they intercept the headwaters of the less powerful streams between them. This, of course, still further favors the already dominant streams. As these master valleys increase in size, their steep side slopes eventually intersect and in their upper courses become separated only by exceedingly narrow "knife" ridges, even though near their mouths the valleys may still be far apart. Thus there develops in the headwater regions a mature topography in which the original shield surface has been completely destroyed, whereas closer to the coast is a series of deep broad valleys separated by triangular segments of the original shield surface dissected only by shallow V-shaped gulches (fig. 113 *A, B*). These flatiron-shaped upland slopes are called planezes. In rainy parts of the island there is still enough water in the streams on the intercanyon facets to cause fairly rapid erosion, and the planezes are gradually dissected into a mature topography consisting of a series of minor canyons and sharp intervening ridges. Eventually even these are largely removed, and the region attains a late mature to old age stage consisting of broad, gently sloping valleys with only narrow, steep-walled ridges and pinnacles between

them. Much of windward Oahu southeast of Waikane Valley and the west-central part of the Waianae Range (fig. 242) are in this stage, although changes in regional base level have resulted in alluviation of the valleys and late (posterosional) eruptions have brought about further alterations to the topography (Stearns and Vaksvik, 1935, p. 24).

Because the amount of rainfall varies greatly from one part of the island to another, all parts of the island do not advance through the erosional cycle at the same rate. For example, as we have just seen, the rainiest parts of Kauai and Oahu have been deeply eroded to a late mature or old age stage (if we disregard the much later, posterosional volcanics), while the driest parts of the islands still remain in a youthful stage of dissection. On the drier,

Figure 109. Approximately 20 plunge pools mark the descent of a captured tributary stream into Waipio Canyon, Hawaii. This type of action contributes to the eventual development of giant vertical grooves on the sides of Hawaiian valleys. The long narrow scars in the vegetation were caused by soil avalanches on the steep canyon sides.

Figure 110. Iao Needle, West Maui. The top of the needle is a prominent point on a winding knife ridge (see fig. 111) left by erosion of caldera-filling lavas. The approximate boundary of the old caldera is the line of great cliffs in the background.

leeward sides of the islands rainfall on the triangular intercanyon facets is insufficient to produce much stream flow. The master streams derive their water from the wet crest region, and these through-flowing streams continue to erode the lower courses of their valleys, while stream erosion is almost nil on the facets between them. Thus we find in these areas broad mature master valleys separated by segments traversed by only very small youthful valleys. This is well displayed in the region northeast of downtown Honolulu, where shallowly dissected triangular facets can be clearly seen between Manoa and Palolo valleys (St. Louis Heights), and between Palolo and Waialae Nui valleys (Wilhelmina Rise).

At first glance the intercanyon facets appear to be almost unaltered remnants of the original shield surface. Actually, however, they have

been appreciably modified by erosion. The faces of the facets have an average slope of about 8°, which is some 3° steeper than the slope of the lava flows that make them up (Stearns and Vaksvik, 1935, p. 23). The entire surface of the facet has been lowered, probably an average of one or two hundred feet, by sheet erosion, which because of the sparse cover of vegetation is far more active in these dry areas than in the areas of higher rainfall.

Another effect of the broadening of valleys is the gradual isolation of certain peaks which eventually stand above the surrounding ridges like glacial matterhorns (figs. 116, 118, 245).

## AMPHITHEATER-HEADED VALLEYS

Some of the most spectacular sights in the erosional architecture of the islands are the

172

precipitous valley walls and near-vertical semi-circular valley heads, commonly termed amphitheater-headed valleys (Hinds, 1931). In general form, these valley heads are strikingly like the cirques formed by valley glaciers; and seen from above, the scalloped pattern of successive amphitheaters cut into the central highland bears a close resemblance to glacial "biscuit-board" topography (fig. 119). Emphatically, the Hawaiian amphitheaters are not the result of glaciation.

The amphitheatral form of the valley heads (figs. 113, 114, 115) is due to the interaction of several factors. The tributaries of the master stream enter the valley as a series of waterfalls that plunge down the steep head wall, and, to a very large extent, the headward growth of the valley is the result of erosion by these waterfalls—the lineal successors of the falls that cut the original youthful gorges. Many of the big valleys have cut back into the dike complex of the volcano and have tapped water held in the

Figure 111. Part of the north wall of Iao Valley, West Maui. From the air it is evident that the Iao Needle, in the center foreground, is a point on a winding knife-edge ridge. The nearly circular valley is the result of erosion scouring out the heart of the West Maui caldera.

Figure 112. Waimea and Olokele canyons, Kauai. Waimea River, on the left side of the photograph, follows the edge of the Makaweli depression, a down-faulted segment of the flank of the Kauai shield volcano. The nearly filled caldera lies in the background. Olokele River, which comes in from the right to join the Waimea River, drains a portion of the Alakai Swamp. Note the distinctive curved skyline of the shield volcano, and the small amount of dissection of the dry southwestern slope at the left of the picture.

interdike compartments (see chap. 15). Thus dike springs issue in the head wall of the valley and at its base, and sapping by these springs, together with the more active stream erosion resulting from the increased volume of water, cause an increased rate of erosion at the foot of the head wall, which contributes to its steepness. The binding effect of vegetation, preventing other means of mass wasting than soil avalanches, also contributes to the steepness of both the head wall and the side walls of the valley.

174

Another very important reason for the steepness of the valley walls is the presence of nearly horizontal beds of alternately greater and lesser resistance to erosion. Even in temperate regions of much lower rainfall, valley heads of very

similar form can develop in low-dipping, alternately hard and soft beds, particularly where the more resistant beds are in the upper part of the succession. More rapid erosion of the less resistant beds, such as the aa clinker layers in the Hawaiian volcanoes, results in undercutting of the more resistant layers. The latter, broken by innumerable, nearly vertical joints, cave away as undercutting reaches a critical amount, producing nearly vertical cliffs. A succession of such cliffs, separated by narrow steep slopes on the less resistant beds, constitutes the walls of the Hawaiian amphitheater-headed valleys.

The valley head gradually expands into its typical rounded form as the headwaters of minor streams on each side are captured by the major stream.

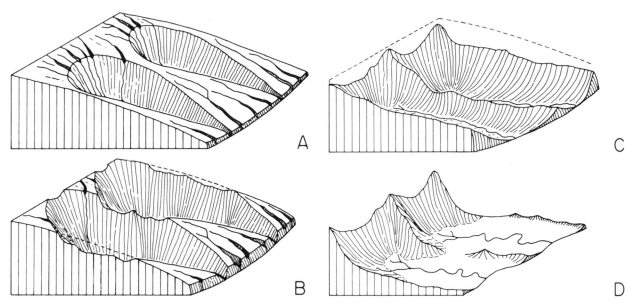

Figure 113. Diagram showing the evolution of amphitheater-headed valleys, and the formation of triangular facets between them (*A* and *B*). With the continued reduction of the ridges between the valleys (*C*), the triangular facets disappear. The further cutting away of the ridges may leave a long scalloped cliff formed by the coalescence of successive amphitheatral valley heads (*D*), such as the Nuuanu Pali. (Modified after Stearns and Vaksvik, 1935.)

175

Figure 114. Hiilawe Falls, Hawaii. Hiilawe is the thumb-like branch on the left side of Waipio Valley seen from a distance in figure 185. Hiilawe is the highest free-fall waterfall in Hawaii and one of the highest in the world, with a vertical drop of about 1,000 feet. The fall is now usually dry because the stream that formerly fed it is diverted by an irrigation ditch. A lava flow from Mauna Kea spilled into Hiilawe Valley, a remnant of it forming the shoulder to the right of the stream in front of the alcove of the fall.

Figure 115. Headwaters of Waihee Stream, West Maui. The amphitheater-headed valley is slowly advancing headward and increasing its width, capturing neighboring streams as it widens. The tiny gullies on the face of the cliff are probably important in the process of maintaining the amphitheater-headed configuration. Note the wastage of the rocks to the right of the head of the falls under the influence of intense chemical weathering.

One of the striking features of all Hawaiian valleys is the succession of huge vertical grooves that make the valley sides look as though they had been raked by giant finger-nails (figs. 118, 231, and photograph on p. 160). These grooves and "vertical valleys" were cut by innumerable waterfalls (fig. 109), some of them on permanent tributary streams, but many of them flowing only during brief periods of heavy rains. Thus the grooves along the Pali south of Waimanalo, nearly always dry, display a succession of spectacular waterfalls during occasional cloudbursts.

Still another striking feature is the extreme narrowness of some of the knife ridges that separate valleys in the more mature stages of dissection (figs. 110, 111, 263). A climber can almost believe he feels the ridge swaying under him in the wind.

## DISCREPANCIES IN EROSIONAL STAGE

In general, the degree and type of erosion are closely related to the amount of rainfall and the age of the region. Certain areas, however, show a degree of erosion that is disharmonious with the present rainfall. Thus the Waianae Range, on Oahu, although receiving only scant rainfall, has been deeply eroded by streams to broad amphitheater-headed valleys typical of high-rainfall regions. The explanation of this apparent discrepancy is not difficult. The present low rainfall on the Waianae Range is the result of shielding by the Koolau Range. The dominant trade winds impinge first upon the Koolau Mountains, and the moisture is largely removed from the air before it reaches the Waianae region. However, the Waianae volcano pre-dates the Koolau, and for a long time the trade winds struck directly against it, after a long sweep across the ocean, and dumped their load of rain upon it. The rainfall was probably even heavier then than the present rainfall on the Koolau Range because the Waianae mountains are higher. Indeed, at that time they may have stood several hundred feet higher above sea level than they do today, bringing their tops well into the zone of maximum rainfall where they may have received

more than 400 inches of rain a year, as does the summit region of Kauai at the present time.

Although the degree of erosion of the Waianae volcano can thus be accounted for, it is not as easy to explain why more erosion has taken place on the western (leeward) side of the range than on the side that would have received more rain, assuming present wind conditions, and consequently would be expected to be more deeply eroded (figs. 242, 244). The suggestion that in earlier geologic times the prevailing winds may have been from the west seems to be negated by the rather small degree of erosion of the western slope of the equally old Kauai volcano. Stearns and Vaksvik (1935, p. 31) believe that collapse of part of the northeastern wall of the Waianae

177

Figure 116. The area of the headwaters of Kahana Valley, Oahu, illustrates erosional three-sided peaks that result from the intersection of several amphitheater-headed valleys. The topographic forms somewhat resemble those that are left by the intersection of glacial cirques and U-shaped valleys, but of course glaciers have played no part in sculpturing Hawaiian valleys.

Figure 117. The Pali along the windward face of the Koolau Range, intersected by the heads of Nuuanu Valley on the left and Kalihi Valley on the right. These two wide valleys were formed by large streams that formerly flowed off the upper slopes of the lofty Koolau volcano, which may have reached nearly the height from which this photograph was taken. Caldera collapse destroyed the top of the volcanic shield, and erosion has removed most of its windward flank and scoured out the caldera, removing the upper sections of the valleys. The city of Honolulu and Pearl Harbor lie beyond the Koolau Range.

caldera at a time when the volcano was still very active allowed lava flows to pour out in that direction and repeatedly cover the northern and eastern slopes, while the western slope was protected from lava flows by the western wall of the caldera. As a result erosion would have begun earlier on the western slope, and the western valleys would of course be larger than those on the other slopes because they are much older. Indeed, if occasional veneering of the northeastern slope by lava flows continued until the Koolau Range attained sufficient size to block off much of the rainfall, the relatively small amount of dissection of the northern and eastern slopes of the Waianae Range is readily understood.

The region at the foot of the Nuuanu Pali, on windward Oahu, is in a late mature to old age stage of dissection, yet it receives much less rainfall than the crest region of the Koolau Range, which has been far less deeply eroded. However, this discrepancy also relates only to present conditions. The region was formerly the highest part of the Koolau shield, rising perhaps 2,000 feet higher than the present crest of the range, and the whole island stood some 1,200 feet higher above sea level than it does today. Thus the crest of the Koolau volcano was once about 6,000 feet above sea level, well within the zone of maximum rainfall. Indeed, if the former shield surface is reconstructed and the probable belts of rainfall intensity are indicated upon it, the region of heaviest rainfall corresponds amazingly well with that of deepest erosion—between Kaaawa and Waimanalo. The present decrease in rainfall has resulted from lowering of the land surface by erosion and sinking of the island in relation to sea level.

Furthermore, in the region between Kaneohe and Waimanalo the rocks occupying the former caldera of the Koolau volcano have been much altered by rising volcanic gases, the original pyroxene of the lavas being changed to chlorite (which gives many of the rocks a distinctly green color). This alteration weakens the rocks and makes them more susceptible to erosion, thus speeding in this area the progress of the geomorphic cycle.

## ORIGIN OF THE KOOLAU PALI

The very spectacular line of cliffs that separates windward Oahu from the crest region of the Koolau Range (figs. 117, 118, 248) was long a subject of geological debate. Early workers attributed this Pali to a fault, on which the northeast side of the island had been dropped down in relation to the crest of the range. Others have suggested that it was cut by ocean waves. Careful study by H. T. Stearns, however, appears to have demonstrated beyond question that it is largely the result of stream erosion (Stearns and Vaksvik, 1935, p. 28). Actually it is a series of closely spaced amphitheatral valley heads (figs. 117, 119) separated by low projecting ridges which are partly

179

Figure 118. Puu Lanihuli, 2,107 feet in elevation, is a prominent point on the Koolau Range at the Pali. It is an example of a peak that is being isolated by the growth of the adjacent valleys, as Nuuanu Valley on the left and Kalihi Valley on the right continue to advance headward and laterally. Plunge-pool erosion is well displayed on the face of the Pali. In the foreground, at the foot of the Pali, are great banks of colluvium.

Figure 119. Haiku Valley is the center of a group of three amphitheater-headed valleys that form huge scallops in the face of the Koolau Pali. Toward the camera from the ends of the steep spurs in the foreground the intervening ridges between the valleys have been largely removed by stream and wave erosion. As the face of the Pali continues to shift westward toward Honolulu the scarp will become lower and individual peaks probably will replace the continuous cliff line. Pearl Harbor, the Ewa coastal plain, and the Waianae Range are visible in the background.

buried in alluvium (fig. 113D). Lava beds in the face of the cliff can be traced out into the ridges without interruption, proving that no fault exists at the present base of the cliff. In part, particularly at the southeastern end, wave erosion has removed the projecting spurs and somewhat straightened the cliff, but by far the dominant agent has been the stream erosion that has cut away a large segment of the volcano and developed the late mature to old age topography east of the base of the cliff (fig. 119).

Although not directly responsible for the present cliff, faulting has played an important part in determining the general position of the

cliff. During the late stages of its history the summit area of the Koolau shield volcano collapsed to form a caldera, 7 or 8 miles long and more than 4 miles wide, stretching from southeast of Waimanalo to beyond Kaneohe. The northeastern boundary lay just beyond the present shoreline at Lanikai and probably crossed the base of the present Mokapu Peninsula in the vicinity of Aikahi School. The southwestern boundary lay close to the present Pali. The caldera was gradually filled with lava flows to a level above the present top of Olomana Peak. The flows were thick and dense, and normally would have been resistant to erosion; but through many centuries gases rose through them from the magma body beneath and gradually altered them to more easily eroded materials (see chap. 8). After volcanism had ceased, both weathering and erosion made rapid headway on the northeast side of the shield, which received heavier rainfall than the southwest side and was directly exposed to the trade-wind waves. The side of the shield beyond the caldera was removed, and erosion ate rapidly into the weakened rocks of the caldera itself. Amphitheater valley heads retreated across the caldera to the far southwestern boundary, but there they encountered the unaltered rocks of the outer part of the volcano and erosion slowed up. Valleys that had been forming less rapidly caught up with those that had cut back fastest, until all the valley heads lay more or less along the same line; and as the rapid erosion of the caldera rocks continued, the amphitheatral heads began to coalesce, forming the present scalloped cliff of the Pali. The cliff has now been cut back a short distance beyond the caldera boundary but is still closely parallel to it. Thus, in a sense, this central part of the Pali *is* the result of faulting. The fault scarp of the caldera boundary was first buried by lava flows but then exhumed by erosion. Technically, the Pali is a resequent fault-line scarp.

Although the removal of the caldera-filling rocks northeast of the Pali was largely the work of stream erosion, waves also played a part at times when sea level was higher than now (see chap. 11). A topographic form that may be a guide to the location of previous wave erosion along and below the Pali, as well as at other cliffs elsewhere in Hawaii, is the "horned crescent ridge ending." A ridge which extends in a straight line toward the coast makes an abrupt sideways bend just before reaching the shore, and both the coastal and the inland sides of this hook are steep cliffs. It is not a natural form to result from weathering or stream erosion. Horned crescent ridges are especially well developed along the windward coast of Molokai (fig. 132) where the ridge endings are now under heavy wave attack. The inside of the hook or crescent commonly presents a high and steep cliff, diminishing in height toward

the end of the crescent. Similar forms are observed well above sea level east of the Pali, beyond Castle Junction, and other examples can be found along the extent of the Pali. The cutting may have been done by the sea when it stood about 25 feet above its present level.

EFFECTS OF CHANGES OF BASE LEVEL

The normal progress of the geomorphic cycle may be interrupted at any stage by a change in general base level—that is, by a change in the position of sea level in relation to the land. In its normal course of development a stream tends to establish a longitudinal profile which is graded to base level and is in equilibrium with the amount of water in the stream and the amount of available load. This "normal" profile is steeper near the headwaters, where the volume of water and available load are less, and flatter in the lower reaches of the valley. Any change of base level, either general or local, will result in either increased erosion or deposition by the stream to reestablish a profile in equilibrium with the new conditions.

Lowering of base level results in an increased stream gradient and consequently in an increased rate of downcutting. If a valley has started to broaden out under the control of one base level, the increase in downcutting which follows lowering of base level may form a steeper-walled valley in the bottom of the older broader valley, leaving a terrace along the valley wall (fig. 120A). This may occur when sea level is lowered in relation to the land. However, in Hawaiian valleys many terraces were caused by destruction of a local base level—cutting away of a hard ledge of rock that for a time limited downcutting by the stream. Although these local terraces may be very conspicuous features for a time, they are ephemeral, lasting perhaps through several generations of humans but disappearing very quickly in the eons of geologic time.

All of the big valleys in the Hawaiian Islands have flat floors built by alluvium deposited by their streams. There are two principal reasons for this alluviation. In some areas, such as the Waianae Range and the part of windward Oahu

181

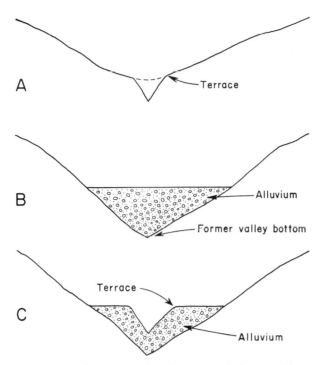

Figure 120. Diagrams showing the effects of changes of base level on the cross profile of a valley. *A*, Lowering of base level, resulting in increased downcutting by the stream and trenching of the valley floor; *B*, rise of base level, resulting in deposition of alluvium in the valley and formation of a flat valley floor; *C*, lowering of base level, resulting in trenching of the flat alluvial floor.

discussed in the section on discrepancies in erosional stage, the deposition of alluvium is caused partly by the decreased volume of water in the streams resulting from a change to a drier climate, and the consequent decrease in transporting power.

A more general cause of alluviation of Hawaiian valleys, however, is a rise of regional base level, which has resulted in a decrease in the total amount of fall of the streams and the establishment of flatter average stream gradients. The rise in base level has occurred as sea level has risen in relation to the land. This rise of sea level has resulted partly from the return of water to the oceans upon the melting of the great continental glaciers, and probably partly from an actual sinking of the islands because of their great weight upon the earth's crust (see chap. 11). During the time when sea level stood relatively lower than it does now, major streams cut the mouths of their canyons down to that level. As sea level rose, the ocean tended to flood the mouths of the deepest

valleys. If the rise was rapid, there may have been produced a deeply indented "shoreline of submergence," with bays extending up the drowned river valleys much as they do on the fjord coasts of Norway or the South Island of New Zealand. However, if even a small bay is formed in this way, the stream entering it immediately begins to fill it with sediment. The reduction of velocity of the stream where it enters the bay causes it to drop its transported load, build a delta out into the bay head, and gradually spread alluvium over the top of the delta.

If the rise of sea level was slow, therefore, deposition by streams may have kept pace with the rise, preventing the formation of bays. In general this probably was the case in Hawaii, an approximate balance between the rising sea level and deposition by streams maintaining a nearly straight coastline without deep marine embayments. Pearl Harbor, on Oahu, is an exception. Its branching lochs (fig. 239) are former stream valleys drowned by the rising ocean, and modified somewhat in form by deposition of sediment and by differential wave erosion of rocks of varying resistance.

Whether the rise in sea level is rapid or slow, the result is a valley in which the walls descend steeply to their junction with a nearly flat floor (fig. 120*B*). Such flat-floored major valleys as Waipio and Waimanu on Hawaii (fig. 185), the great windward valleys of East Molokai, Hanalei and Waimea valleys on Kauai, and Lualualei and Waianae valleys in the Waianae Range of Oahu (fig. 242), all owe their alluviation to a fairly recent relative rise of sea level.

A stream that has built a broad plain in its valley bottom may have its erosive power restored, either by a change of climate with a resulting increase in its volume of water, or by lowering of base level. Such a stream is said to be *rejuvenated*. A sharp inner valley may then be cut into the alluvium, bordered by terraces that represent the former valley floor (fig. 120*C*). Eventually, most or all of the alluvium may be removed from the valley. Many of the large valleys of the Hawaiian Islands contain masses of "older alluvium," some of great volume, which are now being removed by the streams (figs. 207, 273). Whether the rejuvenation of these valleys was caused by change of climate or change of base level, or both, is still uncertain; but it is very probable that relative lowering of sea level was at least partly responsible (see chap. 11).

*Suggested Additional Reading*

Gilluly and others, 1968, pp. 207–245; Holmes, 1965, pp. 468–618; Moberly, 1963*b*; Stearns and Vaksvik, 1935, pp. 22–29; Thornbury, 1962, pp. 99–207; Wentworth, 1927, 1928

*Breccia formed of blocks of tuff, Ulupau Head, Oahu*

# CHAPTER TEN

# Mass Transfer

ONE OF THE GREATEST GEOLOGICAL NUISANCES ENCOUN-tered by mankind today is mass transfer, the movement of material down the side slopes of valleys under the influence of gravity, apart from the transportation of material by tributary streams. It is a normal part of the process of erosion and goes on all the time on all slopes. Generally it is slow, and so long as man's dwellings and most of his activities were confined to flat lands and valley bottoms it seldom caused serious trouble. As houses and roads creep farther and farther up the hillsides, however, as they are doing today, we must expect more and more trouble from natural downslope movements. Loading the soil and subsoil with houses and making steep-sided cuts in them for roads, far from stopping the movement, aggravates it. Future troubles of the general sorts now affecting parts of Palolo Valley and Aina Haina, in Honolulu, must be anticipated.

The processes that bring about mass transfer include the falling, sliding, or rolling of individual rock fragments, washing of loose materials by sheet or rill runoff of rain, soil creep, and landsliding. All of these processes result in a shifting of material downslope—sometimes en masse and sometimes piecemeal, a fragment or a few fragments at a time; sometimes in a rapid spectacular rush, but often so slowly that it escapes notice. By mass transfer the walls of the valley are pushed back, and at the same time the angle of slope of the walls generally is lowered and the valley is widened from a vertical-walled, stream-cut slot to a V shape.

Fragmental material that moves down the sides of valleys as a result of all the processes of mass transfer forms accumulations known as *colluvium*. It should be distinguished from *alluvium,* which is deposited by streams.

A *talus,* which is one variety of colluvium, is a pile of angular rock fragments at the foot of a cliff, formed by the fragments tumbling down the slope. Taluses are well shown at the base of the cliffs surrounding calderas, and along some sea cliffs. The rock fragments making up the pile also are commonly called talus. Colluvial accumulations often consist of material deposited by several differ-ent mechanisms—gravity fall, rain wash, mudflow. The products of the different mechanisms are so intermixed that it is generally impractical to try to identify them.

In Hawaii mass transfer operates somewhat differently in dry regions than it does in very wet regions. Although there is a complete gradation from one sort of region to the other, it is advantageous to consider the two extremes separately.

## MASS TRANSFER IN ARID REGIONS

In regions of relatively low rainfall, such as the leeward side of the Waianae Range on Oahu, gravity fall and sliding of rock fragments, rain wash, and soil creep are all active in mass transfer. Chemical weathering is less effective than in wetter regions, and much of the material in transport is fresh or only moderately decomposed. Soil cover is far less continuous than in wetter regions. The valley sides typically exhibit a series of low cliffs formed on rock layers resistant to erosion—usually the massive centers of aa flows—alternating with slopes commonly averaging 20° or 30°. Many of the slopes, which form on the less resistant beds, are covered with loose rock fragments that have washed or tumbled down from higher up the hillside. This local colluvium is very temporary, representing merely a pause in the movement of the fragments toward the valley bottom.

In these dry areas the action of rain wash is not continuous but is concentrated in the brief infrequent periods of heavy rainfall. The exceptionally heavy rains that occur only every few years may accomplish far more work than is done in all of the time between them. At such times great volumes of water may pour down the valley walls, as sheets or rills, heavily loaded with rock debris. Washing away of the finer material may in turn remove the support from larger fragments, which become unstable and roll or slide down the slopes.

Soil creep is the general en masse downhill movement of the superficial parts of the soil. In temperate regions one of the important causes of soil creep is the repeated freezing and thawing of water in the soil. This occurs in Hawaii only on the uppermost slopes of the highest mountains—Mauna Kea, Mauna Loa, and Haleakala. In those regions water freezing in the soil (which there consists largely of fresh rock fragments) expands, and the soil above it is pushed outward at more or less right angles to the general slope surface. When the ice melts the fragments are released, but because the pull of gravity is directly downward they fall in that direction rather than back toward the points whence they were pushed outward. Thus repeated freezing and melting results in a zigzag downhill movement of the particles in the soil.

Although freezing and thawing do not occur in most of Hawaii, another similar process is widespread in the drier regions. Some clay minerals in the soils have the property of expanding when they are wet and contracting when they dry out. Slight as the expansion and contraction are, repeated wetting and drying of the soil results in a slow steady downslope shifting of the soil due to the zigzag downhill movement of its individual particles. This process is much less active in the wetter parts of the islands, where the soil is almost constantly wet and not subject to alternate wetting and drying.

Landslides are rather rare in the dry areas. When they do occur, they are generally of the rock-slide type, in which a mass of rock becomes loosened and rushes downward, breaking up as it goes, and deposits a heterogeneous heap of angular fragments at the foot of the slope.

All of these movements result in the accumulation of a mass of colluvium at the foot of the valley wall. Some individual colluvial cones can be recognized where a tributary gulch has concentrated the material, or where the downward movement of material has simply been more abundant than elsewhere along the valley wall, or where a recent large landslide has deposited a mass of material not yet dissipated by erosion. For the most part, however, the colluvium forms embankments or aprons that are nearly continuous along long stretches of the valley sides. Two colluvial aprons on opposite sides of the valley may be separated by bedrock or by stream-deposited sediment (alluvium) along the axis of the valley; or in narrow valleys they may join, with the stream running along their junction. The colluvium is unbedded or only very crudely bedded, and is largely unsorted, with angular to partly rounded fragments of rock up to several feet in diameter in a matrix of smaller rock fragments, sand, and silt. Especially near the lower edge of the aprons, occasional masses of stratified and rounded stream-laid gravel are enclosed in the colluvium.

## MASS TRANSFER IN WET REGIONS

In the regions of high rainfall it is immediately apparent, from the long colluvial aprons along the valley sides, that mass transfer is at least as active as in the dry regions. Closer observation soon reveals, however, that the processes of mass transfer which are most effective in arid regions are of comparatively little importance in wet ones. Except on a few steep cliffs, individual fragments are seldom seen to roll or slide down the hillsides. Soil creep resulting from alternate wetting and drying is unimportant because the soil is almost constantly wet. During very heavy rains large amounts of water may flow on the surface at the foot of the valley walls, but where the valley sides above have not been disturbed by construction or artificial clearing, this water is surprisingly clear. It brings down very little silt or clay.

This reduction in the amount of gravity fall, soil creep, and rain wash appears to be very largely due to the anchoring in place of the weathered rock mantle by the roots of plants. The heavily vegetated slopes of the wet regions are markedly steeper than the average detritus-covered slopes in drier regions. Normally, one expects a valley wall with a slope of 50° or more to consist of solid bedrock, but in the wet parts of Hawaii such slopes commonly are covered with soil. Vegetation holds the soil together and binds it to the bedrock, so that it remains on slopes much steeper than the angle of equilibrium in sparsely vegetated areas.

In the wet regions of Hawaii the principal agent of mass transfer is a type of landslide known as a soil avalanche, in which the soil cover on a steep slope simply pulls loose from the underlying bedrock and slides downward. Sometimes the avalanche travels all the way to the foot of the valley wall and adds to the bulk of the colluvial apron, which indeed has been almost wholly built by slides of this sort. At other times it travels only part way down the wall before it is checked. The bright red scars left by these slides (fig. 109) are very conspicu-

ous on the green hillsides for a few months after the avalanche occurs.

Most soil avalanches are small, commonly less than 100 yards long downslope, and less than 100 feet wide, though rarely they may be two or three times that large. The mass of soil removed generally is only 2 or 3 feet thick, but rarely may be as much as 10 feet. Measurements by Wentworth (1943) show that the great majority of them were formed on slopes of 42° to 48°, with a few on slopes as high as 55°

The frequency of soil avalanches is closely related to heaviness of rainfall. Some are caused by nearly every exceptionally heavy rain. The additional weight of the water-saturated soil, possibly together with some degree of lubrication by the water, causes less stable portions of the mantle to pull loose from the bedrock and slide downhill. Sometimes the actual tearing loose seems to be initiated by a single large tree that is leaning outward and pulls the soil loose with its roots. At any one time relatively few soil avalanche scars are visible. Wentworth estimated, however, that in a period of 8 years there were about 200 such avalanches, each one about 1 acre in size, in an area of approximately 15 square miles in the Honolulu watershed. At this rate about 1 foot of material would be removed from the entire area in 400 years. Following a heavy kona storm in 1930, Stearns counted 14 new avalanches in the head of Nuuanu Valley alone.

Occasionally, soil avalanches become transformed into mudflows, which may travel for considerable distances out over the colluvial slope, or even downstream along the axis of the valley.

## TALUS

An abrupt break in slope is usually apparent between the pile of broken rock fragments that constitutes the talus and the cliff face above that is formed of solid rock. The angle that the surface of the talus slope assumes is called the *angle of repose.* It is the steepest angle at which the fragments will stand without sliding or rolling downhill. It is controlled by several factors, including the coarseness and uniformity in size of the fragments, the rapidity of the

*187*

fraying of the cliff above, and the rapidity with which material is removed from the base of the pile. The larger and more uniform the average size of the fragments, the more steeply the slope will stand.

Talus deposits differ from deposits of other types of sedimentary material, laid down in water or distributed by wind, in that there is no orderly arrangement of the fragments; nor are they sorted by size, and bedding or layering is usually not evident. The fragments, which are generally quite angular, lie in jumbled confusion (see photograph, p. 184).

Talus is a temporary place of storage for rock debris. A talus pile does not usually continue to grow in size until it engulfs the cliff that is contributing the material, because there is generally a removal of the talus at the foot of the slope by streams or waves that maintain an efficient policing of the slope. Obviously, where broken rock debris from the cliff tumbles directly into the water of an underloaded stream or into pounding surf no talus will form.

In Hawaii extensive talus deposits have been formed in many areas. The Makapuu Point section of Oahu near Sea Life Park displays a fine apron of talus along the base of the cliff formed of Koolau basalt. Along the north side of the Waianae Range, between Mokuleia and Kaena Point, there are other good examples of talus. Very fresh talus may be observed at the foot of cliffs rimming Kilauea caldera on Hawaii, and within the Halemaumau fire pit itself, broken fragments of the crater wall gather in cone-shaped heaps at the bottom of the wall, to be buried by the next eruption.

Ancient taluses are sometimes very useful in deciphering the history of a region. Masses of old talus plastered against the western wall of Waimea Canyon, on Kauai, have been preserved beneath later lava flows and then revealed by subsequent erosion. They show not only that the cliff was there before the eruption of the lava flows, but also that it was in very nearly its present location. Similarly buried taluses exposed in the upper parts of Waimea Canyon (along Halemanu Stream) and Olokele Canyon indicate the former position of the cliffs that bounded the Kauai caldera.

188

## SOIL SLUMPS

In wet to moderately wet areas in which the larger vegetation has been destroyed and replaced with a cover of grass or other shallow-rooted plants, and to a lesser extent in dry areas, there is an abundant small-scale slumping of the soil resulting in numerous small subhorizontal terraces running along the hillsides. These terraces are commonly referred to as cow paths, and indeed they are extensively utilized by cattle because they are easier to walk on than the sloping surfaces. People also find them useful—farmers, who plant sugar cane and other crops along them, hikers, and geologists—and repeated use by cattle and men accentuates the terraces.

The terraces result from innumerable movements of small bodies of soil on rather poorly defined, curved slip planes, each body tending to rotate back toward the hillside as it slips downward, its upper surface thus producing a nearly flat shelf along the hillside (fig. 121). In

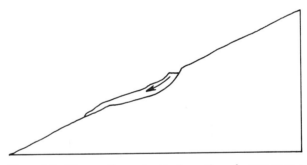

Figure 121. Diagram illustrating the formation of a terrace on a hillside by a soil slump.

general form and behavior they resemble the much larger landslides known as slump slides.

## ROCK FALLS FROM CLIFFS

In regions of any climatic type, where cliffs of fresh rock occur there are occasional falls of single rock fragments or showers of fragments from the cliff face. In Hawaii most of the high cliffs are either fault scarps or sea cliffs; of the fault scarps the most important are the walls bounding calderas and pit craters. Occasional rock falls from the caldera walls occur at any time, but during certain periods they become far more numerous. Both at Kilauea caldera and at Mokuaweoweo, on Mauna Loa, rock falls every few minutes have been observed during some periods of numerous earthquakes. Apparently the rocks are simply shaken loose by the agitation of the ground. At other times, however, frequent rock falls may take place without earthquakes to set them off. Thus, during the latter half of 1954 there were many rock falls from the walls of Halemaumau crater, although the period was one of seismic quiet.

On the high sea cliff between the base of the Kalaupapa peninsula and the mouth of Waikolu Valley, on the north coast of East Molokai, rock falls have been very numerous for many years. During some periods a falling rock can be heard every few minutes, and the passage along the boulder beach from the peninsula to the valley can be somewhat hazardous. Rock falls are less numerous, but nevertheless very common, along other sea cliffs.

## LANDSLIDES

In addition to the fall of individual fragments, large segments occasionally break off the cliff face (fig. 127). As the mass of rock descends, it breaks up into innumerable fragments which form a helter-skelter accumulation at the foot of the cliff. At least some of these falls are initiated by earthquakes. During the Maui earthquake of March 1938, many such slides occurred on the walls of Haleakala Crater.

In April 1868, during the most severe earthquake Hawaii has experienced in historic times, a large part of the sea cliff just northwest of Waimanu Valley, on the island of Hawaii, collapsed and rushed down the slope as a great rock slide which built a peninsula into the ocean at the foot of the cliff. The peninsula is nearly a mile wide, and despite trimming back by the waves, it still extends outward from the cliff base for nearly a quarter of a mile. It is possible, however, that part of the mass had been built by previous slides. A similar, and still larger peninsula lies a little farther southeast, between the mouths of Waimanu and Waipio valleys (fig. 185), and a similar landslide occurred on the sea cliff just southeast of Honopue Valley, also on Kohala

Mountain, during heavy rains in January 1941 (Stearns and Macdonald, 1946, p. 51).

Earth flows, also known as Culebra-type landslides after the famous Culebra Cut in the Panama Canal, have not been observed in Hawaii. These are movements of plastic materials, generally thick claystones or clay shales, in which slumping of the upper part is accompanied by plastic flowage farther downslope. At Culebra the flowage caused the bottom of the cut to bulge upward, blocking the canal. Slides of this sort are common in wet warm countries, such as Central America and the Philippines. The almost total absence of them in Hawaii no doubt can be attributed to the lack of thick accumulations of plastic materials. A small amount of movement of this sort took place during the slide at the Kailua Drive-in Theater, described later.

For convenience, landslides are divided into those that take place in unconsolidated or poorly consolidated material, such as colluvium, sand, clay, or deeply weathered rock, and those that involve a large proportion of solid hard rock. The former are known as *debris slides,* the latter as *rock slides.* Both are characterized by the separation of the overlying material from bedrock by a surface of shearing—the surface along which the slide material moves downhill. The force that drives the slide is simply the weight of the sliding material, pulled downward by gravity.

Before the slide starts, the material is held in place by cohesion within the mass, but as soon as the shear surface develops, the cohesion across the surface is lost and the only restraining force is the friction of movement. Slides form on slopes that are steep enough for the weight of the surficial material to overcome the cohesive force and establish the surface of shearing. In otherwise similar material the tendency to slide increases with increasing steepness of slope. The tendency is also increased by any addition of weight to the surficial material and by any reduction in the cohesiveness of the material. Addition of water to the mass promotes sliding both by adding to the weight and by reducing cohesion. When they are wet, clayey materials may become so plastic that they behave almost like liquids.

*189*

Once the shear surface has developed, water may serve to lubricate it, reducing the frictional resistance to movement.

Water may also abet sliding in another way. If the material above the potential shear surface is saturated to a considerable thickness, the water at the level of the shear surface may be under considerable pressure, owing to the weight of the water above it. Since the pressure in the water is hydrostatic and works in all directions, the water near the shear surface pushes upward, as well as laterally and downward, and tends to lift some of the weight of the overlying rock material. This reduces the pressure of the overlying mass on the shear surface, which in turn reduces the friction involved in movement. In this way, pore pressure of water is an important factor in initiating and maintaining movement of many slides.

The shear surfaces beneath landslides tend to be more or less spoon-shaped, with steep headward parts flattening downslope (fig. 122).

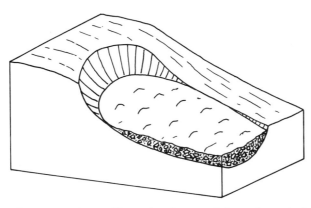

Figure 122. Diagram illustrating the general form of a typical landslide.

Movement of the slide downhill results in a roughly crescentic escarpment at its head. Some slides move more or less as a single block, but many break up internally as they move, and the material becomes much jumbled. Irregular disturbance of the surface of the slide caused by the irregular internal movements results in an irregular hummocky topography on the mass when it comes to rest. Extreme internal shearing of the slide mass may convert wet clayey materials to a highly plastic state and cause the slide to be transformed into a mudflow.

In uniform and structureless materials the shear surface forms a curve from the head to the toe of the slide which approaches a portion of a circle. The mass may simply move downward on that shear surface (fig. 123A), or it may rotate; and if it remains fairly intact the upper surface may be tilted backward at a marked angle to the headwall of the scar (fig. 123B). If the mass breaks up, however, the surface may become exceedingly irregular (fig. 123C). If bedding planes, or other structures

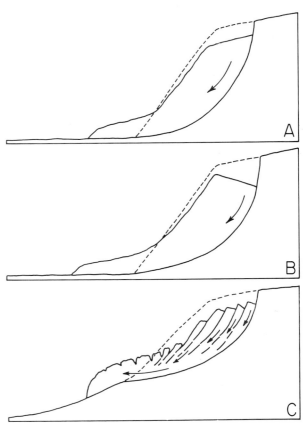

Figure 123. Diagrammatic cross sections illustrating the formation of landslides. Dashed lines represent the configuration of the hillside before the slide.

such as faults or joints, slope toward the toe of the potential slide they commonly modify the form of the shear surface, which then tends to follow these structures. This was the case in the Pali Highway slide of 1965, in which the shear surface followed in part the contact between the bedrock and overlying colluvium. Such structures sloping toward a valley are often planes of weakness, and are conducive to sliding. Many slides move on joint or bedding planes, or surfaces of unconformity coated

190

with clay that becomes slippery when it is wet.

For a number of years recurrent sliding has been a problem in an area on the eastern slope of the Waiomao Branch of Palolo Valley, in Honolulu. The sliding has a maximum width, near Kipona Place, of about 500 feet. The head is approximately at Kuahea Street, and the toe at the bottom of the valley—a total length of a little more than 600 feet. The slow movement has dropped the ground level several feet in places, breaking and tilting road pavements as well as tilting and distorting houses to the point that some of them have had to be abandoned. The slide is of the slump type, and is taking place in colluvium. The shearing surface bounding the slide at the bottom appears to be approximately 60 feet below ground level in the central part of the area.

The Palolo slide probably was caused originally by loading of the colluvium by waste from an old quarry which lies directly upslope, perhaps combined with removal of material at the lower edge by erosion by Waiomao Stream. In this wet valley the sliding certainly is abetted by water, which adds weight to the moving mass, lubricates the shearing surface, and reduces friction on the shearing surface through the lifting effect of pore pressure in the basal part of the mass. Recurrent movements of the slide show close correlation with periods of heavy rainfall. The water is added partly by infiltration of rain falling directly on the mass, and probably in part by drainage from farther upslope. There is little question that some is added also from water pipes broken by internal movements in the sliding mass. Several geologists and engineers who have studied the situation have recommended drainage of the mass itself and diversion of any flow of water from upslope, as measures that may put an end to the movement.

During November 1965, following several weeks of heavy rains, a dramatic and potentially very dangerous landslide took place on the Pali Highway that crosses the Koolau Range of Oahu. Its weight increased and its internal cohesion decreased by the rain water that had seeped into it, some 20,000 tons of rock and mud slid down across the divided highway just above the "horseshoe bend,"

covering all four lanes with a blanket of debris 20 feet thick and surging part way up the opposite bank. The movement actually consisted of three separate slides, each larger than the previous one. The first slide, on November 13, deposited about 1,000 cubic yards of debris on the highway, blocking the Kailua-bound lanes and one of the Honolulu-bound lanes (fig. 124). The second slide occurred the

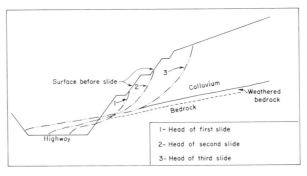

Figure 124. Cross section of the landslide on the Pali Highway during November 1965. (By Philip S. Hubbard.)

next day, and brought down about twice as much material, and the third, on November 16, spread about 8,000 cubic yards of debris completely across all four lanes. Miraculously, no cars were caught by the slides, but traffic was forced to detour for several weeks while the debris was trucked away.

A terraced cut 120 feet high had been made through a ridge composed of colluvium resting on the weathered surface of dense lava rock. The colluvium is itself partly decomposed by weathering, and consists of partly rotten rock fragments in a matrix of clay. When high steep cuts are opened in this type of material the hazard of landsliding is always great. In this case the likelihood was even greater because the contact between the colluvium and the underlying bedrock formed a surface of discontinuity sloping toward the roadcut, and this surface did indeed form part of the shear plane for the landslide.

Only part of the colluvium on the ridge above the highway was removed by the slide of 1965. An attempt was made to drain the remainder by means of horizontal drill holes and a trench across the top of the mass, but it was only partly successful, and the colluvium remained in an unstable position. By early

1966 a new arcuate crack was forming about 30 feet above the head of the landslide scar, showing that the material downslope from it was again moving. Not unexpectedly, then, in August 1967 another slide took place (fig. 125), again blocking the highway for several weeks.

During the same rainy period that caused the first Pali slide another high bank gave way on the north side of the excavation for the Kailua Drive-in Theater. Once again hundreds of tons of mud and rock poured out, this time over part of the parking area of the theater. Fortunately, several weeks before the slide occurred, the opening of large crescentic cracks above the top of the artificial cliff and several minor "back-rotated" slumps had served as warning that a bigger slide was imminent; therefore, part of the parking area was roped off and no one was injured when it came, although it took place during a movie performance. Continued cracking on this hillside probably presages more slides during future wet seasons. The theater excavation is in the very red soil and weathered rock of the Koolau caldera complex, and the slide area is in colluvium resting on bedrock containing numerous dikes. Both colluvium and weathered bedrock were involved in the slide.

MUDFLOWS

Several examples of mudflows, formed in different ways, have been recognized in the Hawaiian Islands, and a few have occurred in historic times. The largest of the historic mudflows took place in 1868, during the same violent earthquake which caused the rock slide near Waimanu Valley. On the slopes just southwest of Wood Valley, in the Ka'u District of the island of Hawaii, a thin bed of basaltic lava lies on a layer of volcanic ash (Pahala ash) as much as 50 feet thick. The ash is partly altered to a mixture of bauxite and iron oxide, and on taking up water exhibits "thixotropic" properties—that is, so long as it is not disturbed

Figure 125. The landslide that blocked the Pali Highway in August 1967.

it behaves as a solid, but if it is agitated it assumes the behavior of a liquid and flows like one. Late March 1868, had been a time of heavy rains, and the ash was thoroughly saturated with water. On April 2 came the great earthquake, and the resulting agitation of the ash transformed it into fluid mud which flowed down the hillside as two main streams, carrying with it fragments of the broken-up overlying lava bed. The smaller of the two branches, which rushed into Wood Valley, was more than a mile long. The larger branch, 2 miles long and more than half a mile wide, swept down a smaller valley just to the southwest. About 500 domestic animals, and a village with 31 persons, are said to have been buried (Alexander, 1899, p. 292). The unsorted debris is well exposed in the walls of a stream gully that crosses the Wood Valley road 3.5 miles north of Pahala.

The largest single mass of mudflow debris thus far discovered in Hawaii crops out near Kaupo, on the south coast of East Maui. The material, most of it buried by more recent lava flows, is more than 350 feet thick and is exposed in patches over a width of 2.5 miles. Some blocks of rock as much as 50 feet across are included in the mass, which consists largely of coarse angular debris, with a very subordinate matrix of sand and silt. The fragments are of rock types exposed in the summit region of Haleakala. The "mud" probably originated in the head region of Kaupo Valley, by mobilization of talus during exceptionally heavy rains, and flowed down the valley to its present position (Stearns and Macdonald, 1942, p. 107).

Many voluminous mudflows contributed to the Palikea formation, which lies against the slopes of the central highland of Kauai. The reason why so many large mudflows occurred in one specific short span of time, represented by the deposition of the Palikea formation, can only be conjectured, but it may be related to the widespread destruction of vegetation and consequent very rapid erosion of the weathered mantle rock at the time of renewed volcanic activity on Kauai (Macdonald, Davis, and Cox, 1960, p. 81).

Of somewhat similar origin is the mudflow that blocked the Pukele and Waiomao tributaries of Palolo Valley, on Oahu. The volcanic activity that accompanied the formation of Kaau Crater, in the head of Palolo Valley, covered the adjacent slopes with unstable pyroclastic material ranging in size from bombs to fine ash. Heavy rains saturated this material with water and transformed it into a slurry that flowed down into and along the valleys on each side of the crater, carrying with it blocks of older Koolau lavas picked up from the ground it traversed. (Stearns and Vaksvik, 1935, p. 124.)

## LANDSLIDE-DAM LAKES

Occasionally a landslide blocks the course of a stream and forms a dam, behind which the water accumulates as a lake. A pond of this sort, 100 yards long, was formed in the West Branch of Honokane Nui Valley in the Kohala Mountains of Hawaii in 1942. A still larger pond was formed by a landslide during prehistoric times in the East Branch of the same valley, and 30 feet of stratified silt and clay was deposited in it. These ponds are always very short lived, being quickly drained as a channel is cut through the loose material of the dam by the stream spilling over it. This drainage itself can be catastrophic. In 1925 a great landslide blocked the Gros Ventre River, in Wyoming, forming a lake nearly 5 miles long. In the spring of 1926 flood waters overflowing the landslide dam eroded it so rapidly that the level of the lake was lowered 50 feet in about 5 hours, and the resulting flood in the valley downstream destroyed part of a town and killed several people. Fortunately, we have had no floods of this sort in Hawaii, nor are they likely to occur.

193

*Suggested Additional Reading*

Eckel, 1958; Legget, 1962, pp. 385–443; Sharpe, 1938; Wentworth, 1943; White, 1949.

*Near Honaunau,*
*Kona Coast, Hawaii*

# CHAPTER ELEVEN

# Work of the Ocean

THE OCEAN SERVES AT ONCE AS A BARRIER AND BUFFER, and as an avenue of communication between the Hawaiian Islands and the rest of the world. It is a source of food for the islands' human population, and makes possible a fair-sized fishing and packing industry. To a large extent it controls the climate, and it contributes the water that falls as rain on the islands, directly or indirectly making agriculture possible. Its beaches are world famous as places of recreation, and for their beauty. Geologically also, the ocean is vitally important to Hawaii. Through the millennia it constantly gnaws away at the island shores, and the great cliffs it thus creates are among the most spectacular features of island landscape. It alone of the natural agents of erosion is ultimately capable of destroying the islands created by the volcanoes; but it will leave in their places memorial monuments of limestone.

The ocean has been aptly termed "the last frontier." It is the least-known part of the earth's surface—a part making up more than 70 percent of the whole. In spite of very rapid advances in our knowledge of the ocean, especially since the second world war, there is still a vast amount to be learned about it, and its study is one of the most fascinating and rewarding lines of geologic research today. However, the shore processes that directly affect the islands have been a subject of intensive study for more than a century, and they are now fairly well understood. It is these shore and near-shore processes that are dealt with in the following pages.

## PRINCIPLES OF MARINE EROSION AND DEPOSITION

As with streams, ice, and wind, the energy by which the ocean erodes the land and transports the debris produced by its own erosional work and carried into it by streams is kinetic energy—the energy of motion. Moving ocean water takes two forms: waves and currents. Currents are simply water flowing from one part of the ocean to another. Some are caused by tides, some by wind, some by the rotation of the earth, some by differences in density of the water from place to place. Differences in density are in turn caused by differences in temperature, in salinity, or in the amount of sediment suspended in the water. Cold water is heavier than warm, salt water is heavier than fresh, and sediment-loaded water is heavier than clear water. Heavier water always tends to get underneath lighter water, and in doing so it creates a density current. Currents of dense, sediment-laden water are believed to move downslope into the deep ocean fast enough to erode submarine canyons, and at times to break submarine telephone cables.

Most ocean currents move slowly—less than half a mile an hour—but some tidal currents attain speeds as great as 10 miles an hour. Currents erode the shore, and more particularly the shallow sea bottom, much as do streams, by abrasion of the bedrock by the rock fragments being transported. As in streams, the fragments may be transported by rolling or sliding along the bottom, by saltation, or in suspension. Currents are more important in transporting and depositing material than in eroding.

Some erosion is accomplished by marine organisms. Such animals as sea urchins and certain types of clams bore holes into hard rock. Roundish holes up to several inches across, made by sea urchins, are conspicuous on some of the benches cut by waves on Hawaiian shores. Seaweeds may aid in the transportation of debris by anchoring themselves onto cobbles and drifting them along as the plant itself is moved by waves and currents. Overall, however, the erosional effects of organisms are minor.

Waves are the most important agent of marine erosion. In the open ocean a wave consists merely of an oscillation in the water. The water particles move in circles a few feet across, and there is no forward movement of the water as a whole. This is demonstrated by the fact that a piece of floating timber merely goes up and down on the passing waves and is not carried along with them. These are known as *waves of oscillation.* In shallow water, however, they are transformed into *waves of translation,* in which the whole mass of water actually moves forward. As it advances toward the shoreline the wave of translation grows steeper and leans forward, as though eager to

reach its destination, and eventually its top curls into the familiar "breaker." Striking shore, the breakers are known as "surf." Even during calm weather the waves strike exposed coasts with considerable energy—a fact to which the annual toll of opihi (limpet) gatherers and shore fishermen along Hawaiian coasts bears tragic witness. And during storms the wave attack becomes far more severe.

Waves erode by *abrasion* and also by *hydraulic action.* The latter is simply the pressure exerted by the water, either directly or through trapped and compressed air. Rapidly moving storm waves may exert pressures of several tons per square foot. Single fragments of rock weighing thousands of tons may be picked up and moved unbroken for distances of several hundred feet, and smaller fragments may be thrown with great force against the land. Probably the most spectacular results of wave action have been reported from the traditionally stormy North Atlantic Ocean. Thus in 1872 at Wick, in Scotland, a storm tore loose a concrete monolith weighing 1,350 tons from the end of the breakwater and dropped it inside the harbor. Annoyed, the Scots replaced it by one weighing 2,600 tons; but this one lasted only 5 years before it too was torn away. At Dunnet Head, in northern Scotland, stones have been thrown through windows 300 feet above sea level. Wave effects on Hawaiian shores are less extreme, but still very impressive. Blocks of lava rock up to 12 feet long, and weighing as much as 15 tons, have been thrown by waves onto the shore platform, in some places 30 feet above sea level, along the Puna coast of the island of Hawaii.

The impact of waves dislodges many rock fragments, some of them large. Also important is the compression by waves of air trapped in fractures and hollows in the rocks. Pressures may develop in this air sufficient to thrust the enclosing rocks apart. The compression of air by waves is well illustrated by the blowholes (more properly called spouting horns) along Hawaiian coasts, which result from compression of air in lava tubes or sea caves at water level as waves move into them. The air escapes through a hole in the roof of the cave, often driving ahead of it a jet of water and carrying spray high into the air. At the Lawai blowhole, on the south shore of Kauai (fig. 133), jets of spray sometimes spout well over 100 feet into the air. A still larger blowhole was destroyed by dynamite, because the spray blowing inland damaged nearby cane fields.

Sand, pebbles, and boulders are moved back and forth by the waves and abrade the underlying rock. At the same time, of course, they are themselves abraded. Most of the rock fragments brought into the ocean by streams are already somewhat rounded, but further rounding is done by the ocean. Those dislodged by hydraulic action by the waves are at first angular, but under constant abrasion they also soon become smoothed and rounded. On beaches exposed to heavy wave action angular rock fragments may become quite well rounded within a year.

Part of the abrasion on a shoreline is the result of rock fragments being thrown against the shore rocks by the waves, the impact chipping or shattering both the fragment and the bedrock. Particularly during storms, fragments may be hurled by the waves with tremendous violence. As in the case of streams, more erosion may be accomplished during the brief interval of one exceptionally heavy storm than in many years of ordinary weather. The rate of wave erosion on exposed coasts may be surprisingly rapid. Part of the southeastern coast of England has been worn back more than a mile since the time of the Norman Conquest. Because Hawaiian rocks are more resistant to wave erosion, the rate of shoreline retreat is much less; but even here, according to estimates by C. K. Wentworth, the island shorelines are being cut back at an average rate of a couple of inches a year.

## SHORELINE FEATURES

The attack of the waves is directed horizontally and concentrated in a zone within a few feet of sea level. As the shoreline is gradually cut back there is produced a relatively smooth surface approximately at sea level at the shoreline and sloping gently seaward. This is the *wave-cut*

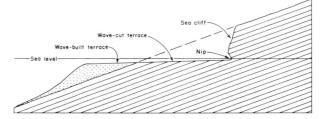

Figure 126. Diagram illustrating wave-cut and wave-built terraces, nip, and sea cliff.

*terrace* (fig. 126). Rock debris is transported across the wave-cut terrace by currents and returning waves, but at its outer edge, where the currents encounter deeper water, their velocity is checked and they tend to deposit their load. In this manner the wave-cut terrace may be extended outward by a bank of sediment known as the *wave-built terrace* (fig. 126). Along many shores where the ocean bottom slopes rapidly into deep water the wave-built terrace is inconspicuous or absent. This is the case around most Hawaiian shores, where we commonly have a wave-cut terrace with little or no wave-built terrace.

The nearly horizontal cutting by the waves into the sloping edge of a land mass produces at the shoreward edge of the wave-cut terrace a *sea cliff* (figs. 126, 127, 130–132). On a gently sloping coast the cliff may remain low even through long periods of wave attack forming a broad wave-cut terrace, whereas along steeper coasts the cliff quickly becomes high. Along the north side of East Molokai the sea cliffs reach a height of 3,600 feet (fig. 232), and those on the northeast side of Kohala Mountain, on Hawaii, are as high as 1,400 feet (figs. 183, 185). Along the northeast coast of Mauna Kea the cliffs are only 50 to 350 feet high, because the mountain is younger and the waves have not had as long to work. Parts of Mauna Loa and Kilauea are so young that almost no cliff has as yet been cut. On the average, sea cliffs are much higher on the northeast side of an island than on the southwest, because the northeast side, facing the prevailing winds, also receives the heaviest wave attack. An exception is the island of Lanai (fig. 222), the northeastern side of which is shielded by Molokai, whereas the southwest side (fig. 127) is exposed to southwesterly ("kona") storms.

197

Figure 127. Kaholo Pali, on the southwest tip of Lanai, is a nearly vertical, wave-cut cliff rising 1,000 feet above the shore. Huge waves generated by southwesterly storms pound this side of the island. Rock slides, like the one in the foreground, are a principal mechanism of cliff retreat. A sill is visible intruded into the lavas in the cliff a short distance to the right of the slide.

The spectacular cliffs of the Napali Coast, on the northwest side of Kauai (figs. 131, 262), are much larger than any along the northeast coast. This is because the northeastern slope has been much more deeply eroded by streams, and also because of the complications resulting from voluminous posterosional volcanic eruptions on the northeastern side.

The concentrated horizontal attack by waves at the base of the sea cliff cuts a notch, known as a *nip* (figs. 126, 128), which ranges from a few inches to 10 or 15 feet deep. As the nip grows larger the rocks higher in the cliff are undermined and from time to time drop off and become fresh tools for the waves. Retreat of the cliff shoreward is very largely the result of caving away of its face, caused by this process of continuous undermining, and the slope remains a steep cliff as long as it is being actively cut back by the waves.

Wave-cut nips are conspicuous along many parts of the Hawaiian shorelines. They are especially well displayed at Laniloa Point, near Laie, on Oahu, where they are cut in limestone formed by the consolidation of dunes of limy beach sand. They are also well shown along former shorelines that have been elevated above sea level, such as those just inland from Waimanalo, on Oahu.

Figure 128. Nip cut by waves about 5 feet above present sea level, near Koko Head, Oahu.

198

On an irregular shoreline wave attack is not uniform, but is stronger on the headlands than in the embayments. As a result the headlands are cut back faster than the bay heads (which often are actually built out by deposition) and the shoreline is gradually straightened. Most Hawaiian shorelines that have been long exposed to wave attack are straight or gently curved.

At points of weakness in the rocks the waves may cut into the base of the cliff more rapidly than into the adjacent rocks, producing a *sea cave* (figs. 136, 216). Many of the sea caves in Hawaii are lava tubes enlarged by wave action. Others get their start in layers of relatively easily eroded aa clinker. Examples of sea caves, cut when ocean level was higher than now, are the wet and dry caves at Haena, Kauai, and the Makua cave on the Waianae coast of Oahu. Occasionally, waves attacking the sides of a promontory or a small island may cut a cave completely through it, producing a *sea arch* (figs. 129, 134). A sea arch extends through an islet off Laniloa Point, Oahu. The Onomea Arch (fig. 135), just north of Hilo on the island of Hawaii, was cut through an old cinder cone

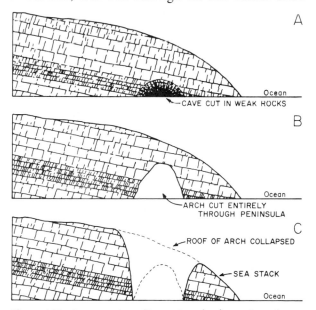

Figure 129. Cross sections illustrating the formation of a sea arch and a sea stack. *A,* Waves cutting more rapidly in an area of weak rock on the side of a peninsula erode out a sea cave; *B,* later the cave may extend all the way through the peninsula, forming a sea arch; and *C,* eventually the roof of the arch collapses, isolating the former end of the peninsula as a sea stack.

on the slope of Mauna Kea when the ocean stood about 25 feet higher than at present, but its top collapsed in 1958. Collapse of the top of a sea arch that is still at sea level may leave the former outer end standing as an isolated rock projecting out of the water. Such rocks, left by the collapse of an arch or simply isolated from the main shoreline by erosion, are referred to as *sea stacks* (figs. 129C, 130, 216, 232). Paoakalani and Mokupuku islands off the Kohala coast of Hawaii; the Mokulua Islands, Mokolii Island, and other small islands and rocks off the coast of Oahu; Mokapu, Mokoleia, and Mokohola islands off the north coast of Molokai, and others off the west coast of Lanai are examples of sea stacks.

It has been pointed out in an earlier chapter that the wearing away of an island (or other land surface) by a stream is limited by the minimum slope that is necessary to give the stream enough velocity to transport solid fragments. Horizontal cutting by waves also has a limiting factor—the loss of energy as the waves cross ever broader areas of shallow wave-cut terrace. However, the width of terrace that waves can cross and still retain cutting and transporting power is very great—certainly many miles. As a result, in time the waves may completely cut away an island, leaving in its place only a shoal. These shallow platforms are ideal environments (other factors being favorable) for the growth of coral, and in tropical and warm temperate regions they are usually covered with coral reef (discussed in a later section of this chapter). A good example in the Hawaiian area of a former island decapitated by wave erosion is the shoal surrounding Kaula Island (fig. 279). The shoal is a nearly horizontal platform, about 7 miles across, and with an average depth of water less than 200 feet, cut across the top of an old shield volcano (fig. 137A). Kaula Island itself (fig. 12) is a tuff cone built on the submerged eroded platform long after the major volcano ceased activity.

Farther northwestward, the Gardner Pinnacles (fig. 283) and La Perouse Rocks rise from similar platforms cut across the tops of ancient shields. The rocks are the tiny remnants of once-large volcanic islands.

Figure 130. Sea cliff and sea stacks on the west shore of the island of Lanai at Nanahoa. The stacks were left as the surrounding rocks were eroded away.

## BEACHES

Along the shoreline, rock fragments form temporary accumulations known as beaches. Due to wave and current action the finer debris (silt and clay) tends to be separated from the coarser fragments and carried outward across the wave-cut terrace to be deposited in deep water beyond. The coarser material (sand and gravel) tends to remain behind and is transported by alongshore currents until it reaches a sheltered place where the strength of the current is reduced. There it is dropped. Where wave action is strong, even the sand is removed leaving beaches of pebbles, cobbles, or boulders. Cobble and boulder beaches lie along the foot of the sea cliffs on the north side of

Molokai and on the Hamakua Coast of Hawaii, and at many other places in the Islands. However, typical beaches consist largely of sand (particles between 0.0025 and 0.06 inch in diameter). On continental beaches the sand is largely the product of erosional attack by waves and streams on the land mass. Some of the sand in Hawaiian beaches also is the product of erosion of the volcanic rocks, but much of it is calcareous (limy, $CaCO_3$) material of organic origin—fragments of coral and mollusk shells torn by the waves from the reefs offshore, and the limy shells of tiny marine animals (particularly the single-celled animals known as foraminifera). The white or cream-colored beaches, such as those of Waikiki and Kailua on Oahu, Poipu on Kauai, Kihei on

Maui, and Hapuna on Hawaii, are very largely calcareous sand.

Some black sand beaches, for example those of Kalapana (Kaimu) and Punaluu on Hawaii, consist of glassy volcanic debris from littoral explosions (see page 44). Others, however, such as some on the south shore of Molokai, consist of the grains of the heavy black minerals magnetite and ilmenite, eroded out of the lava rocks. Still other black, gray, and brownish gray beaches consist largely of fragments of lava rock. A good example of the latter is the beach at the mouth of Waipio Valley on Hawaii.

Beaches composed partly or largely of green sand are found at a few places in the Hawaiian Islands. Perhaps the best examples are at Hanauma Bay on Oahu, and Papakolea, 3 miles northeast of South Point on Hawaii. At both places the sand consists predominantly of green crystals of olivine, separated out of the volcanic rocks by erosion. At Hanauma Bay the rock supplying the olivine is tuff belonging to the row of cones extending northeastward from Koko Head. At Papakolea the olivine is derived from Puu Mahana, a littoral cinder cone formed where an ancient aa lava flow entered the ocean.

Sand being drifted alongshore may build a beach extending out partly across the entrance of a bay, thus forming a *spit*. If, in time, it nearly or completely closes the bay entrance it is called a *bay-mouth bar*. The beach blocking the entrance to Waimea Valley on Oahu is of this sort. Sometimes bars build across from shore to an island, or from one island to

Figure 131. The Napali cliffs, Kauai. A view looking southwestward along the wave-cut cliffs and across stream canyons on the northwestern coast.

Figure 132. Sea cliff on the north coast of East Molokai. The small cove in the tip of a ridge provides an example of a shoreline feature termed here a "horned crescent"—a distinctive concave wave-eroded ridge ending. This type of wave-cut embayment usually occurs along cliffed coastlines that are exposed to heavy wave action. Note the sheer sea cliff at the right, beyond the spur.

another, forming a *tombolo.* The small island, Mokuaeae, at the foot of the sea cliff just west of Kilauea lighthouse on Kauai, is tied to shore by a tombolo.

On very gently sloping shorelines long ridges of sand may accumulate some distance offshore, projecting slightly above sea level and separated from shore by lagoons. These are known as *barrier beaches.* They extend along much of the eastern coast of the United States from Virginia to Florida. We have none in Hawaii, although ridges on the submerged platform northwest of the Napali Coast of Kauai may be old barrier beaches on a wave-cut terrace formed during a former lower level of the sea.

Beaches change constantly in size and profile

with changing conditions of waves and currents. Under storm conditions they generally are cut back. The white sand beach at Kahaluu, 4 miles south of Kailua on the Kona Coast of Hawaii, almost completely disappears during some storms. They may disappear altogether, but more commonly only the outer part is swept away and the seaward slope of the beach is increased. Some of the sand that is carried away remains in shallow water on the wave-cut terrace; the rest is carried beyond it and dropped in deeper water. When calm weather resumes most of the sand on the wave-cut terrace is moved back onto the beach. Thus, beaches are cut away during storms and rebuilt during normal calm weather.

The sand carried out into deep water during

storms is forever lost from the beach. Ordinarily there is a continuous supply of new eroded material or organic debris to replace the sand lost into deep water, and so our beaches remain much the same year after year. But if a continuing supply of material is not available, the beach gradually becomes smaller and smaller as sand is lost, and eventually disappears. This is happening at the beaches of Kalapana and Punaluu, mentioned above, where the black volcanic glass sand originated in each case from a single littoral explosion where an aa lava flow entered the ocean. There is no continuous supply of this sand, and the beaches will eventually disappear. The Kalapana beach has shrunk considerably in the last 20 years. Part of the loss, by sand being washed into deep water and blown inland by the wind, is unavoidable, but removal of sand from the beach for concrete and other such purposes

must be avoided if we are to preserve as long as possible some of our finest scenic attractions.

## CLASSIFICATION OF SHORELINES

A new shoreline may be formed by building of a new land mass, as by a volcano *(constructional shoreline)*, or by uplift of the land and the adjacent sea floor bringing former sea bottom up into a position to be attacked by the waves *(shoreline of emergence)*, or by lowering of the land relative to sea level bringing down into reach of the waves a land surface already dissected into valleys and ridges by stream erosion *(shoreline of submergence)*. Shorelines of emergence are apt to be rather straight and smooth because the surface exposed to wave attack is relatively smooth, near-shore sea bottom, often the surface of a wave-cut terrace. In contrast, shorelines of

Figure 133. Blowhole, or spouting horn, near Lawai, on the south coast of Kauai. The hole is on a wave-cut bench about 5 feet above sea level.

*203*

Figure 134. Sea cliff and sea arches cut by the waves in Kilauea lava flows on the south coast of Hawaii. Note the boulders at the base of the cliff which may serve as "battering rams" during times of very high surf. Some boulders estimated to weigh as much as 15 tons have been thrown onto the flat above the cliff in the distance.

submergence are commonly very irregular because the sea extends as embayments into the stream-eroded valleys, with intervening headlands on the inter-valley ridges. Many shorelines are partly of one origin and partly of another. Thus the California coast includes both shorelines of emergence and shorelines of submergence. In Hawaii we find constructional shorelines on the younger volcanoes, and, on the older ones, complex shorelines in which, however, submergence is dominant. This is because, although ocean level has moved both upward and downward in relation to land in

Hawaii, the greatest change has been a relative sinking of the islands. It should be emphasized that, so far as the shorelines are concerned, it is immaterial whether the land has stood still and the sea level changed, or the sea level has stood still and the islands have moved up or down. It is the *relative* movement of land and sea level that is important.

CHANGES OF SEA LEVEL

The rise or fall of sea level in relation to the land, producing shorelines of emergence or submergence, may result either from actual uplift or lowering of the land mass brought about by deformation of the earth's crust, or from changes in the ocean level itself, caused by increase or decrease in volume of water in the ocean or change of total capacity of the ocean basin. Changes due to deformation of the earth's crust are known as *tectonic* changes, whereas those due to change in ocean level alone are called *eustatic*.

Large parts of the continental coasts have undergone recent tectonic changes of level. In these areas the amount of change differs markedly from place to place, and a former beach line or wave-cut nip uplifted above sea level is often conspicuously tilted. The principal evidences of the former shorelines are wave-cut terraces and nips, and beach deposits containing the remains of marine organisms. The level at which a nip and the inner edge of a wave-cut terrace are originally formed may vary a little from place to place, owing to differences in exposure to the waves, but if the level of an uplifted shoreline varies more than a few feet it must have been tilted or warped by crustal deformation. Correlatively, it is highly unlikely that tectonic deformation of the crust over a very wide area would produce a completely uniform uplift or sinking. Broadly speaking, tectonic changes of level must always be irregular. Therefore, when we find changes of level of the land relative to the sea that are regular and constant over large portions of the earth's surface, we conclude that the changes must be eustatic—the result of change of sea level with the land remaining stationary. We do find evidence of such uniform and constant changes of level over the whole mid-Pacific, and probably over the whole area of the earth's oceans.

Figure 135. Onomea Arch, a sea arch, the top of which has collapsed since the photograph was taken. The arch was cut by waves through a narrow peninsula on the Hamakua Coast north of Hilo. The peninsula is the remains of a cinder cone on the east rift zone of Mauna Kea. The floor of the arch, partly covered by collapsed debris in the picture, was approximately 25 feet above sea level, showing that the arch was cut during the Waimanalo (25-foot) stand of the sea.

Table 12. Ancient shorelines in the Hawaiian Islands

| Approximate altitude (feet) | Name of shoreline | Age Geologic | Years before present (approximate) |
|---|---|---|---|
| 0 | Present | Recent | — |
| −15 | Koko[a] | | — |
| −180[b] | Penguin Bank | | — |
| −350 | Mamala | | 17,000 |
| 2 | Manana | | — |
| 5 | Kapapa | | 26,000 |
| 12 | Ulupau | | 32,000 |
| 25[c] | Waimanalo | | >38,000[d] |
| −40 | Waipio | | — |
| 45 | Waialae | | — |
| 70 | Laie | Pleistocene | — |
| 95 | Kaena | | — |
| −300 | Kahipa | | — |
| 55 | Kahuku | | — |
| 250 | Olowalu | | — |
| 325 | — | | — |
| 375 | — | | — |
| 560 | Manele | | — |
| 625 | Kaluakapo[e] | | — |
| 1,200 | Mahana | | 350,000 |
| −1,300 | Waianae | Pliocene-Pleistocene (?) | 1,500,000 (?) |
| −1,200 to −1,800 | Lualualei | Late Miocene | 11,000,000 |
| −3,000 to −3,600 | Waho | Late Cretaceous (?) | 80,000,000 (?) |

Source: Modified after Stearns, 1966a, p. 17.
a. Named by Easton, 1965.
b. Placed at approximately 160 feet below sea level by Ruhe and others, 1965.
c. Two shorelines, respectively 22 and 27 feet above sea level.
d. Oyster shells at this level near Pearl Harbor give a C-14 age greater than 38,000 years, according to Ruhe and others, 1965.
e. Named by Easton, 1965, but originally recognized by Stearns, 1938.

The most recent list of shorelines in Hawaii, based largely on the work of H. T. Stearns, is given in table 12.

The shorelines which are above present sea level are marked by wave-cut terraces, nips, and sea cliffs, and by deposits of sediment containing marine fossils. Those below sea level are marked by topographic benches representing submerged wave-cut terraces, by sand dunes that could only have been built by the wind above sea level but that now extend below sea level, and by stream-cut valleys now clogged with alluvium that extend far below sea level but which may be accordant in level with one of the submerged benches. Coral reef was encountered in a well in Waikiki, just west of Diamond Head, between 1,005 and 1,055 feet below sea level, and, since reef coral can grow only in water less than 200 feet deep, submergence of the shore of the island must have amounted to at least 800 feet. Other coral reefs lie as much as 1,600 feet below sea level.

Some of the benches that mark the uplifted shorelines around the island of Oahu are the original surfaces of fringing reefs like those now growing around the island (see the following section), very little modified by erosion. This is especially true of the widespread Waimanalo shoreline bench that forms much of the Ewa Plain (fig. 239), the flats near the mouths of Lualualei and Waianae valleys, the coastal plain in the Kahuku area, and the flat surface of the Mokapu Peninsula southwest of Ulupau Head. The same reef surface underlies much of the Honolulu area, but there it has been partly buried by recent volcanic rocks and alluvium.

The benches at the plus-12-foot and plus-5-foot levels (figs. 12, 128, 133) were formed partly by wave erosion of the older Waimanalo reef, and partly also by building of new reef.

On Lanai fossiliferous marine limestone as much as 150 feet thick extends up to 550 feet altitude (in Kawaiu Gulch), and calcareous conglomerate containing many shell fragments and foraminifera is found as high as 561 feet. These deposits are believed to mark a former shoreline at about 560 feet altitude. The evidence of a 625-foot shoreline is less definite. Blocks of marine limestone are found at 625 feet altitude on the southwest side of Kaluakapo Crater on Lanai, and, although the blocks are not in place, they are so numerous that they must be close to their point of origin. Perhaps, as some have suggested, they were carried there by ancient Hawaiians, but they are present in great numbers and there is no apparent indication of any ancient structure in which they might have been used. Although the evidence is inconclusive, it seems probable that the island was at one time submerged to the 625-foot level.

On Kahoolawe a conspicuous change takes place in the character of the surface along an essentially horizontal line approximately 800 feet above sea level (fig. 221). Above that line the surface is covered by a thick capping of soil, but below it the soil has been completely stripped away, presumably by wave erosion during higher stands of the sea. The line is marked by residual boulders and cobbles, apparently rounded by wave action, but no coral limestone or fossils have been found (Stearns, 1940c, p. 147). No definite high-level shoreline has been recognized at 800 feet

207

Figure 136. Koko Head, Oahu. A sea cave is being cut by waves in the fairly soft, thin-bedded tuff. Note the fretwork caused by salt-spray weathering above the zone of active wave erosion.

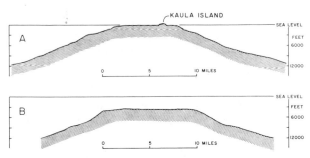

Figure 137. *A*, Profile across Kaula Island, showing the broad wave-cut platform truncating the shield volcano and capped by a tuff cone. It is very similar to *B*, a profile across a typical guyot. (*B* is after Hess, 1946.)

altitude on any of the other islands. This fact suggests that Kahoolawe may have experienced vertical movements of the island mass itself that were independent of any movements that may have taken place on the neighboring islands; and this in turn leads to the speculation that all of the shorelines above 250 feet altitude, and perhaps to a lesser degree those at lower levels, may have been affected by vertical movements of the islands that differed from place to place.

Veinlets of limestone in basalt at 1,069 feet altitude on Lanai contain fragments of coral, coralline algae, and marine mollusk shells (Stearns, 1938). These veinlets probably were formed from calcareous sand which sifted down into cracks in the basalt, and possibly the sand was blown inland from a beach at a lower level, as has happened on parts of the island during recent times. However, when Lanai is viewed from a little distance, a distinct line of demarcation can be seen extending around the island at about the 1,200-foot level. This line separates a zone above in which there is deep red lateritic soil from one below in which the soil has been eroded away leaving numerous boulders and cobbles scattered over the surface. It is believed that the soil was removed by wave erosion when the sea stood about 1,200 feet above its present level (Stearns, 1938). The sharpness and uniformity in altitude of the line are difficult to explain in any other way.

More details of the evidence for the various shorelines above present sea level are given in chapter 19, on geology of the individual islands.

At many places on the coasts of several of the

Hawaiian Islands, lithified sand dunes extend below sea level. Since sand dunes are formed only by wind above water level, these dunes, now under water, must have been formed when sea level was lower than it is today. Usually it is impossible to determine the level of the sea at the time the dunes were formed, but Stearns (1935) has shown that those at Waipio, on Oahu, rise from a platform located about 60 feet below present sea level. Ruhe, Williams, and Hill (1965) state that the platform slopes from present sea level and has an average depth (based on a statistical analysis of 275 submarine profiles around Oahu) of close to 40 feet. The latter figure is adopted in table 12.

A very prominent bench lies approximately 300 to 360 feet below sea level at many places around the Hawaiian Islands. There is little doubt that it is a terrace cut by waves when sea level was about 300 feet lower than now. Furthermore, the bench is so extensive that it is quite likely the product of several different periods of erosion at about the same level. At least two withdrawals of the sea to about minus-300 feet are indicated (table 12). The older of these was named by Stearns (1935) the Kahipa stand of the sea. Evidence for a later withdrawal to the minus-300-foot level was first reported by Ruhe, Williams, and Hill (1965), who named it the Mamala stand. Analysis of offshore profiles around Oahu by the latter workers indicates that the average depth of the shelf is close to 350 feet, and again this figure based on recent precise statistics is used in table 12 rather than the older generalized figure of 300 feet. Stearns (1966) suggests that it be called the Mamala-Kahipa shelf to emphasize that it was formed during two separate periods of erosion.

Another prominent shelf is found around the island of Oahu at depths of about 1,200 to 1,800 feet, and very probably it too is a terrace or succession of terraces cut largely by wave action. It has been called by Stearns (1938) the Lualualei shelf. Analyses of offshore profiles by Ruhe and his co-workers (1965) indicate to them that the shelf varies greatly in depth, rising from more than 2,000 feet off Makapuu Point to only 1,140 feet off the mouth of Pearl Harbor, then descending again to 1,700 feet off

Barbers Point. Thence it rises along the western coast to Kaena Point, and continues to rise along the northern coast until it reaches a high level of only 800 feet off Kahuku Point. If this actually is all the same wave-cut shelf there can be no question that it has been warped by tectonic movements, as these investigators suggest; but the continuity of the shelf between successive profiles is not entirely clear, and it is at least possible that more than one wave-cut terrace, formed during more than one period of erosion, are present. Other evidence points to the existence of at least two former shorelines within this general depth range.

Wells in Waianae and Lualualei valleys penetrated stream-deposited alluvium to depths as great as 1,200 feet before they entered lava rock, indicating that the bottoms of the stream-cut valleys are now approximately 1,200 feet below sea level. Since the valleys could only have been cut above sea level, this is incontrovertible evidence that the island has sunk at least that much relative to sea level since the alluvium was laid down. This fact was recognized as early as 1915 by W. A. Bryan. However, the wells are inland from the present shoreline, and even further inland from the position the shoreline would have if the island were raised 1,200 feet. The valley bottom at the ancient shoreline would have been somewhat lower than at the present position of the well—just how much lower we cannot say, because we do not know how far the shoreline was from the position of the well, nor the slope of the ancient valley bottom. Stearns (1935) correlated the erosion of the alluviated valleys of the Waianae Range with the Lualualei shelf. But reef-forming corals and other fossils of very late Miocene age, 11 or 12 million years old, have been dredged from the Lualualei shelf at depths of 1,550 to 1,700 feet about 5 miles southwest of Honolulu Harbor (Menard, Allison, and Durham, 1962). Since the Waianae valleys are cut into late Pliocene rocks (appendix B), some of which are no more than 2.7 million years old, the cutting of the valleys must have occurred long after the Lualualei shelf was formed. Furthermore, the part of the shelf from which the fossils were collected cannot have been at or very close to sea level at

any time since it was formed, because if it had, the late Miocene reef would have been either eroded away or covered with later reef. It must have remained always below the depth of wave erosion and active coral growth. Thus the shoreline to which the Waianae valleys were cut must lie more than 1,200 feet, but considerably less than 1,600 feet, below present sea level. In table 12 we have indicated it at about minus-1,300 feet, and called the corresponding level the Waianae stand of the sea.

Recent soundings and exploration with a miniature submarine have revealed the presence off the Waianae coast of three well-defined terraces (Brock and Chamberlain, 1968). The two shallower of these, one at a depth of approximately 200 feet, and the other at 230 to 395 feet (fig. 138), are believed to corre-

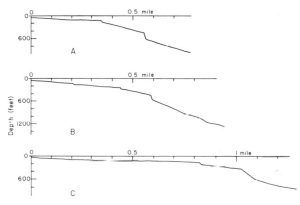

Figure 138. Profiles of the ocean floor off the Waianae coast, showing the terraces believed to correlate with the Penguin Bank, Mamala, and Waianae stands of the sea. (After Brock and Chamberlain, 1968.)

spond to the Penguin Bank and Mamala shorelines in table 12. The third terrace starts at a depth of about 590 feet and slopes seaward beyond the reach of the submarine. It probably is the terrace corresponding to the minus-1,300 foot shoreline. The terraces are cut into rock and are partly covered with sand that appears to be moving gradually seaward into deeper water. They are separated by abrupt escarpments, in places nearly vertical, that certainly represent former sea cliffs. The escarpment separating the Penguin Bank and Mamala terraces is cut in ancient coral reef.

One of the most remarkable features of the submarine topography around the Hawaiian Islands is a broad shelf at a depth of 3,000 to

*209*

3,600 feet. It surrounds all of the islands except the south end of the island of Hawaii, where it may have been buried by lavas of Mauna Loa, Kilauea, and Hualalai. Around the central islands it is several miles wide, reaching a maximum width of 40 miles off the western end of Oahu. It is narrower around Kauai and Niihau, possibly because of the outward encroachment of a mass of sediment around these older islands. This bench has been named by Stearns (1966a, p. 17) the Waho shelf. As he points out, there appears to be no other reasonable explanation for it except wave erosion, and this in turn implies a sinking of the islands by 3,300 to 3,600 feet since the bench was cut. The bench may have been eroded on the same ancient land mass as the Lualualei bench, but it is certainly older, and the age difference may be great.

The Waho shelf is deeper around the northwestern islands than it is further south, suggesting that the sinking which has carried the shelf to its present depth has been greater toward the northwest.

Farther west in the Pacific the Mid-Pacific Mountains is a range of great submarine mountains (seamounts) that extends from near 20° N, 168° W for about 1,500 miles to 18° N, 170° E. At least 20 separate large seamounts can be counted along this range (fig. 289), and most of them have nearly flat tops. Seamounts of this sort are called *guyots* (fig. 137B). Except for the flat tops, their forms are closely similar to those of known volcanic mountains. There is little question that they are shield volcanoes, the tops of which have been cut off by wave erosion. This implies, of course, that the tops of the guyots were once close to sea level, a conclusion supported by the presence of reef-forming corals and other shallow-water fossils, together with rounded cobbles and pebbles of volcanic rocks, on the flat tops (Hamilton, 1956). Since the tops of the Mid-Pacific Mountains now lie 6,000 feet and more below sea level, the volcanoes and the crust on which they rest must have sunk that much since the flat tops were cut. Fossils dredged from the tops of the guyots are of late Cretaceous age (about 80 million years ago). Menard (1964) has found evidence that the

210

Mid-Pacific Mountains lay near the crest of a broad swell on the ocean bottom (the Darwin Rise), and that the swell later sank, carrying the guyots to their present depth.

Reconstruction of the Darwin Rise (fig. 170) indicates that the Hawaiian Islands area lay well out on its eastern flank, in a region that probably sank more than 4,000 feet as a result of the collapse of the Rise. Thus, if the Waho shelf was cut on a land mass that was already present at the time the guyots of the Mid-Pacific Mountains were being truncated by the waves, its sinking can be accounted for as part of the same movements that submerged the guyots. A northeastward bulge of the Darwin Rise in the vicinity of the Mid-Pacific Mountains is nearly at right angles to the trend of the Hawaiian Ridge, and sinking of this bulge would produce a northwestward tilt of the Hawaiian region, thus accounting for the greater depth of the Waho shelf toward the northwest.

If the Waho shelf was cut on an ancient island mass at the same time the guyots were being truncated, land was present in the Hawaiian area long before the late Miocene reef grew on the Lualualei terrace. Both the Waho and the Lualualei shelves need much more study.

Canyons cut into the edge of land masses below sea level have many of the features characteristic of canyons cut on land by streams, and they have sometimes been interpreted as evidence of a former sea level that was lower in relation to the land than is the present one. Submarine canyons are well developed off several coastlines of the Hawaiian Islands, and many others may be found as more numerous and more accurate soundings are made in the vicinity of the islands. At least 14 small canyons notch the submarine slope of northeastern Oahu near Kaneohe Bay. Several are known off the Napali Coast of Kauai, and another group lies off the north coast of East Molokai. The latter, described in detail by Shepard and Dill (1966), are sharply V-shaped, with continuously outward-sloping floors. They are generally less than 800 feet deep, though the one off the mouth of Pelekunu Valley in places is more than 1,100 feet deep. The canyons terminate at a depth of about

6,000 feet, while the slope of the island extends on down to more than 12,000 feet. There is a strikingly close correspondence between the positions of the Molokai submarine canyons and the valleys formed by stream erosion above sea level (fig. 139); all but one of the submarine canyons lie directly off the mouths of big land valleys. The single exception is one north of the Kalaupapa peninsula, and the related land valley probably was deflected comparatively recently by the building of the small posterosional shield volcano that constitutes the peninsula.

The close correspondence of the Molokai submarine canyons with the land valleys suggests a genetic relationship between the two. However, the submarine canyons could hardly have been cut by stream erosion at a time of relatively lower sea level, since this would entail an earlier lowering of sea level of more than 6,000 feet or a subsequent sinking of the island mass by the same amount. Although it is possible that the total isostatic sinking of the

Hawaiian Ridge may have been as great as 6,000 feet (see chap. 18), the canyons must have been cut late in the history of the island. The relative change of sea level of 6,000 feet is far in excess of the late changes of sea level indicated by any other evidence. It is more likely, as Shepard and Dill conclude regarding most submarine canyons all over the world, that the major cause of the Molokai submarine canyons has been erosion beneath sea level by downslope movement of the heavy loads of sediment contributed to the marginal ocean floor by the streams in the island valleys.

It is still impossible to date most of the ancient shorelines in terms of years. There is almost unanimous agreement that the various shorelines between 0 and plus-250 feet, and those 60 and 300 feet below present sea level, are eustatic, the most likely cause of the changes in ocean level being changes in volume of the glaciers during the recently-ended Great Ice Age. During the Pleistocene epoch of geologic time (roughly 2 million to 7,000 years

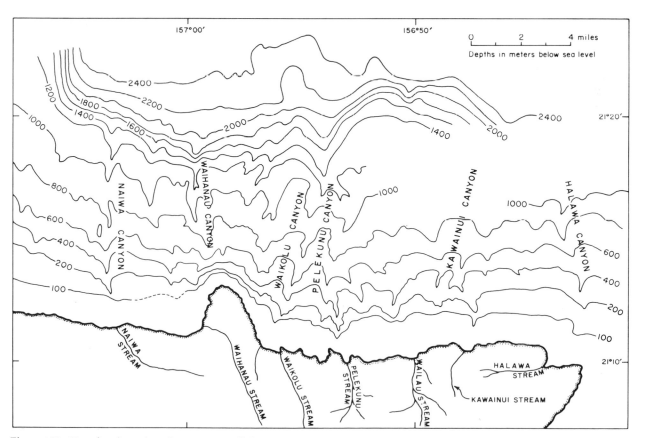

Figure 139. Map showing submarine canyons off the north coast of East Molokai. (After Shepard and Dill, 1966.) Note the close correspondence in position of the land and submarine canyons.

ago) great glaciers covered large parts of the continents. The glaciers were fed by snow, and the water that fell as snow was derived very largely by evaporation from the oceans. Once incorporated in the glacier, it was locked there and prevented from returning to the ocean for many thousands of years. This removal of water must have lowered the level of the oceans. Various geologists have calculated, entirely independently of any considerations of ancient shorelines, that at the maximum stage of glaciation sea level should have been lowered somewhere between 250 and 400 feet—an amount that is in good agreement with the lowest of the Hawaiian eustatic shorelines. Likewise, if all of the ice were melted from the earth's surface, as it probably was before the Ice Age and during some of the interglacial stages, sea level may well have been high enough to account for the Olowalu (plus-250-foot) shoreline. The other, intermediate, shorelines were formed when glaciers were present on continents but were less extensive than at their maximum. During certain periods between the great ice advances the climate was warmer than it is now, and the glaciers were smaller. Shorelines higher than the present one but lower than the Olowalu shoreline presumably were formed during those periods.

During the Ice Age there were five generally recognized successive stages of extensive glaciation in North America. The last of these, the Wisconsin, consists of three substages known, from oldest to youngest, as the Tazewell, Cary, and Mankato.

From evidence in other parts of the world (Shepard, 1961) it appears probable that at the time of the Cary substage, some 17,000 years ago, sea level was about 300 feet below the present one. Since that time there has been a general rise of sea level, with some oscillations, and the present level is close to the maximum attained at any time since then. The period of warm climate known as the "climatic optimum," about 4,000 years ago, during which world climate was somewhat warmer than now, probably lasted too short a time to produce any conspicuous shoreline features that have survived subsequent wave erosion, although it is possible that the plus-2-foot bench recog-

nized at a few places by Stearns was formed at that time. Reef coral on Midway Island, up to 3 feet above sea level, has yielded a carbon-14 age of 2,400 years (Ladd and others, 1967), and, allowing for some error in age determinations, it may also have been formed in a higher stand of the sea during the climatic optimum. It seems unlikely that it is coral from the present-day reef thrown up by storm waves or tsunami. With this exception, sea level does not appear to have been higher than the present at any time since 17,000 years ago, and if this is the case all the Hawaiian shorelines discussed in the foregoing paragraphs must have been formed during or previous to the Cary substage of glaciation.

Ages obtained on shells from the plus-5- and plus-12-foot shorelines by the carbon-14 method (the degree to which original radioactive carbon-14 in them has changed to carbon-12) are respectively 24,100 ($\pm$700) and 31,400 ($\pm$1,200) years before the present (Shepard, 1961). These dates correspond approximately with the interval between the Iowan stage and the first (Tazewell) substage of the Wisconsin. It appears probable that the 12-foot bench was formed during the warmest part of the Iowan-Tazewell interglacial stage, and the 5-foot bench during a pause in the lowering of sea level that accompanied the gradual growth of the Tazewell glaciers.

Coral fragments from the Mahana shoreline have been dated recently by K. Osmond, by a radioactive method involving the breakdown of ionium, as older than 350,000 years (W. H. Easton, quoted by Stearns, 1966a, p. 23).

The shorelines above plus-250 feet and below minus-300 feet almost surely are tectonic, although the cause of the crustal movements remains obscure. It is generally thought that the sinking that caused the submergence and alluviation of the big Waianae valleys was isostatic—the result of loading of the earth's crust with the great weight of rock in the volcanic mountains. The earth's crust seems to be in a state of flotational equilibrium. Just as a board floating in water sinks a bit as a little weight is added to it and rises as the weight is removed, so also does the earth's crust. This principle of flotational equilibrium is called isostasy. We know that under the load of only a few thousand feet of relatively light ice the earth's crust sank beneath the continental glaciers of the Ice Age, and as the ice melted the crust rose again. Surely, then, the crust should sink under the added weight of 20,000 to 30,000 feet of heavy basaltic lava. But reasonable as this argument seems, it is difficult to reconcile with some of the evidence. The sinking of the Waianae Range was not coeval with the slow growth of the volcano; it did not even follow soon afterward. A period of several million years intervened, during which time the great valleys were cut and a large part of the mass that presumably would cause the sinking was removed by erosion. Yet we know that the isostatic response of the continents to the growth and wasting away of the glaciers took place almost immediately, within a few thousand years. Why should the isostatic sinking of Oahu be so long delayed? Actually, by the time of the Waianae shoreline the movement should have been a rise in response to the removal of mass by erosion, not sinking. At best, the hypothesis of isostatic adjustment due to loading of the crust as an explanation for the sinking of the island after the time of the Waianae shoreline must be regarded as doubtful.

A more likely explanation than the hypothesis of isostatic adjustment to loading of the crust is to be found in changes in volume and density of the material in the upper part of the earth's mantle beneath the islands—the same sort of changes that must have been responsible for the initial elevation and later for the sinking of the Darwin Rise. Just what these changes are, we do not know. Perhaps the original elevation of such a rise is due to an increase of heat in the upper mantle, bringing about expansion of the mantle material, perhaps accompanied by a change of state from combinations of dense minerals to combinations of lighter ones (the eclogite-gabbro transformation). In time, molten magma breaks through the crust and builds great volcanoes on the rise; but eventually the excess of heat is dissipated, the mantle material is transformed back into dense materials, and the rise sinks. Some such process must be responsible for the upheaval,

and much later for the sinking, of oceanic rises, but we still have much to learn about them.

This is not to say that the mechanism of isostasy did not operate in the Hawaiian area—only that it does not seem to account for the post-Waianae shoreline submergence. Gravity studies indicate that the Hawaiian area is essentially in isostatic adjustment, and that sinking must therefore have occurred; but it probably took place during and immediately after the period of volcanic activity.

The sinking of Oahu after the Waianae stand of the sea actually was much greater than the 1,300 feet or so necessary to carry the Waianae shoreline to its present depth. We must add the approximately 1,200 feet necessary to bring the Mahana shoreline to sea level. At the time the Mahana shoreline was being cut by wave action, the Waianae shoreline was submerged 2,500 feet, the Lualualei shelf 2,900 feet, and the Waho shelf 4,200 feet. Then, after the sinking and the erosion at the Mahana shoreline, the region was heaved up again some 1,200 feet, with pauses to form the Kaluakapo and Manele shorelines. Why the upheaval? Can we again call on heating and phase changes in the mantle, which led to a renewal of volcanism and eruption of the posterosional (Honolulu and Koloa) volcanism on Oahu and Kauai respectively, and the Lahaina and Hana volcanism on Maui? But this is still simply speculation.

CORAL REEFS

Masses of limestone formed by marine organisms are abundant in tropical and subtropical parts of the oceans. The most conspicuous, and one of the most important, organisms in these masses is coral, and consequently they are generally referred to as *coral reefs*. Actually, many other organisms contribute to the building of reefs. These include mollusks, foraminifera, and bryozoans, all of them animals that secrete limy shells or supporting structures; but most important after corals are calcareous algae—plants that deposit calcium carbonate. Algae sometimes make up the major part of a reef. In general, however, corals provide the framework within and around which the other organic secretions accumulate, and therefore the conditions of growth of the colonial corals govern the conditions under which reefs can form.

Corals are marine animals that extract the dissolved calcium carbonate from sea water and deposit it around themselves to form a limy supporting structure, or external "skeleton." Some corals live individually or in very small groups; others live in colonies of countless thousands, building their skeletons one upon another until a mass of great size may result. Single corals can live in fairly deep water, but the reef-forming colonies can exist only in clear warm water, with salinity normal for ocean water, and less than about 200 feet deep.

Most commonly, coral reefs are classified according to their relationship to a land mass. Reefs that grow directly along the shore, forming platforms that extend out from the shoreline, are called *fringing reefs.* Others, known as *barrier reefs,* lie some distance away from the shore and are separated from it by a belt of shallow water called a lagoon (fig. 140). Still other, roughly circular reefs have no land mass in the center, but only a broad lagoon. These are known as *atolls.* The lagoons of atolls contain numerous "heads" and pinnacles of coral, some of which may reach the surface to

form small islets. All reefs have channels, or "passes," of deeper water leading through them. On fringing or barrier reefs these channels often are opposite the mouths of rivers draining the land mass, where both the fresh water and the sediment brought into the ocean by the river inhibit the growth of the corals. The channels are simply zones in which organic growth was less vigorous than on either side.

The Hawaiian Islands appear to be close to the northern margin of the area in which reefs can grow, probably because of cooling of the surrounding ocean by the Alaska current. The older parts of the major islands are surrounded by rather narrow coral reefs, and in some areas reefs formed during earlier periods have been elevated above sea level by the eustatic movements described in the preceding section. The shores of the younger volcanoes have individual corals and coral heads, but little or no reef growth, simply because sufficient time has not yet elapsed since the last lava flow entered the ocean for reef growth to become established.

The reefs around the major islands are all fringing reefs. At Waikiki the outer edge of the reef determines the line where oscillatory waves change to waves of translation, that supply the motive power to surfboard riders. A narrow nearshore reef at the head of Hanauma

214

Figure 140. Borabora, one of the Society Islands, showing the barrier reef and lagoon, and the deeply embayed form of the island resulting from submergence of a stream-eroded topography.

Bay is often beautifully visible from the top of the cliff above. The fringing reef is nearly a mile wide off Waimanalo and Lanikai, and lies 2 to 3 miles offshore across the mouth of Kaneohe Bay. Fringing reefs are present along many other parts of the shores of Oahu and Kauai, and much of the south coast of Molokai. Probably the best development of living reef is along the north shore of Kauai, from Kalihiwai Bay to Haena.

Both the living and the ancient uplifted reefs of the Hawaiian Islands contain large proportions of algae. In the uplifted reefs spheroidal algal structures up to several feet across often are very conspicuous. The outermost edge of the living reef generally is somewhat higher than the reef surface behind it. This "lithothamnian ridge" is heavily armored with a dense coating of limestone laid down by several different species of algae, and greatly increases the resistance of the reef to the pounding of the surf.

The islands of the northwestern part of the Hawaiian Archipelago are much older than the major islands that make up the southeastern part. Apparently, in a general way, volcanic activity started at the northwestern end of the chain and progressed southeastward. The volcanoes at the southeastern end are still active, but those in the northwest have been extinct so long that streams and ocean waves have carved away their summits, leaving only the stubs of the former mountains. Northwestward from Niihau the remnants of the volcanic mountains projecting above sea level become smaller and smaller. Nihoa Island (area 155 acres) is only a small fragment, less than a mile long, of a once big insular volcanic shield, and Necker Island (41 acres) is even closer to total destruction. The Gardner Pinnacles, two volcanic islets respectively about 200 and 600 feet long, are believed by Palmer (1937) to be the remains of an island that may once have had an area of about 80 square miles.

Between Necker Island and the Gardner Pinnacles (fig. 1), French Frigate Shoals is an atoll, within the lagoon of which projects a tiny volcanic remnant—La Perouse Rocks. Because a bit of the original central volcanic island still exists, the reef can be regarded as transitional between a barrier reef and a true atoll. The northeastern side of the reef is higher than the southwestern and lies close to sea level, with several projecting islets. As on other atolls, the islets are composed of calcareous sand, thrown up by the waves and to some extent blown into dunes by the wind. The greater height of the northeastern side of the reef probably is the result of a greater food supply brought to the corals and other sedentary organisms on that side by waves driven before the prevailing northeast trade winds.

Still farther northwest, Maro Reef, Lisianski Island, Pearl and Hermes Reef, Midway and Kure islands, all are atolls with no volcanic rock showing at the surface of the ocean. The lagoons of Kure and Midway are each about 6 miles in diameter, and that of Pearl and Hermes about 15 miles. Sand islets rise within the lagoons, but not on the ring reefs themselves, which, because they are low and inconspicuous, have been the sites of many shipwrecks. Lisianski is unusual in having a nearly central sand island almost a mile long, itself surrounded by fringing reef. Wake Island, lying southwest of the Hawaiian chain proper, also is an atoll, differing from the others in that a large part of the reef ring is capped by sand islands that rise to a little more than 20 feet above sea level.

Several theories of origin of atolls have been suggested. The one most generally accepted was proposed by the great English biologist Charles Darwin, more than a century ago. According to this theory the process starts with the formation of a fringing reef around an island. As the island slowly sinks and is eroded, the reef builds vertically upward, most actively at the outer edge where the food supply brought by waves and currents is most plentiful. So long as the sinking is slow enough, growth keeps the reef surface close to sea level, within the depth zone in which colonial corals can live. As subsidence and reef growth continue the reef becomes separated from the island by a lagoon, and is transformed into a barrier reef (fig. 141). If sinking continues long enough the central island eventually disappears,

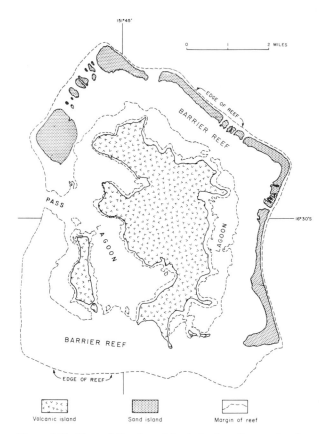

Figure 141. Map of Borabora Island, Society Group, showing the barrier reef and lagoon around the central volcanic island. (Modified after a French Government map.)

and only a circular reef—an atoll—is left (fig. 142).

Darwin pointed to two evidences in support of his theory: (1) examples of all stages of transition from fringing reef to atoll can be seen; and (2) the islands within barrier reefs commonly have embayed shorelines resulting from partial submergence of a stream-eroded topography. Very recently, strong support for Darwin's theory has come from geophysical studies and deep borings on atolls. Formerly, even the existence of volcanic bases beneath the atolls was entirely conjectural, but analysis of artificial earthquake waves traveling through the bases of the islands shows clearly that the reef is only a relatively thin capping on a volcanic base. At Eniwetok, drill holes were sunk through the reef to a depth of about 4,000 feet, where they entered basalt of the volcanic base, thus confirming beyond question the evidence from seismic studies. The lava rock brought up from the bottom of the drill

holes at Eniwetok, and also samples of basalt lava dredged from the lower slopes of the mountain beneath Bikini Atoll, have about the same degree of vesicularity as those poured out above sea level in the Hawaiian and Samoan Islands, and almost certainly, therefore, they were formed, if not above sea level, at least in very shallow water rather than under the high-pressure conditions that exist at great depths in the ocean. If this deduction is correct, the island must indeed have subsided several thousand feet, as Darwin suggested.

Even stronger evidence of sinking is the fact that all of the 4,000 feet of limy sediment at Eniwetok, and also the 2,000 feet of limestone penetrated by drill holes at Bikini, is of shallow-water origin, formed within 200 feet of the ocean surface. This deposition of several thousand feet of shallow-water sediment could have taken place only during a slow sinking of the volcanic base, or a corresponding slow rise of ocean level.

R. A. Daly has suggested that the barrier reefs and atolls grew upward from platforms

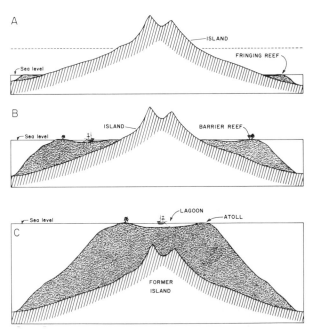

Figure 142. Diagrams illustrating Darwin's theory of coral reef evolution. A, Fringing reef bordering a volcanic island. The dashed line shows the position of sea level in the next diagram. B, Barrier reef, separated from the island by a lagoon. This profile represents the present condition of Borabora Island, in the Society Islands. C, An atoll, with a central lagoon. (After Darwin, 1839.)

216

cut 300 feet below present sea level by wave erosion during the last great glaciation, and, as sea level slowly rose because of the melting of the glaciers, growth of the reef kept pace with it. However, as we have just seen, the depth to the volcanic base beneath several of the mid-Pacific atolls is now known to be far too great to be accounted for in this way. The "glacial control" theory cannot account for the formation of reef cappings several thousand feet thick, but the process suggested by Daly may well have modified reefs that developed predominantly as a result of a larger-scale subsidence.

*Suggested Additional Reading*

Abbott and Pottratz, 1969; Chamberlain, 1968; Moberly, 1968; Moberly and others, 1965; Palmer, 1927; Shepard, 1961; Stearns, 1961; Wentworth, 1938*a,* 1939, 1944; Wiens, 1961.

*Cross-bedded
lithified sand dunes
near Kahului, Maui*

# CHAPTER TWELVE

# Sedimentary Rocks

THE MATERIAL MOVED BY ALL OF THE GEOLOGIC AGENTS of transportation—gravity, streams, wind, ice, ocean currents and waves—both the solid fragments and the material carried in solution, is deposited to form *sedimentary rocks*. Some of the material is deposited on the land surface, to form *nonmarine* sediment, and some is deposited in the ocean as *marine* sediment. Most of the deposits on the land are very temporary, however, and are soon moved on again. The ocean is the final resting place of most transported debris; the great bulk of sedimentary rocks are marine. Over broad areas these marine sediments have been lifted up to form the surface of the land. Sedimentary rocks (and *metamorphic* rocks derived by the alteration of sedimentary rocks) comprise only about 5 percent of the earth's crust as a whole, but they make up 75 percent of the surface of the continents.

The marine sedimentary rocks now exposed on the continents were not formed in the deep ocean. They were deposited in shallow seas that covered submerged parts of the continents. Rarely, if ever, has deep ocean bottom been upheaved to form land.

Most sedimentary rocks are deposited in layers, or *strata* (singular: stratum), and they are therefore said to be *stratified* (fig. 144). An individual layer may be less than an inch, or several tens (rarely hundreds) of feet thick. Most of these layers (strata, or "beds") were originally laid down in nearly horizontal positions, but later deformation of the earth's crust has broken and tilted them until they may now lie in any position—steeply inclined, on end, or even upside down.

Many sedimentary rocks contain fossils—the remains of ancient life buried in the sand of the ancient sea bottom or under the shifting sands of ancient deserts. Most fossils are the remains of the hard parts of the actual organisms, the shells or bones of animals, or the trunks or leaves of trees. Some, however, are other sorts of organic traces—filled-in worm borings or trails of animals. For instance, during the Triassic period, some 180,000,000 years ago, dinosaurs wandered about the mudflats of what is now the Connecticut River Valley, in New England, and left their footprints, which were buried by succeeding layers of mud and sand and preserved as fossils. These are truly "footprints in the sands of time."

Sedimentary rocks are classified first as to the manner in which the material composing them was transported and deposited. Erosional detritus transported and deposited as solid fragments forms *detrital* (sometimes called clastic, or fragmental) sediment. Material carried in solution may be deposited directly as a chemical precipitate *(chemical sediment),* or it may be taken out of solution by organisms and used to build their shells or skeletons, or simply deposited around them *(organic sediments).* These organic remains may accumulate in such abundance as to form layers of rock, such as coral reefs. Sediments commonly are deposited as masses of loose material such as loose sand or soft mud, but in time they become consolidated (hardened, or "lithified") into hard rock by deposition of cement between the grains or by compaction resulting from the weight of accumulating sediment above. The cementing material is usually silica $(SiO_2)$, calcite $(CaCO_3)$, iron oxide, or clay, but more rarely it may be other things.

The materials composing detrital sedimentary rocks are obtained by erosion of land masses, and they consist of whatever rocks and minerals are present in the land mass, or of the products of their breakdown by chemical weathering. This process tends to destroy the less resistant minerals, so that the debris finally reaching the ocean is somewhat enriched in the more resistant minerals and in the products of weathering (largely clay and iron oxides) of the less resistant ones. On continents the sediments contain much quartz, which is the most resistant to chemical weathering of the ordinary abundant rock-forming minerals. Quartz grains tend to be concentrated in the sand portion, and the finer material consists predominantly of clay. In Hawaii the detrital material is, of course, largely basaltic. Usually quartz is almost wholly lacking, as is the case also in the basaltic source rocks. A small amount is present in the sediments in a few areas, derived by erosion of soil in which it originated, as explained in chapter 8. However, a special case is represented at Kaneohe Bay. In the region around the bay the rocks that filled the caldera of the Koolau volcano were much altered by rising gases, which changed the original pyroxene to chlorite, and the silica released by this alteration was deposited in cavities as quartz crystals. Erosion of these rocks has contributed both chlorite and quartz to the sediments accumulating in the bay.

During transportation by water or wind there is a separation of fragments of different sizes (sorting). Currents of a given strength can carry fragments only up to a certain size, and larger ones are left behind. Thus there is a sorting of the material, of various degrees of perfection, and the deposited sediment commonly consists of fragments mostly within a restricted size range: that is, for instance, it is very largely sand, or very largely mud. In a general way, the farther materials are transported the better they are sorted; and marine waves and currents produce a higher degree of sorting than do streams.

Detrital sediments are classified according to the predominant size of the fragments composing them and whether or not they have been consolidated into coherent rock. Further subdivision may be based on the shapes of the predominant fragments, or special structures such as platiness (fissility). Table 13 shows the common major divisions.

If the fragments of a consolidated gravel are rounded (ordinary pebbles, cobbles, and boulders) the rock is called *conglomerate,* but if the fragments are angular it is called *breccia.* If a consolidated silt or clay is composed of many thin plates (commonly a fraction of an inch to a couple of inches thick) it is called *shale,* but

Table 13. Classification of detrital sediments

| Size of fragments | Unconsolidated sediment | Consolidated rock |
|---|---|---|
| Greater than 2 mm (.08 inch) | Gravel | Conglomerate or breccia |
| 2–.06 mm (.08–.002 inch) | Sand | Sandstone |
| Less than .06 mm (.002 inch) | Silt and clay (mud or dust) | Mudstone or shale |

if it is massive it is *mudstone* or *siltstone.* Deposits of wind-blown dust receive the special name, *loess.*

Detrital sedimentary rocks are of only minor importance in the Hawaiian Islands. Masses of poorly sorted breccia, consisting of angular rock fragments in a finer matrix of partly weathered material, form heaps (taluses) at the foot of cliffs and steep valley walls (fig. 118), deposited there by gravity, soil avalanching, and other means of mass transfer. These are temporary deposits, continually being eroded and their materials carried seaward by streams, but continually reforming. Conical piles of alluvium, known as alluvial cones or alluvial fans (fig. 143), are deposited at the mouths of some valleys as a result of reduction of stream gradient or loss of stream water by seepage into permeable underlying rock.

Bars of gravel and sand formed along stream courses likewise are very temporary and are shifted seaward with every flood. More permanent, however, are the deposits of poorly sorted conglomerate, sandstone, and mudstone that are formed on the floors of the big valleys that are being alluviated as a result of rise of base level. The accumulation of alluvium is largely the result of stream deposition, but to some extent, particularly in the drier regions, it is the result of mud flows. Toward the mouths of the valleys the nonmarine alluvium grades into marine deposits of stratified clay, with lenticular beds of sand and gravel. None of them are very well lithified. In Lualualei Valley, on the Waianae coast of Oahu, deposits of this sort are more than 1,200 feet thick. Poorly lithified, brown, stratified siltstones and claystones with lenses of gravel, containing some marine fossils, are exposed in the vicinity of Pearl Harbor. Poorly consolidated deposits of clay and mudstone, calcareous sand and gravel, and marl (a mixture of calcareous material and clay) on the Mana Plain between Kekaha and Barking Sands, Kauai (fig. 275), appear to have been deposited in a shallow lagoon between the sea cliff and a calcareous barrier beach ridge formed on a gently sloping, wave-cut terrace.

Figure 143. South coast of West Maui. Big alluvial fans lie at the mouths of Olowalu *(left)* and Ukumehame canyons. The streams are now cutting down into their own fans. Steep dips of the làva beds of the Wailuku Volcanic Series can be seen in the sides of the valleys.

Figure 144. Well-bedded beach rock, cemented calcareous beach sand, at Halena Beach, West Molokai.

Detrital sediments may consist of organic debris. Thus the constituents of common white Hawaiian beach sands are fragments of coral and shells eroded from the offshore reefs by waves and washed up onto shore, mixed with the shells of single-celled animals and plants, especially foraminifera, washed in by currents. Nearly all of this material is $CaCO_3$ in the form of the minerals calcite and aragonite. This calcareous (limy) sand may become consolidated by cementation into a hard calcareous sandstone, known locally as beach rock.

The beachrock generally shows stratification dipping (sloping) seaward at approximately the same angle as the bedding in the loose sand of the adjacent beach (fig. 144). Beachrock extends from a few feet above sea level to a few feet below. Where it is exposed at the surface it is generally being eroded by both solution and wave action, and therefore it was probably formed under earlier conditions, different from those of the present. Precisely what those

conditions were, and what brought about the deposition of the calcareous cement between the sand grains, is not known.

At a few places, notably in the beach and dunes just south of the cliffs of the Napali Coast, on Kauai, there is found a rather fine grained calcareous sand, containing a small proportion of grains of basaltic lava rock, that under just the right conditions of dampness emits a peculiar squeaking or barking sound when walked on or squeezed between the hands. These are known as sonorous sands, or more commonly, barking sands.

Chemical sediments are formed when solutions become oversaturated—that is, when they contain more of a dissolved substance than they can continue to hold. The excess amount of the substance is then precipitated out of solution. Oversaturation may result from change in composition of the solution, from chemical reactions in the solution, or from evaporation of the water making the solution

222

more concentrated. Complete evaporation of the water in desert lakes, called playas (fig. 278), results in precipitation of all its dissolved material, including even the very soluble sodium chloride (NaCl, ordinary table salt). Commonly, however, only the less soluble materials are deposited. Rocks resulting from evaporation are often referred to as evaporites, and include rock salt (NaCL), gypsum ($CaSO_4 \cdot 2H_2O$), limestone ($CaCO_3$), and dolomite ($CaMg(CO_3)_2$).

Another factor that contributes greatly to the deposition of the carbonate rocks (limestone and dolomite) is the loss of carbon dioxide from the solution. You will recall from the discussion of chemical weathering that carbon dioxide combines with water to form carbonic acid ($H_2O + CO_2 = H_2CO_3$), and in the presence of this weak acid the carbonates of calcium and magnesium are fairly soluble, but when $CO_2$ evaporates leaving the solution somewhat less acid, the calcium carbonate becomes rather insoluble and part of it is precipitated as limestone. Most of the deposits (including stalactites, stalagmites, and sheet deposits) in limestone caves, such as the Mammoth Cave in Kentucky and the Carlsbad Caverns in New Mexico, are formed as a combined result of evaporation of water and loss of $CO_2$ from the solution. Similar "dripstones" have been formed in Hawaii where water has seeped through limestone, either raised coral reefs or dunes of calcareous sand, and has dissolved calcium carbonate from the rock. Emerging onto the roof of a cave or the face of a cliff, a drop of water partly evaporates and loses $CO_2$, and deposits some of its contained calcium carbonate, gradually building up a limestone pendant known as a stalactite. Dripping onto the cave floor below, it evaporates further and again deposits calcium carbonate, building up a stalagmite. Limestone stalactites and stalagmites can be seen on the faces of cliffs and in shallow caves eroded in calcareous dune sandstone just inland from the

highway northwest of Kahuku, and also near the eastern edge of Kailua, both on Oahu.

Although some dolomites are formed by direct chemical precipitation, most of them result from secondary alteration of either chemical or organic limestone by circulating ground water or heated water partly of magmatic origin rising from deeper within the earth. Dolomite is not known to be important in the reefs of the major Hawaiian islands, but in the calcareous reefs of other parts of the Pacific it makes up a large proportion of the lower part of the reef masses. It appears to have formed by alteration of reef limestone by magnesium-bearing ground water moving downward through the limestone at some stage of the reef's history.

The most important organic sediments are *limestone* and *coal*. The masses of organic limestone known as coral reefs are discussed in the chapter on the work of the ocean. Other types of organic limestone, not found in Hawaii, are of great importance in continental areas (such as the great sheets of limestone, formed largely of the remains of crinoids and brachiopod shells, that are found in central United States).

Coal is composed of the carbonized remains of plants, generally accumulated in swamps. The alteration of vegetable material into coal consists in the gradual elimination of other materials, particularly volatile ones, and the concentration of carbon. The first stage in the alteration is a woody brown substance known as *peat*. As more and more of the material other than carbon is removed the peat passes into *lignite* ("brown coal"), and then into true coal. No lignite or true coal has been found in Hawaii, but deposits of peat occur in several of the swampy regions high in the mountains, and in a few other places. For instance, peat occurs in the crater of Kilohana cone west of Lihue, and just north of the Knudsen Gap through which the highway runs from Lihue to the south side of the island of Kauai.

223

*Suggested Additional Reading*

Edmondson, 1928; Leet and Judson, 1965, pp. 91–110; Moberly, 1963a; Pollock, 1928; Stearns and Vaksvik, 1935, pp. 165–172

*Longitudinal sand dunes, Niihau*

# CHAPTER THIRTEEN

# Work of the Wind

IN SOME REGIONS WIND IS AN IMPORTANT AGENT OF transportation, erosion, and deposition, but it is of comparatively little significance in most of Hawaii. Its effects are greatest in arid regions, but even there, if a reasonably complete vegetative cover exists, wind is geologically rather ineffective. The great dust storms of the southwest-central part of the United States are the result not only of aridity, but of the loss of plant cover due to poor agricultural practices. In past years a great plume of red dust quite often could be seen blowing out to sea from the island of Kahoolawe, where the vegetation had been destroyed by overgrazing.

Dust (consisting of mineral grains less than .06 mm, or .002 inch, in diameter) is lifted by wind high into the air, and transported in suspension as a cloud that may obscure all visibility to a height of several thousand feet. Sand (grains greater than .06 mm), on the other hand, is transported in a zone within a few feet of the ground surface. A man standing in a sandstorm may not be able to see the ground around him, while at the same time, his vision at eye level may be almost unobstructed. Very little of the sand is carried any long distance in suspension. Most of it is rolled along the ground surface, or moves in a series of jumps of a few inches or a few feet (saltation).

Erosion by wind consists of abrasion of rock surfaces by moving sand (dust particles are too small to abrade much), and of the blowing away of material. The latter process is known as *deflation.* Blowing sand abrades any object in its path. The bases of telephone poles in desert regions often become so abraded that the poles have to be replaced. Sand abrasion likewise may cut away the base of a rock outcrop or a boulder, leaving a "wine-glass rock" or "pedestal rock." It also results in grooved and often finely polished rock surfaces on or close to the general ground surface. Examples can be seen on the western parts of the islands of Molokai and Lanai, where devegetated soil surface is scarified in long grooves a few inches deep and from a few inches to several feet wide, lying parallel to the direction of the prevailing wind. The grooves and other wind-formed features give a marked parallel grain to the land surface as seen from the air.

Rock fragments exposed to sand blasting commonly become faceted, the abrasion flattening out one or two surfaces, depending on the angle at which the fragment lies with respect to the prevailing wind. These fragments may later roll over and expose other sides to the blast, and these in turn become flattened and polished. Wind-faceted fragments are known as *ventifacts* ("wind-created" stones, in contrast to man-made "artifacts," which they sometimes resemble). No good ventifacts have been found in Hawaii, although some rather poorly developed ones have been found on Molokai.

Deflation removes large amounts of dust from some arid regions, and this dust sometimes is carried for hundreds of miles. Sand movement is more restricted. Hollows (called *blowouts*), generally only a fraction of a mile across and a few feet deep, may be excavated as loose sand is blown away. In Hawaii, deflation is largely confined to the blowing away of beach sand, but as much as 7 to 8 feet of soil has been removed from the summit region of Kahoolawe, largely by the wind. Small blowouts, up to about 200 feet across and 2 feet deep, are present at the northeast end of the "desert strip" of West Molokai, southwest of Moomomi Beach.

Dust deposited by wind may form extensive thick deposits, known as *loess*. Yellow loess, blown from the Gobi Desert, forms a blanket as thick as several hundred feet over much of north China, and similar loess from the Sahara Desert occurs in the Sudan, in Africa. A belt of loess extends across much of central United States and Europe, but there the dust was derived not from deserts in the ordinary sense but from broad stream-deposited plains, nearly devoid of vegetation, that spread southward beyond the limits of the continental glaciers during the Ice Age. Almost all of the dust blown from Kahoolawe is dropped in the ocean, and it contributes to marine sediments forming around the islands. On the island of Hawaii, however, sizable dust storms are quite often seen in the Ka'u Desert, southwest of Kilauea caldera, and in the vicinity of South Point there are extensive deposits of yellow loess formed of dust-size volcanic ash blown by the wind from the Ka'u Desert.

Blowing sand commonly encounters obstacles, such as boulders or bushes, and piles up around them to form hillocks, called *dunes.* Part of the sand is dropped in front of the obstacle, but most of it falls just behind it, where an eddy occurs and the velocity of the wind is checked. Once formed, the dune itself constitutes an obstacle to the wind, and promotes further sand deposition, with the result that dunes tend to grow by the accumulation of more and more sand. Most dunes remain fairly small—only a few feet to a few tens of feet high—but some as much as 700 feet high are known.

As they grow, many dunes migrate. Sand is blown from the windward side of the dune up across the top, and dropped on the leeward side, thus causing a constant shifting of the mass to leeward. In this way dunes may march across the countryside for many miles.

The side of a dune facing the wind has a gentle slope, but as the sand is carried over the top of the dune and is dropped, it slides down to an equilibrium slope that generally is inclined about 34°—much steeper than the windward slope. Some dunes are crescent-shaped as viewed from above, with the horns of the crescent pointing to leeward. These (known as *barchans*) commonly are considered typical sand dunes. However, in many areas they are far less common than other types, which form as fairly straight ridges either parallel or at right angles to the direction of the wind. Ridges parallel to the wind, known as longitudinal dunes, are common in Hawaii. Examples of such dunes are shown in figure 145 and in the photograph on page 224.

Cross sections of dunes commonly show conspicuous stratification, with the layers inclined at an angle to the surface on which the dune rests, each layer having been formed parallel to the lee side of the dune. This cross-bedding is characteristic of sand dune deposits, and allows us to recognize ancient sandstones formed by the wind. Successive major beds of wind-deposited sandstone may have the internal cross-bedding sloping in dif-

Figure 145. Strong prevailing northeast trade winds have swept beach sand inland from Moomomi Beach on West Molokai. Some dunes near shore are migrating, whereas others farther from their source and of an older generation have become lithified and form low ridges of sandstone.

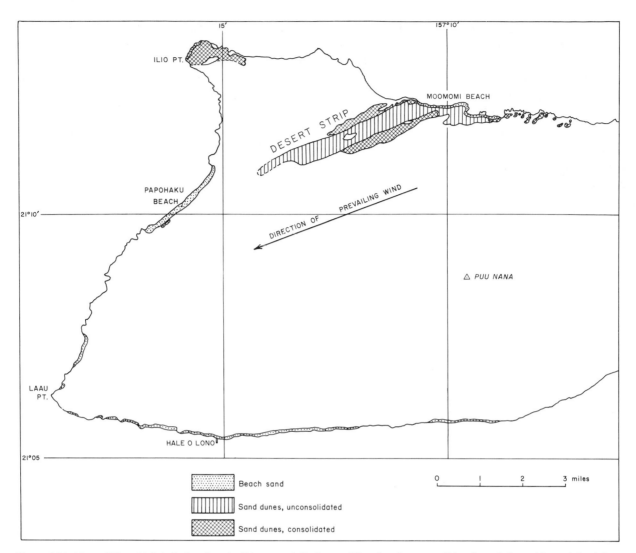

Figure 146. Map of West Molokai, showing the "desert strip" of consolidated and unconsolidated sand dunes blown inland from Moomomi Beach, and the location of the principal beaches. (After Stearns and Macdonald, 1947.)

ferent directions, owing to changes in the direction of the wind. (See photograph on page 218.)

Where sandy beaches face the prevailing winds, the loose sand frequently is blown inland. Just beyond the inner edge of the beach, the sand often encounters vegetation that serves as a barrier to its movement, and a belt of dunes roughly parallel to the beach is formed. This has happened at many places in Hawaii. For example, dunes of this sort are well developed along the coast of Maui near Kahului, where some of the dunes are as much as 200 feet high, and near Mana on Kauai behind the Barking Sands beach. On Molokai sand has blown inland from Moomomi Beach

228

(fig. 145) and formed a belt of dunes and irregular deposits, partly cemented to calcareous sandstone, up to a mile in width and extending southwestward almost across West Molokai (fig. 146). (This is sometimes referred to as the desert strip of Molokai.)

On the slopes of Mauna Kea sand-size volcanic ash deposited on the upper slopes of the mountain by some of the late-stage eruptions has been picked up by the wind and built into dunes of black sand on the lee (southwest) side of the mountain. Cross sections of such dunes are well shown in road cuts on the Saddle Road 2 to 3 miles southeast of Waikii. Recently, trucks and tanks running over the surface in the Humuula Saddle area during military maneuvers have destroyed large amounts of vegetation and caused a great increase in wind erosion in that area. Great clouds of yellow dust can be seen blowing to leeward during maneuvers, and new sand dunes are forming, particularly in the area 3 to 6 miles west of Pohakuloa.

During the past two centuries much sand has blown southwestward from the black sand beach at Kaimu, on the island of Hawaii, and accumulated as dunes in the vicinity of the village of Kalapana. This is part of the cause of the recession of the beach, mentioned on page 203. Because there is no continuing source for its sand, the Kaimu (Kalapana) beach eventually will disappear completely, but its final disappearance can be delayed by artifically returning to the beach some of the sand from the Kalapana dunes.

## Suggested Additional Reading

Leet and Judson, 1965, pp. 208–218; Stearns, 1940*c*, pp. 13–15, 52, 126–127; 1947, pp. 24–25; Stearns and Macdonald, 1947, p. 28; Wentworth, 1925*b*.

*Mauna Kea*
*(White Mountain)*

# CHAPTER FOURTEEN

# Work of Ice

ONE ORDINARILY THINKS OF HAWAII AS A TROPICAL RE-gion, where snow and ice do not exist, but this is not entirely true. Climate depends on altitude as well as latitude, and even at the equator high mountains are crowned with ice and snow. Snow falls on the high mountains of the island of Hawaii practically every winter, and some years it remains on Mauna Kea well into the succeeding summer. On Mauna Kea, as in most other regions of the world, one winter's snow melts or evaporates completely away before the snowfalls of the next winter. However, there are places on the earth where some snow remains throughout the year and is buried by the next year's snow. Under such conditions it accumu-lates to greater and greater thickness, and beneath the surface the delicate and beautifully-shaped snowflakes are gradually trans-formed, first into granular snow, known as *névé* or *firn,* consisting of little lumps of ice, and finally into dense blue ice. The latter is a true crystalline solid—actually a rock.

We do not ordinarily think of a solid as being capable of flowing, but actually all solids can flow under certain conditions, and ice flows more readily than most. When the accumulating ice reaches a sufficient thickness it starts to move under the influence of gravity and its own weight, just as a little heap of tar placed on a table top will gradually spread outward in all directions. A mass of flowing ice is a *glacier.* (In some areas recent changes have brought about a thinning of the ice and once-flowing glaciers have ceased to move. These we still call glaciers, though we really should modify the term and call them stagnant glaciers.)

The glacial ice flows slowly outward, away from the center of accumulation. Only in the lower part of the glacier is the pressure from the weight of the overlying ice great enough to cause flowage. The lower zone of flow is overlain by an upper zone of fracture in which the ice is merely carried along on the lower flowing portion, and breaks up as it moves. In this respect it resembles a moving aa lava flow. Large cracks in the zone of fracture, known as *crevasses,* constitute both an annoyance and a serious hazard to persons crossing the ice.

Outside the zone of accumulation the ice begins to lose volume. This region is known as the zone of wastage. There are several ways in which a glacier wastes away. The most obvious is by melting of the ice at and near its margin. This may produce a large volume of meltwater that trickles down through fractures and accumulates beneath the glacier, pouring forth as streams from beneath the ice. Some of the ice evaporates directly into the atmosphere without passing through an intermediate liquid stage. (The combination of melting and evaporation in the zone of wastage is called ablation.) Where glaciers flow into the ocean or into lakes, chunks of ice may break off the end and float away. This process is known as *calving,* and the floating chunks of ice are *icebergs.*

So long as accumulation (alimentation) exceeds wastage, the front of the glacier continues to advance. When accumulation and wastage are equal the front is stationary, and when wastage is greater than accumulation the glacier grows thinner and the front recedes. In the last case the glacier is sometimes said to be receding or retreating, but since the ice must flow downhill or outward from a thick center of accumulation, the glacier cannot actually retreat in the sense of moving backward. The ice continues to flow outward; only the position of the ice front recedes.

## CLASSIFICATION OF GLACIERS

Glaciers are grouped in three general categories. *Valley glaciers,* also known as Alpine glaciers because they are widespread in the Alps and were first studied there, are streams of ice that form in mountainous areas and flow down the valleys. Some are very short—only a fraction of a mile in length—but others attain lengths of many tens of miles. Very small ones, occupying niches high on the mountainside, are sometimes called *hanging glaciers (cliff glaciers,* or *glacierettes).*

A series of valley glaciers may emerge at the edge of a mountain range and coalesce to form a broad apron of ice known as a *piedmont glacier.*

In contrast to valley glaciers that move outward only along the line of a single valley, broad mounds of ice that tend to spread radially in all directions are known as *ice caps*

if they are small, and *ice sheets* if they are large. The great mass of ice covering most of Greenland is an ice sheet. Very large glaciers covering many thousands of square miles are sometimes called *continental glaciers.* They may be either ice sheets or a series of partly coalescing valley glaciers.

## EROSION AND DEPOSITION BY GLACIERS

A glacier carries rock debris on its surface, and frozen within and at the base of the ice. Above valley glaciers rise mountain peaks and steep valley walls on which frost action breaks the rocks into fragments, which are then brought down onto the surface of the glacier by landslides and snow avalanches. The surface load of rock debris tends to sink into the ice as the ice melts beneath the fragments and refreezes above them. Other fragments tumble down into crevasses. The base of the moving ice quickly scrapes away the soil and subsoil, and soon rests on fresh rock. Where the rock has been fractured the ice freezes onto individual fragments, pulls them free, and carries them away. This process is called *plucking.* The bottom load of a glacier (the rock debris carried in the base of the ice) is acquired largely by picking up already loose fragments and by plucking, though some material sinks all the way through the ice from the surface. In ice caps and ice sheets the load is almost entirely bottom load, because few rock masses rise above them to supply debris to the surface.

The rock fragments carried in the base of the ice grind against the underlying bedrock, smoothing, scratching, and polishing the bedrock and becoming partly or completely worn away themselves. A rock surface over which a glacier has moved commonly is highly polished and marked by a series of long scratches, or *striations* (fig. 147), that indicate the direction of movement of the ice. In being worn away, the fragments frozen in the ice become faceted, with one or more flat sides that also may be polished and striated. These faceted, polished, and striated rock fragments constitute one of the principal means by which we identify ancient glacial deposits.

At the head of a valley glacier there is

formed a steep-walled amphitheater known as a *cirque.* (It has already been mentioned, in chapter 9, that the amphitheater heads of some of our Hawaiian valleys resemble surprisingly these glacial cirques, although glaciers had nothing to do with their origin.) The cirque seems to be formed largely by plucking of rock fragments by the ice. Around the head of the glacier is an arcuate fracture (the *bergschrund*) formed as the glacier pulls away from the ice frozen to the headwall of the valley. Meltwater often pours down into the bergschrund during the daytime, and may penetrate into the fractures in the rock beneath the ice and freeze there, prying rock fragments loose and freezing them to the base of the glacier.

Two cirques cutting into a ridge from opposite sides may modify it into a jagged knife-edged wall known as an *arête.* Once again, arêtes find their morphological counterparts in Hawaiian land forms—the knife ridges that result largely from soil avalanching. Three or more cirques cutting into a mountain may produce a very steep sided pyramidal or tooth-shaped peak known as a *horn.* The classic example is the famous Matterhorn, in Switzerland. Seen from some directions, Olomana Peak, on the windward side of the island of Oahu, bears striking resemblance to a glacial horn.

Farther downstream the valley glacier changes the cross section of the valley from the sharp V shape typical of youthful or submature stream-cut valleys to a U shape, and straightens the valley by eroding off the ends of spurs that project into it. Yosemite and Tenaya Canyons, in Yosemite National Park, are good examples of U-shaped glaciated valleys. Along the course of the valley some zones of rock are more fractured, and therefore more easily eroded by the glacier, than others. At the more easily eroded places the glacier cuts deeper, and when the ice melts away the valley floor has a series of basins that become lakes. A lake *(tarn)* also often occupies the floor of the cirque after the glacier has melted away.

The glacier occupying a major valley is commonly much thicker than those occupying

233

Figure 147. Glacial striae on a ledge of andesite (hawaiite) at an altitude of 11,750 feet on the southern slope of Mauna Kea. The loose rubble in the foreground is ground moraine, dropped by the ice when the glacier melted.

tributary valleys, and it cuts down the floor of its valley much more deeply. When the ice disappears, the mouths of the tributary valleys arc lcft hanging, sometimes far above the floor of the major valley. The tributary streams drop from the hanging valleys into the master valley as waterfalls; good examples are the Bridalveil Fall at Yosemite, and Yosemite Fall itself. Again, the hanging tributary valleys with their plunging waterfalls in the heads of many deep Hawaiian valleys, such as Nuuanu and Manoa in Honolulu, are surprisingly reminiscent of glaciated topography.

Beneath ice sheets and ice caps, valleys trending parallel to the direction of ice movement may be modified in a similar manner, though U-shaped valleys are far less characteristic of them than of valley glaciers. Hills of bedrock often are eroded into rounded knobs with a gently sloping side facing the direction from which the ice came, smoothed and polished by the abrasion of the rock-loaded ice, and with the other side steeper and often rather irregular due to plucking out of rock fragments by the ice. These rounded hills are known as *roches moutonnées,* because of a fancied resemblance to the rounded backs of sheep.

At the terminus of the glacier the melting ice deposits its load of rock fragments in a long ridge, parallel to the ice front, known as a *terminal moraine.* If the climate becomes warmer and the front of the glacier gradually recedes as the ice melts away, a sheet of rock debris *(ground moraine* or *till)* is left on the uncovered surface. Pauses in the recession of the ice front may result in a series of debris ridges called *recessional moraines.* Along the edges of valley glaciers, or of tongues protruding beyond the general margin of an ice sheet, there may be deposited ridges of rock fragments called *lateral moraines.*

Melting glacial ice may release large volumes of water that flows out away from the ice front, carrying with it part of the rock debris dropped by the glacier. This material may be deposited as a broad apron, or *outwash plain,* beyond the terminal moraine, or if it is less regular in its distribution it may be called simply *outwash.* There are also many special types of water-laid deposits associated with glaciers ("fluvio-glacial" deposits).

THE GREAT ICE AGE

Today's glaciers are restricted to the arctic regions and to high mountains in the temperate zone and very high mountains in the tropics. But several times in the history of the earth the climate has grown colder and the glacier-covered areas have expanded. The last of these Ice Ages was in the Pleistocene epoch of geologic time, roughly from 2,500,000 to 8,000 years ago. During part of that time the ice covered essentially all of North America down to the latitude of New York City. (In fact, Long Island is part of the terminal moraine left by the glacier.) All of northern Europe and much of northern Asia also were ice covered. South of the general ice margin, mountain glaciers were much more extensive than they are today. The thickness of the great ice sheets reached 5,000 feet or even more, and the glaciers covered at least 30 percent of the total present area of the continents.

In North America the ice spread out from three general centers, one in the Canadian Rockies, one in central Canada, and one in the vicinity of Labrador, though the maximum accumulation of ice was not simultaneous in all three. Not only once, but five times, major southward advances of the ice occurred. These advances were separated by periods of warmer climate, during at least some of which the climate was actually a good deal warmer than it is now, with subtropical plants growing as far north as southern Canada. There are some people who suggest that we may now be in another interglacial period—that the Great Ice Age may not be over, and that in the geologically near future ice may again spread over much of the earth.

GLACIATION ON MAUNA KEA

Hawaii also had its glacier. During the Pleistocene epoch an ice cap existed on the top of Mauna Kea (fig. 148). Actually, the amount of climatic change necessary to bring this about is not very great. Even now snow sometimes

234

persists on Mauna Kea through the summer and as late as September, and A. H. Woodcock has recently shown that permanent ice exists in the cinder of the summit cones a few feet below the surface. It has been estimated that an increase in average rainfall of only 2 inches a year, or a drop in average temperature of only a few degrees, would result in a year-round snow cap, and the accumulation of snow from year to year would soon form a glacier.

The Mauna Kea ice cap was quite small. It covered an area of 28 square miles, and, although the ice had a maximum thickness of about 350 feet, it probably averaged less than 200 feet, and many of the large cinder cones in the summit area protruded through the ice. The edge of the ice cap was irregular, several lobes extending down to about the 11,000-foot

level, and one (on the south side of the mountain, above Pohakuloa) extending to 10,500 feet.

Evidence that the glacier existed is found in the presence of several of the erosional and depositional features characteristic of glacial action.

Within the limits of the glacier, many areas were scraped bare of ash and cinder, the clinkery tops of aa flows were removed, and the lower slopes of the cinder cones were eroded and steepened. Denser ledges of rock were sculptured into roches moutonnées, and locally small areas of rock surface were left well polished and striated (fig. 147). In general, however, glacial erosion was minor, in keeping with the thinness of the ice.

Broad areas of thin ground moraine were

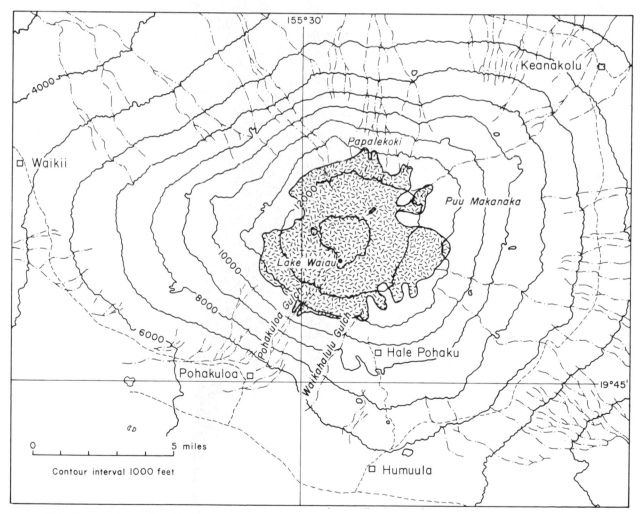

Figure 148. Map of Mauna Kea, island of Hawaii, showing the extent of the ice cap that occupied the summit of the mountain during the Ice Age. (After Wentworth and Powers, 1941.)

235

Figure 149. The summit and south flank of Mauna Kea, showing the terminal moraine left by the glacier. Near the center is a V-shaped loop of moraine deposited by the Pohakuloa lobe of the glacier. From it the main moraine ridge extends both to the left and the right. Below the Pohakuloa loop the prominent gorge of Pohakuloa Stream was excavated primarily by meltwater from the lobe of the glacier.

left, and around its edges the ice deposited ridges of terminal moraine, visible from a distance as a light-colored band extending around the mountain a short distance below its summit. Lateral moraines were formed along the edges of some of the ice lobes. By far the most prominent lobe was located on the slope above Pohakuloa; its former position is clearly indicated by a conspicuous loop of terminal and lateral moraine, easily seen from the Humuula Saddle between Mauna Kea and Mauna Loa (fig. 149). Abundant meltwater from this lobe eroded the present gorge of Pohakuloa Stream, which together with lesser streams from other lobes, such as that at the head of Waikahalulu Gulch, spread a mass of outwash gravel across the base of the mountain out onto the Humuula Saddle.

Glaciers may also have existed on Mauna Loa, but if so all evidence of them has been buried by more recent lava flows. Snowfall melts off Mauna Loa very much faster than from Mauna Kea, perhaps partly owing to escape of volcanic heat, but probably largely because the many small pinnacles of very black rock that project through the snow absorb more solar heat than do the lighter-colored rocks of Mauna Kea. None of the other Hawaiian mountains is as high as the lowest point reached by the glacier on Mauna Kea, and consequently they were not cold enough to have had ice caps.

Just when the ice cap disappeared from Mauna Kea is not known for certain. It probably occurred somewhat earlier than the disappearance of the great continental glaciers, however, because the mountain is nearer the Equator and its ice was thinner. The date may be tentatively placed between 10,000 and 20,000 years ago.

## POSSIBLE MULTIPLE GLACIATION OF MAUNA KEA

In gulches on the south slope of Mauna Kea, a series of older deposits of fragmental rock debris are exposed beneath the generally recognized glacial moraine. These were interpreted by Wentworth and Powers (1941) as glacial moraine laid down during four different stages of glaciation, corresponding with four of the major glacial advances (Nebraskan, Kansan, Illinoisan, and Wisconsin stages) in North America. However, the deposits were later studied by Stearns, who reached quite different conclusions. He agrees, as do all other workers, that the latest deposits (Makanaka moraines) are indeed of glacial origin. But he believes that the two earliest deposits, called by Wentworth and Powers the Pohakuloa and pre-Pohakuloa drift, were laid down by volcanic explosions; and he considers the next-to-youngest unit, the Waihu formation, to be water-laid breccia and conglomerate, largely deposited by mudflows during floods caused by sudden catastrophic melting of the ice cap by volcanic eruptions beneath it. Such glacial floods are well known in Iceland, where they attain enormous volumes and have caused great damage.

The Pohakuloa and pre-Pohakuloa deposits do indeed resemble known explosion deposits in other areas, and they contain cinder. However, a glacier on the summit of Mauna Kea certainly would pick up large amounts of cinder, and deposit it in the moraines. The fact that blocks have not been found in the deposits that are definitely either faceted or striated has been used by Stearns as evidence against their glacial origin, but it must be pointed out that such blocks are exceedingly rare and never very well developed even in the definite Makanaka moraines. It is not likely that the thin ice on Mauna Kea, transporting rock fragments such a short distance, would have produced much faceting or striation on the hard lava blocks it carried. Well faceted and striated blocks are rare in recent moraines of short glaciers in other regions of hard rocks, such as the Cascade Range of Oregon.

For the present, the question of multiple glaciation of Mauna Kea must be left open.

*Suggested Additional Reading*

Gregory and Wentworth, 1937; Holmes, 1965, pp. 619–672; Jaggar, 1925; Leet and Judson, 1965, pp. 176–200; Stearns, 1945; Stearns and Macdonald, 1946, pp. 57–58, 166–167; Wentworth and Powers, 1941, 1943

*A perched spring,*
*near Laupahoehoe,*
*Hawaii*

# CHAPTER FIFTEEN

# Ground Water

WATER OCCUPYING THE VOIDS IN THE ROCKS BELOW THE level of the soil is called *ground water*. It is essential to the development of many areas, and today in Hawaii its importance is increasing steadily as community developments requiring more and better water spread to the drier parts of the islands, where surface water supplies are inadequate. This vital resource is subject to depletion through overproduction, but it differs from such other natural resources as metallic ores and oil in that beyond a certain critical point its depletion cannot be tolerated. It cannot be allowed to be "mined out," because without water the community cannot continue to exist. It is subject also to organic pollution and chemical contamination, through improper development methods, excessive production, or improper disposal of wastes into the ground, making it unfit for its principal vital uses. Overproduction and spoilage of our ground water must be avoided, and as population and land use increase this demands an ever-better understanding—particularly by engineers and geologists, but also by the general population—of the location, dimensions, and source of our ground-water bodies, and of the behavior of water within them.

The source of practically all ground water is rainfall. Although it probably is true that all of the water on and near the surface of the earth has been released from deep within the earth at volcanoes, its accumulation has required all of the billions of years of geologic time. During any one year the amount added to the ground-water body from deep sources is only an infinitesimal part of that added from rainfall. Careful geologic studies indicate that claims of great amounts of "pristine" water rising from the earth's interior, in any region, are false.

The addition of water to ground-water bodies is known as *recharge*. Knowledge of the amount of recharge is essential in planning and controlling the use of ground water, because if, for more than a short period of time, the amount of ground water removed by combined natural and artificial means (such as pumping from wells) exceeds the recharge, serious depletion of the ground-water body will result.

Unfortunately, the amount of recharge generally is very difficult

239

to determine accurately. Not all of the water that falls as rain infiltrates the rocks to become ground water, so that simply measuring the amount of rainfall does not give us the amount of recharge. Some of the water runs off over the surface as streams, most of which eventually find their way to the ocean. Another portion evaporates back into the air, and still another is taken up by plants, used in their metabolism, and eventually transpired back into the atmosphere. This "hydrologic cycle" is illustrated in figure 150.

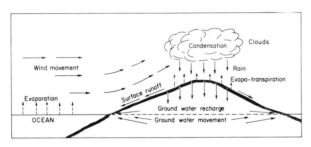

Figure 150. Diagram illustrating the "hydrologic cycle." Water evaporated from the ocean is condensed and falls as rain on land. Part of the rainfall runs back to the ocean as streams, part is evaporated back into the atmosphere, part is transpired back by plants, and part sinks into the ground to recharge the ground water bodies.

The amount of water flowing in streams is measured at many places, but the vast majority of streams are still unmeasured. Furthermore, it is often difficult or impossible to tell how much of the water in a stream is direct surface runoff, and how much has been fed into it by

springs and thus actually represents a leakage from the ground-water body. In most regions we can still only estimate the amount of rainfall that is disposed of by surface runoff. The amounts lost by evaporation and transpiration are even more difficult to determine. Table 14 gives some measurements made at two localities on Oahu. A great deal of work is being done by such agencies as the Pineapple Research Institute, the Hawaiian Sugar Planters' Association Experiment Station, the Agricultural Experiment Station and Water Resources Research Center of the University of Hawaii, and the United States Geological Survey to determine the extent of these losses in Hawaii. Direct measurements of the amount of water seeping into the ground also have been and are being attempted, principally by placing broad pans (lycimeters) below the surface of the ground to intercept the downward-moving water. The accuracy of these measurements in indicating the amount of infiltration in the region as a whole is open to serious question, however, because it is impossible to place a lycimeter in the ground without disturbing the ground above it and thereby altering its water-transmitting properties. For practical purposes, perhaps the most effective way to determine the amount of recharge is to observe the behavior of the water level in the rocks as various amounts of ground water are removed from them by pumping.

Table 14. Evaporation and consumptive use (evaporation plus transpiration) at two stations on Oahu

| Station | Year | Rainfall, inches | Evaporation inches | Evaporation % of rainfall | Consumptive use Panicum pan inches | Consumptive use Panicum pan % of rainfall | Consumptive use Fern pan inches | Consumptive use Fern pan % of rainfall |
|---|---|---|---|---|---|---|---|---|
| Lower Luakaha (890 feet altitude in Nuuanu Valley) | 1931 | 144.7 | 47.6 | 33 | 50.0 | 34 | — | — |
| | 1932 | 180.2 | 34.2 | 19 | — | — | — | — |
| | 1933 | 96.9 | 33.3 | 34 | 44.8 | 46 | — | — |
| Kaukonahua (1,250 feet altitude on side of Koolau Range above Schofield Plateau) | 1932 | 289.4 | 16.3 | 6 | 21.3 | 7 | 24.5 | 8 |
| | 1933 | 173.3 | 14.6 | 8 | 27.6 | 16 | 24.6 | 14 |

*Source:* Data from Stearns and Vaksvik, 1935.

240

Table 15. Estimates of disposition of rainfall in some areas in Hawaii (% of rainfall)

| Area | Consumptive use | Runoff | Infiltration |
|---|---|---|---|
| Honolulu-Pearl Harbor | 40 | 22 | 38 |
| Kaukonahua | 12 | 30 | 58 |
| Hilo-Puna | 20 | 0 | 80 |

In general, in the Hawaiian Islands as elsewhere, the relative proportions of recharge, runoff, evaporation, and transpiration can only be estimated. In dry regions evaporation and transpiration may greatly exceed surface runoff and infiltration. In the wetter regions of the Islands, however, they appear to amount to only about one-third of the total rainfall, the other two-thirds being divided between runoff and infiltration. As we have already seen (page 164), during the initial stage of the stream-erosion cycle in Hawaii there is essentially no runoff; and in these young regions, such as the southern half of the island of Hawaii, essentially all of the rain water that is not evaporated or transpired infiltrates into the ground. Over the Islands as a whole, except in the very dry areas and the very young areas, the rainfall appears to be distributed approximately one-third to stream runoff, one-third to evaporation plus transpiration, and the remaining third to recharge of the ground-water bodies. Table 15 gives a rough estimate of the proportion of the rainfall that reaches the ground-water body in several areas of differing climate and rock permeability.

The infiltrating water percolates downward through openings in the rocks until it encounters some obstacle, which may be simply an increasing tightness of the rocks. With increasing depth within the earth the rising pressure squeezes the rocks, thus closing most of the openings in them. Other openings are filled by deposition of various minerals, the materials for which were carried in solution in circulating water. The downward movement of ground water is thus limited to the uppermost part of the earth's crust, probably usually to the top few thousand feet, and other types of barriers may reduce or stop the downward movement at even lesser depths.

The openings in rocks through which water moves include spaces between the grains in sedimentary rocks, caves dissolved out by solution in limestone, and cracks (joints) in all kinds of rocks. In the lava rocks of Hawaii they include also lava tubes, openings between flow layers (fig. 151), and the spaces between the fragments in aa clinker. Bubble holes (vesicles) commonly are filled with water, but it is doubtful that communication between them is usually free enough to allow much active circulation of water through them.

Within the zone through which ground water is moving downward, the openings in the rocks remain partly filled with air, and consequently this is known as the *zone of aeration* (fig. 152). Water within the zone of aeration is sometimes called *vadose* water. The zone of aeration is of great practical importance because within it oxidation destroys or renders harmless much of the organic material that is carried into the rocks with the infiltrating water. Thanks to the zone of aeration, even with the large number of cesspools that are scattered over the hills behind Honolulu, there has never been any known contamination of the deep ground-water body that lies beneath the area. Certain shallow ground-water bodies are suspect, however, and when they are used as a source of domestic water they must be constantly monitored for pollution.

When the downward movement is halted by some barrier, the water accumulates in the rock, filling all the openings. The zone in which this occurs is a *zone of saturation,* and the upper surface of the water in it is called the *water table* (fig. 152). Above it water rises in narrow cracks to form the "capillary fringe." The water table is not a completely flat surface, as it would be on a mass of stagnant water in a pan, but slopes in one direction or another and commonly has humps and depressions. Indeed, it is often, to some extent, a

241

Figure 151. Ground water issuing from interflow spaces in pahoehoe lava into the development tunnel of the Halawa well, Honolulu, Oahu.

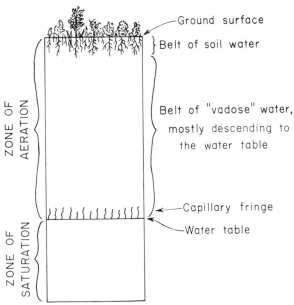

Figure 152. Diagram illustrating the zones of aeration and saturation, the belt of soil water, the water table, and the capillary fringe.

242

subdued counterpart of the irregular land surface above it. At places the land surface intersects the water table, and there the water flows out as *springs* (fig. 153). Draining out of ground water at a spring causes the water table to slope toward the spring. Similarly, when water is pumped from a well the level of the water table is lowered at that point, causing it to slope inward toward the well from all directions (fig. 153).

The amount by which the water level is lowered at a well by pumping is known technically as the *drawdown*, and the cone-shaped form of the water table around the well as the *cone of drawdown*. The amount of drawdown is governed by the rate at which water is pumped from the well and by the permeability (the freeness with which water can pass through the surrounding rocks). Consequently, when a well produces water at a high rate with only a small drawdown it is generally a good indication of a successful high-yielding well. A recently completed well near Hilo yielded water at a rate of nearly 7,000,000 gallons daily with only 4 inches of

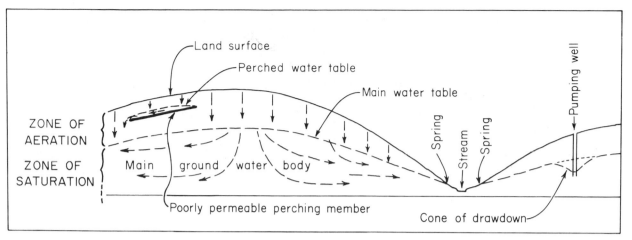

Figure 153. Diagram illustrating the movement of water beneath the ground surface.

drawdown, and therefore it is estimated that the well is capable of producing several times that much water.

## BASAL GROUND WATER

In the Hawaiian Islands and other oceanic islands, and along many continental coastlines, there is a special type of lower barrier to the movement of fresh ground water. Most of the base of the island is quite permeable—that is, it allows water to pass through it readily. As a result, all of the base of the island below sea level (except that within the rift zones of Hawaiian shield volcanoes, to be discussed later) becomes saturated with ocean salt water. Fresh water moving downward encounters the salt water in the rocks, and, because it is lighter than the salt water, floats upon it, just as a block of wood or a drop of oil floats on water. Thus the downward movement of fresh ground water is limited by the top of the salt water that saturates the base of the island.

The fresh water resting on salt water is known as the *basal* ground water body, and its upper surface is the *basal water table* (fig. 154*B*). The fresh basal water moves outward and escapes into the ocean at the edge of the island, but because of frictional resistance to its movement through the rocks it does not move with complete freedom, but piles up within the island until it attains sufficient hydraulic head to overcome the friction. In this way the basal water table acquires a slope toward the shore-

line, the amount of slope depending on the permeability of the rocks and the amount of recharge. In Hawaii the slope generally is between 1 and 8 feet per mile. In very permeable rocks the slope is less than in tighter

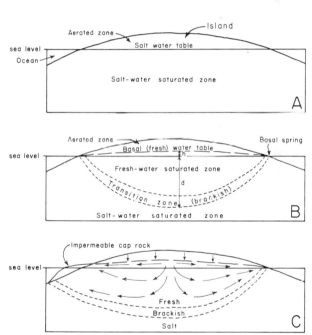

Figure 154. Diagrams illustrating the development of the Ghyben-Herzberg lens of fresh water within an oceanic island. *A,* An island without rainfall, saturated up to sea level with salt ocean water. *B,* An island in which water that fell as rain on the surface has descended through the rocks until it encountered salt water. The fresh water, being lighter, forms a lens-shaped body floating on the heavier salt water. *C,* An island with a relatively impermeable cap rock on one side. The cap rock raises the level of the water table inland from it and increases the thickness of the underlying fresh-water lens. The arrows show approximate directions of movement of water in the lens. (Modified after Palmer, 1946.)

243

rocks because the water can flow outward more readily (if it were not restrained at all, of course, it would quickly assume a completely horizontal surface); and in regions of high recharge the slope is greater than in drier regions because there is more water to flow outward through the available openings and a higher hydraulic head is necessary to drive it all through the rock within the same unit of time. The ideal shape of the basal water table in an island of homogeneous rocks is a broad flat dome (fig. 154B).

Springs formed by the escape of basal water are called *basal springs.* They are very numerous around some parts of the Hawaiian Islands, either just above or just below sea level. Some of them discharge several million gallons of water daily. Submarine basal springs often are noticed by swimmers as zones of unusually cold water. In the old days Hawaiians sometimes obtained drinking water from submarine basal springs by diving down into them and filling gourds with the fresh water issuing beneath the ocean.

Fresh basal water floating on salt water presses down on the salt water, just as any floating object presses down on the water beneath it. In this way the surface of the salt water beneath the island is pushed down below sea level. How far it is pushed down depends on the weight—that is, on the thickness—of the body of fresh water above it, and since the layer of fresh water is thicker in the central part of the island, the salt water boundary is depressed more there than near the coast. The precise amount it is pushed down depends on the ratio of the specific gravities (densities) of fresh and salt water. Using the average specific gravity of ocean water, we calculate that the top of the salt water should be pushed down approximately 40 feet below sea level for every 1 foot that the basal water table rises above sea level. Thus, starting at sea level at the coast, the bottom of the basal fresh-water body slopes downward toward the center of the island about 40 times as fast as its top rises. As a result, the basal ground water body has the form of a biconvex lens, with the bottom surface bulging much more than the top

surface. The principle of flotation of fresh ground water on salt water in coastal regions was discovered independently by a Dutch scientist, Baden-Ghyben, and a German named Herzberg, and the basal ground-water body is commonly referred to as the *Ghyben-Herzberg lens* (or as the *basal lens*).

The paths of outward movement of water through the basal lens vary greatly from place to place, depending on the location of the paths of greatest permeability through the rocks. Some water moves seaward along the surface of the lens, but most of it follows paths through the lens at various distances below its surface (fig. 154C). In the southeast Hilo area, for example, the surficial part of the basal lens is nearly stagnant, and the major movement of water is through the upper quarter of the lens below the stagnant upper layer. This zone of major movement of water is identified by the very low salt content of the water and by its low temperature, reflecting the low surface temperature in the region of infiltration higher up the side of the mountain.

The fresh basal water is not sharply separated from the underlying salt water. The lower part of the basal lens is a transition zone (fig. 154B, C) in which the water is brackish, due to upward diffusion of the salt and to physical mixing resulting from the up-and-down movements of the base of the lens, caused largely by tidal action. The thickness of the transition zone varies greatly from place to place. In areas of high recharge the transition zone tends to be thin because the high rate of movement of water through the lens flushes out the salt that has diffused upward. In dry areas, where the rate of recharge is low, the brackish water may extend completely through the basal lens, as it does in the seaward portion of the northern part of the Kona District on Hawaii.

Water can be obtained from the basal lens by sinking wells into it. It is the source of all the major production of ground water in Hawaii. Obviously, however, since the lens is underlain by salt water, care must be taken not to sink the wells too deep, or they will produce salt water. They cannot even reach close to the bottom of the lens because of the brackish transition zone. The amount of water a well will yield in a given length of time depends partly on the area of its surface within the zone of saturation. In order to obtain a large area below the water table without going too deep into the basal lens, many of the biggest wells in Hawaii are so-called Maui-type wells (first used on the island of Maui), consisting of a shaft leading from the ground surface to a nearly horizontal tunnel in the upper part of the basal lens. The present trend, however, is to drill several vertical wells penetrating only the uppermost part of the water body, instead of the more costly shaft and tunnel of the Maui-type well.

The amount of basal ground water available in some parts of the Hawaiian Islands is tremendous. Some single Maui-type wells have potential yields of as much as 40 million gallons a day. Calculations based on the probable rate of recharge indicate that about 3.5 billion gallons of water is escaping into the ocean each day around the east coast of the island of Hawaii from Hilo to Cape Kumukahi. Much of this water can be recovered whenever it is needed.

## ARTESIAN WATER

When a well encounters ordinary ground water, such as the basal water in Hawaii, the water does not rise in the well above the top of the zone of saturation. In the wells in some areas, however, the water rises to levels above the top of the saturated zone. These are called *artesian wells,* and the water is *artesian water.* The water may rise only part way, or it may come all the way to the ground surface as a flowing artesian well. Very commonly, as the ground water is depleted in a region that once had flowing artesian wells, the "head" of water diminishes until the water no longer overflows at the surface but rises only part way in the well.

The fact that water rises in the well indicates that it is under pressure, and in order to be under pressure it must be confined. In most artesian areas the water is found in a permeable stratum (called an *aquifer*) lying between two relatively impermeable strata (called *confining members*). The artesian aquifer receives its recharge in an area where it is exposed at the surface. From the recharge area it slopes downward, between the confining members, and may continue either gently sloping or horizontal, or even in part bent upward again, for many tens or even hundreds of miles. For instance, an aquifer of this sort that is exposed and recharged along the edge of the Rocky Mountains extends eastward beneath the Great Plains and is tapped by artesian wells as far away as 400 miles to the east. The water moves through the aquifer very slowly, and water being produced today by wells in the Mississippi Valley entered the aquifer along the Rocky Mountain front hundreds of years ago.

The level to which water will rise in an artesian well is governed by the level of the water table in the recharge area, minus a little for frictional losses in moving through the aquifer. Thus, if the water table in the intake (recharge) area is at an altitude of 1,000 feet, in a well a few tens of miles away the water will not rise quite to the 1,000-foot level because some of the potential energy of the water in the recharge end of the aquifer has been used up in overcoming friction in driving the water through the openings in the aquifer.

In Hawaii artesian conditions are present in the Honolulu-Pearl Harbor area and other smaller areas on the island of Oahu, and in a small area on the east side of the island of Kauai, but they are somewhat different from the more usual ones described above. In these areas the aquifer is permeable basaltic lava rock (in the Honolulu area, flows of the ancient Koolau volcano). The upper confining member (the so-called cap rock) is relatively impermeable soil and alluvial deposits, with some marine sediments, including poorly sorted gravel, sand, and silt washed down from the mountains (fig. 154C). Strictly speaking, there is no lower confining member. At the bottom the fresh water is confined by salt water, just as it is in the ordinary Ghyben-Herzberg (basal) lens (fig. 156a, p. 249).

Artesian water was first found in the area west of Pearl Harbor in 1879, and in 1880 near

245

the mouth of Manoa Valley. Many other wells were soon drilled throughout the Honolulu area. At first the wells were flowing, but production of the water gradually lowered the artesian head (the height to which water rises in the well, or in an extension of the well above ground level) until in recent years nearly all of these wells have had to be pumped to bring the water to the surface. The artesian head is governed by the height of the basal water table in the area inland from the edge of the upper confining member. The Honolulu-Pearl Harbor artesian district is divided into a series of smaller areas, in each of which the artesian head is a little different from that in the adjacent areas (fig. 155). The areas appear to

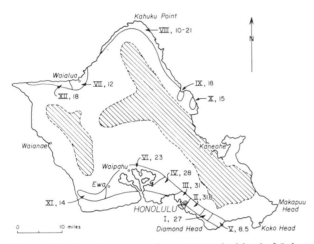

Figure 155. Map showing artesian areas on the island of Oahu. The roman numerals identify the areas mentioned in the text; the arabic numerals give the approximate height of the water table and the artesian head (in feet above sea level) in each area in 1930. The areas with diagonal shading are those in which water is confined between dikes. (After Stearns and Vaksvik, 1935.)

be separated primarily by wedges of alluvium deposited in valleys that were cut at a lower stand of the sea and extend well below present sea level (such as Nuuanu, Kalihi, and Manoa valleys). In each area the amount and conditions of recharge and the perfection of the confinement of the water are a little different and result in differences in the artesian head. Thus, in 1932 the head in artesian area II, between Manoa and Nuuanu valleys, was 31.8 feet above sea level, whereas that in artesian area I, between Manoa and Palolo valleys, was only 27 feet.

246

Since the first discovery of artesian water in Honolulu, both the artesian head and the basal water table have gone down as a result of heavy pumpage, and, as fresh water has been removed, the bottom of the basal lens has risen, bringing the transition zone and underlying salt water closer to the surface. Many of the older wells, drilled too deep into the aquifer, have become brackish. Great care is being exercised in new developments to prevent too great a lowering of water levels and salting up of the basal lens.

In places, water escapes through leaks in the upper confining member as artesian springs. The great Pearl Harbor springs, with a discharge of more than 80,000,000 gallons a day, are not artesian springs, however. They issue directly from Koolau lava rock at low points along the edge of the upper confining member, and thus really represent overflows of the artesian basin.

PERCHED WATER

Not all of the water percolating downward through the zone of aeration goes directly to the basal water table. In some areas there are beds that are much less permeable than those above and below them, and when the descending water encounters such an impermeable bed it tends to run off along it instead of sinking through (figs. 153, 156). In this way there is formed a saturated zone directly above the impermeable bed. Water thus held up, with an unsaturated zone between it and the basal water table, is known as *perched ground water.*

The impermeable bed, known as the *perching member,* may be a bed of dense lava, but more commonly it is a layer of either alluvium or volcanic ash. In the Ka'u area on Hawaii several layers of partly weathered volcanic ash are interbedded with the lavas and act as perching members. Where erosional valleys cut across the perching members the perched water may run out as springs. Extensive tunnels have been driven at the base of the saturated zone to obtain water for plantations. This water is especially valuable because it occurs naturally at high levels and does not have to be pumped to the level at which it is used, as does basal water.

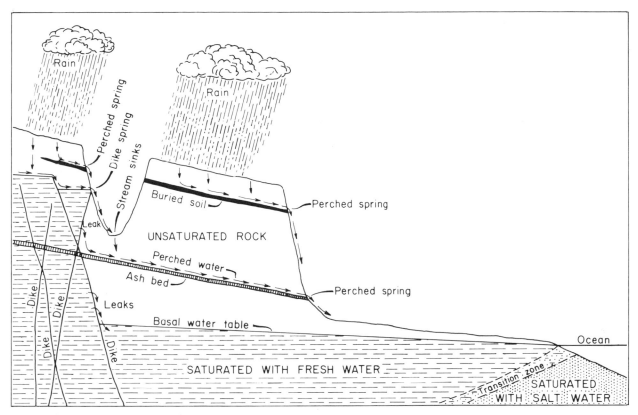

Figure 156. Diagram showing perched water, water confined between dikes, basal water, and perched and basal springs. (Modified after Stearns and Macdonald, 1946.)

In the Nahiku area, on the north slope of East Maui Mountain, streams of perched ground water follow ancient valleys filled with later lava flows. The water is tapped by driving tunnels along the base of the valley-filling lava flows until the old valley axis is reached.

Similar perched water bodies exist in many other parts of the islands, but they are all small as compared with the basal water lens.

## DIKE WATER

The numerous dikes in the rift zones of Hawaiian volcanoes constitute nearly vertical walls that are less permeable than the masses of ordinary lava flows between them. Water sinking into the rocks between the dikes is retarded in its escape toward the coast by the relatively tight dikes, and it accumulates in the compartments of more permeable rocks between the dikes (fig. 156), gradually rising in level until the leaks through the dikes equal the amount of recharge from above. In some of the dike complexes water is held between the dikes to a

height of more than 2,000 feet above sea level. In general, this water is probably not floating on salt water like the basal lens, but is simply held up by denser intrusive rock beneath.

On Lanai, wells have been drilled from the overlying surface into the saturated rocks between dikes. On Oahu, and in the Kohala Mountains of Hawaii, tunnels have been driven nearly horizontally into the dike complexes. The water obtained in this way from sources at high altitudes is, like perched water, more valuable than basal water because it will flow to the place it is needed without being pumped.

As in other types of ground-water development, if a permanent yield of dike water is desired, care must be taken not to remove water over any long period of time at a rate greater than the supply is renewed by recharge. The drainage of water from a tunnel penetrating the dike complex can be controlled by building a water-tight bulkhead across the tunnel where it crosses a tight dike and limiting the amount of flow through the bulkhead by means of a valve.

247

In some areas ground water encounters hot igneous rocks, becomes heated up, and issues as hot springs. These are commonest in volcanic areas. Elsewhere, ground water may penetrate to unusually great depths and become heated up by the normal temperature increase within the earth. Springs of this latter sort generally occur along faults—great fractures along which the rocks of the earth's crust have been broken and displaced.

Natural warm water is found at a few localities in the Hawaiian Islands. On the east rift zone of Kilauea volcano warm basal springs issue in the beaches between Cape Kumukahi and Opihikao village. Before it was buried by the 1960 lava flow, Warm Spring, just east of the village of Kapoho, was a pool of basal water in a crack along a fault cliff. In 1868 the water temperature of Warm Spring was 90° F., but by 1941 it had dropped to about 70°. The heat probably was derived from hot lava injected into the rift zone during the eruption of 1840. On West Molokai a hole drilled to test the quality of the ground water encountered brackish water with a temperature of 93° F. Water in a Maui-type well at the mouth of Ukumehame Canyon, on West Maui, has a temperature of 95° F. Steam vents on Kilauea and Mauna Loa volcanoes have already been described (page 46).

All thermal water in Hawaii undoubtedly owes its heat to volcanic sources. That there are not more hot springs in the islands is probably due to the great abundance of cold ground water, which effectively keeps down the temperature of the ground water body as a whole.

Geysers are simply spouting hot springs. Water accumulates in a subsurface cavity and becomes heated until that at the bottom of the cavity is above its boiling point at surface pressure, but it is prevented from boiling by the pressure resulting from the weight of the water above it. Slight expansion of the heated water may cause some of it to overflow at the surface, slightly reducing the pressure on the water at the bottom. This allows some of the hottest water to change into steam, the expansion of which causes further surface overflow, the further reduction in pressure allowing more water to flash into steam, and so on in a rapidly progressing chain reaction, the expanding steam blowing the overlying water into the air and producing a geyser eruption. Some geysers, such as Old Faithful in Yellowstone National Park, are very regular in their action, the period between eruptions being governed by the length of time required for the geyser tube to refill with water and heat the water to the boiling point. No geysers are known in Hawaii, probably because of the absence of suitable cavity structures and also the lack of water heated appreciably above surface boiling temperature.

In some parts of the earth steam accumulates under high pressure in tight rocks within a few hundred feet of the surface. This steam can be tapped by wells, and both the heat and the pressure can be utilized. Steam is now being produced from wells for the generation of electricity in New Zealand, Italy, Mexico, Central America, Iceland, and California. An unsuccessful attempt was made to develop similar steam wells along the east rift zone of Kilauea volcano. One well encountered boiling water, and hot water was found in other wells also, but it was generally below the boiling point. In Hawaii the rocks are too permeable to permit steam to accumulate under pressure at levels close to the surface, and the abundance of cold ground water makes unlikely the occurrence in most areas of steam at a temperature appreciably above the boiling point. In general, where heat sources exist they will merely heat more of the plentiful ground water to boiling point, rather than raising a smaller amount of steam to higher temperature. The steam issuing from vents in the wet Kilauea summit region is, with rare exceptions, always at or slightly below boiling temperature for that altitude. However, steam issuing in the caldera of Mauna Loa is slightly superheated (above boiling temperature), possibly because of the smaller amount of ground water available in that dry area.

Despite the lack of promising prospects for

the development of steam wells by the methods employed in other parts of the world, much volcanic heat unquestionably exists at relatively shallow depths in Hawaii. Methods are under consideration for the utilization of this heat, and there is every likelihood that ways will eventually be found to put to work the tremendous resources of heat energy that exist within the volcanoes.

*Suggested Additional Reading*
Cox, 1954; Davis and DeWiest, 1966; Meinzer, 1923*a*, 1923*b*; Palmer, 1946, 1957; Stearns, 1942*b*; Stearns and Clark, 1930, pp. 172–191; Stearns and Macdonald, 1947, pp. 53–79; Stearns and Vaksvik, 1935, pp. 235–272; Watson, 1955; Wentworth, 1951, pp. 59–102

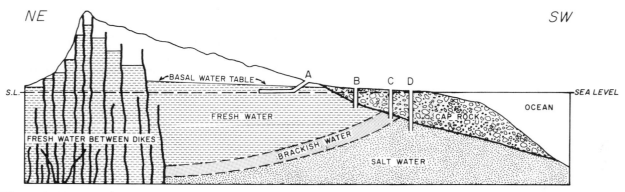

Figure 156a. Diagrammatic cross section of the Koolau Range at Honolulu (not to scale), showing general ground-water conditions and the nature of the Honolulu artesian system. *A,* Maui-type well producing fresh water; *B,* artesian well producing fresh water; *C,* artesian well producing brackish water; *D,* artesian well producing salt water. (Modified after Watson, 1955.)

249

*Hilo, Hawaii,*
*after the tsunami*
*of April 1, 1946*

# CHAPTER SIXTEEN

# Earthquakes and Tsunamis

## EARTHQUAKES

Deformation of the earth's crust causes the rock to bend until it reaches the breaking point. When breaking occurs, elastic energy stored in the rock during the bending is suddenly released, causing a vibration or trembling of the earth known as an *earthquake.* Some earthquakes are caused in other ways, such as by the collapse of cave roofs and volcanic explosions; but most earthquakes, and all of the major ones, result from breaking of the rocks under stress. Breaking of the rock, and movement of the rock on one side of the break past that on the other side, forms a *fault.* Many faults are known on which the displacement has amounted to thousands of feet; on some it has been several miles. There are even a few, such as the great San Andreas Fault in California, where one side has shifted past the other as much as several hundred miles. However, most faults, and certainly all of the very large ones, form by repeated small movements rather than by a single large one. Each of the small movements may cause an earthquake, because the frictional resistance involved in the slipping of rocks along a fault causes a considerable amount of elastic deformation of the rocks before the resistance is finally overcome and fault movement occurs. Thus each small movement on the fault may amount essentially to a fresh break, and cause an earthquake.

Earthquakes range in violence from those so small that they can be detected only by means of sensitive instruments, to those that cause great disturbance of the ground surface and extensive damage to man-made structures. Two different types of scales are now in general use to express the strength of earthquakes, and this has resulted in some confusion.

One type of scale is based on the amount of damage caused by the earthquake. Damage is greatest, of course, near the place of origin of the quake, where the shaking is strongest, and less at greater distances. Most earthquakes originate at some depth within the earth, rather than directly at the earth's surface, and the point on the surface above the point of origin of the quake is known as the *epicenter.* In the type of scale based on damage by the quake, known as an *intensity* scale, the number expressing the size of the quake is greatest at the epicenter, and becomes smaller at increasing distances. For example, the Kona earthquake of 1951 had an intensity of 6.5 near its epicenter offshore near Kealakekua, and intensities of 5 at Hilo, 56 miles away, and 4 at Hawi, about the same distance away but in a different direction. Lines drawn on a map through points of equal intensity of the earthquake are known as *isoseismals*. They are shown for the Kona earthquake in figure 157.

The other type of scale, known as a *magnitude* scale, is a measure of the actual amount of energy released during the earthquake, and does not depend on the kind or amount of damage done by the quake. The magnitude scale that is in common use was devised by Beno Gutenberg and C. F. Richter, and is often called the Richter scale. The magnitude value is determined by means of standard recording instruments (seismographs), taking into consideration the distance of origin of the quake from the instrument; theoretically the value should be the same everywhere. The figures usually reported in newspapers are magnitude values. There is no upper limit to this scale, but an earthquake of magnitude 7 or higher is a very big earthquake. The Kona earthquake of 1951 had a magnitude of about 6.5; the San Francisco earthquake of 1906 had a magnitude of 7.25. (The scale is logarithmic, not linear, so that a quake with magnitude 4 is not twice as large as one with magnitude 2, but 100 times as large). The highest magnitudes thus far reported have been between 8.5 and 9.

The intensity scale in commonest use today is the modified Mercalli scale, with a range from 1 to 12. An earthquake of intensity 1 is barely perceptible, whereas one of intensity 12 results in total destruction of man-made structures. Very few earthquakes have an intensity greater than 7, even in the epicentral area. Because man-made structures are generally more uniform in their reaction to earthquakes than are natural objects, the scale is based largely on the effects produced on artificial structures. The Mercalli scale is given on page 254, as modified by H. O. Wood (who was the first seismologist at the Hawaiian Volcano Observatory and later founded the superb net of seismograph stations of the California Institute of Technology), and further modified to include effects that are especially applicable in Hawaii, such as the slopping of water out of tanks. The table will also serve as an enumeration of the sorts of effects commonly caused by earthquakes.

Earthquakes can occur anywhere, as the quake that damaged buildings in New England in 1755 and the very violent quakes that shook the Mississippi Valley in 1811 and 1812 bear eloquent witness. However, the vast majority originate in certain zones of the earth's crust in which the rocks are being deformed, bent, and broken, as mountains are being built. One of the two principal mountain-building belts extends eastward from the Mediterranean region through the Himalayas and Indonesia and joins the other belt, which extends around the edge of the Pacific Ocean. In these belts, where the earth's crust is still being folded and faulted, earthquakes are frequent.

### Hawaiian Earthquakes

Except for the island of Hawaii itself, the Hawaiian Islands are not a highly seismic area; that is, they are not subject to numerous earthquakes. Even on Hawaii, most of the quakes are small and do little or no damage.

In 1929 a swarm of several thousand earthquakes came from a source beneath the north flank of Hualalai, the most severe of them causing minor damage in central Kona. The same area, and also South Kona and part of Ka'u, still further south, were somewhat more severely damaged by the earthquake of August 1951, that originated on the Kealakekua fault west of Kealakekua Bay (fig. 157).

A few earthquakes originating in other parts of the Hawaiian Islands have done small to moderate amounts of damage. The Maui earthquake of 1938, which had its epicenter about 25 miles north of Pauwela Point on the north coast of Haleakala, damaged roads and buildings on Maui and Molokai, and did minor damage even in Honolulu. A somewhat smaller earthquake, from a source somewhere in the area of Oahu, broke windows in downtown Honolulu in the spring of 1948.

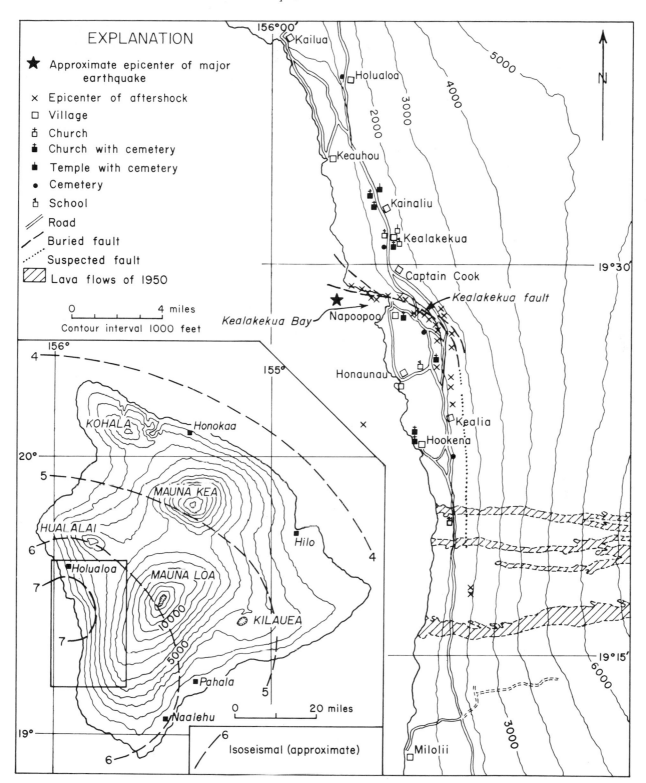

Figure 157. Map of the western slope of Mauna Loa, island of Hawaii, showing the epicenters of the Kona earthquake of August 1951 and of some of the aftershocks. The inset map of the island shows the isoseismal lines for the major earthquake. (After Macdonald and Wentworth, 1952.)

A few really large earthquakes have occurred during historic time on the island of Hawaii. That of 1868, which originated near the southern tip of the island, was an exceptionally strong one, with an intensity of 10 or more in the epicentral area. It is said that every

European-style building in the Ka'u District was destroyed. The earthquake was accompanied, and no doubt caused by, displacement on several faults. One of these, just west of Waiohinu, crossed a wagon road and offset the broken ends of the road about 12 feet. The violent shaking of the ground brought about the great mudflow at Wood Valley, described in chapter 10; and the crustal displacements in the ocean caused the tsunami mentioned in the following section.

Many areas in the vicinity of continental volcanoes experience earthquakes caused by volcanic explosions. However, as might be expected from the very slight explosiveness of Hawaiian-type eruptions, explosion earthquakes are rare in Hawaii. Some did accompany the violent phreatic explosions at Kilauea in 1924, but such episodes are very infrequent. Thousands of earthquakes stemming from the active volcanoes are recorded each year on the island of Hawaii, but they are the result of shifting of the rocks in the volcanic structures as the volcanoes swell and shrink, not of explosion. Eruptions of Kilauea and Mauna Loa are generally preceded by earthquake swarms consisting of thousands of individual earthquakes. Most of the quakes are too weak to be felt, even in their epicentral regions, and it is seldom that one is severe enough to do any serious damage. Some of them are accompanied by rumbling and booming noises from the ground. The quakes appear to result from minor fault movements as the top of the volcano is pushed upward by inflation of the magma reservoir beneath it (fig. 6). Often the earthquake swarms preceding flank eruptions culminate with the opening of fissures in the ground, and surface faulting that may produce grabens (troughs formed by sinking of the ground surface between two faults) several feet deep along the rift zone. The graben may be half a mile or more wide and several miles long. The point or points at which eruption takes place (generally within a few hours of the formation of the graben) is usually within the graben rather than along the fault at its edge.

The number of earthquakes which accompany the pre-eruption swelling of the mountain is almost unbelievable. During the earthquake prelude to the 1955 eruption 400 to 600 distinct quakes per day could be counted on the records (seismograms) from a rather insensitive seismograph over a period of a week or so before the eruption, and commonly they came so close together that their records overlapped, making it difficult to count individual quakes. But generally, when eruption starts the earthquakes suddenly cease. The volcano is no longer swelling, and the opening of the eruptive fissures has been completed.

Similar earthquake swarms accompany the shrinking of the volcano. For instance, in 1955, earthquakes preceding the eruption came largely from the Puna area, 20 miles east of the summit of Kilauea, where the rift zone was inflating preparatory to the flank eruption. But when, as a result of drainage of magma from beneath the mountain top during the eruption, the summit began to sink, earthquake activity shifted to the summit region. Although most of the quakes in the swarms accompanying subsidence also are small, they tend to be somewhat stronger than those accompanying swelling, and some are large enough to do damage.

Thus earthquakes are useful indicators of changes in the volume of the volcano, and constitute one of the principal sources of information used in predicting eruptions. But since earthquakes can accompany either swelling or shrinking of the mountain, the additional evidence supplied by tiltmeters is necessary to determine whether swelling or shrinking is taking place, and consequently whether the situation is building up toward an eruption, or vice versa.

## MODIFIED MERCALLI SCALE OF EARTHQUAKE INTENSITY

1. Not felt except by very few favorably situated persons.

2. Felt only on upper floors, by a few persons at rest; swinging of some suspended objects.

3. Quite noticeable indoors, especially on upper floors, but many persons fail to recognize it as an earthquake; standing automobiles may sway; vibrations feel like those of a passing truck.

4. Felt indoors by many during day, outdoors by

254

few; if at night, awakens some; dishes, windows, and doors rattle, walls creak; standing automobiles may rock noticeably; sensation like heavy truck striking building; water may slop from tanks.

5. Felt by nearly all, many awakened; some fragile objects broken, and unstable objects overturned; a little cracked plaster; trees and poles notably disturbed; pendulum clocks may stop; some damage to stone walls.

6. Felt by all; many run outdoors; slight damage; heavy furniture moved; some fallen plaster; some water tanks damaged; extensive damage to poorly built stone walls.

7. Nearly everyone runs outdoors; slight damage to moderately well built structures, negligible to substantially built, but considerable to poorly built; some chimneys broken; noticed by drivers of moving automobiles; water tanks extensively damaged or destroyed.

8. Damage slight in well-built structures, considerable in ordinary substantial buildings, with some collapse, and great in poorly built structures; panels thrown out of line in frame structures; chimneys and monuments thrown down; heavy furniture overturned; some sand and mud ejected; water levels changed in wells; automobile drivers disturbed.

9. Damage considerable even in well-designed and well-constructed buildings; frame structures thrown out of plumb; substantial buildings greatly damaged, shifted off foundations, partially collapsed; conspicuous ground cracks; buried pipes broken.

10. Some well-built wooden structures destroyed; most masonry and frame structures destroyed or knocked off their foundations; rails bent; ground cracked; landslides on steep slopes and river banks; water slopped out of rivers.

11. Few if any masonry structures left standing; bridges destroyed; underground pipes completely out of service; rails bent greatly; broad cracks in ground, and earth slumps and landslides in soft ground.

12. Damage total; waves left in ground surface, and lines of sight disturbed; objects thrown upward into the air.

TSUNAMIS

Probably the most feared natural scourge to which Hawaii is subject is the tsunami (figs. 158–160). A *tsunami* or *seismic sea wave*, is a series of elastic waves in the water of the ocean, caused by a sudden displacement of

255

Figure 158. Front of one of the waves of the tsunami of April 1, 1946, advancing as a bore past the former railroad bridge at the mouth of the Wailuku River in Hilo. Note the steep front, the turbulence of the water behind it, and the placidity of the water in front. The right-hand span of the bridge was destroyed by an earlier wave of the same tsunami.

Figure 159. Tsunami of April 1, 1946, flooding Hakalau Gulch, north of Hilo. The wreckage of the Hakalau sugar mill is in the foreground.

rock in the ocean. The generated waves spread outward in all directions from the source, commonly for thousands of miles, to become what are popularly, but wrongly, called a tidal wave.

Three different types of displacement may cause tsunamis. Some result from submarine volcanic activity, either explosions or the collapse of the top of a volcanic mountain during the formation of a caldera. During the eruption of Krakatoa volcano in 1883 the top of the mountain caved in, causing a tsunami that took many thousands of lives in coastal villages on the neighboring islands of Java and Sumatra. Small ones were set in motion by submarine volcanic explosions during the eruption at Myojin Reef, south of Japan, in 1952. Other tsunamis originate from landslides, either falling into the ocean, or occurring below sea level. Such tsunamis have been reported at Oshima Island in Japan, and at Yakutat Bay in Alaska. In July 1958, a landslide in Lituya Bay, also in Alaska, caused the water to sweep up to a height of 1,740 feet on the shore of the bay.

Tsunamis caused by volcanic action or landslides may be large near their points of origin, but generally they possess relatively little energy and decrease rapidly in size, so that they are small or undetectable at any great distance. Most tsunamis, particularly those that have sufficient energy to remain large to great distances, probably result from sudden fault movements on the ocean floor. A block of rock is suddenly elevated, forcing the water violently outward; or a block drops down, allowing the water to rush into the void thus created and rebound outward; or, perhaps most usually, an area of sloping sea bottom is shifted suddenly sideways giving the water a violent push in one direction. These are possiblities, but it should be emphasized that the precise cause is still not known; it is currently the subject of intensive study, at the University of Hawaii and elsewhere.

Whatever the precise mechanism by which the tsunami is generated, unquestionably it is associated with sudden crustal movements, and probably nearly always, if not always, with

256

faulting. Fault movements, such as displacement of segments of the ocean bottom, are accompanied by earthquakes, and therefore the generation of a tsunami is usually associated with a severe earthquake. It should be understood, however, that the earthquake does not cause the tsunami. Rather, both result from fault movement.

Only a small proportion of the earthquakes originating beneath the ocean are accompanied by tsunamis of sufficient size to be readily observed, and in many instances no tsunami can be detected even with sensitive instruments. Probably this is because most fault movements take place below the surface of the rocky crust and do not cause displacement of rocks at the surface. Because such a small proportion of even fairly strong earthquakes are accompanied by tsunamis, the occurrence of a submarine earthquake is not, in itself, an

adequate basis for the prediction of a tsunami; and tsunami warnings based solely on reports of submarine earthquakes lead to so many false alarms that all such warnings become useless. (In the 1920s the Hawaiian Volcano Observatory issued several warnings of tsunamis, at least one of which, in 1923, saved small boats in Hilo harbor from considerable damage; but it was soon realized that these warnings, based wholly on the occurrence of earthquakes, were undesirable.) Observations as to whether or not actual water waves have been formed are required for consistently successful prediction, and they are the only sound basis for issuing a warning, although the earthquake can, and does, alert the agencies responsible, which then start the investigation to determine whether a warning is necessary.

Slight movements which produce only small earthquakes do not cause tsunamis that are

Figure 160. Hilo waterfront, devastated by the tsunami of April 1, 1946.

large enough to be dangerous at great distances. The records of past tsunamis and earthquakes indicate that quakes of magnitude less than 6 in the border regions of the Pacific are unlikely to cause tsunamis of dangerous size in Hawaii.

"Tsunami" is a Japanese word which means "long wave in a harbor." The wave is indeed long, in the sense of distance between one wave crest and the next. In the open ocean this distance is of the order of 100 miles. If we couple this with the fact that in the deep ocean the wave is probably only a foot or two high, it is obvious that the wave cannot be seen by persons on shipboard far from shore. Reports of observation of tsunamis in the open ocean are erroneous.

Tsunamis are not ordinary water waves, like the familiar waves generated by wind. Rather, they resemble earthquake waves, traveling across the deep ocean at a speed approaching 500 miles an hour. The speed of the waves depends on the depth of the water:

$$\text{velocity} = \sqrt{\text{coefficient of gravity} \times \text{depth of water.}}$$

As they enter shallow water the waves slow up greatly, at the same time gaining in height. Depending on the character of the shore below sea level and the direction in which the wave approaches, the tsunami may produce a relatively gentle rise of water level and flooding of the shore zone, or it may be transformed into a wave of translation with a steep front and much turbulence that strikes the shoreline with great violence. Commonly the waves are less severe in bays and more severe on headlands, for the same reason that ordinary storm waves strike the headlands more violently. Under other conditions, however, the waves may be larger in bays than elsewhere. Some of the greatest observed wave heights have been near the heads of long, funnel-shaped bays.

The height to which the water rises on shore depends also on the size and shape of the land mass it encounters. Commonly the water rises much less on small islands, such as Wake or Midway, than it does on the large main Hawaiian islands and it is not safe to assume that because a wave is small on Midway, it will be small also on the major islands. Perhaps with the accumulation of more knowledge it will someday be possible to say that a wave one

258

foot high, for instance, at Midway will not rise more than some specified number of feet on the larger islands, but this time has not yet come, and the only safe procedure still is to assume that a tsunami of any size observed on the small islands is a potential danger to the main islands.

At some places a seismic wave advancing through shallow water is transformed into a

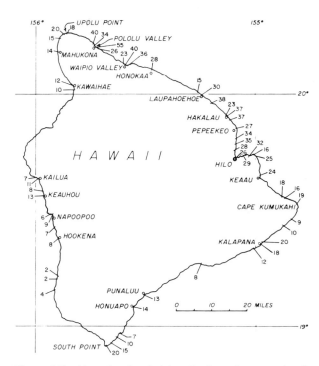

Figure 161. Map showing heights (in feet above sea level) reached by the water on the island of Hawaii during the tsunami of April 1, 1946. (After Macdonald, Shepard, and Cox, 1947.)

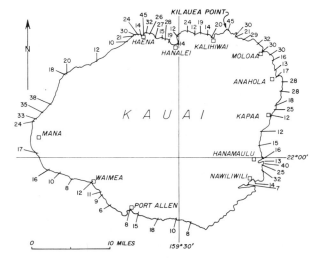

Figure 162. Map showing heights (in feet above sea level) reached by the water on the island of Kauai during the tsunami of April 1, 1946. (After Macdonald, Shepard, and Cox, 1947.)

phenomenon known as a *bore* (fig. 158). (Bores are also formed in some other parts of the world, such as the long narrow lochs of Scotland, by water entering during the ordinary rise of the tide, and their characteristics are therefore well known.) In essence, the oncoming wave breaks free and advances over the surface of the water in front of it, instead of involving this water in its movement; thus it attains a speed much greater than it would otherwise have had in shallow water. A bore formed in Hilo Bay during the tsunami of May 1960 struck the shore with a velocity of about 40 miles an hour, and with such force that the 2-inch pipes supporting parking meters along the waterfront were bent over parallel to the ground.

Generally there is a succession of waves during a tsunami. At each crest there is a rise of water level, flooding the shore; at each trough, a lowering of water level and withdrawal of water from shore, exposing part of the shallow sea bottom. The length of time between crests is usually between 12 and 20 minutes, averaging around 15. In many instances the first visible sign of a tsunami is the withdrawal of water from shore, but at other times the first apparent change is a rise of water level. Tsunami waves may dash far up onto shore. The tsunami of April 1, 1946, reached at least 115 feet above sea level on Unimak Island, near its origin, and heights up to 55 feet above sea level were measured in the Hawaiian Islands. Figures 161 and 162 indicate the heights reached by the water at that time on the islands of Hawaii and Kauai. The tsunami of April 2, 1868, which originated just south of the island of Hawaii, is reported to have come in "over the tops of the coconut trees" on the south shore of the island.

It cannot be too-much emphasized that the first wave of the series commonly is not the largest, nor is the wave that is largest at one place necessarily the largest at other places. During the 1946 tsunami each of the first eight waves was found to have been the largest at one place or another. Therefore it must not be assumed that because one wave of the series has come and gone without causing damage it is safe for people to return to evacuated coastal areas. Later waves may be more severe. Furthermore, the waves may be highest at one

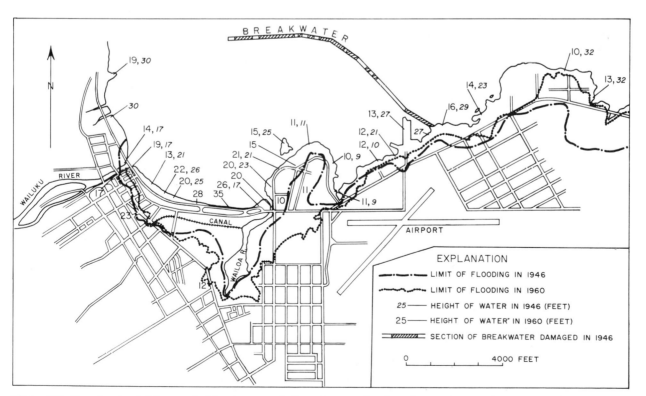

Figure 163. Map showing heights reached by the water (in feet above sea level), and areas devastated at Hilo, Hawaii, during the tsunamis of April 1, 1946, and May 23, 1960. (After Macdonald, Shepard, and Cox, 1947; and Eaton, Richter, and Ault, 1961.)

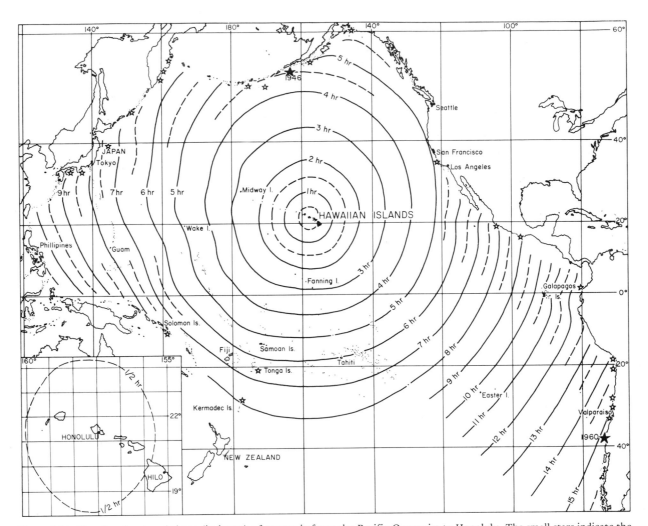

Figure 164. Map showing travel times (in hours) of tsunamis from the Pacific Ocean rim to Honolulu. The small stars indicate the approximate points of origin of some tsunamis that have affected Hawaii. The origins of the disastrous tsunamis of 1946 and 1960 are shown by the two larger stars. (Modified after Zetler, 1947.)

place during one tsunami, and at another place during a different one; a place that escaped damage during one severe tsunami may be heavily damaged by another. Many lives were lost in Hilo in 1960 because people believed the Waiakea area, unharmed by the violent waves of 1946, would again be safe from damage (see fig. 163.)

Most of the tsunamis that affect the Hawaiian Islands originate in the belts of mountain building which surround the Pacific Ocean (fig. 164). In the past, waves that did moderate to severe damage have come from the Aleutian Islands, South America, and Kamchatka. Table 16 lists the principal tsunamis recorded in the Hawaiian Islands during historic time. Thus far waves from Japan have caused only minor damage, but it is possible that future ones may

be more destructive. No waves from the southwest Pacific have been large in Hawaii, possibly because of the screening effect of the numerous islands in that part of the ocean, as well as the greater distance. Although they have not occurred as yet, tsunamis of dangerous proportions from the west coast of North America (including Mexico and Central America) are a distinct possibility.

Regrettable as has been the loss of life and property in Hawaii from tsunamis, it should be pointed out that other areas have suffered far more than has Hawaii. Perhaps most seriously hit has been the northwestern coast of the main island (Honshu) of Japan, where tsunamis are frequent, due to crustal movements that occur in and near the great trough (Japan Deep) that lies offshore, and that are related to the growth of the great complex mountain ridge of which the Japanese islands are the top. In the Sanriku area alone, some 30,000 persons were drowned by the tsunami of 1896, and several thousand more in 1933. Furthermore, the sources of these tsunamis are so close that commonly the populace has a warning of less than half an hour. In Hawaii, except for the

very rare tsunamis of local origin, a span of several hours is available in which to evaluate the likelihood of a tsunami, issue the warning, and evacuate the endangered areas (fig. 164). With increasing knowledge it should eventually be possible to eliminate all casualties from tsunamis of distant origin in Hawaii.

Table 16. Principal tsunamis during historic time in Hawaii

| Date | Source | Damage in Hawaii | Average speed of waves (miles per hour) |
|---|---|---|---|
| 1819, April 12 | Unknown | Unknown | — |
| 1837, Nov. 7 | South America | Severe | — |
| 1841, May 17 | Kamchatka | Small | — |
| 1868, April 2 | Hawaii | Severe | — |
| 1868, Aug. 13 | South America | Severe | — |
| 1869, July 25 | South America (?) | Moderate | — |
| 1872, Aug. 23 | Hawaii | Small | — |
| 1877, May 10 | South America | Severe | — |
| 1883, Aug. 26 | East Indies | Small | — |
| 1896, June 15 | Japan | None | 478 |
| 1901, Aug. 9 | Japan (?) | None | — |
| 1906, Jan. 31 | Unknown | None | — |
| 1906, Aug. 16 | South America | Small | — |
| 1918, Sept. 7 | Kamchatka | Small | 456 |
| 1919, April 30 | Unknown | None | — |
| 1922, Nov. 11 | South America | None | 450 |
| 1923, Feb. 3 | Kamchatka | Moderate | 432 |
| 1923, April 13 | Kamchatka | None | 438 |
| 1927, Nov. 4 | California | None | 462 |
| 1927, Dec. 28 | Kamchatka | None | 438 |
| 1928, June 16 | Mexico | None | 462 |
| 1929, March 6 | Aleutian Islands | None | 492 |
| 1931, Oct. 3 | Solomon Islands | None | 447 |
| 1933, March 2 | Japan | Small | 477 |
| 1938, Nov. 10 | Alaska | None | 496 |
| 1944, Dec. 7 | Japan | None | 425 |
| 1946, April 1 | Aleutian Islands | Severe | 490 |
| 1952, Nov. 4 | Kamchatka | None | 434 |
| 1957, March 9 | Aleutian Islands | Small | 497 |
| 1960, May 22 | South America | Severe | 442 |
| 1964, March 28 | Alaska | None | — |

*Source:* Modified after Macdonald, Shepard, and Cox, 1947.

## Suggested Additional Reading

Eaton, Richter, and Ault, 1961; Hodgson, 1964; Leet and Judson, 1965, pp. 291–306; Macdonald, Shepard, and Cox, 1947; Macdonald and Wentworth, 1952; Richter, 1958; Shepard, Macdonald, and Cox, 1950; Wood, 1914, 1933

*Lualualei Valley, Oahu*

# CHAPTER SEVENTEEN

# Age of the Islands

UNTIL RECENTLY THERE WAS NO DIRECT MEANS OF MEAS-
uring the age of the rocks of the Hawaiian Islands—or anywhere.
Ages were estimated from geological evidence, but large errors were
always possible. Only in the last two decades have we had anything
approaching a real clock to measure geologic time. The clock is a
geochemical one, depending upon the breakdown of radioactive
materials in the rocks or in substances buried by the lava flows or ash
beds at the time of their formation. It has been gratifying to find
that, in general, the radioactive dates for the rocks confirm the
general succession of events and agree, at least approximately, with
the ages that have been deduced from geological reasoning.

GEOLOGIC EVIDENCE

Until the discovery of the radioactive clock the usual way of dating
rocks was by means of fossils contained in them or in related rocks.
Early in the 19th century it was recognized that in the formation of
sedimentary rocks, layer was deposited upon layer, so that any given
layer must be younger than the one beneath it and older than the
one above it. Furthermore, William Smith, working in England,
found that different layers contained different characteristic fossils,
and that by applying the principle of superposition of layers the
relative ages of different fossils could be established. It was soon
found that the same fossils and successions of fossils were present in
rocks over large parts of the earth, and there was thus developed a
means of correlating, or establishing the relative ages of, rocks in
different regions. It was also observed, however, that fossils of
different types of organisms were present in rocks of the same age
that had accumulated in different environments. Deep-water sediments contain different fossils from those in sandstones formed
along ancient beaches, and land-laid sediments contain still different
ones; but there is enough intermingling of life forms, and interbed-
ding of sediments formed in different environments, to establish the
general succession of fossil types. Furthermore, it was discovered
that often an evolutionary sequence could be traced within groups of
fossils, earlier forms showing features that are ancestral to those in
later forms.

263

Certain groups of rocks, containing fossils of similar age, were found to be separated from other groups by unconformities—surfaces marking the absence of rocks of intermediate age, which either have never been deposited at that particular locality or have been eroded away after they were deposited and before the area was covered by younger rocks. In many instances the older rocks have been folded or tilted, and beveled by erosion, and the younger rocks laid down across the edges of the older beds, showing that a period of deformation of the earth's crust (commonly a period of mountain building) intervened at the time marked by the unconformity. In this way a general time scale was gradually established, in which the rocks were assigned to different units formed during different eras, periods, and epochs of the earth's history. But although there was an increasing realization of the immense spans of time involved, educated guesses remained the only means of assigning ages in terms of years to the different units of geologic time. The geologic time scale is given in appendix B, with ages in years that are derived by the radioactivity methods described later.

Fossils are rare in lava flows everywhere, and in Hawaii the only ones that have been found are molds of tree trunks that, although they show the rocks to be geologically very young, are not useful for close dating. In other regions lava flows have been dated from fossils found in beds of sedimentary or pyroclastic rock between the flows, but in the Hawaiian Islands the only such find is the bones of a goose and some plant remains in an ash layer in the Ka'u District on Hawaii, and again these show only that the rocks are geologically young. To be sure, fossils are abundant in the coral reefs that cling to the shores of the older Hawaiian islands, but these are so young that many of the species are still living either in Hawaiian waters or in warmer waters farther south, and they tell us little about the age of the underlying volcanic rocks except that they are older than the late part of the Pleistocene ("glacial") epoch (see appendix B).

Of necessity, therefore, estimates of the ages of the Hawaiian Islands have been based on nonpaleontological grounds, particularly on the degree of weathering and erosion that the surface of each volcano has undergone since it was formed, and on the overlapping onto one volcano of the lava flows from another. Estimates of the length of time required for the construction of the individual volcanoes can be made using the assumption that the rate of building was about the same as that at Mauna Loa and Kilauea during historic time. However, such estimates give only a very crude minimum for the duration of the volcanic activity, because the presence of erosional unconformities and sedimentary beds between successive lava flows during the late stages of building of some of the volcanoes, such as Haleakala and the Waianae volcano, show that eruptions during those stages were far less frequent than they are at the present active volcanoes.

Further uncertainty results from the fact that the rate of lava outpouring by different volcanoes is quite variable even in the same general stage of building, and from the possibility that building during the earlier stages may have been even faster than that during historic activity. Nevertheless, these figures do suggest some limiting values. At the historic rate, the formation of the entire mass of Mauna Loa, from the sea floor up, could have taken place within the last 1,500,000 years; but since we know that erosional interruptions occurred in some places (see chap. 19), it is very probable that Mauna Loa's activity has extended over a considerably longer period—perhaps 3,000,000 years or even more.

Mauna Kea was already built to its present size by the time the glaciers disappeared, at the end of the Pleistocene, and if we make allowance for the slower pace of volcanism during its late stages it seems almost certain that the activity of Mauna Kea must also have begun several million years ago.

Comparison of the degree of weathering and erosion of the older Hawaiian volcanoes, such as the Waianae and Koolau volcanoes on Oahu, and the Kauai and Niihau volcanoes, to that of similar rocks of known geologic ages in other parts of the world suggests that these volcanoes must have ceased activity well over a million years ago. Using the rate of burial of the trunks of kiawe trees on the island of Lanai by debris eroded from the higher parts of the island since the introduction of the trees in 1837, Wentworth (1925, 1927) determined an average rate of erosion of the uplands of Lanai of about 1 foot in 2,900 years, and allowing for other factors he estimated that the general rate in the Hawaiian Islands was probably about 1 foot in 5,000 years. On this basis he concluded that the Lanai volcano finished building about 100,000 to 150,000 years ago. (The age appears to us too young.) The Koolau volcano, on Oahu, and the Kauai volcano he estimated respectively to have ceased activity 1,000,000 and 2,000,000 or more years ago.

Considering the same sort of evidence, and in addition the relationship of coral reefs to the big valleys cut into the Koolau Range and the fact that Koolau lavas overlap the edge of the Waianae volcano, showing that the Waianae is the older of the two, H. T. Stearns also estimated ages of the volcanoes in terms of the geologic time scale (appendix B). The Waianae volcano, he concluded, is of late Pliocene, the Koolau volcano of late Pliocene or early Pleistocene, and the later Honolulu Volcanic Series of Pleistocene age (Stearns and Vaksvik, 1935, p. 67).

For a long time it was believed that the boundary between the Pleistocene and Pliocene epochs was about 1,000,000 years ago. During the last few years, however, largely as a result of a redefinition of the Pleistocene on the basis of climatic change as well as the percentage of living species among the fossils found in the rocks, the boundary has been pushed back to about 2,500,000 years before the present. Thus some rocks that formerly would have been called late Pliocene are now classified as of Pleistocene age. This must be borne in mind in comparing the older with the recent geologic age designations for Hawaiian rocks.

Also using the degree of weathering and erosion, the main Kauai shield volcano has been estimated (Macdonald, Davis, and Cox, 1960, pp. 22–23) to have finished building about 4 million years ago, during the Pliocene epoch, and to have started building during the same epoch, some 2 to 4 million years earlier.

Still more recently, Moberly (1963*b*) has estimated the rate of erosion in the Koolau Range above Kaneohe Bay to be about 1 foot in 2,300 years, based on the amount of calcium in solution in the stream waters draining from that area, and he concludes that erosion of the area by streams alone to its present stage of dissection would have required about 5 million years. This estimate, however, leaves out of consideration the possibility of more rapid erosion of part of the area by waves during higher levels of the sea and of more rapid erosion because of higher rainfall before the general level of the range was lowered by isostatic (?) sinking and erosion (see chap.10). Making all possible allowances, Moberly concludes that the beginning of the present cycle of erosion of the Koolau volcano must have been at least 1.3 million years ago.

The deep weathering and erosion of the West Maui volcano, East and West Molokai volcanoes, and the Kohala volcano on Hawaii, has led to the conclusion that all became extinct (except for small, very local, much later eruptions on West Maui and East Molokai) during middle to late Pliocene time. In view of the revision of the Pliocene-Pleistocene boundary, the extinction of these volcanoes would now be placed within the Pleistocene epoch. Moreover, radioactive methods, described in a later paragraph, indicate that the rocks of the Kohala volcano are considerably younger than was formerly believed. To account for the degree of erosion exhibited by Kohala volcano it is necessary to revise upward our previous estimates of the speed of erosion in Hawaii.

One further bit of geological evidence should be mentioned. In 1961, dredging on a submerged terrace 1,600 to 1,700 feet below sea level southwest of Honolulu brought up fossils, including shallow-water corals, that are probably of very late Miocene age (Menard, Allison, and Durham, 1962). The terrace is separated from Oahu by a shallow basin, and it must be the remains of a flat-topped, reef-crowned seamount that existed during Miocene time, more than 10 million years ago. The present Hawaiian Islands may be built on an older ridge, perhaps contemporaneous with a chain of islands that existed in middle and early Tertiary time farther northwest, in the region of the Leeward Islands.

EVIDENCE FROM GEOMAGNETISM

When a lava flow solidifies each ferromagnetic grain in it (mainly magnetite and ilmenite) becomes a tiny magnet with its field oriented parallel to that of the magnetic field of the earth at that time. If at a later time there is a change of the orientation of the earth's magnetic field, or of the orientation of the rock within the earth's field, the rock to some extent takes on a new direction of magnetism, but also to a considerable extent the old direction remains "frozen in." Measurement of this "remanent magnetization" has become an important tool in the attempt to decipher the changes of position of different parts of the earth's crust relative to each other. Differences in the direction of magnetization of rocks of the same age in different parts of the world constitute some of the most important evidence for "continental drift"—the shifting of one large segment of the crust in relation to others. In Hawaii the horizontal component of magnetism is found to be essentially parallel to the present magnetic field of the earth (allowing for relatively small cyclical variations), and consequently there is no evidence from magnetism that the position of the islands has changed appreciably since they were formed.

Another, even more remarkable, sort of change is found, however. Study of well-dated rocks in other parts of the world has shown that there have been several abrupt reversals in the polarity of the field during the last few million years. Although the horizontal direction of the magnetic force lines has remained unchanged, the position of the north and south magnetic poles has been reversed. The north pole of the magnet in one set of rocks points in exactly the opposite direction to that in adjacent sets of rocks formed at a slightly different time. The last of these reversals took place about 700,000 years ago. Another reversal, to the "normal" position coinciding with the present one, took place about 2.5 million years ago (Cox and Dalrymple, 1967). Study of the magnetism of the rocks in Hawaii is still in a very early stage. Thus far the results indicate that some of the rocks have "normal" magnetism, but others have the polarity reversed. From this it can be concluded that those with normal polarity were formed either within the last 700,000 years or more than 2.5 million years ago, and those with reversed polarity, between 700,000 and 2.5 million years ago or during very brief periods of reversal that existed about 900,000 and 3 million years ago. Rocks of the main Kauai shield volcano have normal magnetism, and are presumed to have formed more than 2.5 million years ago, since some of the later Koloa (posterosional) lavas show reversed magnetism. On Oahu, most of the rocks of the Waianae Range have normal magnetism, and probably formed more than 2.5 million years ago, whereas the rocks of the younger Koolau volcano have reversed magnetism and presumably formed between 700,000 and 2.5 million years ago. The directions of magnetic polarity, so far as they are known, are

266

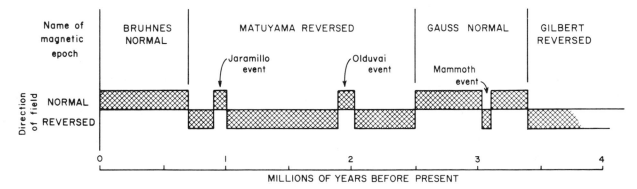

Figure 165. Magnetic polarity epochs during recent geologic time. (Modified after Cox and Dalrymple, 1967.)

shown in figure 165. Most of the ages of Hawaiian rocks as interpreted from magnetism are in reasonable agreement with other geologic evidence.

## GEOCHEMICAL EVIDENCE

The most accurate methods of determining the ages of rocks in terms of years make use of the constant rates of decay of radioactive substances in the rocks or in associated materials. Various methods exist, such as those based on the change of uranium and thorium to certain isotopes of lead, and radioactive rubidium to strontium, but for rocks of very young geologic age, like those of Hawaii, the most useful are those involving the change of potassium-40 to argon-40, and carbon-14 to nitrogen-14.

Organisms incorporate into themselves carbon from the environment, either air or water, in which they live. Plants, for instance, take in carbon from the atmosphere in the form of carbon dioxide. Most of the carbon is the isotope $C^{12}$, but cosmic rays from outer space bombard the nitrogen of the air and transform some of it to $C^{14}$. The ratio of $C^{14}$ to $C^{12}$ in the air is quite constant (though some changes are known to have occurred in past time), and the ratio of the two carbon isotopes incorporated into the organisms is the same as that in the atmosphere. However, after the death of the organism the $C^{14}$ gradually disintegrates, at a constant rate, while the $C^{12}$ remains the same in amount if it is not affected by outside processes. Thus, in fossil materials that have not been contaminated by addition of new carbon, the age of the material can be ascertained by determining the ratio of $C^{14}$ to $C^{12}$ in it. Because $C^{14}$ breaks down quite rapidly (its half-life, or the span of time in which half of it is destroyed, is only about 5,570 years), the carbon method is good only for very young rocks, less than 40,000 years old.

The carbon method has yielded useful dates of a few lava flows and ash beds in Hawaii, through the analysis of bits of charcoal buried by the volcanic rocks. It also has given the ages of a few marine fossils. Shells in two elevated coral reefs on the north shore of Oahu, 5 and 12 feet above sea level, are found by this method of dating to be about 24,000 and 31,000 years old, respectively, thus giving approximate ages in terms of years to the late Pleistocene reefs in which they occur (Shepard, 1961). Fragments of coral limestone in the tuff at Koko Head on Oahu also have been investigated, but because of the limitation of the method we can say only that the fragments are more than 33,000 years old. Even this limiting date applies only to the submerged reef from which the fragments were torn. The explosion that formed the tuff probably was very much more recent.

Very recent work has shown that corals and algae from the base of the reef in the head of Hanauma Bay are 6,000 years old (Easton, 1968); and since this reef rests not only on the tuff of the Hanauma cone, but also in places on younger black ash from the eruption of Koko Crater, both of these eruptions must have taken place more than 6,000 years ago.

Potassium-40 makes up a small but constant part of the total potassium in the original constitution of the minerals of rocks. Its breakdown generates argon-40, which is trapped within the structure of the mineral. Careful determinations of the amounts of potassium and argon-40 in the mineral make it possible to calculate the age of the mineral and of the rock that contains it. The same method has been extended to the treatment of fine-grained volcanic rocks, which are analyzed as a whole rather than as separate minerals.

Several factors may affect the accuracy of the determined ages, in addition to the obvious one of accuracy of the analyses. One is the slow leakage of $A^{40}$ from the mineral and rock, reducing the amount present and thereby the apparent age of the rock. Some minerals appear to lose argon more readily than others, feldspar being generally regarded as one of the most leaky. Because, in Hawaiian rocks, feldspar is the mineral that contains most of the potassium, the leakage factor must be kept in mind. Most workers in this field feel, however, that in dealing with very young rocks the amount of leakage is too small to have any important effect on the determined age of the rocks. Possibly of greater importance is the fact that some minerals have been found to contain more $A^{40}$ than should have been generated by the decay of the $K^{40}$ in them, giving an apparent age that is too great. This excess $A^{40}$ is probably magmatic gas trapped in the mineral during its formation. Some of the minerals in which it is found contain small fluid inclusions that in turn enclose tiny bubbles of gas. One of these minerals is biotite, which previously has been regarded as one of the best minerals for K-A age determination because it is one of the least likely to leak argon. Actually, the amount of excess argon is so small that it would have an important effect on the apparent age only of very young rocks in which the amount of $A^{40}$ generated from $K^{40}$ within the mineral is still very small.

The rocks of Hawaii fall within the age range in which argon leakage is probably unimportant, but in which the inclusion of magmatic $A^{40}$ may cause serious errors in age determinations. An apparent example is the age determined for the rhyodacite at Mauna Kuwale, in the Waianae Range. Ian McDougall, of the Australian National University, who published the first K-A determinations of the ages of Hawaiian rocks, found ages of 3.6 to 2.7 million years for most of the Waianae rocks, by means of whole-rock analyses. In the case of the rhyodacite, however, he carefully removed the biotite and worked on it separately. The age obtained for the biotite was 8.4 million years—more than twice that found for the other rocks. To explain this, he suggested that the rhyodacite is the top of a much older volcano that had been buried by the Waianae lavas (a suggestion first made by Stearns and

Table 17. Ages of Hawaiian volcanic rocks as determined by the potassium-argon method

| Island | Volcano | Volcanic series or formation | Orientation of the remanent magnetic field (R = reversed, N = normal) | Age (millions of years) | Source of information[a] |
|--------|---------|------------------------------|------------------------------------------------------------------------|-------------------------|--------------------------|
| Hawaii | Mauna Loa | Ninole | N | 0.1–0.5 | A, B |
| | Kohala | Pololu | N | 0.33–0.45 | E |
| | | Hawi | N | 0.06–0.25 | E |
| Maui | Haleakala | Kula | N | 0.4–0.8 | A, B |
| | West Maui | Honolua | R | 1.15–1.17 | A, B |
| | | Wailuku | R | 1.27–1.30 | A, B |
| Molokai | East Molokai | Upper member | R | 1.3–1.5 | A, B |
| | | Lower member | R | 1.5 | A, B |
| | West Molokai | West Molokai | R | 1.8 | A, B |
| Oahu | Posterosional | Honolulu | — | 0.1–0.9 | C, D |
| | Koolau | Koolau | R | 2.2–2.6 | A, B, C |
| | Waianae | Upper member | R and N | 2.7–2.8 | A, B |
| | | Middle member | R and N | 2.5–3.0 | A, B |
| | | Lower member | R and N | 2.9–3.3 | A, B |
| Kauai | Posterosional | Koloa | N and R | 0.6–1.4 | A, B, D |
| | Kauai | Makaweli | — | 3.3–4.0 | A, B |
| | | Napali | N | 4.5–5.6 | A, B |
| Nihoa | Nihoa | — | — | 7.5 | C |
| Necker | Necker | — | — | 11.3 | C |

268

a. Sources of information: A, McDougall and Tarling, 1963; B, McDougall, 1964; C, Funkhouser, Barnes, and Naughton, 1968; D, unpublished determinations, courtesy of J. Dymond and N. J. Hubbard, Lamont Geological Observatory, Columbia University; E, McDougall and Swanson, 1972 (Bull. Geol. Soc. Am., vol. 83, pp. 3731–3738).

Vaksvik, in 1935, as a result of a wholly different line of reasoning).

The explanation would be a reasonable one, in view of the demonstrated existence of a seamount or island in the area of Oahu in Miocene time, were it not that careful field studies of Mauna Kuwale reveal no indication of weathering or erosion between the rhyodacite and the overlying basalt, as there certainly should have been if the rhyodacite were 5 million years older than the basalt. More recently, Funkhouser and other investigators at the University of Hawaii determined the ages of the rocks underlying the rhyodacite and of the feldspar phenocrysts in the flow overlying the rhyodacite, as well as of the rhyodacite itself, by means of whole-rock analyses. The underlying rocks gave ages of 2.3 and 3.1 million years, and the rhyodacite 2.3 million years, in good agreement with the ages found by McDougall for rocks of the middle member of the Waianae Series. The feldspar from the overlying flow gave an age of 4.3 million years, somewhat older than even the whole-rock determinations on the flows of the lower member. It appears almost certain that the apparent age of the feldspar crystals is too great, although there is a possibility that they were brought up by the rising magma from an accumulation of older crystals at depth. Moreover, it has been demonstrated (Funkhouser and others, 1966) that there is a considerable range in ages determined from different samples of biotite separated from the rhyodacite, and that fluid inclusions are present in the biotite in variable amount. The whole-rock age of the rhyodacite appears to be more nearly correct than those derived from the biotite.

Other inconsistencies still exist in the ages determined by the potassium-argon method. For instance, determinations by two different laboratories on the Sugarloaf lava flow of the Honolulu Volcanic Series yield respectively 100,000 (±80,000) years, and 900,000 (±500,000) years. In contrast, the very small amount of weathering and erosion of the flow suggest that it is probably only a few thousand years old, and preliminary results of K-A age determinations now being made by John Gramlich at the University of Hawaii seem to confirm that the flow is indeed less than 20,000 years old. In general, however, the K-A ages almost surely are of the right order of magnitude, and they serve to confirm the estimates made on wholly geological grounds.

Potassium-argon ages of Hawaiian rocks are given in table 17, together with magnetic polarities. Even though there may still be some error in the determined ages, it appears clear that the oldest rocks of the major Hawaiian islands now exposed above sea level were formed during the Pliocene epoch. Although they seem very old in terms of human life, they are very young indeed in terms of the history of the earth.

*Suggested Additional Reading*

Doell and Cox, 1965; Funkhouser, Barnes, and Naughton, 1968; McDougall, 1964; McDougall and Tarling, 1963; Menard, Allison, and Durham, 1962; Moberly, 1963*b*; Shepard, 1961

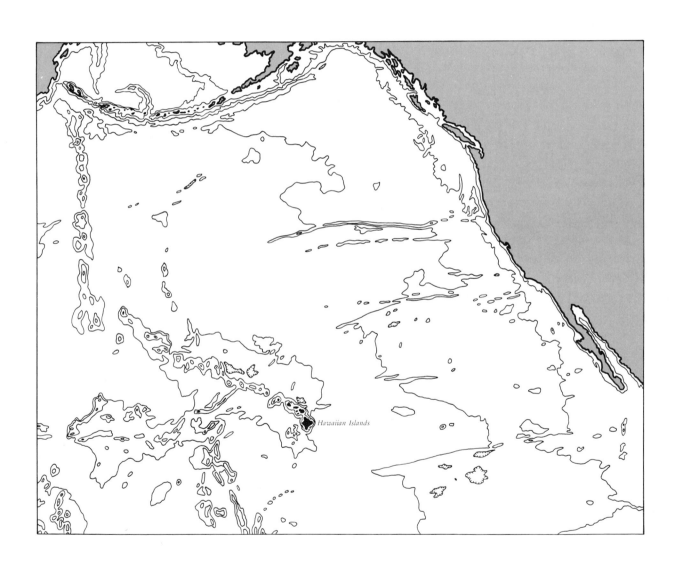

Hawaiian Islands

# CHAPTER EIGHTEEN

# Structural Setting of the Islands

TO PLACE THE HAWAIIAN ISLANDS IN THEIR GEOLOGICAL environment we must consider their relation both to the earth as a whole and to the Pacific Ocean basin in which they are located. Volcanism, such as built the Islands, originates within the earth, and we must therefore consider very briefly the internal structure and composition of the earth. The location and distribution of the Hawaiian volcanoes has been determined by the structure of the earth's crust in the Pacific basin, and certain differences between Hawaiian volcanism and that of the continental borders surrounding the basin are probably related to differences between the crust beneath the basin and that of the continents.

## GENERAL STRUCTURE OF THE EARTH

It is generally accepted today that the earth consists of several concentric shells, or layers, having different physical properties (fig. 166). At the center of the earth is a mass known as the *core*, having a radius of approximately 2,200 miles and a density (specific gravity) of 10.7. Surrounding the core is a shell known as the *mantle*, approximately 1,800 miles thick, with an average density of about 4.5. Finally, outside the mantle is the earth's *crust*, with an average density of about 2.8. Since the radius of the earth is approximately 4,000 miles, simple addition of the thicknesses of the core and mantle will show that the crust of the earth must be very thin. In fact, it is about 20 to 45 miles thick beneath the continents, but only about 3 miles thick beneath much of the ocean basins.

Despite its thinness in relation to the earth as a whole, the crust is the only part of the earth of which we have any certain direct knowledge, because it is the only part exposed to direct observation. The deepest mines extend downward only about a mile and a half, and the deepest oil wells only between 4 and 5 miles. In some parts of the continents folding of the crust and subsequent erosion have exposed rocks that once were 10 or 12 miles below the surface, but these are still part of the crust. In a few places rising magma has carried to the surface small fragments of rocks that certainly originated below the crust; but while they give us excellent clues as

*271*

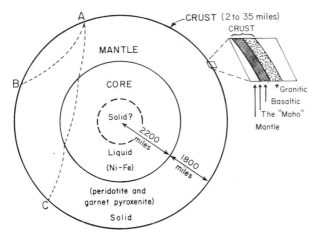

Figure 166. Diagrammatic cross section of the earth. The enlarged inset on the right shows the general structure of the crust of continental regions.

to the nature of the mantle they probably do not represent the nature of the mantle as a whole. In other places large masses of the olivine-rich rock, peridotite (see chap. 4), or of serpentine formed by the hydration of peridotite, have been brought to the surface by faults. Although many geologists believe that these may be portions of the outermost mantle, we cannot yet be certain. With the exception of these possible samples, all of the vast interior of the earth is invisible to us. How, then, do we know anything about it?

There are several lines of evidence. For instance, it has been possible, by measuring the earth's gravitational attraction for other bodies, to "weigh" the earth. In this way we know that the average density of the earth is about 5.5. However, since the visible rocks of the earth's crust average less than 3 in density, there must be much denser material within the earth to compensate for the lighter rocks at the surface. Studies of the changes produced in the earth's rotation by external forces tell us that this denser material is not uniformly distributed, but is concentrated near the center. The most precise information, however, comes from earthquake waves that have traveled through the earth.

272

*Seismic Evidence*

The general nature and causes of earthquakes have been discussed briefly in chapter 16. Most earthquakes are believed to result from a sudden breaking, or sudden movement along a

fault plane, of rocks that have been under elastic strain. This releases energy that travels outward in all directions as elastic vibrations, or waves. There are several types of earthquake waves, but the only ones we need consider now are the so-called P and S waves. The P (primary) waves consist of an alternate compression and rarefaction of matter along the path of propagation of the wave, resembling ordinary sound waves in air. The S (secondary) waves are shear waves, in which particles of matter vibrate back and forth across the path of the wave (fig. 167). The P waves have been

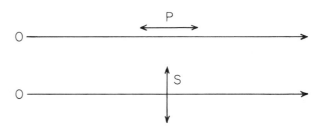

Figure 167. Diagram illustrating the direction of particle vibration in relation to the direction of propagation in P and S earthquake waves.

referred to as "push-pull" waves, and the S waves as "shake" waves. P waves travel faster than S waves. Thus, if both types start simultaneously from a given point, say point A in figure 166, the P waves will arrive at any other point (say point B) before the S waves. Furthermore, the speeds of both types of waves differ in different materials, depending on the elastic properties of the material. In general, S waves travel at about three-fifths the speed of P waves.

Thus, at point B (fig. 166), representing a seismograph station, the P wave will be recorded somewhat before the S wave; and if we know (from the records of stations near the origin of the earthquake) the time and place of origin, we can calculate the average speed with which each wave traveled through the earth to station B. By comparing these average speeds at many stations it is possible to reconstruct the paths of the waves through the earth, and variations in speed along different parts of the paths. It is found, the world over, that abrupt changes in velocity, caused by changes in the physical properties of the rocks, occur at certain quite consistent depths within the earth. These levels, where abrupt changes occur, are known as *discontinuities.*

The highest of these discontinuities is found at a depth of 20 to 45 miles beneath the continents and 3 to 5 miles beneath the floor of the deep ocean basins. In the rocks just above this discontinuity the average speed of P waves is about 6.5 kilometers per second. Below the discontinuity the speed is markedly higher, in some places as low as about 7.3 km/sec, but usually close to 8 km/sec. This discontinuity is named after its discoverer, the Yugoslavian seismologist A. Mohorovicic; but the cumbersome term, Mohorovicic discontinuity, is commonly contracted to "the Moho." You will recognize that the Moho was taken in a previous paragraph as the base of the earth's crust.

A few years ago there was much talk of an exploratory hole to be drilled by the United States Government through the Moho into the mantle—the "Mohole." The plan was to drill it in the ocean basin because the crust there is so much thinner than beneath the continents, and a site about 130 miles north of the island of Maui (fig. 171) had been chosen as the most favorable. In 1967 the project was shelved because of its very high cost, but the knowledge to be gained from the Mohole is so great that almost surely the project someday will be revived. Samples from the Mohole will give us our first direct look at the earth's mantle.

Another even more pronounced break marks the boundary of the core (fig. 168). Downward

through the mantle the speed of P waves gradually increases to about 13.5 km/sec, but at the boundary of the core their speed drops abruptly to a little more than 8 km/sec. What is even more striking, the S waves disappear altogether. One of the characteristics of S (shear) waves is that they are not transmitted through an ordinary liquid, and even in a liquid of extreme viscosity they become very much weakened—hence the obvious deduction that the core of the earth is liquid. However, recent studies indicate that S waves probably are transmitted through the innermost part of the core, and suggest that the inner core (inside the dashed line in fig. 166) is probably solid. A correlative conclusion is that, since the S waves travel through the earth all the way to the boundary of the core, the earth must be solid down to that level. The picture geologists once held, of a thin solid crust resting on a molten, liquid interior, is now generally abandoned.

It is possible that small bodies of liquid, of dimensions less than the wave length of ordinary earthquake waves (about 30 km) might escape detection, but there cannot be any completely liquid shells, or even very large bodies of liquid, at a level above the core boundary. It should be noted that a few geologists, notably W. Kuhn and A. Rittmann, believe that the mantle is actually liquid, but is so viscous and has such a long relaxation time that it will transmit S waves. Under very high pressures the distinction between solids and very viscous liquids becomes vague, and it is possible that the deeper part of the mantle is, after all, a very viscous liquid. If so, the viscosity must change abruptly at the core boundary to account for the disappearance of S waves there, and the viscosity of the liquid lower mantle must be so great that for practical purposes it can be regarded as a solid.

### Composition of the Earth's Interior

Thus we arrive at a general picture of the physical properties of the materials in the successive shells of the earth. But do we know what the actual materials are? The answer is no, but we can make some guesses. At present the almost unanimous choices are: a core of nickel-iron, molten at the outside but probably

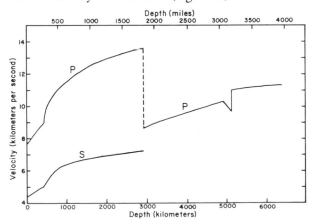

Figure 168. Graph showing discontinuities in velocity of P and S earthquake waves within the earth. The very conspicuous break in the curves at a depth of approximately 1,800 miles marks the boundary of the core. (After Bullen, 1963.)

solid at the center; a solid mantle of peridotite, probably varying somewhat in composition from place to place and containing small bodies of material of non-peridotitic composition, and probably undergoing a "phase" change at some depth from an assemblage of minerals stable under relatively low pressures to an assemblage stable under higher pressures; and a crust of light silicate rocks among which basalt (or gabbro) and granitic rocks (granite and granodiorite) are predominant. (The rock types are described in chapter 4.)

The evidence on which these materials are chosen is twofold. The first consideration is that they must correspond in density and elastic properties to those that have been shown by geophysical methods to exist within the earth. The second approach is, in a sense, astronomical. The thousands of meteorites that strike the earth each year are generally believed to be either fragments of a disrupted planet (or planets) that originally resembled the earth, or bits of cosmic "dust" such as originally were gathered together to form the earth. Under either hypothesis, they may logically be assumed to give us an idea of what the interior of the earth must be like. Meteorite materials are predominantly of two sorts: stony meteorites, resembling in various degrees the rock peridotite; and metallic meteorites, composed chiefly of an alloy of nickel and iron. The former presumably indicate the general nature of the earth's mantle. As long ago as 1866, the French geologist Daubrée took the latter as an indication that the core of the earth is probably nickel-iron.

We all know that the earth behaves as a great magnet, and this fact has sometimes been cited as evidence of a nickel-iron core. Actually, however, it is not. For many decades we have known that the temperature increases downward within the earth, until at the boundary of the core it almost certainly exceeds 2,500° C. With increasing temperature the strength of ordinary ferromagnetism decreases, until at a level known as the Curie point it disappears altogether. Under ordinary pressures the Curie point of iron is close to 500° C., and in the laboratory it is little affected by changes in pressure. It is very unlikely, therefore, that the earth's magnetic field is due to simple ferromagnetism in a nickel-iron core. More probably the earth's field is electro-magnetic, the result of electrical currents in the core.

To illustrate the fact that geological science is in a dynamic state of development, and knowledge of the earth is still far from complete, we should point out that there is not complete agreement on the nature of the core. For instance, Kuhn and Rittmann, prominent Swiss geochemist and volcanologist respectively, have suggested that the central part of the earth consists not of nickel-iron, but of solar matter rich in hydrogen, in a highly condensed state. At first glance this might seem impossible, because the earth's core is so dense; but solar, or stellar, matter can be very dense indeed. Kuhn suggests that the atoms in the core are "stripped"—that is, bereft of their electron rings. Matter believed to be of this sort in some of the "white dwarf" stars may attain the fantastic density of 4,000—more than 300 times as heavy as lead. Just a little atom stripping might well boost the density of solar matter to the value of 10.7 calculated for the earth's core. Kuhn and Rittmann's idea is very interesting, but few other geologists or cosmologists have accepted it. Most believe that even at the earth's center the pressure is insufficient to bring about such a change in the condition of matter.

Other geologists have suggested that some of the discontinuities within the earth are caused by phase changes in silicate minerals, light mineral assemblages stable under low pressures being replaced by assemblages of heavier minerals stable under higher pressures. Thus the augite and feldspar of gabbro, stable under low pressures, are transformed at high pressure to a mixture of garnet and omphacite (an augite-like pyroxene containing sodium) to form a heavy rock called eclogite. Phase change may be a possibility for the deeper part of the Moho beneath the continents, but few geophysicists are willing to admit that it could occur at levels as shallow as that of the Moho beneath the ocean basins. At present it appears most likely that the Moho, even beneath the continents, marks a change in composition, possibly in some areas combined with a phase change; and the most probable compositions for the mantle and core, respectively, appear to be a rock resembling peridotite in chemical composition, and an alloy of nickel and iron.

In Hawaii, lavas of both the late phase of the principal period of volcanism and of the posterosional period have brought up from depth inclusions of peridotite, and at Salt Lake Crater on Oahu nephelinite of the posterosional period contains inclusions of garnet-pyroxene rock (eclogite). In the late lavas of the principal period, the minerals of the inclusions form an assemblage that is stable under fairly low pressure, but in the 1801 lava of Hualalai volcano the inclusions contain microscopic bubbles of carbon dioxide in which the gas pressure is so high that the rock must have come from the uppermost part of the mantle. The peridotite inclusions in the posterosional lavas contain augite that is abnormally rich in aluminum, indicating that it was formed under high pressure; and the eclogite inclusions at Salt Lake Crater are a high-pressure assemblage. The inclusions in the posterosional lavas must have come from the upper part of the mantle, probably from a somewhat lower level than did the inclusions in the late lavas of the principal period. The latter probably represent fragments from gabbroic intrusive bodies in which the olivine and pyroxene crystals had sunk toward the bottom of the body and accumulated there to form layers of peridotite. They do not appear to represent original mantle material. The Salt Lake Crater eclogite also appears to be a high-pressure phase of olivine-rich gabbro, and probably represents only a relatively small mass within the mantle. The peridotite inclusions in the posterosional lavas, on the other hand, may represent fairly typical mantle material—but mantle material from which a certain proportion of material of basaltic composition had been removed to supply the basalt magmas of the Hawaiian volcanoes. (See chap. 2.)

## The Continental and Oceanic Crust

Beneath the continents the earth's crust is composed of two parts. The upper part, which we can see, consists of a great variety of rocks—igneous (volcanic and plutonic), sedimentary, and metamorphic. The sedimentary and metamorphic rocks form a relatively minor proportion, however. Igneous rocks make up about 95 percent of the upper 10 miles of the crust, the proportion of plutonic rocks increasing at deeper levels. Among the oldest (pre-Cambrian) rocks, the most abundant single type is granite. In the younger mountain-building (orogenic) belts the plutonic igneous rocks are on the average somewhat poorer in silica and alkalies, approximating granodiorite. There is relatively little gabbro, but great volumes of basaltic volcanic rocks are present in some areas, such as the flood-basalt accumulations of the pre-Cambrian Keweenawan Series in Michigan, or the much later (Tertiary) Columbia River basalts of Washington and Oregon. Considering all of the various rock types, the average composition of the upper continental crust is about that of granodiorite or quartz diorite. It is commonly called the *granitic* layer, but a better term is *sialic* layer, indicating that its rocks are rich in silicon (Si) and aluminum (Al).

The velocities of P waves in the sialic layer vary greatly, according to the particular type of rock, but average about 4 km/sec. At depths generally ranging from about 15 to 30 miles beneath the surface of the continents the velocity of P waves commonly undergoes a sudden increase, to about 6 km/sec. This increase marks the top of a lower portion of the crust, which is believed to consist of igneous rocks fairly rich in magnesium and iron ("mafic" rocks), and is commonly called the *basaltic* layer. The term "basaltic" layer refers only to the general composition of the material, since magma of basaltic composition crystallized slowly at depth would yield a coarse-grained gabbro, not a basalt. Thus the continental crust consists of an upper granitic (sialic) layer and a lower basaltic (mafic) layer. In some areas, however, the typical basaltic layer appears to be absent, and the lower part of the crust has properties intermediate between those of the typical basaltic and granitic layers.

The crust beneath the ocean basins is markedly different from that under the continents. The sialic layer is absent. The seismic proper-

275

ties of the suboceanic crust are approximately the same as those of the basaltic layer beneath the continents, and the basaltic layer is generally believed to be present over essentially the entire earth. Beneath the continents this layer is invisible, and it is called basaltic only because of theoretical considerations largely related to its physical properties as revealed by earthquake waves. Beneath the oceans there is more direct evidence that it is basaltic. All of the specimens of igneous rocks dredged from the deep ocean floor are basalt.

The presence of the sialic layer appears to be the fundamental reason for the existence of the continents. Sialic rocks are light. Studies of gravity over the face of the earth indicate that, on a moderate to large scale, the earth's crust is in a condition resembling flotational equilibrium. Areas of light rocks stand high, and areas of heavy rocks are low, as though they were floating in a denser fluid beneath. According to this principle of flotational equilibrium, known as *isostasy,* the continents are relatively high because they are light, and the ocean basins are low because they are heavy.

Figure 169 shows diagrammatically the conditions at a continental boundary, with the thick light continental mass bordered by the thin basaltic crust of the ocean basin. It should be pointed out, however, that the edges of the

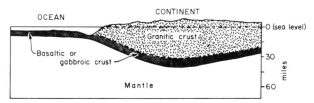

Figure 169. Diagram illustrating the change in crustal thickness and composition at the boundary between the continents and ocean basins.

oceans do not coincide with the edges of the true oceanic basins. There is too much water in the oceans to be contained entirely in the ocean basins, and some of it extends over the edges of the continents. The submerged edges, known as the continental shelves, geologically are parts of the continents, not the ocean basins.

*Geosynclines and Mountain-Building*

In the geologic past there have been formed repeatedly along the edges of the continents down-wrinkles in the earth's crust, hundreds of miles long, known as *geosynclines.* In these troughs sediments and lava flows were deposited, and as the layers accumulated, the bottom of the trough gradually sank. Thus, although the water in the trough seldom, if ever, was very deep, the total thickness of rocks—conglomerate, sandstone, shale, limestone, and lavas—reached 40,000 feet or even more. Eventually, forces of unknown origin crushed the edges of the geosyncline together, folding the sedimentary layers and raising them into a range of mountains. The upheaval was probably isostatic, due to the thickening of the mass of light rocks. During and after the folding, granitic magmas were generated at depth and rose into the rocks above them, forming great *batholiths* of granitic rocks such as the core of the Sierra Nevada in California, or the great complex of granitic bodies that forms the core of the Coast Range of British Columbia. The folded and granitized geosyncline was welded to the edge of the continent and became part of it. The development of these belts of mountains from the geosyncline is known as *orogeny.*

During the growth of the folded mountain belt, and for millions of years afterward, magma escaped from time to time at the surface as volcanoes. The entire border of the Pacific basin consists of folded mountain belts, and everywhere along it volcanoes either are still active or have been active in the geologically recent past. Most of the volcanoes are quite different from those of Hawaii. The eruptions are more explosive (Strombolian, Peléean, or even Plinian), and the rocks are predominantly andesites instead of basalts. This latter difference is expressed by the "andesite line" (fig. 170), which marks the limit of the predominantly andesitic volcanoes around the Pacific. For the most part, the andesite line marks the true limit of the continents, the thin edges of which are partly under water. In some places, however, as in the Tonga and Mariana islands, narrow belts of andesitic volcanoes are separated from the true continent by broad areas of oceanic-type crust on which the sialic layer is nearly or completely absent. What the reason for this may be, we cannot yet say.

Volcanism of Hawaiian type is found in various parts of the continents, and the continental flood basalts are even more voluminous and their eruptions less explosive than are Hawaiian eruptions. But with the exception of a few belts such as the Tonga and Mariana archipelagoes, which are associated with long troughs (foredeeps) that may be related in origin to geosynclines, the andesitic volcanism of the orogenic belts is absent from the ocean basins. Some geologists believe that the continental andesites are the result of melting and/or dissolving of part of the continental sial layer by rising basaltic magma; others believe that they are formed by melting of the base of the sial beneath the sinking geosyncline, or even in part by melting of the lower portion of the geosynclinal sediments. The belts such as the Tonga and Mariana islands that are isolated from the continental sial masses do not fit well either of these pictures. The magmas of these andesites most probably formed, like basalt magmas, by partial melting under appropriate physical conditions in the upper part of the earth's mantle. On the other hand, the general spatial relationship between continental andesites and continental sial strongly suggests some sort of genetic relationship. Over geologic eons, have the processes that generate andesites—whatever they may be—also generated the sial masses of the continents?

## STRUCTURE OF THE PACIFIC BASIN FLOOR

Three prominent structural trends can be recognized in the Pacific basin. One is a northwest-southeast trend that is quickly apparent from the alignment of rows of islands on a map. The Hawaiian Islands are one such row; but there are many others, such as the Society Islands and Austral Islands, farther south. A nearly north-south trend is much less obvious,

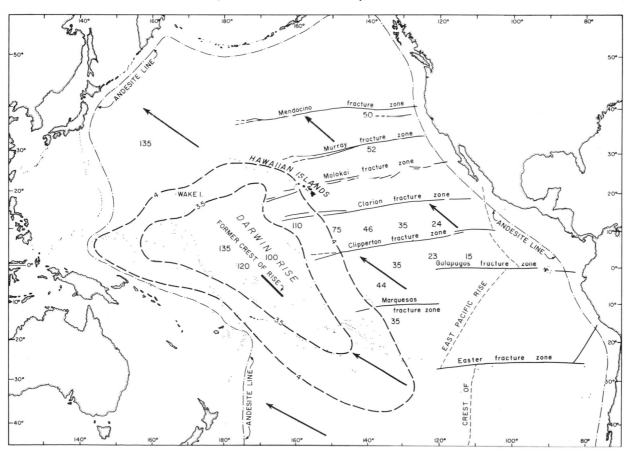

Figure 170. Map of the Pacific Ocean, showing the principal fracture zones, the course of the East Pacific Rise, and the position of the former Darwin Rise. The dashed contours on the Darwin Rise show the depth of water (in miles) over the Rise about 100,000,000 years ago (after Menard, 1964). The scattered numbers give the approximate age, in millions of years, of the lowest sediment resting on the basalt crust (after Winterer, 1973, Bull. Am. Assoc. Petr. Geol., vol. 57, pp. 265–282). The heavy arrows show the present direction of movement of the Pacific plate.

277

but appears on bathymetric maps in the alignment of seamounts and some other lineaments, particularly in the part of the ocean north of the Hawaiian Islands. The most prominent of the north-south lines is that of the Emperor Seamounts, that extend for 1,500 miles from a point west of Kure Island at the western tip of the Hawaiian Archipelago northward to near the junction of the Aleutian Ridge and Kamchatka. A less prominent line of seamounts extends northward from near Pearl and Hermes Reef to the deep trough just south of the Aleutian Islands. A very abrupt north-south trench, almost surely a graben, lies just east and northeast of Gardner Island.

The third trend is defined by a series of great "fracture zones" (Menard, 1964, p. 41) that extend in a direction somewhat south of west from the coasts of North and South America. They are tremendous fault zones that separate major blocks of the sub-Pacific crust. Thirteen of these fracture zones are now known. Individual ones can be traced as far as 2,000 miles, and are as much as 120 miles wide. They separate parts of the ocean floor that may differ by several thousand feet in average depth and may be markedly different in other properties. Thus the Mendocino fracture zone (fig. 170), which trends westward from Cape Mendocino in California, separates sea floor to the north from an area to the south that averages about 4,000 feet deeper. Furthermore, the sea floor to the north is littered with seamounts (submarine volcanoes), while there are very few in the area to the south. The basaltic crust is much thinner north of the Mendocino fracture zone than south of it, and the velocity of earthquake waves in the uppermost part of the mantle is slower (only about 7.5 km/sec). Farther south, trending westward from the Santa Barbara Islands of Southern California, is the Murray fracture zone, and south of it the sea floor is more like that north of the Mendocino fracture zone than that between the two fracture zones. The Murray fracture zone intersects the Hawaiian Ridge near Necker Island, and its trend is continued southwestward by the Mid-Pacific Mountains (figs. 170, 489). Still farther south, trending westward

from near Cedros Island off the coast of Baja California, is the Molokai fracture zone (fig. 170), which heads for the major Hawaiian islands. Near its western end it splays out, one strand heading for a steep north-facing slope northeast of Molokai, and another toward the eastern end of Maui. Recent evidence indicates that, although the Molokai fracture zone is apparently interrupted by the Hawaiian Ridge and Deep (see next section), it continues southwest of the islands (Malahoff and Woollard, 1966).

Evidence suggests that at least some of the fracture zones have had very large horizontal fault offsets. Roughly north-south belts of magnetic anomalies on the sea floor are broken and offset by the fracture zones. On the Mendocino fracture zone the offset is about 700 miles, with the block on the north shifted westward (Vacquier and others, 1961). The offset may be even larger on the Molokai fracture zone, in the same direction.

All three of the major structural trends recognized in the sub-Pacific crust are also evident on the Hawaiian volcanoes. Not only are the islands aligned northwestward, but the rift zones of the Waianae and Koolau volcanoes on Oahu, and the principal rift zone of Kohala volcano on Hawaii trend in the same direction. The principal rift zones of Kauai and Niihau, West and East Molokai, Haleakala, Mauna Loa, and Kilauea are at least approximately parallel to the west-southwest trend of the great fracture zones, as also is a row of seamounts northeast of Oahu and another west of Hawaii. The following pairs, and one triplet, of volcanoes also show the west-southwesterly trend: Mauna Kea–Hualalai, Haleakala–Kahoolawe, West Maui–Lanai, East Molokai–West Molokai, and Kauai–Niihau–Kaula (figs. 1, 279). For the pairs Kilauea–Mauna Loa, and Koolau–Waianae, the trend is slightly north of west. The nearly north trend is shown by alignment of some of the posterosional vents on Oahu and Kauai, though others show a northeast alignment. It is evident that many volcanic structures on the islands have been influenced, if not governed, by the older structures of the sea floor.

278

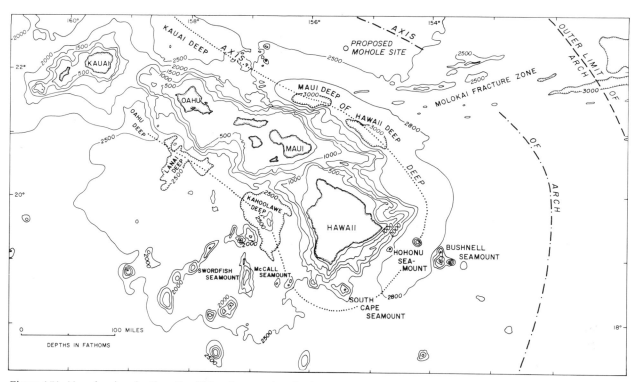

Figure 171. Map showing the Hawaiian Ridge, Deep, and Arch, the position of the Molokai fracture zone, and the proposed site for the Mohole. (Modified after Malahoff and Woollard, 1968.)

## CRUSTAL STRUCTURE
## OF THE HAWAIIAN REGION

The Hawaiian Islands are the projecting tops of a volcanic mountain range known as the Hawaiian Ridge. Partly surrounding the Ridge is a moatlike depression known as the Hawaiian Deep, and outside of that is a broad low upbowing of the ocean floor which is called the Hawaiian Arch (figs. 171, 172). The Hawaiian Deep is conspicuous northeast of Molokai and Maui (the Maui Deep), and swings around the eastern side of Hawaii (the Hawaii Deep). Southwest of the islands it is far less conspicuous, but it can be traced as far as Oahu. On the northeastern side of the islands it becomes shallower north of West Molokai, and it is interrupted off Oahu by a north-northeast-trending belt of seamounts (figs. 171, 246) that are probably volcanic cones at least in part equivalent in age to the posterosional Honolulu volcanics on Oahu. North of Oahu the Deep again appears (fig. 171) and can be followed as a series of basins separated by shallower saddles for more than 600 miles, nearly to Gardner Island (fig. 1). Northeast of Maui and Hawaii the Deep reaches depths greater than 18,000 feet below sea level, and its axis lies 50 to 70 miles from the shoreline of the islands.

Northeast of the Deep, the Arch is about

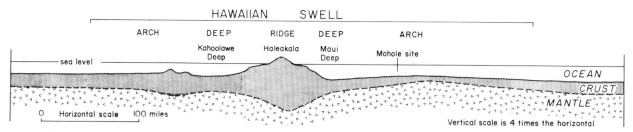

Figure 172. Profile and cross section running NE—SW across the Hawaiian Swell, showing the thinning of the crust beneath the Arch and the thickening beneath the Ridge.

279

200 miles across, and its crest has an average depth of approximately 15,000 feet below sea level. Still farther northeast, beyond the Arch, the general depth of the ocean floor is about 18,000 feet. Thus, the Arch rises about 3,000 feet above the general level of the ocean floor and about the same amount above the bottom of the Deep; but note that the Deep, although reaching depths of 3,000 feet below the crest of the Arch, actually does not descend below general oceanic depths.

Southwest of the major islands the Arch is less well defined than it is to the northeast. The sea bottom rises fairly abruptly southwest of the Deep to about 15,000 feet below sea level, but then descends as a broad, gently sloping plain 100 miles or so wide to merge with general oceanic depths of about 16,000 feet.

The whole, generally raised portion of the ocean floor lying between the outer limits of the Arch both to the northeast and the southwest of the Islands is known as the Hawaiian Swell. It is continuous for the full length of the Hawaiian Archipelago, from Hawaii to Ocean Island. The Ridge, Deep, and Arch are all parts of the Swell, which thus is 500 to 600 miles across and 1,500 miles long. Quite likely, the Swell was formed in much the same way as other broad uplifts of the ocean floor, such as the East Pacific Rise and the Darwin Rise (see chap. 11). The Ridge was formed by volcanic accumulation along the axis of the Rise, probably over a period of 100,000,000 years or more; and the Deep is generally regarded as the result of isostatic downbowing of the sea floor immediately adjacent to the Ridge under the weight of the volcanic accumulations. The Arch may have been raised partly or wholly as material of the earth's mantle was squeezed out from beneath the sinking Ridge and Deep.

On the shelves around the islands and in the Deep, sedimentary rocks reach thicknesses of a few hundred to about 2,000 feet, in general becoming thinner with increasing distance from the islands. Below that layer, however, the near-surface rocks of the entire region appear to be basaltic volcanics. Earthquake velocities indicate that lavas with about the same porosity as those formed above sea level extend to a depth of about 7,000 to 8,000 feet below sea level. Since the porosity of lava flows erupted beneath sea level is known to be only a few percent at depths of 3,000 feet, this indicates that the islands have sunk, carrying subaerial lavas to at least a mile below sea level. We have already seen (chap. 11) that the Waho shelf indicates a sinking of at least 3,600 feet. Two or three thousand feet of sinking must already have occurred before the Waho shelf was cut, presumably due to isostatic adjustment to the weight of the pre-Waho volcanic mountains.

Studies of both natural earthquakes and shock waves from artificial blasts show that the earth's crust beneath the Hawaiian Ridge is much thicker than that beneath most of the Pacific basin. The general sub-Pacific crust is about 3 miles thick. Studies by the U.S. Geological Survey have shown that the crust beneath Mauna Loa and Mauna Kea, on Hawaii, is some 12 miles thick, thinning to 9 miles beneath Hualalai. In the Oahu area, studies by the Hawaii Institute of Geophysics show the crust under the Waianae Range to be about 12 miles thick, and under the Koolau Range about 9 miles. Thus the Hawaiian Ridge in the vicinity of the major islands consists of a great welt of crustal material 9 to 12 miles thick (fig. 172). Interestingly, the uppermost part of the mantle beneath the Waianae Range also appears to be somewhat different from that under the Koolaus. Beneath the Waianae Range, P waves traveling just below the Moho have a velocity of more than 8 km/sec, whereas beneath the thinner crust of the Koolau Range their velocity is a little less than 8 km/sec. One possible explanation is that the mantle under the younger Koolau volcano is a little hotter, and the higher temperature has decreased the rigidity of the rocks and consequently the velocity of the seismic waves.

Beneath the Hawaiian Arch the crust is unusually thin—only about 2 miles (fig. 172). (It was for that reason that the site on the Arch northeast of Maui was chosen for the Mohole.) This means that the Arch cannot be simply a portion of the normal oceanic crust that has been elevated by mantle material squeezed out from beneath the sinking Ridge and Trough. Neither can the Arch be the result of isostatic

280

adjustment, since a thin crust should float with its surface lower than that of a thicker crust with the same density. Something has operated to cause both thinning and upheaval of the crust beneath the Arch. From the inner edge of the Arch the crust gradually thickens beneath the Trough to the great thickness found beneath the islands (fig. 172).

East of Molokai the Arch is broken by innumerable faults, most or all of which are related to the Molokai fracture zone. The faults disappear beneath the Deep and Ridge, presumably because they have been buried by younger rocks, both lava flows and sediments. This suggests that the Molokai fracture zone is older than the present Hawaiian Ridge, and probably is now essentially inactive, although it may have had an important effect on the general configuration of the Ridge. In 1849, J. D. Dana pointed out that the major Hawaiian volcanoes lie along two parallel northwest-trending lines. These lines, he believed, marked great deep-seated fractures in the earth's crust, termed by

him "fundamental rifts," that guided the magma to the surface. He believed that the major volcanoes had developed at points where the fundamental rifts intersected fractures that trend northeastward, in the direction now recognized as that of the great Pacific fracture zones.

Our present view differs little from Dana's. At two places, however, the conspicuous northwestward alignment of the volcanoes is interrupted. The two major volcanoes of Oahu do not lie directly on the projection of the alignments of the more southeasterly volcanoes, but are offset westward from it about 55 miles. One of the strands of the Molokai fracture zone heads directly for the steep submarine escarpment northeast of Molokai (figs. 171, 173), and magnetic patterns (discussed in a later paragraph) strongly suggest that it continues westward north of Molokai and south of Oahu. It appears quite likely that the fundamental rifts have been offset westward by movement on this fault. If Kauai and

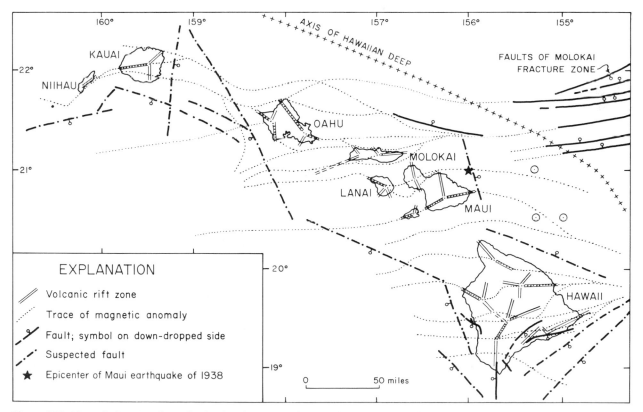

Figure 173. Map of the Hawaiian Islands, showing magnetic anomaly trends, volcanic rift zones, and faults. The star marks the approximate epicenter of the Maui earthquake of 1938, and the circled dots the epicenters of other large nonvolcanic earthquakes of recent years. The locations of the epicenters are accurate only within about half a degree. (Magnetic trend lines and faults of the Molokai fracture zone from Malahoff and Woollard, 1968; earthquake epicenter locations after U.S. Coast and Geodetic Survey.)

Niihau are taken as a pair of volcanoes built one on each of the two fundamental rifts, a similar offset of the rifts has taken place between Oahu and Kauai. These great fault offsets of the Hawaiian fundamental rifts and Ridge are highly speculative, but are consonant with what is believed to have been the movement of the Molokai fracture zone farther northeast. It should be noted that the offset of the Ridge does not mean necessarily that the fault movement occurred later than the building of the Ridge, which may have been built by eruptions along the previously offset fundamental rifts.

The steep escarpment northeast of Molokai almost surely is a fault scarp resulting from downward movement of the sea floor to the north relative to that south of the escarpment. The height of the scarp suggests that the vertical displacement was more than 13,000 feet. Such vertical displacement does not mean that there could not have been even greater horizontal offset on the same fault, since precisely the same behavior is known to have taken place on other great faults. This escarpment is only one of many on the submerged slopes of the Hawaiian Ridge, and probably most or all of the others also are fault scarps. In each case the apparent movement has been downward on the oceanward side, or upward on the landward side. The movement was similar to that on the tangential faults on land, discussed in chapter 2. Some of the more conspicuous of these suspected submarine faults are shown in figure 173, in which the fault northeast of Molokai is indicated as definite.

If these and other suspected faults are actual, it is apparent that the Hawaiian Ridge has in part been elevated as a great horst in relation to the surrounding ocean floor. Whether this means an absolute elevation of the Ridge, or simply that the Ridge has sunk less than its surroundings, we do not now know. Certainly the latest large movements of the Ridge in relation to sea level have been predominantly downward (see chap. 11). The evidence is strong, however, that the sea floor in the area of the Darwin Rise, west of Hawaii, first was elevated and then sank. Is the Hawaiian Swell undergoing a similar history? Both rises are studded with great volcanoes. Beneath both, the temperature of the mantle must have been raised sufficiently to bring about partial melting and the generation of magma. Perhaps the rise was forced up when the underlying mantle rocks expanded as a result of heating, perhaps above the crest of a rising convection current deeper in the mantle, that operated for a few tens of millions of years and then ceased, allowing the mantle to cool and shrink. These are highly conjectural possible answers to questions that are among the many fascinating problems of geology still to be solved.

Still other questions relate to the explanation of magnetic and gravity anomalies in the central Pacific that have been mapped in recent years. A magnetic anomaly can be simply defined as a departure in the strength or orientation of the magnetic field from that of the general earth field around it. The anomalies in the central Pacific are "dipoles," with one side stronger and the other weaker than the general field, as though the effects of a local magnet were superposed on the field as a whole. The maps (Malahoff and Woollard, 1968) show that these dipole anomalies form long belts or lineaments generally arranged in a northeast-southwest pattern, but with interesting variations in the vicinity of the Hawaiian Islands.

East of the Hawaiian Swell many of the anomaly belts coincide closely in position and orientation with strands of the Molokai fracture zone. The magnetic lineaments continue across the Hawaiian Deep and Ridge, where the faults presumably are hidden beneath later rocks. As they reach the islands, some of the anomaly lines merge with the volcanic rift zones. Thus, one lineament merges with the rift zone on the long submarine ridge that extends eastward from Haleakala volcano, continues along the northeast and southwest rift zones of Haleakala, and farther west coincides with the principal rift zone of Kahoolawe (fig. 173). Even more striking is the behavior of one of the magnetic lineaments that approaches Oahu from the east, splits, and bends abruptly northwestward to coincide with the two principal rift zones of the Koolau and Waianae volcanoes. It has been suggested that the local increase in magnetic intensity that causes the anomalies, not only along the rift zones but across the ocean floor remote from the volcanoes, is the result of large numbers of dikes with high magnetic susceptibility.

It should be noted, however, that not all the magnetic lineaments coincide with known, or even probable, volcanic rift zones. Lineaments crossing West Maui bear no apparent relationship to the rift zones of the West Maui volcano. A lineament approaching the island of Hawaii from the east merges with the submarine portion of the east rift zone of Kilauea and runs along it almost to the coast line, but then departs from it to run west-northwestward across the slope of Mauna Loa. Although this latter portion of the lineament is far from the present rift zone of Mauna Loa, it may coincide with an ancient, long inactive rift of Mauna Loa that has been largely buried beneath lavas of Kilauea and the present active center of Mauna Loa. However, another lineament, farther south, also bears no apparent relation to the east rift of Kilauea, though it does follow the Hilina fault system. Another, still farther south, coincides with no known volcanic structures. At the northern end of the island a magnetic lineament coincides with the southeast rift zone of Kohala volcano, but then bears to the westward, departing completely from the prominent northwest rift zone of the volcano.

Other examples, both of coincidence and lack of coincidence with volcanic rift zones, could be cited. At the present stage it appears best to regard the cause of the magnetic lineaments as still unknown.

Also still partly in doubt is the explanation of gravity anomalies that have been found on the Hawaiian Islands. A gravity anomaly is a local departure in the strength of the earth's gravitational attraction from that which would be expected if conditions were uniform. (Allowances must be made for local topographic effects and the sorts of rocks known to be exposed at the surface.) Surveys of the Islands have revealed strong positive anomalies (areas in which the gravitative attraction is greater than normal) in close association with the

summit or main vent areas of most of the major volcanoes (Strange, Machesky, and Woollard, 1965; Kinoshita, 1965; Kinoshita and Okamura, 1965). In the case of Oahu, very strong positive anomalies are centered in the calderas of the Koolau and Waianae volcanoes (figs. 94, 174). On the island of Hawaii a strong

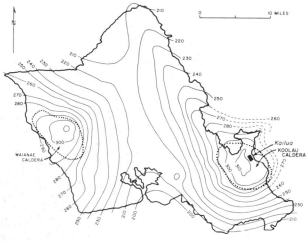

Figure 174. Map of the island of Oahu, showing the relation of strong positive gravity anomalies to the calderas of the Koolau and Waianae volcanoes. The dotted lines show the approximate outlines of the calderas. The values for gravity are Bouger anomalies, in milligals. (After Strange, Machesky, and Woollard, 1965.)

positive anomaly lies just southwest of Kilauea caldera, another about 5 miles southeast of the caldera of Mauna Loa, and another just south of the summit of Mauna Kea (fig. 175). The close relationship to those volcanic centers is clear. On Kohala the anomaly is offset further from the volcanic center, but lies on the southeast rift zone of the volcano; and on Hualalai it lies far to the south of the summit, on no known volcanic structure.

The probable explanation of the strong positive gravity anomalies is the presence of abnormally dense bodies of rock that exert a stronger gravitative pull than do the less dense rocks of the main body of the volcano. Analysis of the measurements on the Oahu anomalies suggests the presence of a body of rock with a density of about 3.2 grams per cubic centimeter (about the same as that of some peridotites) close to the surface near the center of the Koolau caldera. The gravity findings have been confirmed by studies of artificial earthquake waves, which show that a

283

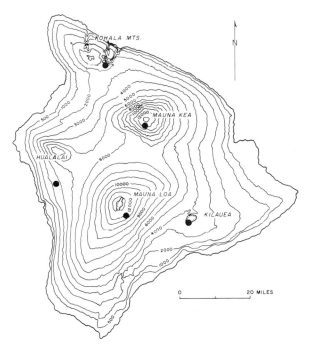

Figure 175. Map showing the centers of strong positive gravity anomalies (*solid black dots*) of the major volcanoes on the island of Hawaii.

body of rock with P-wave velocities greater than 7 km/sec (as compared with 4.6 km/sec velocities in the adjacent caldera-filling lava flows) lies beneath Kawainui Swamp, near the center of the caldera, with its top only about a mile below the surface. This dense, high-velocity "plug" is 3 to 3.5 miles in diameter, and appears to extend downward all the way to the Moho (Adams and Furumoto, 1965; Furumoto and others, 1965).

This much seems well established, but the nature of the dense mass remains uncertain. The density of the rock is similar to that believed to exist in the upper mantle, and the seismic velocity is within the range of velocities found in the upper mantle. The mass may be an upward projection of the mantle into the core of the volcano. Alternatively, it is at least as probable that it is a mass of olivine and perhaps other heavy crystals that have lagged behind in basaltic magma rising toward the surface, and have accumulated to form a plug of peridotite that resembles mantle rock in its physical properties. The uncertainty remains as still another challenge to future workers.

Study of the floor of the oceans has revealed a worldwide system of broad ridges, or "rises,"

284

the tops of which are elevated a mile or more above the general level of the ocean bottom. On the two sides of these ridges symmetrical patterns of bands have alternately stronger and weaker magnetic attraction than the average of the ocean floor. These magnetic anomaly bands are believed to result from normal and reverse polarity in the rocks (p. 266), respectively adding to and subtracting from the strength of the earth's present magnetic field. The rocks of the anomaly bands are believed to have formed during periods of normal and reversed polarity of the earth's field, and to have taken on the orientation of the field at the time they were formed. To explain this symmetrical arrangement of magnetic anomaly bands it has been suggested that the earth's crust beneath the ocean is moving outward away from the rises, and as it does so new magma rises from lower in the mantle to take its place. Thus, new crust is constantly being generated at the rises, and taking on a polarity governed by the earth's field at that time, only to be torn apart and carried away approximately at right angles to the crest of the rise and replaced by still more new material, and so on.

Under this concept of "sea-floor spreading" the ocean floor should become progressively older away from the crest of the rise, and dating of the sea-floor rocks by the potassium-argon method (p. 267), and by means of fossils in the sediments immediately overlying the basaltic floor, shows that this is indeed the case (fig. 170). It appears that the outward-moving material includes not only the crust, but also the upper part of the mantle, down to a depth of 30 to 50 miles, and that this material forms a relatively rigid plate, referred to as *lithosphere*. It moves on more plastic material beneath it, known as the *asthenosphere*.

However, if new lithosphere is constantly being generated and moving outward away from the ridges, either the circumference of the earth must be getting constantly greater, to make room for it, or lithosphere must be being destroyed at some other places at the same rate it is being formed at the rises. There is good evidence that the earth is remaining about the same in size, and consequently that lithosphere is being destroyed. This destruction is believed to be accomplished by sinking of the lithosphere plate into the underlying asthenosphere, where it is eventually assimilated into the mantle. The "subduction" of the lithosphere takes place along narrow zones, generally marked by deep sea-floor trenches bordered on the side away from the sinking plate by mountain belts and rows of volcanoes.

The surface of the earth appears to be made up of about ten major plates, which move independently of each other. Each plate is being created at one edge, destroyed at the opposite edge, and slipping past other plates at the sides. The Hawaiian Islands lie on the Pacific plate, which is moving west-northwestward away from the East Pacific Rise (fig. 170) at a rate of about 5 cm (2 inches) a year. The plate is being destroyed along the series of trenches and island arcs (Philippines, Japan, Kamchatka) along the western side of the Pacific Ocean. The sea floor in the vicinity of the Hawaiian Islands was formed at the East Pacific Rise about 80 million years ago, but the islands themselves have been built on the surface of the plate much more recently (chap. 17).

A correlative of this "plate tectonic" theory is the suggestion that the Hawaiian volcanoes were formed successively as the lithosphere plate moved northwestward over a "hot spot," near the island of Hawaii, where magma is being generated in the asthenosphere. As each volcano was built and carried northwestward on the moving lithosphere, a new one was formed behind it over the hot spot. The general increase in age of the volcanoes toward the northwest (p. 268) fits the idea, but there are some ages that do not fit, and other lines of evidence also are difficult to reconcile with it. For instance, the generation of the magmas of the Honolulu Volcanic Series on the island of Oahu after a period of volcanic quiet of more than a million years, 300 miles or more away from the probable position of the hot spot, and apparently from a depth in the mantle considerably greater than that at which the earlier Koolau magmas were formed, is incompatible with the hot-spot hypothesis. During the autumn of 1973 investigators on a cruise of the National Science Foundation's Deep Sea Drilling Project found that there is no appreciable difference in age of the seamounts at the two ends of the Line Islands chain, one of the northwest-trending volcanic chains south of Hawaii. This seems to indicate that the chain was not formed sequentially by drift over a hot spot, and this conclusion presumably should apply also to the other similar linear island groups in the Pacific, including the Hawaiian chain. The subject is being intensively investigated, and it would be premature to draw a conclusion regarding it at present.

*Suggested Additional Reading*

Eaton, 1962; Eaton and Murata, 1960; Hodgson, 1964; Macdonald, 1961; Malahoff and Woollard, 1966; Smith and Menard, 1965; Takeuchi and others, 1967

*Waimea Canyon,
Kauai*

# CHAPTER NINETEEN

# Regional Geology of the Individual Islands

THE GEOLOGIC HISTORY OF THE HAWAIIAN ARCHIPELAGO is probably a good deal more complex than that revealed by the parts of the islands exposed to view above sea level. Far more exploration of the submerged portions, by both submarine geological studies and deep drilling, are needed before we will even begin to have the complete story. Only by lucky chance, during a demonstration cruise of a vessel of the Scripps Institution of Oceanography for members of the Pacific Science Congress, in 1961, were fragments of very ancient limestone dredged from off southern Oahu which showed for the first time that a coral reef had existed in that area before the present island was built. And since coral reef can grow only in shallow water, a still older volcanic ridge must have been present. It may have been a southeastward continuation of the ridge that underlies the Leeward Islands, but this is only conjecture, and its actual extent remains to be determined. Did it, and the reefs that crowned it, sink as a result of the subsidence of the Darwin Rise, that carried downward the guyots of the Mid-Pacific Mountains to the west of Hawaii? How much of the submarine portions of the Hawaiian mountains consists of pillow lavas, how much of hyaloclastite, how much of tephra formed by explosions where rising lava encountered sea water? Are the heavy masses that underlie Kailua on Oahu, and other caldera areas, accumulations of sunken olivine crystals, or are they portions of the earth's mantle that somehow have been shoved upward into the cores of the volcanoes? Only deep drill holes will supply the answers. The account of the geology of the individual Hawaiian islands on the following pages is only a progress report based very largely on the uppermost, visible parts of the Hawaiian volcanic range.

Even in their visible tops alone, the northwestern islands of the Hawaiian chain have had a considerably longer and more complex geological history than those to the southeast. They show more clearly the effects of outside events such as the changes of sea level during the glacial period, and their original forms have been far more extensively altered, so that the interpretation of their geology is more difficult. In this chapter the islands are described in turn, from southeast to northwest, in the order of increasing age and generally of increasing geological complexity.

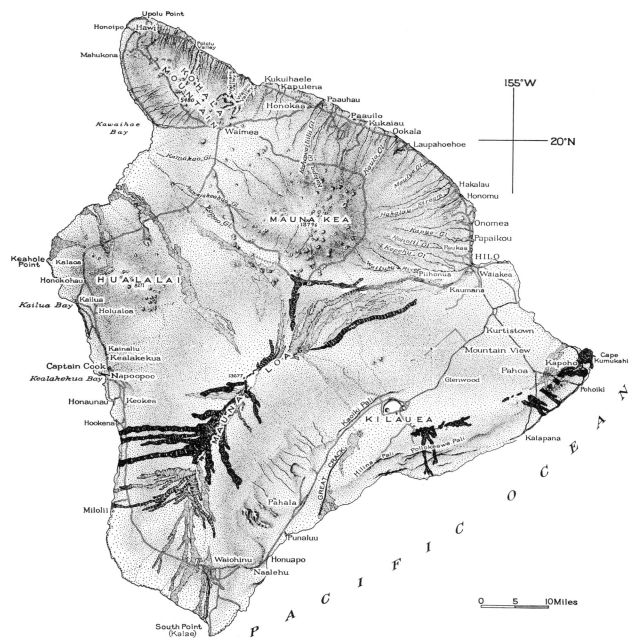

Figure 176. Topographic map of the island of Hawaii. (Modified after Stearns and Macdonald, 1946.)

Only the more salient features of their geology can be summarized here. For more details, both in facts and in the evidence for the interpretations presented, the reader should consult the original publications which are listed at the end of the chapter.

## Hawaii

The island of Hawaii today consists of five volcanic mountains. All of them are very young, and three of the volcanoes have been active in historic time. At least two other volcanoes which helped to build the island have been buried by more recent ones. Kohala Mountain, which forms the northern end of the island (fig. 176), appears to be the oldest. Rocks exposed in the cliffs on its northeastern side have been shown by the potassium-argon method to be about 700,000 years old (table 18). The Kohala volcano had already reached nearly its final size before its southern flank was buried beneath lava flows from Mauna Kea, next to the south.

About the same time as the birth of the Kohala volcano, or shortly afterward, another vent opened on the sea floor beneath the present south end of the island of Hawaii. Lavas from this southern vent built a shield volcano that rose to a height of about 8,000 feet above sea level in the region northwest of the present town of Naalehu. Potassium-argon determinations indicate that its rocks are about 100,000 to 500,000 years old. The rocks are known as the Ninole Volcanic Series after the now-abandoned village of Ninole, near which they are exposed. The Ninole volcano ceased activity perhaps as long ago as 100,000 years, and stream erosion cut deeply into the shield, forming a series of large valleys separated by

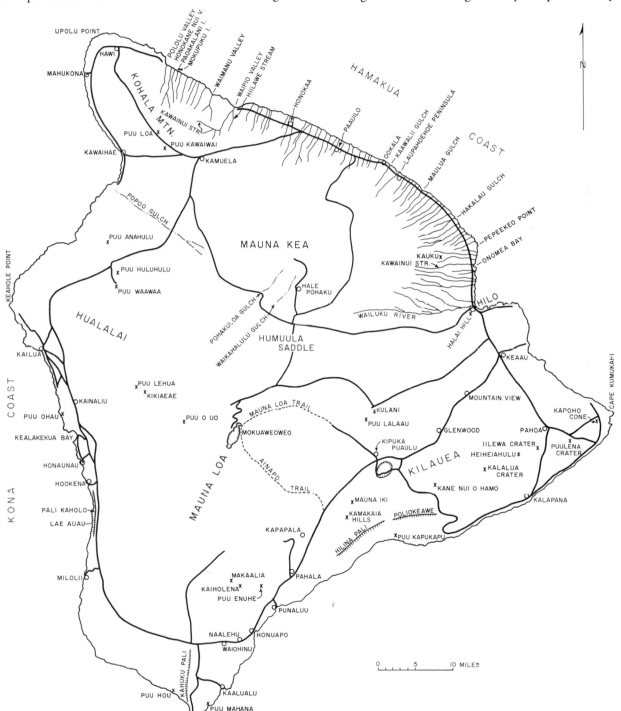

Figure 177. Map of the island of Hawaii, showing the location of places mentioned in the text.

289

GEOLOGIC MAP
OF THE
ISLAND OF HAWAII
Geology by
G. A. Macdonald and H. T. Stearns
1924, 1940-1943
Prepared in cooperation with the
Division of Hydrography,
Territory of Hawaii

Contour interval 1000 feet
Datum is mean sea level

290

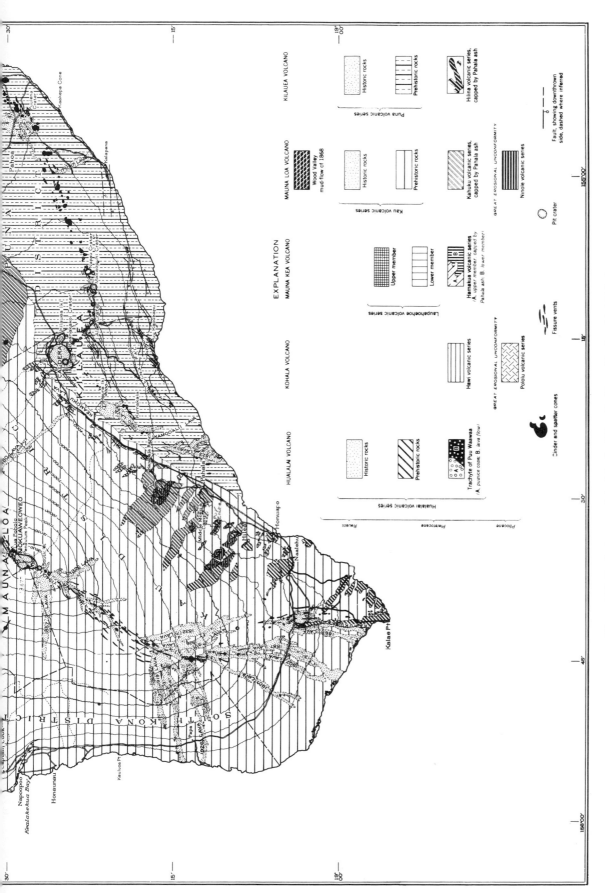

Figure 178. Geologic map of the island of Hawaii. (From Macdonald, 1949b.)

flat-topped ridges. This old topography has been almost completely buried by the present Mauna Loa volcano, but the tops of some of the ridges can be seen projecting through the Mauna Loa lavas north of Naalehu, where they form the prominent hills called Kaiholena and Puu Enuhe, near Pahala (figs. 177 and 178).

The rocks of Mauna Kea overlap those of Kohala Mountain and must, therefore, be younger. The oldest of them that are now exposed above sea level are in the sea cliff at Laupahoehoe, on the Hamakua (northeastern) Coast (fig. 187). All of the visible portion of Mauna Kea must be more recent than middle Pleistocene. Nevertheless, Mauna Kea had already reached its present size by the time the last Pleistocene glacier disappeared from its summit, 15,000 or so years ago (see chap. 14). Only three lava flows are known that are later than the glacial moraines. There are no Hawaiian traditions of eruptions of Mauna Kea, and it probably has not been active during the last 2,000 years, but it is quite impossible to say whether the volcano is extinct or only dormant. Occasional earthquakes originate beneath it and emphasize the possibility that it may someday erupt again.

Hualalai, on the west side of the island, is a dormant volcano that last erupted in the years 1800–1801. At that time two lava flows poured into the sea, burying the old Hawaiian trail near the coast and destroying at least one village. The rocks of Hualalai are in a less advanced stage of magmatic evolution than those of Mauna Kea, and a more advanced stage than those of Mauna Loa. Consequently, the mountain probably started to grow upward from the sea floor at a time between the beginnings of its two larger neighbors, though probably not before the start of the Ninole volcano. We do not know as yet how uniform are the rates of growth of individual volcanoes or how uniform are the rates of evolutionary change of magma composition.

292 Since 1801 Hualalai has been dormant, but a series of several thousand earthquakes came from beneath its northern flank in 1929, and occasional other earthquakes indicate that its magmatic hearth is still alive. We may expect the volcano to erupt again, but perhaps not for many decades. The appearance of the lava flows on the flank of Hualalai suggests that, during the latter part of the life of the volcano, eruptions have been rather infrequent, perhaps at intervals of a few hundred years.

When the Ninole volcano became extinct, activity appears to have shifted to two other centers, one beneath the present summit of Mauna Loa and the other 18 miles or so to the east, around both of which broad shield volcanoes were built. Toward the end of the history of the eastern shield volcano, the explosiveness of the eruptions increased somewhat and large tephra cones were built. These cones, Kulani cone and Puu Lalaau, stand near the summit of the eastern shield, which can be referred to as the Kulani shield volcano. The activity of the Kulani shield came to an end probably not more than a few tens of thousands of years ago. The more westerly shield has continued active and has built the present summit of Mauna Loa. Its lavas have now nearly buried the Ninole and Kulani volcanoes. Although Mauna Loa has continued to erupt frequently during the 19th and first half of the 20th centuries, it must already have grown to nearly its present size by the end of the Ice Age. Lava flows from Mauna Kea that are earlier than the end of the last glacial period descend southward into the Humuula Saddle between Mauna Kea and Mauna Loa and there turn sharply eastward or westward, showing that they must have been diverted by the obstacle of Mauna Loa, and this in turn indicates that the slope of Mauna Loa must have been in essentially its present position when these flows were erupted.

The volume of Mauna Loa, including the buried Ninole shield, is on the order of 10,000 cubic miles, making it probably the largest volcanic mountain on earth. However, at the rate that lava has been poured out by Mauna Loa during historic time, this entire bulk, from the ocean floor up, could have been produced in a little more than a million years.

Kilauea volcano forms only a slight protuberance on the southeast flank of Mauna Loa (fig. 189), but it appears to be a completely independent volcano. Along the boundary northeast of the summit of Kilauea its lavas interfinger with those from the present Mauna

Loa. At a depth of only a few tens to hundreds of feet, however, they probably rest on the flank of the Kulani shield. There has been little correlation between the activities of Kilauea and Mauna Loa during historic time, and the magmas erupted at the two centers have reached slightly different stages of evolution, indicating that the two volcanoes are fed from different magma bodies at depth. True, there have been a few instances when the level of lava in the Halemaumau lake dropped abruptly at the time of an outbreak of Mauna Loa, and in 1868 the two volcanoes erupted nearly simultaneously, but in most instances eruption of one has caused no apparent reaction at the other. More than a century ago the great American geologist, James Dwight Dana, pointed out that the great difference in level of the two eruptive centers, one 13,000 feet above sea level and the other less than 4,000, precludes any direct connection of the two in a common magma chamber at shallow depth. If any connection exists, it must be deep and indirect. Recent unpublished work by J. G. Moore indicates that the average rate of production of lava in the southern Hawaii region has remained about the same despite the fact that from 1934 to 1950 all of the eruptive activity was at Mauna Loa, and since then all of it has been at Kilauea. This suggests that magma is being formed at a fairly constant rate in the earth's mantle, but that for some unknown reason the upward feeding of magma has changed, for the time being, from the Mauna Loa reservoir to that of Kilauea.

PAHALA ASH

Near the town of Pahala the remnants of the Ninole volcano are covered with more than 50 feet of yellowish volcanic ash, known as Pahala ash. What appears to be the same ash is found in many parts of the island—just below the top of the Hilina Pali on the southern slope of Kilauea, buried by younger Kilauea lavas; capping lavas of the Kahuku Volcanic Series on the slope of the Kulani shield northeast of Kilauea, where it has been buried by later lavas of Mauna Loa; and on Mauna Kea and Kohala Mountain (figs. 178, 179).

Most of the Pahala ash consisted originally of vitric ash—bits of Pele's hair, Pele's tears, and fragments of pumice thrown into the air by lava fountains and carried by the wind, like the material produced by the great lava fountains of the 1949 eruption of Mauna Loa and the 1959 eruption of Kilauea. Throughout most of its extent the once-glassy ash has been largely altered by weathering to a mixture of clay minerals and hydrated oxides of aluminum and iron. (The reddish brown to yellow, altered material somewhat resembles that of the tuff cones of Oahu, and both have been called palagonite.) As a result of alteration it is usually impossible to determine the original composition of the material, and so the source of much of this ash is uncertain. Furthermore, it probably was erupted over a considerable span of time, and Pahala ash in one area may not be strictly coeval with that in another area.

Several different sources probably have contributed to the deposits of Pahala ash. On the flanks of Mauna Kea just north of the Wailuku River, the ash layer is more than 20 feet thick, while across the eastern and northern flanks of the mountain it becomes progressively thinner.

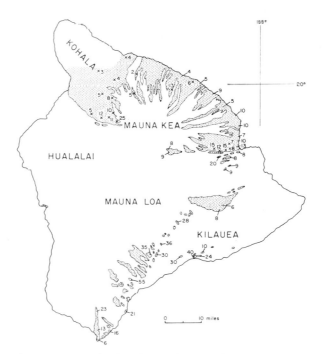

Figure 179. Map showing the distribution of Pahala ash on the island of Hawaii. The numerals indicate the approximate thickness of the ash at different points. (After Stearns and Macdonald, 1946.)

293

On the dry northwestern flank near Kamuela, some of the ash is quite fresh and has the same composition as late lavas of Mauna Kea. It is believed that most of the ash on Mauna Kea came from eruptions on that mountain, and that the ash layer becomes thinner toward the north because the trade winds carried most of the ash southwestward from the erupting vents. South of the Wailuku River, where it is found in water-development tunnels beneath the Mauna Loa lavas, the ash layer again becomes thinner, and on the flanks of the Kulani shield near Glenwood it is only 8 feet thick. These facts suggest very strongly that much of the ash on the east slope of Mauna Loa also came from vents on Mauna Kea. However, to the leeward of Kilauea, near Pahala, the ash is much thicker than to windward, and it seems certain that much of the ash southwest of Kilauea caldera came from Kilauea. Undoubtedly, some contributions also have been made by eruptions on Mauna Loa and Hualalai.

The Pahala ash is the only rock formation that is found on more than one of the volcanic mountains, and consequently, until more potassium-argon ages are obtained, it provides the only direct means of correlating the times of activity of the different volcanoes. Unfortunately, however, there is no certainty that the Pahala ash is everywhere of exactly the same age. It represents the total accumulation of all ash falling in a given area during an interval that was not interrupted by lava flows, and, in places where it has not been covered, its thickness is still increasing. For example, thin ash layers that overlie lavas of the Laupahoehoe Series on Mauna Kea form the top of the Pahala ash in adjacent areas that have not been covered by the Laupahoehoe lavas, and yet they are generally indistinguishable from the underlying ash. Areas mapped as Pahala ash on the eastern slopes of Mauna Kea and Mauna Loa are distinguished from adjacent ash-covered areas only by the greater thickness of the ash. There does, however, seem to have been a general period of abundant ash formation, when most of the Pahala ash was deposited essentially simultaneously over much or all of the island. This probably was the period of greatest activity of the late Hamakua and Laupahoehoe cones on Mauna Kea, and strong lava fountains at Kilauea, while fault scarps shielded the southern slope of Kilauea and the southeastern slope of Mauna Loa from lava

294

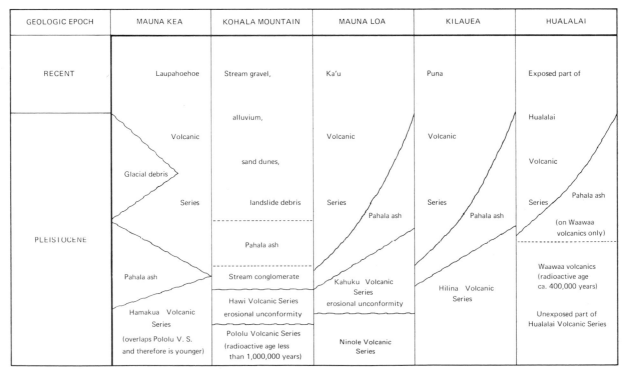

Figure 180. Diagram showing principal rock units on the island of Hawaii. All rock units thus far investigated have normal magnetization, and the oldest radioactive dates determined are less than 1,000,000 years, so that it appears unlikely that any exposed rocks are older than Pleistocene.

flows, and only occasional flows invaded the eastern slope of the quiescent Kulani shield.

All of the rocks of the Kohala and Ninole volcanoes are older than the Pahala ash. On Mauna Kea most of the ash is older than the late lavas (Laupahoehoe Volcanic Series), but a thin layer of ash covers even the late lavas, as would be expected if the eruptions that produced these lavas were also the source of the ash. On the east flank of Mauna Loa the ash rests on, and is therefore younger than, the rocks of the Kulani shield. The lavas of the present Mauna Loa appear to be mostly younger than the ash, although in the region northeast of Mountain View, layers of ash are interleaved with lava flows, and the earliest Mauna Loa lavas in that area are probably coeval with the main period of formation of the Pahala ash. Ash resembling the Pahala ash is exposed also in the sea cliff at Kealakekua Bay, on the west flank of Mauna Loa, resting on pre-Mauna Loa lavas that may belong to either the Ninole volcano or an early phase of Hualalai volcano. Pahala ash is totally absent on Hualalai itself; all of the surface lavas there are younger than the ash. The Pahala ash which is exposed in the Hilina Pali on the south slope of Kilauea, is overlain by a cap of Kilauea lavas generally less than 100 feet thick. Other thin beds of ash lie between the underlying, older lava beds. A layer of ash formerly exposed at the base of the west wall of Kilauea caldera (but buried by the lava flow of 1919) probably was also Pahala ash. If so, the present Kilauea lavas form a cap, with a maximum thickness of only about 400 feet, overlying the Pahala ash. Thus, the main period of deposition of the Pahala ash was later than the Kohala, Ninole, and Kulani volcanoes; it was contemporaneous with the late lavas of Mauna Kea and some of those of Mauna Loa; and it appears to mark the beginning of the present phase of Kilauea.

The approximate time relationships of the rocks of the different volcanoes are shown in figure 180.

## KOHALA MOUNTAIN

The northern end of the island of Hawaii, known as Kohala Mountain (fig. 176), consists of an oval shield volcano built around two rift zones that trend northwestward and southeastward from the summit region (figs. 182, 189). The last eruptions were moderately explosive and formed a series of large cinder cones that stud the surface of the shield. The top of the highest cone is 5,480 feet above sea level.

On the western and northern sides of Kohala Mountain erosion has made little headway (figs. 181, 184). The deepest stream-cut gulches are less than 100 feet deep, and the mountainside is essentially the original land surface built by the volcano. The sea cliff along its margin is less than 50 feet high. This gentle topography stands in marked contrast to the topography of the northeastern side. There, much of the original constructional slope has been cut away, and the shield is truncated by a series of great sea cliffs reaching heights of 1,400 feet, notched by canyons 1,000 to 2,500 feet deep (figs. 181, 183, 185). The canyons were once even deeper. Near their mouths the bottoms are now flat floors of alluvium, deposited as a result of a recent rise of sea level relative to the land. The thickness of the alluvial fill is not known, but near the valley mouths it is probably about 300 feet. Paoakalani and Mokupuku islets, off the northeastern shore of Kohala, are sea stacks left by wave erosion of the surrounding rocks. The northeastern shore of Kohala between Waipio and Pololu valleys appears to have been cut back nearly a mile by wave erosion. The difference in the degree of erosion of the northeastern and southwestern slopes of Kohala is explained partly by the heavier rainfall and stronger wave attack on the northeastern side, but largely by the structure of the volcano, described later.

The rocks of the Kohala volcano have been divided into two series (Stearns and Macdonald, 1946, pp. 170–180). The older, Pololu Volcanic Series consists very largely of flows of basalt, each one usually 5 to 20 feet thick, and rarely reaching as much as 50 feet. Successions of these lavas 2,000 feet thick are exposed in the walls of the deep canyons (fig. 178), and they extend on down below sea level probably all the way to the ocean floor. Through most of the succession, ash layers are rarely found, but near the top of the series they become more numerous. At about the same level another change occurs. The lower, main part of

*295*

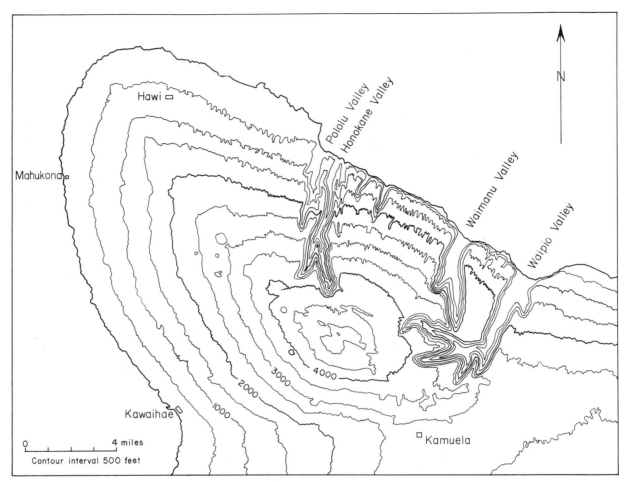

Figure 181. Map of Kohala Mountain, showing the huge valleys cut into the windward (northeastern) slope, and the small degree of dissection of the leeward slope.

the series is made up of tholeiitic basalt, tholeiitic olivine basalt, and oceanite, but in the upper part alkalic olivine basalts appear. Some of the flows in the upper part of the tholeiitic part of the series contain very abundant large crystals (phenocrysts) of labradorite feldspar. The lava beds of the Pololu Series are especially well exposed along the trail down the southeast side of Waipio Valley.

The younger, Hawi Volcanic Series is separated from the rocks of the Pololu Series by an eroded surface (unconformity), with valleys as deep as 300 feet, covered with red soil and in places underlain by as much as 50 feet of decomposed, weathered rock. The unconformity marks a long period of time during which the volcano was either completely inactive or erupted only very infrequently. The lavas of the Hawi Series, above the unconformity, are mostly mugearite, but a few flows and domes

of trachyte are known and one flow of hawaiite has been found at the northwestern shore, near Mahukona. (Some of the rocks are transitional between mugearite and trachyte, and have been referred to as "benmoreite.") Thin beds of ash, partly weathered to soil, lie between flows of the Hawi Series at some places, and elsewhere the flows are separated by lenticular deposits of gravel laid down by streams in shallow gullies. Much longer intervals of time must have elapsed between eruptions of the Hawi Series than between those of the main part of the Pololu Series. The rocks of the Hawi Series are best seen along the highway from Kamuela to Hawi.

The deep highway cut just southwest of Puu Loa, 5 miles beyond the junction with the road to Kawaihae, is in a thick short flow of "benmoreite," erupted at Puu Loa, which is a cinder cone built at the vent. Puu Kawaiwai, on

Figure 182. Andesitic cinder cones of the Hawi Volcanic Series mark the trend of the northwest rift zone on Kohala Mountain, Hawaii. The largest cone is about 700 feet high.

Figure 183. Sea cliff cut across lavas of the Pololu Volcanic Series on the northeastern side of Kohala Volcano, Hawaii. The straight line of the cliff and the faceted spurs suggest a fault origin for this coastline, but there is no other evidence that it is not entirely the result of erosion. The floors of the major valleys are close to sea level, but the smaller valleys are "hanging" and their streams enter the ocean as waterfalls. Beyond Pololu Valley, at the right-hand end of the high sea cliff, the long gentle slope is formed by later lavas of the Hawi Volcanic Series. Note the much smaller degree of dissection of the Hawi lavas as compared with the Pololu lavas in the center of the photograph.

Figure 184. Northwestern slope of Kohala volcano, Hawaii. Contrast the much smaller degree of dissection on this drier and younger side of Kohala with that on the wet northeastern side of the mountain shown in figures 183 and 185. The port of Kawaihae lies in the foreground, and the summit of Haleakala, on Maui, looms in the distance.

the lower side of the road 3 miles beyond the same junction, is a cinder cone of hawaiite composition. The magma for the Kawaiwai eruption may have come from the reservoir beneath Mauna Kea, and been guided to the surface by a fissure that cut the flank of the adjacent Kohala volcano; or, like the lava flow near Mahukona, it may be simply a variant of the magma from the Kohala reservoir.

The focus of the Kohala volcano lay near the present summit of the mountain. A series of curved faults in that region suggests that a caldera probably was formed in the summit of the shield near the end of the eruption of the Pololu Series, but it has been entirely buried by lavas of the Hawi Series. However, the massive filling of a pit crater a mile in diameter can be seen near the head of the canyon of Kawainui

Stream. The exposed part of the crater fill is 2,000 feet thick and consists mostly of dense flows of tholeiitic basalt up to 100 feet thick. At the western side the lavas rest on talus breccia that accumulated at the foot of the crater wall.

The geologic map (fig. 178) shows that lavas of the Hawi Series are absent for a stretch of 7 miles on the northeast slope of the mountain between Waimanu and Honokane Nui valleys. There, the Hawi lavas were prevented from flowing northeastward by a series of fault scarps in the summit region. Either at the end of the formation of the main shield or early in the period of eruption of the Hawi lavas, a northwest-trending graben 6 miles long and 1 to 3 miles wide formed across the top of the volcano. Lavas erupted within this graben

Figure 185. Waipio Valley, Hawaii. The wide flat floor of the valley is the result of submergence of a stream-cut valley due to relative rise of sea level, and the filling of the valley by alluvium and possibly marine sediments. Sea level probably was about 300 feet lower than now when the valley was cut. The head of Waipio Valley, following a fault zone, bends sharply to the right and has captured the principal former tributaries of Waimanu Stream. The head of Waimanu Valley and the streamless notch (wind gap) separating it from the head of Waipio Valley are visible at the right of the photograph. Note the lateral capture by Waipio Stream of parallel consequent streams, and the vertical grooves cut in the side of the main valley by the captured streams plunging down to the valley floor. The low peninsula at the foot of the sea cliff to the right of Waipio Valley was formed by a landslide from the cliff.

gradually filled it, escaping from it at the northwest and southeast ends and eventually overflowing much of its southwestern rim, but on the northeast side the bounding fault scarps were higher and were never overtopped by the flows. Consequently, although on the western and northern slopes erosion was constantly interrupted and its damages repaired by repeated lava flows of the Hawi Series, on the northeastern slope it went on undisturbed, and huge valleys were cut. To this long period of uninterrupted erosion, even more than to the higher rainfall, is due the much greater erosional dissection of the northeast side of Kohala Mountain. Eventually, however, lava spilled over the fault scarp at the northern end of the graben and poured into the head of Pololu Valley (fig. 186).

The great valleys of the northeastern side of Kohala Mountain differ in two marked respects from the usual radial valleys of the Hawaiian Islands. Waipio Valley extends inland southwestward, and then makes an abrupt right-angle bend toward the northwest (fig. 185). Waimanu Valley also extends inland southwestward, and ends in a deep, streamless, V-shaped notch (a wind gap) high on the northeastern

Figure 186. Sketch of Pololu Valley, Kohala Mountain, showing the flow of mugearite that poured into the head of the valley. (From Stearns and Macdonald, 1946.)

*299*

wall of the northwest-trending portion of Waipio Valley (fig. 181). The notch is clearly the result of cutting by Waimanu Stream during earlier times, and formerly the stream and canyon must have extended farther inland. Kawainui Stream was once the headwaters of Waimanu, but has been captured by the faster-cutting Waipio Stream, robbing Waimanu Stream of much of its water. Mass transfer (chap. 10) continues on the walls of upper Waimanu Canyon, but the stream is no longer adequate to carry away all the debris that reaches the valley bottom, and the head of the canyon is becoming clogged with landslide and other debris. The northwesterly trend of the upper part of Waipio Canyon is the result of the same series of graben-boundary faults mentioned in the last paragraph, which prevented the streams, as well as the lava flows, from flowing northeastward.

The late activity of Kohala volcano was contemporaneous with the later part of the building of the main shield of Mauna Kea. Lava flows of the Hawi Series interfinger with lavas of Mauna Kea in the sea cliff east of the mouth of Waipio Valley. However, Mauna Kea continued active long after the death of the Kohala volcano. The canyons of Kohala Mountain had already reached nearly their present size by the time of eruption of the late lava flows of Mauna Kea (lava flows that were considerably more recent than the building of the main Mauna Kea shield). One of the late Mauna Kea flows poured over the headwall of Hiilawe Canyon, a tributary of Waipio Valley, and filled the canyon to a depth of about 150 feet. The flow has since been cut through by the stream to the level of the old canyon floor.

MAUNA KEA

Mauna Kea volcano passed through the primitive shield-building stage into the late stage, and produced a cap of differentiated lavas that almost completely buried the original shield above sea level. The rocks have been divided into an older Hamakua Volcanic Series and a younger Laupahoehoe Volcanic Series (fig. 178), and the former has been further divided into upper and lower members. The lower member of the Hamakua Series comprises the tholeiitic basalts, olivine basalts, and oceanites of the early shield-building stage. It is exposed only in the lower part of the sea cliffs along the Hamakua Coast north of Hilo (fig. 187). The rocks are the usual thin beds of pahoehoe and aa, like those of Mauna Loa and Kilauea. They pass upward gradationally into alkalic olivine basalts, hawaiites, and ankaramites that make up the upper member of the series. The rocks of the upper member are well exposed in highway cuts along the Hamakua Coast. They are covered by a layer of Pahala ash that reaches a thickness of 25 feet along the Wailuku River above Hilo but gradually thins northward to about 6 feet near Paauilo (fig. 179). Along the Hamakua Coast the ash is much altered, to a mixture of clay minerals and aluminum and iron oxides. The structure of the fragments composing the ash is still sufficiently preserved, however, to show that they were originally fragments of pumice and cinder and finer glassy debris from Strombolian- or Hawaiian-type eruptions. Layers of similar ash, a few inches thick, lie between the lava beds of the underlying upper member of the Hamakua Series.

The maximum thickness of the Pahala ash is found only on the lower slopes of Mauna Kea. On the upper slopes its accumulation was repeatedly interrupted by lava flows of the Laupahoehoe Series, and the ash consists of several thin beds between the flows. The top of Mauna Kea consists almost wholly of rocks of the Laupahoehoe Series (figs. 178, 188). Rocks of the Hamakua Series are exposed only in Waikahalulu and Pohakuloa gulches on the south flank of the mountain (fig. 148), and in another, smaller gulch 2 miles farther west. The Laupahoehoe Series consists predominantly of hawaiite, with lesser amounts of alkalic olivine basalt and ankaramite. Hawaiite flows are well exposed along the highway between Honokaa and Kamuela, and south of Popoo Gulch on the road to Kona. They are thick, with very hummocky tops. In the highway cuts, the massive flow centers are seen to be underlain and overlain by fragmental portions in which the fragments are less spinose and more regularly shaped than those of typical aa, and grade toward the shapes characteristic of block lava flows. The three latest flows show no signs of having been erupted beneath ice, and hence must have been erupted within the last 15,000 years. The last eruption was about 3,600 years ago.

Eruptions of the Laupahoehoe Series were almost wholly restricted to the upper slopes of the volcano. One small eruption of alkalic basalt on the lower slope, near Ookala, formed a small dome and a stubby lava flow. Most of the eruptions were of Strombolian type and

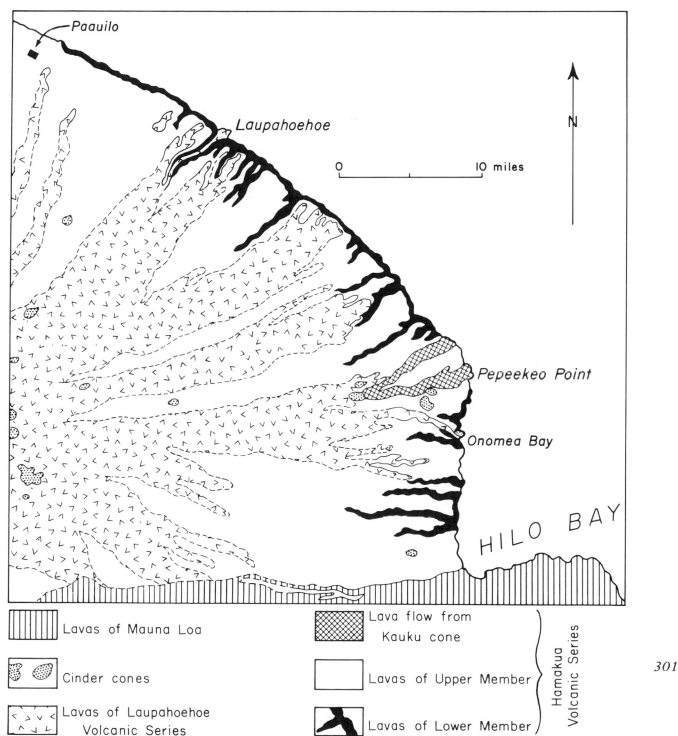

Figure 187. Map showing distribution of rock series on the eastern slope of Mauna Kea. (After Stearns and Macdonald, 1946.)

301

Figure 188. Looking westward toward the summit of Mauna Kea. Lava has issued from the flank of the cinder cone in the foreground and spread out fanwise. A similar fan of lava issued from the base of the Summit Cone in the distance.

built big cinder cones, some of them more than a mile across at the base and several hundred feet high. Good exposures of the internal structure of cinder cones are provided by cinder quarries in some of the cones along the road through the Humuula Saddle, between Mauna Kea and Mauna Loa. Large cones south of the road are Mauna Kea cinder cones projecting through later lava flows of Mauna Loa. The cones just west of Hale Pohaku, at 9,500 feet altitude on the south slope of Mauna Kea, and some cones elsewhere on the mountain, contain numerous cored bombs in which the cores are fragments of peridotite and gabbro brought up from depth.

The broad shield volcano built in the early stages of Mauna Kea is surmounted by a steeper-sided cone formed by the rocks of the Laupahoehoe Series and the upper member of the Hamakua Series. The general form of the shield is still clearly visible on the lower slopes, particularly from viewpoints on Halai Hill and along the shore east of Hilo. Whether or not a caldera was present at the top of the shield is uncertain. However, some of the cones of the Laupahoehoe Series are situated along lines which curve around the summit of the mountain and probably were controlled by concentric fractures similar to those that bound calderas (fig. 87). Probably a former caldera does lie buried beneath the later lavas.

The rift zones now visible are less pronounced than they probably were in the earlier shield, but westerly and southerly rift zones are suggested by alignments of cinder cones of both the Laupahoehoe and Hamakua Series. An easterly rift zone (fig. 189) is suggested by a few late cones of the Hamakua Series, including the cones that have been extensively quarried near Pepeekeo and the cone remnant at the north side of Onomea Bay that formerly contained the Onomea sea arch (fig. 135). It is still more clearly indicated by the broad ridge that extends eastward from Pepeekeo Point all the way to the deep ocean floor. Recent studies show that the lavas on this deep-sea ridge are tholeiitic basalts of the early Mauna Kea shield. Kauku ("the louse") is an appropriately named parasitic cone on the east rift zone from which issued a flow of ankaramite. The flow divided into two branches, both of which reached the sea. The more southerly branch forms Pepeekeo Point (fig. 187). The Kauku eruption was one of the last of the Hamakua Series.

Mauna Kea furnishes an excellent example of the influence of climate on erosion. The dry western slope is almost unscarred by erosion. In contrast, the northeastern slope, which is exposed to the trade-wind rains, has in the same span of time become trenched by canyons several hundred feet deep. Maulua Gulch is 650 feet deep a mile inland from its mouth, and Kaawalii Gulch is nearly as deep. Both were formerly deeper. The mouths of Maulua, Kaawalii, and Hakalau gulches are filled with

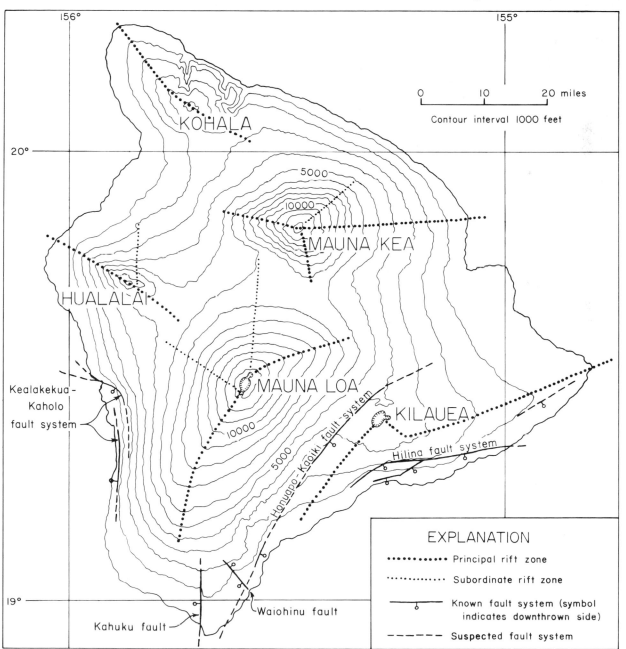

Figure 189. Map showing volcanic rift zones and faults on the island of Hawaii.

accumulations of stream-laid alluvium, deposited as a result of a recent rise of sea level. Laupahoehoe Gulch, cut into the rocks of the Hamakua Series, has been partly filled by a later flow of hawaiite of the Laupahoehoe Series. At the canyon mouth the flow spread out to form the flat-topped, delta-like Laupahoehoe Peninsula. A similar gulch once extended inland from Onomea Bay, along the course of the present Kawainui Stream. It also has been filled by a late lava flow of hawaiite. The channel of Kawainui Stream, with its magnificent display of potholes, follows approximately the former channel of the lava flow, and the hillocks adjacent to the stream mark the former levees along the lava river.

Between the occasional large gulches, the northeastern flank of Mauna Kea is barely beginning to show the scars of erosion (fig. 102). Even the largest gulches extend inland only 3 or 4 miles. The streams plunge into them over high waterfalls, the erosive action of which is gradually extending the canyons headward. Between the major gulches broad surfaces of Pahala ash are crossed only by shallow gullies. The large number of small streams are consequent on the original constructional slopes, and integration of the drainage into a smaller number of master stream systems is just beginning. Excellent examples of near capture of one stream by another can be seen from the air during flights between Hilo and Honolulu.

## HUALALAI

The present shield of Hualalai volcano was built by eruptions from a well-defined rift zone that trends approximately N50°W across the summit (fig. 189). About 3 miles east of the summit this is intersected by a less well defined rift zone that trends nearly north-south, fanning out on the northern flank of the mountain. More than 100 small cinder and spatter cones (fig. 190) are scattered along these rift zones. The eruptions were much less explosive than the late eruptions of Mauna Kea and Kohala, and the cones built by them are smaller and contain a larger proportion of spatter. However, they were more explosive than those of Mauna Loa and Kilauea, as evidenced both by larger cones at the vents and

by more abundant ash. At the summit of the mountain a collapse crater about a third of a mile across occupies the apex of a small lava shield. Scattered around it are angular blocks thrown out by a recent phreatic explosion. There is no direct evidence that a caldera ever existed on Hualalai. If it did, it has been completely buried by later lavas.

Hualalai appears to have just entered the late stage of the eruptive cycle. Nearly all of its lavas are alkalic olivine basalt, but a few flows are gradational to hawaiite. No tholeiitic basalts have been found, though presumably a tholeiitic shield lies buried beneath the alkalic cap. The flows are of both pahoehoe and aa types. They are especially well exposed in cuts along the highway between Kainaliu and Kailua. On this steep southwestern slope of the volcano the flows are thin, averaging only 4 or 5 feet in thickness. They must have been very fluid, and have flowed very rapidly.

Many of the lava flows contain scattered angular inclusions of bright green to brown peridotite, from a fraction of an inch to a foot across. They are especially abundant in the flow of 1800–1801, where they can be seen in cuts at the highway; but near the telephone relay station, about 2 miles upslope from the highway, they are present in tens of thousands. There the flow spread over a comparatively flat area and lost velocity, and the heavy inclusions tended to settle out like pebbles being dropped from a slowing stream of water. Most of the liquid drained out from between them, but left each fragment coated with a thin black layer of lava. The resulting accumulation of partly rounded, black to brown masses, each a few inches across, resembles more than anything else a heap of potatoes buried by the lava. In addition to peridotite there are also fragments of gabbro and rocks intermediate between gabbro and peridotite (picrite).

On the southern slope of Hualalai most of the lava flows are covered with a layer of ash. At the level of the highway it ranges from a few inches to about 2 feet thick, but it thickens upslope, and, southeast of the summit, there are areas several square miles in extent covered with ash and pumiceous cinder averaging 2 to 3 feet in depth.

On the north slope of Hualalai, 6 miles from

the summit, Puu Waawaa is a large cone of trachyte pumice, more than a mile in diameter (figs. 178, 191*A*). A trachyte lava flow more than 900 feet thick extends 6 miles northward from the cone. The pumice of the cone (figs. 191, 192) is being quarried for lightweight concrete aggregate and insulating material. Scattered through the pumice are many blocks of black trachyte obsidian. The lava flow consisted of several flow units, each 250 to 500 feet thick. Seen from the southwest, the edges of some of these units form conspicuous terraces. The western edge is a steep escarpment 500 feet high. The eastern side of the flow has been completely buried by later basalt flows of Hualalai and Mauna Loa, and some of these have spread across the surface of the trachyte and cascaded down its western escarpment. The closest exposure of the trachyte flow is separated from its source cone by more

than a mile of these later flows of basalt (fig. 191*B*). Where the surface of the trachyte flow is still exposed, it is very irregular and hummocky, with crescentic flow ridges up to 50 feet high, convex downstream. Puu Anahulu and Puu Huluhulu are simply high points on the irregular flow surface.

The Puu Waawaa trachyte represents a stage of magmatic differentiation far more advanced than that of any other rock of Hualalai. However, the cone lies squarely on the north rift zone of Hualalai and more than 20 miles from the nearest other occurrences of trachyte—on Kohala Mountain. The Waawaa trachyte was probably formed by differentiation in a relatively small magma chamber belonging to Hualalai, but isolated from the main magma chamber of that volcano.

During the latter part of its history, Hualalai was active simultaneously with Mauna Loa, and

Figure 190. Pit crater *(foreground)* and spatter cone on the north slope of Hualalai volcano. A collapsed lava tube extends to the left between them.

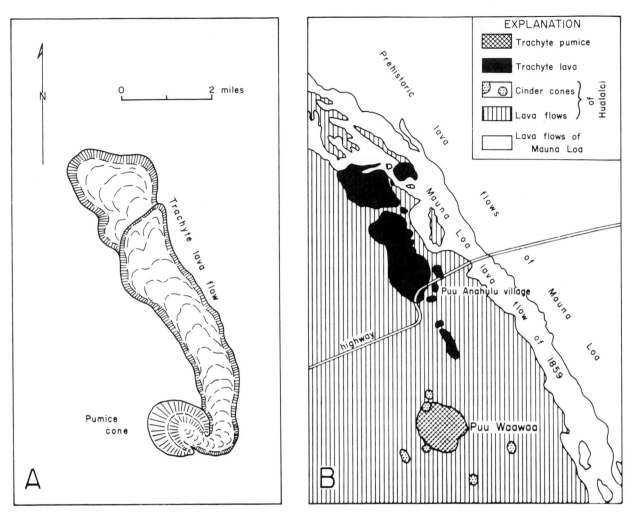

Figure 191. *A*, Diagram of the trachyte pumice cone of Puu Waawaa, on the northern slope of Hualalai Volcano, and the lava flow from it. *B*, Geologic map showing the relationship of the Puu Waawaa cone and flow to the later lava flows of Hualalai and Mauna Loa. (After Stearns and Macdonald, 1946.)

the lava flows of the two volcanoes interfinger on both the southern and northeastern flanks of Hualalai. A short distance south of the edge of the Hualalai lavas (fig. 178), the small cones of Puu Lehua and Kikiaeae, at an altitude of

5,000–5,500 feet, may possibly be cinder and spatter cones on the north-south rift of Hualalai, surrounded by later lavas of Mauna Loa. However, since the eruptive fissures on which the cones were built trend N60°W and line up

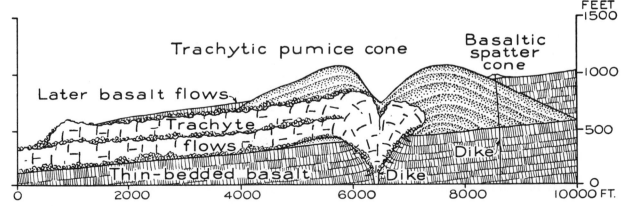

Figure 192. Diagrammatic cross section showing the relationship of the Puu Waawaa pumice cone and trachyte lava flow. (After Stearns and Macdonald, 1946.)

with another cone (Puu o Uo) still higher up the flank of Mauna Loa, the cones probably belong to Mauna Loa.

## MAUNA LOA

As noted on a previous page, the present mass of Mauna Loa appears to be made up of at least three huge shield volcanoes built around three separate eruptive centers. The lavas of all of them are tholeiitic basalts, olivine basalts, and oceanites. It is not possible to determine the position of the rift zones of the Ninole shield, or whether it had a caldera, because all but a few small remnants of its southeastern flank have been buried by lavas from the present southwest rift of Mauna Loa.

The remnants of the Ninole shield (fig. 178) form a series of steep-sided hills projecting high above the surrounding gently sloping lava surface. The rocks, known as the Ninole Volcanic Series, consist of thin layers of pahoehoe and aa exposed in the sides of the hills. In one, Puu Enuhe, a layer of vitric tuff 2 to 12 feet thick is interbedded with the lavas about 500 feet below the top. A thickness of 2,100 feet of Ninole lavas is exposed in Makaalia hill. The hills represent the high parts of ridges that lay between valleys cut by streams into the Ninole shield. Their tops have remained unburied by lava flows ever since the extinction of the Ninole volcano, and they have received the full accumulation of Pahala ash, which reaches a thickness of 55 feet on Puu Enuhe and Kaiholena hill.

Little more is known about the structure of the Kulani shield, which likewise is nearly buried by lavas from Mauna Loa's northeast rift. Part of the eastern flank of the Kulani shield is still exposed, however, and a row of cones and pit craters extending nearly eastward from Kulani Cone appears to delineate part of its old rift zone. It has been found also that a zone of magnetic anomalies extending westward from the ocean east of Cape Kumukahi (East Cape) follows the projection of this rift zone—not the present rift of Kilauea (fig. 173). Much of the eastward bulge of the island of Hawaii may be due to the Kulani shield, with only a thin cover of Kilauea lavas.

The main evidence for the existence of the Kulani shield is physiographic. Viewed from the summit of Kilauea, the flank of Mauna Loa exhibits a distinct terrace, or bench, which slopes gradually southwestward from a point south of Kulani Cone and finally disappears near Kapapala. The southeastern edge of the terrace is the Kaoiki fault scarp, and the northwestern edge is the beginning of the steeper rise of the present Mauna Loa shield. There seems to be no way to account for the terrace except as a remnant of an older, buried topography. It is crossed by many recent lavas of Mauna Loa, which, however, have left many kipukas ("islands") of older rock, capped by Pahala ash. Kipuka Puaulu (Bird Park), 2 miles northwest of Kilauea caldera, is one of these.

The large triangular kipuka east of Kulani Cone (fig. 178) is part of the flank of the Kulani shield that has been sheltered from later lava flows by the cone, and a similar, but smaller, kipuka lies just to the north. These segments of the shield surface also are covered with Pahala ash.

On the southern slope of Mauna Loa, from Kapapala to South Point, a succession of large and small kipukas consist of tholeiitic lava flows covered by Pahala ash. These lavas and associated cones have been named the Kahuku Volcanic Series. They have buried the erosional topography formed on the older Ninole shield. The lavas are best seen in the Kahuku Pali (figs. 36, 193), a fault scarp that extends inland from South Point, where about 600 feet of interbedded aa and pahoehoe flows is exposed. Near the coast, the base of the series lies an unknown distance below sea level. The Kahuku lavas south of Kapapala appear to have come from the present southwest rift zone of Mauna Loa, and the rocks of the Kulani shield also are included in the Kahuku Series.

The rocks of the present Mauna Loa shield are known as the Ka'u Volcanic Series, which includes lavas of present-day eruptions. Nowhere has erosion cut deeply into the volcano, but the upper 600 feet of the series is visible in the wall of Mokuaweoweo caldera, at the summit of the mountain. Ash is almost absent in the exposed sections, which consist of interbedded pahoehoe and aa flow units averaging about 15 feet in individual thickness. The flows in the caldera walls slope gently

Figure 193. Three nearly circular pit craters lie along the crest of the Kahuku fault scarp 9 miles north of South Point, Hawaii. The closest crater is Lua Puali, the second is Lua Poai, and the third is unnamed. The fresh lava at the base of the scarp is part of the 1868 lava flow of Mauna Loa.

outward, showing that they came from vents within the area of the present caldera which have been dropped down out of sight by the caldera collapse.

A blanket of pumice as much as 5 feet thick extends half a mile to leeward of the cinder and spatter cone of the 1949 eruption, on the southwest rim of the caldera (fig. 52); and small areas of pumice and vitric ash lie to leeward of some of the other recent cones. Thin layers of angular lava blocks and lithic ash, thrown out by small phreatic explosions, lie on the surface on the east rim of the caldera near the rest house, and on the northwest rim (fig. 47B). Another small patch of phreatic explosion debris lies along the Ainapo Trail 2.3 miles south of the rest house.

Mokuaweoweo caldera is 2.5 miles long and 1.5 miles wide (fig. 47). The cliff bounding it on the west is 600 feet high, but the one on the east is only 225 feet, and both decrease in height northward to only about 10 feet at the northeastern edge. At the south end the caldera merges with the adjacent pit crater known as South Pit (fig. 32), and during the 1949 eruption, lava flowed from the caldera into South Pit, filling it to overflowing. Two other pit craters, Lua Hohonu and Lua Hou, lie on the rift zone 0.3 and 0.7 mile southwest of South Pit.

Mokuaweoweo was first mapped by Lt. Charles Wilkes, the commander of the U.S. Exploring Expedition, in 1841. At that time it had a nearly circular central pit, 800 feet deep on the west side, bordered by crescentic benches on both the north and south. Still farther north the present North Bay was a separate pit crater. Since then eruptions in the caldera have gradually filled the central pit and overflowed the north and south "lunate platforms," extending the floor of the caldera across North Bay (fig. 47). Wilkes's map (fig. 47A) shows neither the East Bay nor Lua Poholo, and south of the caldera it shows only two pit craters instead of the present three: both East Bay and Lua Hohonu have been formed since 1841. (Lua Hohonu is now the one that should be named Lua Hou—"new pit.") Lua Poholo still did not exist when the area was mapped by J. M. Lydgate in 1874, but it had been formed by 1885, when Mokuaweoweo was again mapped by J. M. Alexander. The present condition of Mokuaweoweo is shown in figure 47*B*.

The Mauna Loa shield has been built principally by eruptions along two rift zones that extend southwestward and east-northeastward from the caldera (fig. 189). A less well defined rift zone, which fans out on the northern slope, gave vent to the great lava flow of 1859. Another, even less prominent alignment of vents extends northwestward down the western slope of the mountain toward the summit of Hualalai. The rift zones are marked by scores of open cracks, an inch or two to 10 feet wide, along many of which spurting lava has built spatter ramparts from 1 to 20 feet high. Where the eruptions centralized in big lava fountains, the ramparts merge into cinder-and-spatter cones. Most of the cones are less than 100 feet high, but the lava cone of the 1940 eruption, in Mokuaweoweo, rises about 170 feet above its base, which is buried by flows from the same and later vents. More than 160 fissures and cones have been mapped on Mauna Loa.

The Halai Hills, in Hilo, are a row of three small cinder cones. They are covered with Pahala ash, and are correlated with the Kahuku Volcanic Series. The middle hill has been almost completely dug away for cinder used in the surfacing of secondary roads. It has been suggested that the Halai cones belong to Mauna Kea, but it appears more probable that they are early cones of Mauna Loa, since they are aligned parallel to the northeast rift zone of Mauna Loa and lie almost directly on the projection of that rift zone. Most of downtown Hilo lies on an ash-covered kipuka that has been sheltered from later lava flows by the Halai Hills.

Few eruptions have taken place along the lower 20 miles of the northeast rift zone during the time of formation of the Ka'u Series. However, one small spatter cone lies on the rift in the Waiakea Homesteads area, 4 miles southwest of Halai Hill. Spatter from this cone buried tree ferns and converted them into charcoal. Carbon-14 determinations on the charcoal show that the eruption occurred about 2,000 years ago.

Along the south shore of Hawaii, for about 14 miles west of South Point, a row of cinder and ash cones (fig. 178) has been built by littoral explosions where lava flows from the southwest rift of Mauna Loa entered the sea. The most recent of the cones (fig. 42) is an unusually large one, 240 feet high, known as Puu Hou (new hill), formed by the lava flow of 1868. Except in this area, only a few littoral cones are found around the shore of Mauna Loa. There are two on the Kona Coast, Puu Ohau and a small cone at Lae Auau, and 3 miles northeast of South Point, Puu Mahana is a littoral cone belonging to the Kahuku Series. Wave erosion has destroyed the seaward side of Puu Mahana, and grains of olivine from the cone have been concentrated in a small bay to form one of the best green sand beaches in Hawaii.

Near the coast the southwest rift zone is bounded along the eastern side by the north-south-trending Kahuku fault (fig. 189), along which the western side has been dropped down relative to the eastern side. The resulting fault scarp is 600 feet high at the coast. It can be traced inland some 10 miles, gradually decreasing in height until it disappears a short distance beyond the highway. Seaward, the scarp extends at least 18 miles beyond South Point. About 6 miles farther east the southeast-trending Waiohinu fault is displaced in the opposite direction. It extends 4.5 miles from Waiohinu to the sea, and has produced an eastward-facing scarp, generally less than 50 feet high. In 1868 lateral movement of several feet took place on the Waiohinu fault.

Along the southeastern base of Mauna Loa, the Kaoiki fault system (fig. 189) is a series of echelon tangential faults along which the mountainward side has moved relatively upward. Step faulting has produced a series of terraces separated by southeast-facing scarps that can be traced for 18 miles southwestward from Bird Park, near Kilauea caldera, to beyond Kapapala. The Kaoiki faults appear to provide a sort of expansion joint between Mauna Loa and Kilauea. When one or the other volcano is swelling or shrinking, adjustments

Figure 194. Kealakekua Bay, with Napoopoo in the foreground. The cliff on the right-hand side of the bay is a fault scarp. In the background, lava flows of Mauna Loa have poured over the cliff and spread out to form the Kaawaloa peninsula, on the near side of which Capt. James Cook was killed.

take place in the form of movements on the faults, with resulting earthquakes. In places, the total height of the escarpment is more than 500 feet, but along part of its length the escarpment is veneered by later lava flows, and the visible height represents only part of the displacement on the faults, since its base is buried by later lava flows of both Mauna Loa and Kilauea. One of the flows is the very late prehistoric Keamoku flow, which originated from a vent at about 9,000 feet altitude on the northeast rift zone of Mauna Loa, flowed southeastward 10 miles down the flank of the mountain and over the Kaoiki fault scarp, then turned and flowed southwestward 5 miles along the base of the escarpment.

A series of older tangential faults extends

southwestward from the Kaoiki escarpment to beyond Honuapo. They have been partly eroded and largely hidden by a cover of ash and later lavas, but near Honuapo the recent lavas have been offset a few feet by more recent fault movement.

Along the west slope of Mauna Loa, the escarpment at the northeastern edge of Keala-kekua Bay (fig. 194) is the expression of the Kealakekua fault (figs. 157, 189). The damaging Kona earthquake of 1951 was caused by movement on this fault beneath the ocean, about a mile southwest of the tip of the peninsula north of the bay. A branch of the fault diverges northward and is responsible for the steep lava-mantled escarpment at the inland edge of the peninsula. Southeastward, the fault

310

extends diagonally inland for about 3 miles beneath a cover of later lavas. Beyond there, it appears to bend southward, parallel to the coast, and to be responsible for the abnormal steepness of the lava-covered slope of that part of the mountain.

Farther south, the Kaholo Pali lies just inland from the shore for about 16 miles. This west-facing escarpment is almost entirely mantled by later lava flows. South of Hookena, the height of the escarpment is about 500 feet, but it decreases both northward and southward, and over most of its length it averages about 250 feet. There appears to be little question that the Kaholo Pali is primarily a buried fault scarp, although it may have been modified to some small extent by marine erosion.

At many places near the shores of the island, the Ka'u Series is only a few flows thick. Kipukas of ash-covered Kahuku lavas are found southwest of Hilo, in addition to those already mentioned along the south slope near Pahala. Pahala ash is also encountered at shallow levels in water-development tunnels southwest of Hilo and near Pahala and Naalehu, where it perches ground water in the overlying Ka'u lavas. Ash which may be correlative with the Pahala ash is exposed in the cliff at the north side of Kealakekua Bay, overlain by 25 feet of Ka'u lavas. The rocks beneath the ash may belong to Hualalai, but more probably they are a part of the Kahuku Series.

In some of the kipukas on the south slope of Mauna Loa, such as that at South Point, the Pahala ash is as much as 40 feet thick, which is nearly the maximum thickness found on the remnants of the Ninole shield. These parts of the Kahuku Series must have remained unburied by lava flows through most of the time of accumulation of the Pahala ash. It is not now possible to tell whether the Mauna Loa shield remained inactive throughout this time, or whether its southeastern flank was shielded from lava flows by a high caldera wall and fault scarps paralleling the rift zone. More probably, however, activity shifted for a time almost wholly to the Kulani shield. That activity probably did not cease entirely at all vents is indicated by the fact that the ash on the

eastern slope, between Keaau and Mountain View, although it totals about 15 feet in thickness, is shown by drill holes to be divided into several beds by intervening lava flows.

A low cliff, largely mantled with later lava flows, which extends for several miles northward from Kealakekua Bay, is probably a sea cliff. The lava flows beneath the later lava veneer are capped by a layer of ash 6 to 12 inches thick. The ash must have accumulated, and the sea cliff probably was cut, during a time when the Mauna Loa shield was inactive or when that part of the slope for some other reason received no lava flows.

Thus the evidence suggests that, at various intervals in its history, there were periods of at least several centuries during which lava flows did not reach some sectors of the volcano. Nevertheless, there is almost no sign of erosion between successive flows, probably because of the high permeability of the lava surface which allows water to sink in instead of running off. Only one noteworthy exception is known. Above Hilo, the Wailuku River has been crowded northward against the slope of Mauna Kea by successive lava flows of Mauna Loa, and it now follows approximately the boundary between the two volcanoes. Despite loss of water into the Mauna Loa lavas, the river is kept flowing by many tributaries that enter it from the ash-covered slopes of Mauna Kea. At some time within the last few millenia, the river was able to cut a shallow gorge, seldom more than 50 feet deep, along the edge of the Mauna Loa lavas. Still later, the gorge was refilled by another lava flow, which in turn is now being eroded. A cross section of the flow can be seen at Rainbow Falls, in Hilo, where at the north edge of the plunge pool the old valley floor can be seen sloping southward, overlain by the later lava. Along the south side of the plunge pool, springs issue at the bottom of the lava flow. The water for the springs is derived by leakage from the river farther upstream. Chilling of the lava against the valley side caused columnar jointing, which can be seen in the face of the fall. The cross sections of columnar joint columns are well exposed also at the Boiling Pots—a series of potholes cut

311

by the river about a mile upstream from Rainbow Falls. Through most of its lower course the Wailuku River follows the former feeding channel of the late lava flow. Maui's Canoe is a remnant of the flow downstream from the falls.

## KILAUEA

The rocks of Kilauea are divided into the older Hilina Volcanic Series and the younger Puna Volcanic Series (fig. 178). The Hilina Series is capped by Pahala ash, and is believed to be coeval with the Kahuku Series on Mauna Loa. The Puna Series overlies the Pahala ash and is correlative with the Ka'u Series of Mauna Loa. Lava flows of the Puna and Ka'u series interfinger along the boundary between the two volcanoes. Both the Hilina and the Puna series consist of pahoehoe and aa lava flows of tholeiitic basalt, olivine basalt, and oceanite, and associated cinder-and-spatter cones and ash deposits.

The Hilina Series is exposed only in fault scarps along the southern coast. In the Hilina Pali a succession of about 1,000 feet of thin lava flows is exposed. A few thin beds of vitric tuff lie between the flows. A similar section is exposed in the seaward face of Puu Kapukapu.

The Pahala ash is 30 feet thick in Hilina Pali and 40 feet thick on Puu Kapukapu (fig. 179). Most of it is sandy to silty yellow vitric ash, resembling most of the outcrops on Mauna Loa, but in Hilina Pali coarser beds containing pumiceous lapilli and Pele's tears are abundant. They must have been derived from vents relatively close by, probably not farther away than Kilauea caldera. As on Mauna Loa, the thick deposits of Pahala ash in the Hilina Pali area mark an interval during which no lava flows reached that part of the volcano. However, the coarse lapilli at Hilina Pali, as well as the great thickening of the ash in the Pahala region, indicate that Kilauea volcano was active during this interval. The Hilina Pali area was probably protected from lava flows by north-facing cliffs formed by collapse around the caldera and along the adjacent rift zones.

The lavas of the Puna Volcanic Series were erupted almost entirely from vents in the area of the present Kilauea caldera and along two rift zones that extend eastward and southwestward from the summit of the volcano (fig. 189). At the top of the Hilina fault scarp they are only one or two flows thick, but in the cliff at the western side of the caldera lava flows totaling about 420 feet thick are exposed. Before 1919 a thick bed of yellow ash (the Uwekahuna ash) was exposed at the base of the cliff, and if it is the same as the similar Pahala ash the total thickness of the Ka'u lavas at that place is only a little more than the thickness now visible in the cliff. Elsewhere their base is not exposed, and their thickness is unknown.

Kilauea caldera consists of a central, well-defined, oval sunken area 3 miles long, 2 miles wide, and a little more than 400 feet deep at its western edge, surrounded by a series of benches which have not sunk as much (figs. 57, 195). The inner caldera is bounded throughout most of its circumference by abrupt fault scarps. In two places, however, the collapse has been in the form of step faulting. The step-fault blocks beneath the Volcano House, at the northeastern edge of the caldera, are irregular, and a graben 50 feet deep lies behind one of the blocks. Those at Uwekahuna, on the west side of the caldera, are quite regular. Near the southern edge of the caldera a horst (the "Sand Spit") projects from the eastern wall. The southeastern wall is low and banked with ash from eruptions within the caldera. In 1921 a short lava flow escaped through a low gap in the southeastern wall of the caldera (fig. 195). With the exception of the Sand Spit, the entire inner caldera floor is covered with lava erupted within the last century. Halemaumau, in the southwestern part of the caldera floor, is a collapse crater that occupies the top of a very flat cone or shield which slopes off in all directions to the edge of the caldera floor. Through the last 150 years Halemaumau has been the main focus of surface activity of Kilauea.

Surrounding the central caldera is a series of benches. For the most part they also are bounded by well-defined fault scarps, but in some places the faults pass into sharp monoclinal down-bends of the surficial lavas. Individual faults tend to die out and be replaced by others arranged en echelon (fig. 195). Particularly on the southwest side, the marginal area gives the impression of having sagged toward the inner caldera. The diameters of the outer sunken area are approximately 3 by 4 miles.

On the southwest the caldera merges with the rift zone, which is there bounded on the south by a series of faults trending parallel to it and down-thrown on the north side (fig. 195). Another series of less conspicuous faults extends southwestward from the caldera along the north side of the rift zone, which thus lies within a shallow graben. Farther southwest the fault scarps along the south side, if they ever existed, have been buried by lava flows; but a series of south-facing fault scarps lies along the north edge parallel to the nearby Kaoiki fault system on Mauna Loa.

The east rift zone of Kilauea extends east-northeastward, from the faults bounding the south edge of the southwest rift zone, 28 miles to Cape Kumukahi, and at least 70 miles farther beneath the ocean. It is linked to the caldera by a row of seven pit craters that trends southeastward to Pauahi Crater (fig. 62). This line of pit craters is crossed almost at right angles by both the outer boundary faults of the caldera and the faults bounding the southwest

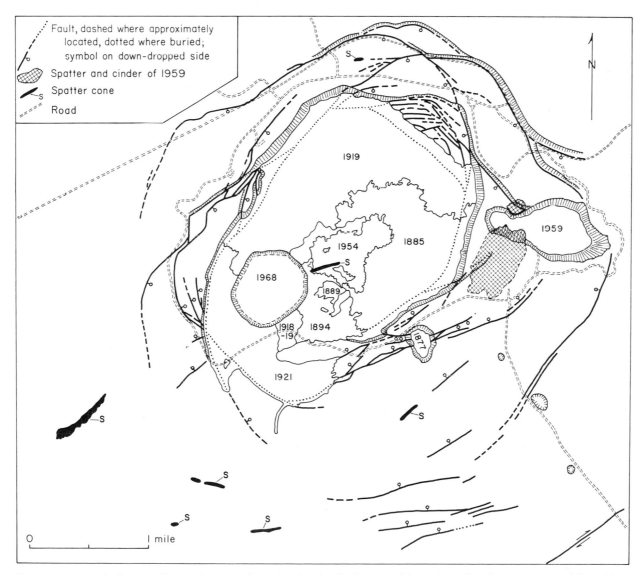

Figure 195. Map of Kilauea caldera and surroundings, showing the distribution of faults bounding the sunken area of the caldera, and the dates of some historic lava flows on the caldera floor. The more prominent fault scarps are shown by hachures. (Modified after Peterson, 1967.)

rift zone, and it is noteworthy that the eruptive fissure of the 1962 eruption, near Aloi Crater, was parallel to these faults, not to the line of pit craters. Between 1938 and 1942, horizontal movement on these faults offset the Chain of Craters Road about 4 feet, the northern side of the fault moving eastward. Similar movements occurred again during the 1962 and 1963 eruptions, again with the north side moving eastward on several faults for an aggregate displacement of several feet. This was accompanied by vertical movement on one fault, with the north side moving upward nearly 3 feet.

In the 7 miles east of Pauahi Crater six other pit craters lie along the rift zone (figs. 62, 77). Beyond that point, however, there are no others for 16 miles to the east. This region of numerous pit craters must mark a zone of large intrusions from the central magma body of the volcano that have approached close to the surface.

Two small pit craters on the southwest rift zone, 3 miles from the caldera, lie above a large lava tube and were probably formed by collapse of the roof of the tube. Puulena, Pawai, and Kahuwai craters, on the east rift zone 7 miles southwest of Cape Kumukahi, are small pit craters perforating the summits of small lava shields.

Lava beds in the walls of Kilauea caldera slope away from the caldera and project into the air where the caldera is now located. This indicates that the caldera must once have been much smaller, or perhaps it did not exist at all. The summit of the Kilauea shield was complex, consisting of several coalescing smaller shields. One occupied the position of the present caldera. Another is still clearly discernible on the east edge of the caldera. Its summit collapsed to form Kilauea Iki Crater. Another small shield on the side of the Kilauea Iki shield is indented by the two small Twin Craters, one of which was drained by the Thurston Lava Tube.

Other small parasitic shields lie on the rift zones. Kane Nui o Hamo and Heiheiahulu, on the east rift zone (figs. 62, 68), and Mauna Iki, formed by the 1920 eruption on the southwest rift zone, are examples. However, most of the discrete cones on the rift zones are spatter or spatter-and-cinder cones. Among the largest of these are the Kamakaia Hills, 100 feet high, on the southwest rift (fig. 58), and Kalalua and Iilewa craters on the east rift (figs. 68, 76). Kapoho Cone, 2 miles inland from Cape Kumukahi (fig. 74), is a tuff cone formed by a phreatomagmatic eruption. The cone of the 1960 eruption, just north of the Kapoho Cone, also consists in large part of ash formed by phreatomagmatic explosions. The basal ground water lies only about 80 feet below the surface in this area. Phreatic explosions have thrown out small amounts of lithic ash and blocks around Aloi Crater, Puulena, and Pawai Crater. Near the rim of Puulena Crater some of the blocks weigh several tons. The phreatic explosions that took place at Halemaumau in 1924 have already been described (page 77). The area around Halemaumau is heavily covered with blocks and ash from that eruption.

In the immediate vicinity of the caldera, the surface of the lava flows is mantled with a deposit of ash. On the Sand Spit, within the caldera, the ash is more than 30 feet thick, and along much of the southwestern rim of the caldera, more than 10 feet. The deposit thins in all directions away from the caldera (fig. 196). It is well exposed in the fault scarp along the east side of the caldera south of Keanakakoi Crater, in road cuts and cracks around the south side of the caldera, and in cuts along the Mamalahoa Highway north of the caldera. Near the caldera the lower several inches of the deposit generally consists of pumice (reticulite) similar to that formed in abundance during the Kilauea Iki eruption of 1959. In many places the pumice rests on a layer of red ashy soil, which suggests that, for a considerable period, weathering of the surface of this part of the volcano was not interrupted by volcanic activity—a period of quiescence of at least this part of Kilauea that may have lasted several hundred years. Above the pumice layer, most of the deposit consists of vitric ash with occasional pumice lapilli, derived from the spray of lava fountains; but occasional layers of lithic ash and lapilli were formed by phreatic explosions. Along the southwest edge of the caldera the lithic ash of the 1924 explosions forms a layer a few inches thick, overlain only by scattered pumice of later eruptions. The 1924 ash is separated by several inches of vitric ash from the lithic debris of the 1790 eruption. Especially in the area near Keanakakoi Crater, the lower part of the 1790 deposit contains many cored bombs and lapilli with shells of magmatic material, indicating that magma was involved at least during the beginning of the eruption. Coarse pumice overlying the 1790 ash came from big lava fountains during the return of activity following the 1790 collapse.

The tangential faults of the Hilina fault system lie along the southern slope of Kilauea. The system consists of a series of subparallel and en echelon faults (fig. 197) with a total

downthrow on the seaward side of more than 2,000 feet. The Hilina Pali has a maximum height of 1,500 feet. Westward it becomes lower and then disappears beneath recent lava flows from the southwest rift zone. To the east it merges with the Poliokeawe Pali, which decreases in height eastward and disappears about 4 miles west of Kalapana. Throughout most of its length the Hilina-Poliokeawe escarpment is mantled with late lava flows (fig. 198), but lavas of the Hilina Series are exposed in windows. The seaward face of Puu Kapukapu is a fault scarp more than 1,000 feet high, plunging directly into the ocean. Puu Kapukapu is a small horst lying between the main

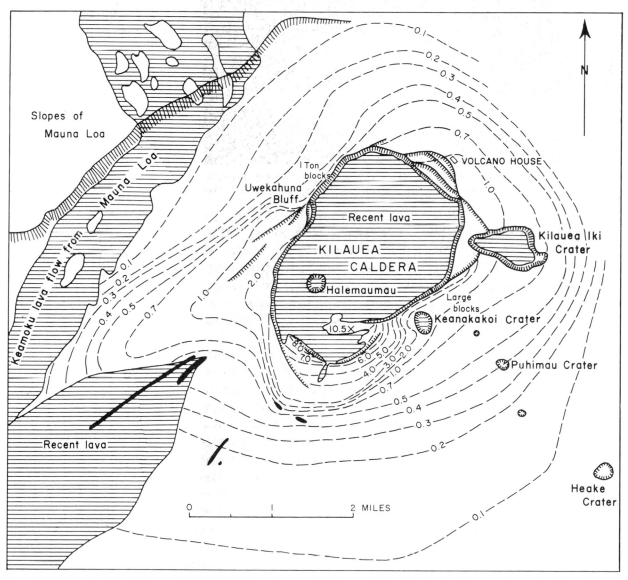

Figure 196. Map showing thickness of surficial ash deposits around Kilauea caldera. The dashed lines are isopachs (lines of equal thickness) indicating the depth of the ash cover, in meters. (After Stearns and Clark, 1930.)

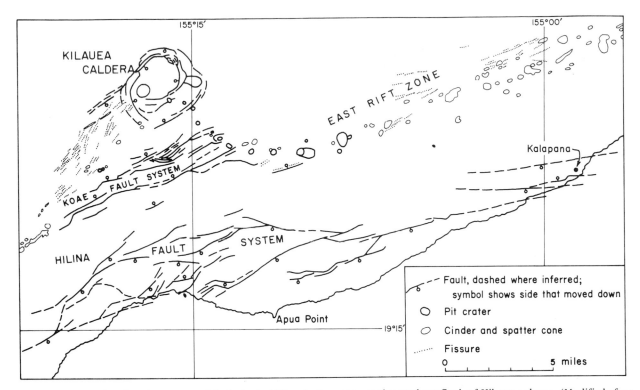

Figure 197. Map showing the pattern of faults in the Hilina fault system, on the southern flank of Kilauea volcano. (Modified after Stearns and Macdonald, 1946.)

316

Figure 198. View eastward along Poliokeawe, one of the fault scarps of the Hilina system on the south slope of Kilauea. Converging with it from the right is another of the fault scarps, the Holei Pali. The black lava flows in the foreground, between the camera and the Kalapana road, were formed in 1969. Other older flows can be seen mantling the scarp.

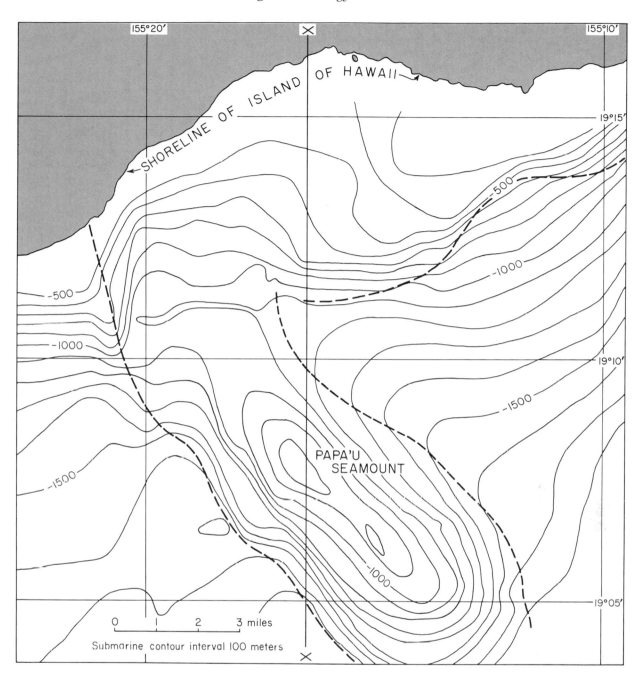

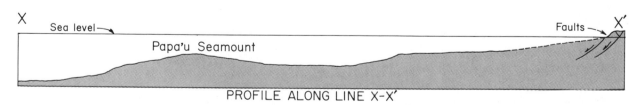

PROFILE ALONG LINE X-X'

Figure 199. Map and profile of Papa'u Seamount, just south of the southern coastline of Kilauea volcano. (After Moore and Peck, 1965.) Moore and Peck consider the areas bounded by the heavy dashed lines to be huge landslides that have slid southward from the flank of Kilauea volcano; but it appears more probable that Papa'u Seamount is a shield volcano built along a rift zone with the same trend as that of Hualalai, with its flanks modified by faulting.

Kapukapu fault and a fault of smaller, opposite displacement lying about half a mile inland. The low ridge seaward of Kalapana village also is a horst, and the village lies within a graben. Movement has continued on the Hilina fault system until recent times. Along the foot of the escarpment west of Kalapana the veneer of recent lavas is broken by a scarplet 10 feet high. During the violent earthquakes of 1868 the floor of the Kalapana graben subsided several feet, allowing ocean water to invade the old churchyard.

The southern slope of the east rift zone east of the Kalapana-Pahoa road undoubtedly owes its abnormal steepness to fault scarps that have been buried by later lava flows. Similar steep slopes are found in the ocean floor both south of the island and along the southern slope of the rift-zone ridge east of the island; they are almost certainly the result of submarine faulting.

A broad submarine ridge south of the island of Hawaii (fig. 199) has been interpreted by Moore (1964) as a huge landslide block that has moved outward and downward from the southern flank of Kilauea volcano, leaving the Hilina and associated fault scarps as part of the scar at the head of the slide. Alternatively, Macdonald (1955) interpreted the mass as another shield volcano built against the southern slope of Kilauea in the same way that Kilauea lies against the slope of Mauna Loa. The crest of the supposed submarine shield lies on the prolongation of the principal rift zone of Hualalai and the minor northwestern rift that crosses the western slope of Mauna Loa. At present, neither hypothesis for the origin of the submarine mass can be proved.

# Maui

The island of Maui is part of a huge volcanic massif consisting of at least six major and one minor volcanoes (fig. 200). At present the

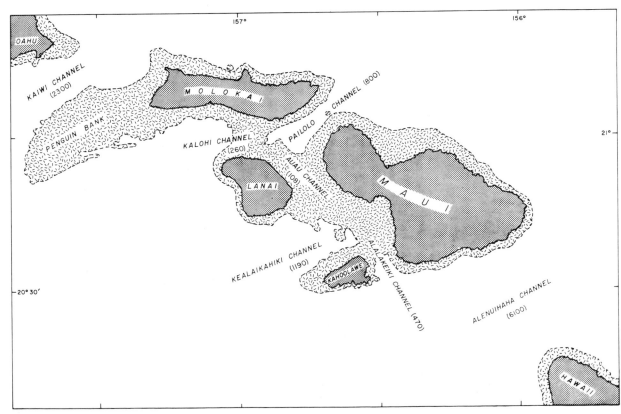

Figure 200. Map of the islands of the Maui group. The dashed line at the outer edge of the shaded area is the 100-fathom (600-foot) submarine contour. The figures in parentheses indicate the approximate depth of water in each of the channels separating the islands.

318

lower saddles between the volcanoes are flooded by shallow sea water, dividing the massif into four separate islands—Maui, Kahoolawe, Lanai, and Molokai; but at times of lower sea level in the geologically recent past all have been united as a single large island. This island, which we may call Maui Nui, had an area of about 2,000 square miles—roughly half the present area of the island of Hawaii.

Maui itself consists of two major volcanoes (figs. 201, 202). The older one is West Maui. Lavas of Haleakala (East Maui) volcano have banked against the already-existing slope of West Maui to form the broad, gently sloping plain of the Maui Isthmus. West Maui volcano may be extinct. It has passed through the principal stages of Hawaiian volcanism, and has produced four small posterosional eruptions. Haleakala volcano erupted most recently less than two centuries ago and must be regarded as dormant. The potentiality for future eruptions certainly exists.

WEST MAUI

The volcanic rocks of West Maui have been divided (Stearns and Macdonald, 1942) into three series (fig. 203). The oldest is the Wailuku Volcanic Series—the basaltic lava flows and associated pyroclastic and intrusive rocks that built the major shield volcano. It was covered by a thin, discontinuous "frosting" of andesitic and trachytic flows, domes, and pyroclastic deposits termed the Honolua Volcanic Series. Then, after a long period of erosion, came the eruptions that produced the flows and cones of the Lahaina Volcanic Series.

The Wailuku Volcanic Series consists predominantly of thin pahoehoe and aa lava flows of tholeiite, olivine tholeiite, and occanite; but in its uppermost part these grade into alkalic olivine basalt. Very little explosion accompanied the building of the shield, though some thin layers of vitric and lithic tuff are intercalated with the flows. The amount of

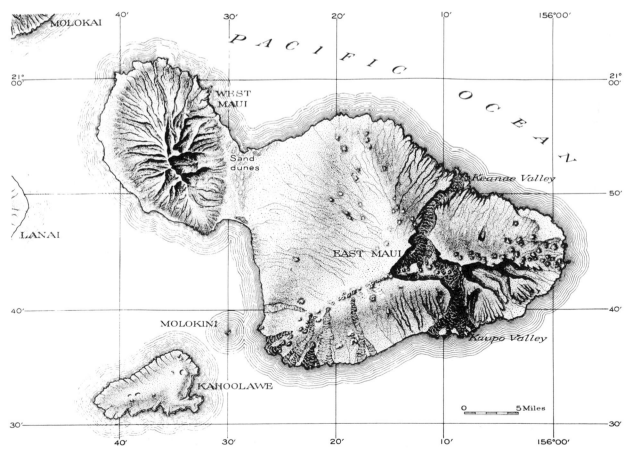

Figure 201. Shaded relief map of the islands of Maui and Kahoolawe. (After Stearns and Macdonald, 1942.)

*319*

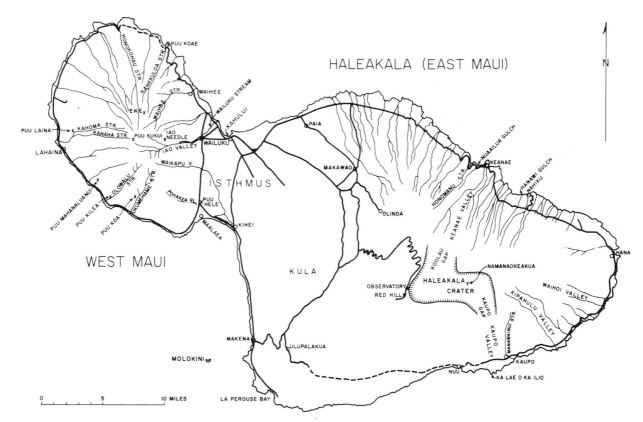

Figure 202. Map of Maui, showing the location of places mentioned in the text.

explosion increased somewhat toward the end of the eruptive period, and tuff beds are more numerous in the uppermost part of the series. Some phreatic explosions were violent enough to produce localized beds of lithic breccia.

The shield eventually reached a height of about 7,000 feet above sea level (more than 20,000 feet above its base at the ocean floor) before its top collapsed to form a caldera about 2 miles across (figs. 204, 205). Outside the caldera thin lava flows dip (slope) outward at angles of 10° to 20°, somewhat steeper than is usual in Hawaiian volcanoes. Within the caldera thick flows that were once nearly horizontal have been much disturbed—tilted, fractured, and brecciated by repeated collapses—and have been altered by rising gases. Alteration of original pyroxene in the caldera-filling rocks to chlorite has released silica, which has been redeposited as nodules of opal and chalcedony. These, eroded out of the rocks, can be found along Wailuku Stream, and along the beaches on the west coast. They are known locally as moonstones. Much of Iao Valley has been cut into the rocks of the caldera fill by stream

erosion (figs. 206, 207). Part of the boundary of the caldera can be seen high on the opposite wall of the valley and just upstream from the parking area near the Iao Needle (figs. 110, 111). The Needle itself is simply an erosional residual in the rocks of the caldera fill. The caldera boundary is exposed also in the head walls of Waikapu and Ukumehame valleys.

The rift zones of West Maui are much less well defined than are those of most Hawaiian volcanoes. There is a tendency for dikes to radiate in all directions from the summit, and this arrangement of the fissures and vents that fed the lava flows is responsible for the nearly circular ground plan of the volcano. There is, however, some concentration of dikes into two general zones. The more pronounced of these crosses the mountain in a nearly north-south direction; the other trends northeastward in the northeast part of the mountain (fig. 204). In each, hundreds of dikes are exposed, most of them trending parallel to the zone in which they are situated. Some sills also are present.

Small stocks of gabbro and basalt porphyry cut the Wailuku lavas in several of the canyons.

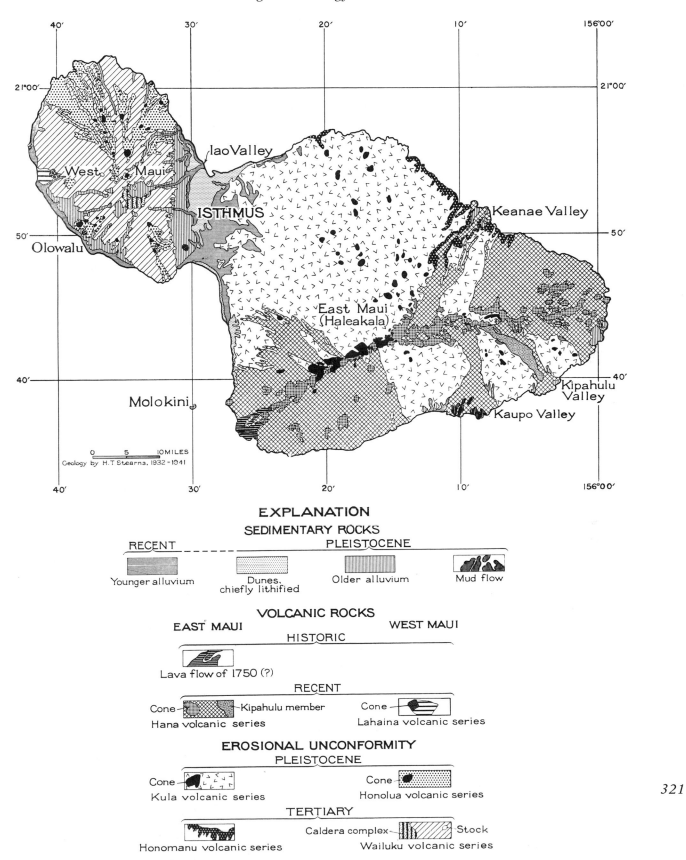

**EXPLANATION**

**SEDIMENTARY ROCKS**

RECENT ———————— PLEISTOCENE

Younger alluvium · Dunes, chiefly lithified · Older alluvium · Mud flow

**VOLCANIC ROCKS**

EAST MAUI · WEST MAUI

HISTORIC

Lava flow of 1750 (?)

RECENT

Cone—Kipahulu member · Cone—Lahaina volcanic series
Hana volcanic series

**EROSIONAL UNCONFORMITY**

PLEISTOCENE

Cone—Kula volcanic series · Cone—Honolua volcanic series

TERTIARY

Caldera complex—Stock
Honomanu volcanic series · Wailuku volcanic series

321

Figure 203. Geologic map of the island of Maui. (After Stearns, 1946.)

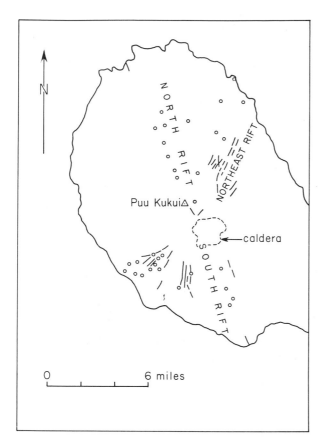

Figure 204. Map of West Maui, showing the principal volcanic rift zones, dikes (*short lines*), and vents (*circles*), of the Honolua Volcanic Series, and the approximate boundary of the caldera (*dashed line*). (Modified after Stearns and Macdonald, 1942.) Dikes of the Wailuku Volcanic Series are much more abundant than those of the Honolua Series.

Most of them are roughly circular in ground plan (bosses) and only a few hundred feet across. Two can be seen in Black Gorge, just to the east of the Iao Needle (fig. 88), and larger ones are exposed in Ukumehame, Kahoma, and Kahakuloa valleys. The one in the latter valley is bisected by the stream, 1.5 miles from the shore, and is approximately 0.4 mile long and 0.15 mile across. The diameter in some cases has been found to decrease slightly downward. Collapse of the roof above some of these stocks may have formed pit craters at the surface.

Two former pit craters, later filled, have been exposed by erosion. One, about 0.3 mile across, lies near the head of Waihee Valley; the other, slightly smaller, is in Kanaha Valley near Lahaina. The latter crater was partly filled by talus breccia and hillwash, but at one stage it must have contained a pond of water in which layers of mud accumulated to form shale. This sedimentary crater filling is capped by massive ponded lava flows (fig. 208).

The formation of the Wailuku Volcanic Series was followed by a period of weathering and erosion. In some places the Wailuku lavas are separated from those of the overlying Honolua Volcanic Series by a layer of soil as much as 5 feet thick, and in other places the two are separated by hillwash and stream-deposited conglomerate. Elsewhere, however, the Honolua lavas rest directly on the Wailuku lavas. Furthermore, the soil between the two series is commonly ashy. It is quite likely, therefore, that there was no long period when the volcano was totally inactive; but eruptions were infrequent, and many parts of the mountain were invaded by lava only at long intervals.

The lavas of the Honolua Series are mostly mugearite, with less abundant trachyte and a little hawaiite. Some of the flows are pahoehoe, but most are aa, commonly transitional to block lava. The capping of Honolua lavas was incomplete, and over much of the mountain it consists of only one or two flows. Its average thickness is about 75 feet, and the maximum, on the northeastern slope, is about 750 feet. Individual flows of mugearite average about 40 feet thick. Trachyte flows are thicker, averaging about 150 feet.

In some eruptions of trachyte the magma was so viscous that it piled up over the vent to form domes (fig. 209). On the upper east wall of the head of Honokohau Canyon a mass of trachyte appears to consist of three superposed flows, each more than 200 feet thick, surmounted by the dome of Mt. Eke. Puu Koae, southeast of the mouth of Kahakuloa Canyon, is a trachyte dome that was built in the crater of a pumice cone. Around the base of the hill the crumble breccia of the dome can be seen resting on and against the remnants of the easily eroded pumice cone. Three miles southeast of Lahaina, Puu Mahanalua Nui (Launiupoko Hill) and the smaller hill just inland from it are trachyte domes. The Mahanalua Nui dome ruptured during a late stage of formation and the seaward portion moved outward and downward as a short thick lava flow.

Dikes of the Honolua Series follow the same general trends as the earlier dikes, and the vents lie mostly on the same major rift zones. Mugearite dikes are generally thin, but those of trachyte range up to 25 feet in thickness, and some of them have been traced for 2 miles. Some dikes expand close to the surface to form shallow-seated plugs, which may in turn protrude as domes. An example of this is well displayed at Puu Koai, on the northwest wall of Ukumehame Canyon, where a 25-foot dike expands into a plug and dome 600 feet high. A small crater appears to have been blasted out by explosion as the magma reached the surface. The viscous lava expanding in this crater thrust aside the bordering Wailuku basalts, brecciating them, and tilting them upward to an angle of 75°. These expanded upper parts of dikes

filling craters at the surface have been referred to as "buds" (Wentworth and Jones, 1940).

The end of the Honolua eruptions ushered in a long period of erosion, which is still continuing. It was interrupted, however, by a brief interval of renewed volcanism during which four small eruptions took place on the western slope of the mountain (fig. 203). The rocks of this posterosional volcanism, known as the Lahaina Volcanic Series, are very different from those of the Honolua Series. The largest of these eruptions took place on the alluvial fan of Kahoma and Kanaha streams, 1.5 miles northeast of Lahaina, forming the cinder cone of Puu Laina. The lava flow is picrite-basalt containing moderately abundant phenocrysts

Figure 205. The summit area of West Maui is a great scoured-out, once-filled caldera. Remnants of the slopes encircling the caldera are clearly recognizable and one can easily make out the general form of the volcanic shield and its summit caldera. Most of the shield is in a submature stage of dissection, but the dry leeward slope in the right foreground is in a late youthful stage.

Figure 206. Puu Kukui, the rounded knob against the sky, is the highest point on West Maui. It rises 5,788 feet above sea level, and is usually shrouded by heavy banks of clouds. The scoured-out West Maui caldera lies to the left of the cliffs near the left edge of the photograph.

of brownish green olivine. It blocked the mouth of Kahoma Valley and displaced Kahoma Stream southward more than half a mile from its former course.

Another cinder-and-spatter cone of the Lahaina Series forms Kekaa Point, on which part of the Sheraton-Maui Hotel is built. Still another, also of picrite-basalt, is Puu Hele, on the alluvial fan of Pohakea Stream 1.5 miles northeast of Maalaea. Puu Hele has been almost wholly quarried away for cinder to use on roads. A hole now marks the position of the former 60-foot hill.

On the south bank of Olowalu Stream, half a mile from the coast, Puu Kilea is a small cinder cone from which erupted a flow of basanite that rests on alluvium of Olowalu Stream. This lava contains nepheline that is visible under the microscope, hence it is compositionally related to the posterosional lavas of Oahu and Kauai.

The picrite-basalts of the Lahaina Series contain no recognizable nepheline, but, like those of the Hana Series on Haleakala, they are strongly undersaturated in silica, and their normative (calculated theoretical) composition includes more than 10 percent nepheline.

Stream erosion of West Maui volcano has reached a late youthful to submature stage. Because of the thick armor of Honolua flows, the rainy northeastern slope has reached a less advanced stage of dissection than might otherwise be expected, and broad surfaces that have not been lowered much below the original surface lie between the deep canyons. In contrast, the drier southwestern slope has been much more deeply dissected, leaving sharp-crested ridges between the valleys.

Broad alluvial fans fringe the eastern and southwestern sides of the mountain. Those along the eastern side are at least partly due to

324

loss of water from the streams to the permeable lavas of Haleakala that have built the Maui Isthmus, but along the southwestern side they have built out the shoreline with debris transported and deposited by streams in greater volume than it is removed by waves and ocean currents in the quiet channel between Maui and Lanai. A well near the mouth of Waikapu Valley penetrated 240 feet of alluvial gravels, and at other places the alluvial fans on the isthmus may be more than 1,000 feet thick.

As elsewhere in the Hawaiian Islands, the heads of the canyons are choked with great deposits of colluvium and alluvium, commonly as deep as 200 feet, which are now being removed by the streams. The cause of this accumulation of material in the canyons was probably a reduction in gradient, and conse-

quently of transporting power of the streams, owing to a rise of regional base level (sea level). Much of the alluviation must have taken place during the Olowalu stand of the sea, about 250 feet above present sea level. In places, however, terraces appear to be graded to lower stands of the sea. Many streams are bordered by terraces, about 25 feet above the present stream bed, that were probably deposited during the Waimanalo (plus-25-foot) stand of the sea.

Several of the high stands of the sea are marked on West Maui also by fossiliferous beach deposits. The type locality of the Olowalu stand is a conglomerate of lava boulders and cobbles cemented by calcareous material at 240 feet altitude in a small gulch on the south side of a cinder cone 1.25 miles north of Olowalu. Other similar deposits have been

325

Figure 207. Iao Valley, West Maui. The sheer cliffs that nearly enclose Iao Valley approximately coincide with the former boundary of the caldera. Lavas which formerly filled the caldera have been partly gouged out by stream erosion. The rough terrain on the sides and floor of the valley is composed of moderately well consolidated colluvium and alluvium deposited at a former stage of higher base level and now being removed by renewed stream erosion.

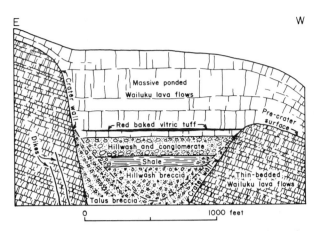

E                                     W

Massive ponded
Wailuku lava flows

Red baked vitric tuff

Pre-crater surface

Hillwash and conglomerate

Shale

Hillwash breccia

Thin-bedded Wailuku lava flows

Talus breccia

Crater wall

Dikes

0              1000 feet

Figure 208. Cross section of a pit crater in Kanaha Valley. The vertical scale is exaggerated 1.5 times as compared to the horizontal. (Modified after Stearns and Macdonald, 1942.)

found 250, 100, and 69 feet above sea level on the south and west sides of Puu Mahanalua Nui. No actual elevated coral reef, like that on Oahu, has been reported, however. Stripping of the soil below a level of 1,200 feet altitude, particularly on the southwest side of the mountain, and even more pronounced stripping below 560 feet, have been attributed by Stearns (1935) to stands of the sea respectively 1,200 and 560 feet above the present level.

Lithified calcareous sand dunes rest on the alluvial fans near the shore between Kahului and Waihee (see photograph, p. 218), and extend inland almost across the western edge of the isthmus. Near the north coast some of the dunes are as much as 200 feet high. A test boring 1.5 miles north of the mouth of the Wailuku River shows that the dunes extend below present sea level. They were formed by wind blowing sand inland from wide beaches exposed during a stand of the sea lower than the present sea level—probably the minus-40-foot stand. Less consolidated to totally unconsolidated dunes are of later date, and are still forming.

## HALEAKALA

The primitive shield of Haleakala volcano is composed of pahoehoe and aa flows of tholeiite, tholeiitic olivine basalt, and oceanite averaging about 15 feet in thickness, with which are associated very minor amounts of pyroclastic materials. The assemblage is known as the Honomanu Volcanic Series. Above sea level the shield has been almost wholly buried

by later lavas. It is now exposed only in sea cliffs along part of the north shore (figs. 203, 210), in the walls of Honomanu and Keanae valleys, along a short stretch of the north wall of Kipahulu Valley on the eastern slope, near the head of the deep Manawainui Canyon on the southeastern slope, and possibly in the lower part of the south wall of Haleakala Crater. In Manawainui and Kipahulu valleys several cinder and spatter cones and thin layers of vitric tuff are interbedded with the lavas, and a thin tuff bed is exposed in the northern sea cliff near Hanawi Gulch. A peculiar breccia, 100 feet thick, in the eastern wall of Nuaailua Valley near Keanae may be the filling of some sort of depression, perhaps a pit crater. The lower part appears to be an ordinary talus breccia, but the upper part is aa clinker containing many accretionary lava balls, probably formed by a lava flow pouring into the upper part of the depression.

The Honomanu Series is overlain by the Kula Volcanic Series, composed predominantly of hawaiite with lesser amounts of alkalic olivine basalt and ankaramite. In the hill just north of the observatory, at the western rim of Haleakala Crater, hawaiite lava contains phenocrysts of hornblende—a mineral which is rare in Hawaiian lavas. The Kula flows are well exposed in cross section in the walls of Haleakala Crater and in Keanae and Kaupo valleys, and they form the surface over most of the northwestern and southeastern segments of the mountain. Eruptions were more explosive than those of the earlier series, and many large cinder cones were formed. Beds of poorly consolidated vitric ash are common, especially near the cones, where they are as much as 30 feet thick. The flows are characteristically thicker than those of the Honomanu Series. Individual flows average 20 feet in thickness near the summit of the mountain and 50 feet near the coast. Aa, commonly with a tendency toward block lava, is predominant, but some pahoehoe is present near the vents. A few small domes are exposed in cross section in the walls of the crater. Near the summit of the mountain the Kula Series is at least 2,500 feet thick, but near the shore it is only 50 to 200 feet thick, at many places consisting only of a single flow.

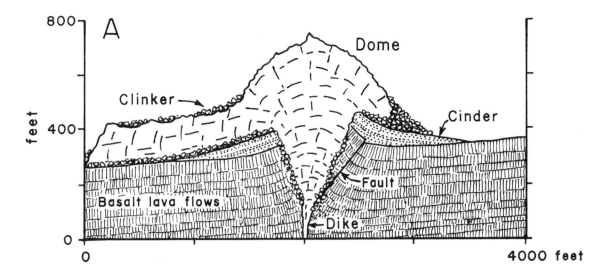

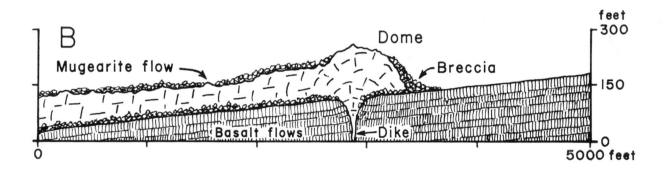

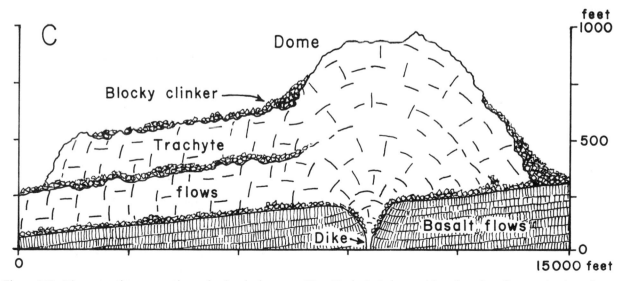

327

Figure 209. Diagrammatic cross sections of volcanic domes on West Maui. *A,* A dome with a short lava flow resting in and on a cinder cone built in the early stages of the same eruption. A dome of this sort is Puu Koai, in the west wall of Ukumehame Canyon near Olowalu. At the edge of the vent, basalt lava flows are tilted upward and outward, faulted, and brecciated. *B,* A small dome with a long lava flow and no cinder cone. An example is Puu Anu, on the ridge 3.5 miles southwest of Waikapu. *C,* A large dome with two thick stubby lava flows and little or no associated cinders, such as Puu Eke, 2 miles north of the summit of West Maui. (After Stearns and Macdonald, 1942.)

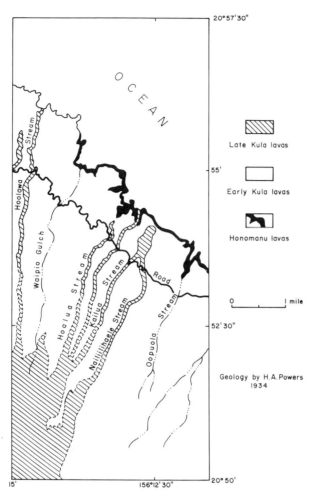

OCEAN

Late Kula lavas

Early Kula lavas

Honomanu lavas

0 ————— 1 mile

Geology by H. A. Powers
1934

Figure 210. Map of an area on the northwestern slope of Haleakala (East Maui) volcano, east of Haiku, showing late Kula lava flows filling valleys cut into earlier Kula lavas. (After Stearns and Macdonald, 1942.)

Occasional lenticular beds of hillwash debris, and stream-laid gravel occupying shallow gulches, are found between lavas of the Kula Series, indicating that at least locally some moderately long intervals occurred between eruptions.

In the sea cliff 500 feet east of Hanawi Gulch, lava of the Kula Series fills a gulch cut into the Honomanu lavas, but in general there is little evidence of erosion at the contact. At some localities the Kula lavas are separated from the underlying Honomanu lavas by a layer of red ashy soil up to about 6 inches thick, but at many places there is no obvious break between the two series. Commonly they pass into each other through a transition zone 50 to 200 feet thick in which the rocks are

328

indeterminate in appearance, both in outcrop and under the microscope, and alkalic olivine basalts are interbedded with tholeiitic basalts.

In the lower part of the south wall of Haleakala Crater, exposed in windows through much later volcanics that mantle most of the lower part of the cliff, is a series of thin pahoehoe flows of basalt that dip southward about 8°. These beds are truncated by an erosional surface that slopes southward 25° to 35°. The erosional surface is covered with a deposit up to 2 feet thick of talus or hillwash breccia composed of blocks of pahoehoe similar to the underlying flows, mixed with red ashy soil, and the upper few inches of the underlying rock is slightly weathered. The breccia is overlain by aa flows, typical of the Kula series, dipping southward at about 35°. This structure has been interpreted (Stearns and Macdonald, 1942, p. 72) as an eroded tangential fault scarp buried by Kula lavas, and the thin pahoehoe flows cut by the scarp were thought probably to belong to the Honomanu Series. Later chemical studies show, however, that the basalt is markedly alkalic, and it now appears probable that the lavas are a part of the Kula Series.

The Kula eruptions took place from three well-defined rift zones (figs. 211–213). Most prominent are those extending southwestward and east-northeastward from the summit, forming a nearly straight line across the mountain. The third rift zone, extending north-northwestward from the summit, is much less prominent, but is clearly marked by the row of cinder cones that extends almost to the coast. Many large cones of the Kula Series lie along the upper part of the southwest rift zone and present an impressive display as seen from the air. Other cones lower on the rift may be Kula cones mantled with later cinder and ash. Still other, buried cones are exposed in cross section in the walls of the crater. Most of the east rift zone has been buried by later volcanics, but the pronounced east-northeasterly bulge of the mountain demonstrates that the rift zone was active in Kula time. Many dikes parallel to the rift zones are exposed in the walls of the crater and of the big valleys. It is presumed that the same rift zones were the sources of the Honomanu lavas, though these are everywhere buried and invisible.

Toward the end of the Kula period, the intervals between eruptions increased, and local erosional unconformities, soil beds, partly weathered beds of ash, and stream-laid conglomerates are commonly found between the flows. Some of the canyons cut in Kula lavas and filled by later Kula flows are several hundred feet deep and must indicate time intervals of hundreds, if not thousands, of years. Figure 210 shows an area adjacent to the northwest rift zone in which late Kula lavas have filled gulches eroded into earlier Kula lavas.

In addition to the numerous dikes, a few larger intrusive masses of Kula age protrude from beneath the general cover of later ash in the west wall of Haleakala Crater. One of them, below Kilohana Peak, is 0.75 mile long and 0.25 mile across. All are fine to moderately fine grained; no coarse gabbros have been found. Some of the very fine grained, dense intrusive rock was used by the ancient Hawaiians for the manufacture of adzes.

Probably the most intriguing feature of East Maui, both geologically and scenically, is Haleakala Crater (fig. 211). Seven miles long, 2 miles wide, and half a mile deep, it is commonly said to be the largest extinct volcanic crater in the world. The statement is completely inaccurate. Haleakala is far smaller than many volcanic craters (calderas); there is an excellent chance

Figure 211. The summit of Haleakala, Maui. The cinder cones in the lower part of the photograph lie along the southwest rift zone of Haleakala, which continues on across the crater in the background. Koolau Gap is on the upper left and Kaupo Gap on the upper right. Haleakala Crater is the result of erosion by the streams that cut the Koolau and Kaupo valleys.

Figure 212. A view of Haleakala Crater looking southwestward along the rift zone shows the continuity of the rift inside and outside the crater. Kaupo Gap is on the lower left and Koolau Gap on the far right.

that it is not extinct, but only dormant; and strictly speaking it is not of volcanic origin, beyond the fact that it exists in a volcanic mountain. There is, however, no disputing the spectacular beauty of its scenery, and the story of its origin is more interesting than that of the formation of most volcanic craters and calderas.

At the end of the accumulation of the Kula Series, volcanic activity became very infrequent, if it did not cease altogether, and stream erosion began to make real progress in wearing away the mountain. All the way around, radial streams started to cut valleys, but erosion was most rapid on the rainy northern and eastern sides. Some streams cut faster than others, capturing neighboring drainage and acquiring greater supplies of ground water, and eventually these streams carved master valleys. Several were particularly large: Keanae Valley on the north, Waihoi and Kipahulu valleys on the east, and Kaupo Valley on the southeast. Keanae and Kaupo valleys had the shortest and steep-

est courses, and in time cut headward into the very heart of the mountain (fig. 214). There the rate of erosion was probably accelerated by a greater abundance of easily removed pyroclastic material, and possibly by lavas that had been softened and made more easily erodable by gas alteration (like the rocks of the West Maui, East Molokai, and Koolau calderas). The heads of the two valleys expanded and merged, forming a single huge depression that extended entirely across the mountaintop, divided only by a relatively low and narrow ridge between the two drainage basins. Kipahulu Valley likewise extended itself headward, nearly merging with the head of Kaupo Valley. Finally, after this period of erosion, which was long enough for streams to cut canyons several thousand feet deep all the way into the core of the mountain, volcanic activity resumed. The principal rift zones reopened right across the depression at the top of the mountain (fig. 212) and on down the flanks, and the great erosional valleys were flooded with lava flows and filled with cinder cones. Thus Haleakala Crater, with its magnificent display of volcanic features, is primarily erosional in origin (Stearns, 1942*a*). The Koolau and Kaupo gaps, leading out of the crater, are the upper parts of great stream-cut canyons, now partly refilled with later lava flows (fig. 215).

The erosion that produced Haleakala Crater also destroyed the former top of the mountain. By the end of the Kula eruptions, Haleakala volcano rose to a summit possibly more than 3,000 feet higher than the present crater rim—a summit undoubtedly studded with large cinder cones and resembling in profile that of Mauna Kea on Hawaii.

The lava flows and associated cinder cones and ash deposits erupted after the long period of erosion are named the Hana Volcanic Series. The rock types are the same as those in the Kula Series, but alkalic olivine basalts and basaltic hawaiites are predominant over the more siliceous types. Many of the flows contain phenocrysts of plagioclase feldspar, and in some, the plagioclase grains reach as much as an inch in length and may be nearly as clear as window glass. Large black phenocrysts of augite are abundant in ankaramites of the Hana Series, and also in similar rocks of the Kula Series. One of the outstanding characteristics of the Hana lavas is a marked deficiency in silica. In theoretical (normative) mineral compositions calculated from chemical analyses, a deficiency in silica results in the presence of nepheline, but no nepheline can be detected in the actual rocks. The Hana lavas resemble the posterosional, nepheline-bearing lavas of Oahu and Kauai in containing little silica, although the deficiency is less extreme in the Hana lavas.

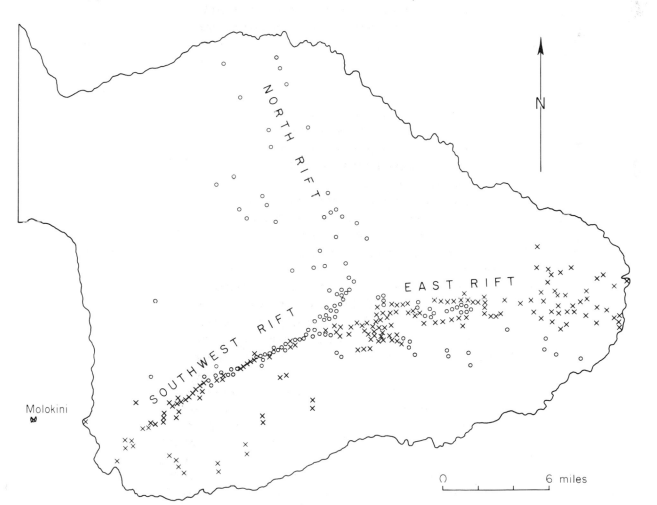

Figure 213. Map of Haleakala volcano, showing vents of the Kula *(circles)* and Hana *(crosses)* Volcanic Series. Molokini Islet is a tuff cone on the southwest rift zone of Haleakala. (After Stearns and Macdonald, 1942.)

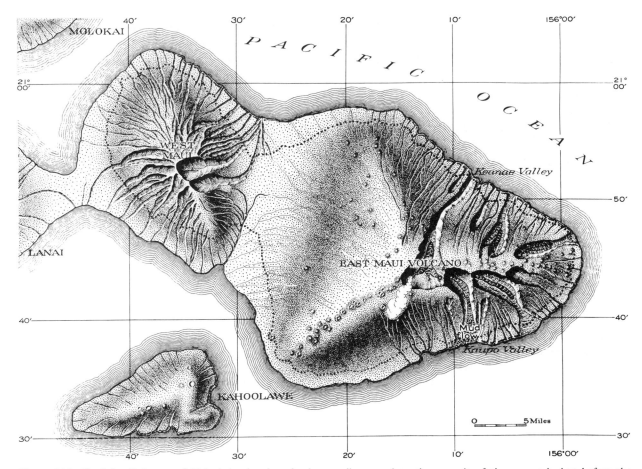

Figure 214. Shaded relief map of Haleakala showing the large valleys cut into the summit of the mountain just before the amphitheaters at their heads started to merge, and before they were partly filled with Hana lavas and cinder cones. The shoreline is that of one of the low sea levels of early Pleistocence time. The present outline of the island is shown by the dotted line. (From Stearns and Macdonald, 1942.)

The distribution of the Hana Series is controlled by two factors: the location of the vents, and the nature of the preceding topography. The north rift zone of the Kula Series did not reopen; consequently, Hana lavas are absent on the entire northwestern sector of the mountain. Hana eruptions on the southwest and east rift zones poured lava flows over those sectors, but the flows followed the valleys, and some higher parts of the earlier erosional topography remained unburied. Across the top of the mountain the rift zones lay entirely within the erosional depression, the high walls of which shielded the northern and southern slopes of the mountain from lava flows, except for those that poured down the Keanae and Kaupo valleys. The floor of the depression itself was deeply buried by both lava and cinder. The thickness of Hana volcanics in the summit depression and adjacent valley heads is

unknown, but it is probably in the vicinity of 3,000 feet. On the eastern flank of the mountain, lava flows from the rift zone poured into Kipahulu and Waihoi valleys, but left uncovered the high triangular sectors between them, and also the one between Kipahulu and Kaupo valleys. Farther north, lava poured into smaller amphitheater-headed valleys upslope from Hana, and, along the present Kuhiwa Stream south of Nahiku, they almost completely buried another large amphitheater.

As in late Kula time, at any one place Hana eruptions often were separated by intervals long enough to allow weathering and erosion of the lava surface before it was buried by the next flow (fig. 216). Where they have been studied in detail, as around the mouth of Keanae Valley and in the Nahiku district 3 miles to the east, the Hana rocks consist of a series of flows filling shallow valleys eroded

332

into the preceding flows (fig. 217). At Keanae the lavas rest on a thick accumulation of alluvium (fig. 218), deposited in the Keanae canyon as a result of relative rise of base level after the canyon had been cut to a sea level lower than the present one.

A much longer interruption in flooding by lava flows is indicated in Kipahulu Valley. There, after the valley was filled with lava to a depth of more than 1,000 feet, stream erosion was able to cut a new valley half a mile wide and probably more than 1,500 feet deep along the northern boundary of the earlier valley-filling lavas. Still later, Hana lavas again flooded the valley and partly refilled the canyon (fig. 219). Near the head of the valley the late lava fill is only about half the depth of the canyon, leaving the surface of the earlier Hana lavas standing as a broad terrace 750 feet above it; but near the coast the later valley was entirely refilled, and new lavas spread southward as a broad fan across the lower end of the terrace of older lavas.

In general the Hana lavas were erupted in a more fluid condition than were the Kula lavas, and the flows tend to be thinner and less massive. Some of the lavas were so fluid that, in pouring down the steep, stream-cut gulches, the central part of the flow drained away and left only a thin plaster of lava adhering to the valley walls. These plasters are commonly only a few feet thick, and sometimes only a few inches, but they extend unbroken along the valley walls sometimes for several miles, and mark a former lava level as much as 50 feet above the remaining portion of the flow.

The cones built by Hana eruptions range from small spatter cones and ramparts a few feet high to big cinder cones more than a mile across at the base and 600 feet high. Some of the latter contain well-formed spherical to fusiform bombs. Molokini Islet, 3 miles offshore, is a cone of palagonite tuff formed by eruption in the ocean on the southwest rift zone. Deposits of vitric ash and pumice as much as 20 feet thick are spread to leeward of

Figure 215. Kaupo Gap, Haleakala volcano, East Maui. Lava flows of the Hana Volcanic Series poured seaward through an erosional valley that formed during a long period of volcanic inactivity. Mudflows containing huge jumbled blocks are associated with these lava flows. Manawainui Valley is on the right.

Figure 216. Keanae peninsula on East Maui. Pauwalu Point, in the foreground, is composed of early lavas of the Hana Volcanic Series and shows the effects of prolonged wave erosion. Note the sea cave and sea stacks. Keanae Point, in the center of the photograph, is formed of late Hana lavas and shows only slight effect of waves. The cliffed coastline in the distance is formed on older lavas of the Kula Volcanic Series with still older lavas of the Honomanu Volcanic Series at the base.

some of the cones along the southwest rift zone and cover a belt up to 2 miles wide along much of the northern edge of the east rift zone. Great banks of ash lie against the lower walls of Haleakala Crater, and in places layers

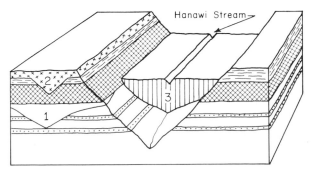

Figure 217. Diagram showing a succession of valleys filled by lava flows at Nahiku, on the northern slope of Haleakala. Valley *1* was cut and filled by a lava flow, and then buried by two more lava flows before valley *2* was cut and subsequently filled. Finally the present Hanawi Gulch was cut, and later partly refilled by the last lava flow in the area. Hanawi Stream is now excavating a valley in this last lava flow. (Modified after Stearns and Macdonald, 1942.)

of ash are interstratified with talus. Near Manaokeakua, a cinder cone built by an eruption of ankaramite in the central part of the crater, well-formed crystals of augite half an inch long are scattered through the ash. Similar augite crystals can be picked up from the ash cover near the road just west of the rim of the crater.

Phreatic explosions from a series of five craters on the southwest rift between 7,500 and 9,250 feet altitude threw out showers of blocks, some of which are several feet across. The deposits are as much as 10 feet deep close to the rims of the craters, but are restricted to a radius of about 200 feet. These rather mild explosions must have occurred when rising magma encountered ground water confined between dikes of the rift zone.

Fossiliferous marine conglomerate, deposited during a higher stand of the sea, has been found 40 feet above sea level in a drill hole on the isthmus 3 miles south of Kahului, and also

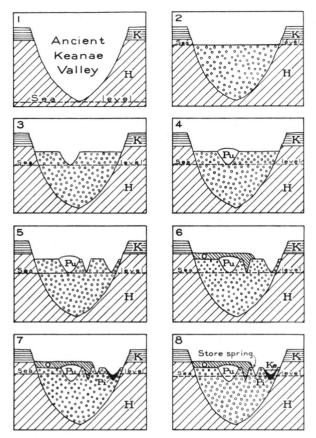

Figure 218. Stages in the development of the lower Keanae Valley, on the northern flank of Haleakala. The original valley (diagram *1*) was partly filled with alluvium as a result of rise of base level (*2*), and then the alluvial floor was trenched by the stream as base level fell again (*3*). The small valley thus formed was filled by the Pauwalu lava flow (*Pu* in *4*), two more small valleys were cut (*5*), one of these valleys and the Pauwalu lava flow were buried by the Ohia lava flow (*O* in *6*), the second valley was partly plastered by the Piinaau lava flow (*Pi* in *7*) and another small valley was cut at the edge of the Ohia lava, and finally the Keanae lava flow (*Ke* in *8*) poured down the gulch on top of the Piinaau lava and spread out at the shore to form the present Keanae Peninsula. (After Stearns and Macdonald, 1942.)

up to about 50 feet above sea level in a gulch 4.5 miles southeast of Kihei. Calcareous sand dunes lie along the north shore between Kahului and Paia, and along the southwest shore

between Kihei and Makena (figs. 201, 202). Some of the consolidated dunes along the northern edge of the isthmus show traces of a wave-cut nip 25 feet above sea level, showing that the dunes and the Kula lavas on which they rest are older than the Waimanalo stand of the sea. If, as appears probable, the dunes were formed during the minus-40-foot stand of the sea, the Kula lavas are older than that stand also.

An interesting cap of limestone a few inches to 2 feet thick, rests on Ka Lae o ka Ilio, at the mouth of Kaupo Valley, 73 feet above sea level. It contains shells of the marine gastropod *Littorina*. However, it is not regarded as evidence of a stand of the sea at that level. H. T. Stearns found living snails climbing the sea cliff, and he concluded that the limestone is forming today, incorporating the shells, as a result of the evaporation of ocean spray (Stearns and Macdonald, 1942, p. 109). This illustrates the care which must be used in interpreting apparent marine deposits above present sea level.

A large alluvial fan in the mouth of Keanae Valley probably was graded to the plus-100-foot stand of the sea. A similar fan in the mouth of Kaupo Valley has been largely buried by lava flows. Deposits of old, partly consolidated alluvium are found also near the mouths of Waihoi and Kipahulu valleys, and in patches resting on Kula lavas west of the mouth of Kaupo Valley. Streams descending the western slope of Haleakala are depositing alluvial fans where they issue onto the gentler slopes of the isthmus, and are burying older alluvium in various stages of consolidation.

A large mass of mudflow breccia lies in the mouth of Kaupo Valley. The deposit is more than 300 feet thick, with its base below sea

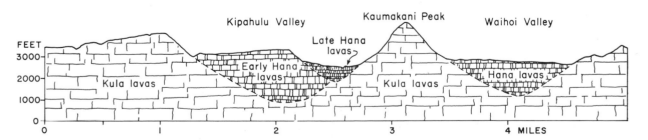

Figure 219. Geologic section across upper Kipahulu and Waihoi valleys on the southeastern slope of Haleakala.

335

level. It is partly buried by Hana lavas and is cut by small stream valleys which also extend below sea level. The valleys have been filled with stream gravels which appear to have been graded to the plus-100-foot stand of the sea, and thus the Kaupo mudflow is probably older than that stand. It probably formed at a time when sea level was about 300 feet lower than now, or during the interval when the sea was rising from that low level. The mudflow deposit consists of angular blocks of lava of types common in the Kula Series, some of them several feet across, in an earthy matrix. Occasional blocks of vitric tuff are found, and one mass of tuff is 50 feet across. The deposit is totally unsorted and unstratified. The material is identical with that found in talus fans along the walls of Kaupo Valley and Haleakala Crater. There appears to be no question that the mudflow moved down Kaupo Valley from a source or sources high on the mountain, but its underlying cause is unknown. Perhaps taluses in the upper part of Kaupo Valley were set in motion by exceptionally heavy rains, strong earthquakes, or both.

# Kahoolawe

Kahoolawe is a single shield volcano with a caldera, about 3 miles in diameter, at its present eastern end (figs. 201, 221). During later stages the caldera was completely buried beneath a cap of later volcanics. The shield was built by eruptions along a prominent rift zone that extends west-southwestward from the summit, and two less prominent rift zones with eastward and northward trends. The west and east rift zones are essentially a single continuous belt of fractures. The east rift has little present topographic expression, but is marked by a swarm of 40 dikes in the sea cliff. The west rift is marked by a broad ridge from which the lava flows slope both northward and southward. The former vent structures have been mostly eradicated by erosion, but two cinder cone remnants, Puu Moiwi and Puu Kamama, lie on its crest (fig. 220), and Lua Kealialalo is the remains of a collapse crater at the summit of a broad parasitic lava shield. The north rift zone is indicated only by a north-

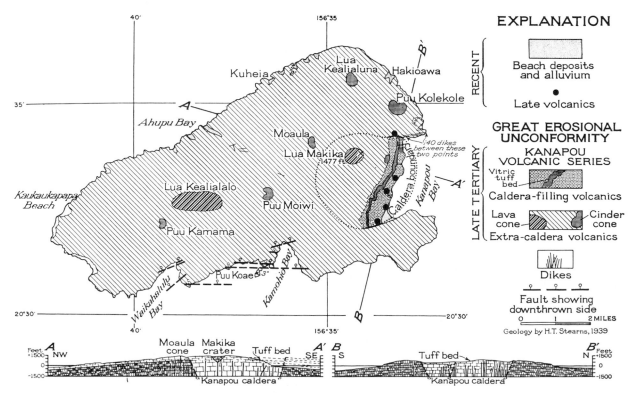

Figure 220. Geologic map and cross sections of the island of Kahoolawe. (From Stearns, 1940c.)

Figure 221. The southern side of the island of Kahoolawe, a dome-shaped shield volcano that reaches an altitude of 1,477 feet. Note the many thin lava flows that built the shield. The island is uninhabited, waterless, and windswept. The fact that the reddish soil, visible as a mantle on the upper slopes, does not extend appreciably below the 800-foot contour is believed to indicate that the sea was once that high and washed away the loose soil. In the distance the domical form of the West Maui shield is clearly visible.

ward bulge of the island and a single cinder-cone remnant, Lua Kealialuna. Moaula and Puu Kolekole also are cinder-cone remnants. Lua Makika is the eroded crater of a lava shield that covers an area of about 10 square miles, and constitutes the highest part of the island.

In the southern sea cliff are several faults on which the downdropped side is to the north. Most of these faults trend east-northeastward, toward the caldera, and suggest the presence at one time of a graben that trended west-southwestward from the summit of the shield.

Wave erosion has cut cliffs 800 feet high on the eastern end of the island, exposing the structure of the shield, the east rift zone, and a cross section of the eastern edge of the caldera (fig. 221). Kanapou Bay is nearly coincident with the eastern edge of the caldera, and the erosion that formed the bay was undoubtedly guided by structures associated with the caldera.

The precaldera lavas have been positively identified only in the eastern sea cliff. They are thin pahoehoe and aa flows of tholeiite and

olivine tholeiite. In contrast, the caldera-filling flows exposed in the head of Kanapou Bay are massive, thick, and nearly horizontal, with frequent pronounced columnar jointing. They have not been studied in detail petrographically, but the lower lavas appear to be tholeiitic, whereas the upper ones resemble in composition the postcaldera lavas on the outer slopes of the volcano. The latter range in composition from alkalic olivine basalt to hawaiite. Beds of yellow palagonite tuff lie between the lavas in the upper part of the caldera fill.

Streams have cut only shallow gulches into the carapace of the island, but weathering of the rocks to a depth of 30 to 50 feet suggests that the volcano is quite old. In the absence of more accurate dating, its age is estimated to be early Pleistocene—perhaps about 1.5 million years.

After a long period of quiet, volcanism returned in a series of five small eruptions along the sea cliff in Kanapou Bay. These formed masses of spatter and cinder, accom-

*337*

panied by very little flowing lava, which mantle the sea cliff and in places overlie talus. Associated dikes cut the talus. One of the dikes is remarkable in that it partly expanded to form pumice. Two of the eruptions were at the edges of the former caldera, and the magma probably was guided to the surface by the caldera-boundary faults. The rocks appear to be alkalic olivine basalts. These eruptions must have been very recent, because the deposits have not been eroded. Kahoolawe lies almost directly on the prolongation of the southwest rift zone of Haleakala volcano, and the tiny eruptions in Kanapou Bay may actually be related to the Hana Volcanic Series of Haleakala.

Among the most striking geologic features of Kahoolawe are the effects of wind erosion. In earlier times, the top of the island was covered with grass and bushes and scattered small trees growing on a thick layer of soil, although the lower slopes probably had been swept largely bare of soil by wave erosion during high stands of the sea. The introduction of sheep in the mid-19th century resulted in destruction of the plant cover and exposed the soil to wind action. At present the top of the island is largely bare. Even before the island was used as an artillery and bombing target, great plumes of red dust often could be seen blowing off the top of the island far to sea. Soil is also being removed by rain wash during occasional heavy rains. (Although the average annual rainfall on the summit of Kahoolawe is only about 25 inches, individual storms sometimes produce as much as 4 inches of rain.) Over much of the top of the island the upper soil zone has now been completely removed, exposing the hard surface of the B horizon (see chap. 8). The stripping has been as deep as 8 feet, and averages about 5. In places, small flatiron-shaped residuals stand 5 to 8 feet above the general level of the stripped surface, and parallel grooves 6 inches deep have been cut into the exposed surface by wind erosion.

338 Despite the low rainfall, Kahoolawe once supported a population of 30 to 80 persons, and except during extreme droughts potable brackish water was obtained from shallow wells in some of the gulches. Water suitable for animals probably existed until about 1900, but no potable water is known to exist today. The decrease in fresh water is probably due to the introduction of kiawe (algaroba) trees (Stearns, 1940, p. 131), which intercept and transpire back into the atmosphere much of the fresh rain water that formerly filtered down to the water table. However, the island is not unusable. Previous to the second world war, when Kahoolawe was taken over by the Navy, the Haleakala Ranch Company had made considerable progress in restoring it to a useful condition.

# Lanai

The island of Lanai is a shield volcano built by eruptions at the summit and along three rift zones (figs. 222, 224). The principal rift zone trends northwestward as a broad ridge, and is responsible for the conspicuous elongation of the island in that direction. A less conspicuous bulge on the southern side of the island is the result of building on the southwest rift zone. The summit of the shield collapsed to form a caldera (fig. 223) from which a shallow graben, bordered by en echelon step faults, extends south-southeastward toward Manele Bay. Numerous dikes exposed in the sea cliff indicate that this Manele graben lies along another rift zone.

The caldera was largely, but not completely, filled by lava flows, and the present Palawai Basin is a remnant of the caldera. The basin is roughly circular, with a diameter of about 2.5 miles. Just to the west of it, Miki Basin, with an average diameter of about a mile, is a nearly filled pit crater. The top of the ridge between them is 144 feet above the floor of Palawai Basin, but only 39 feet above the floor of Miki Basin. On the south, the floor of Palawai Basin merges with that of the Manele graben, through which the last lava flows in the caldera overflowed onto the outer slope of the volcano. The highest wall of Miki Basin rises about 75 feet. On the northeast side, Palawai Basin is bordered by a steep slope about 500 feet high, beyond which lies a nearly level bench about a mile wide. The thick, massive character of the

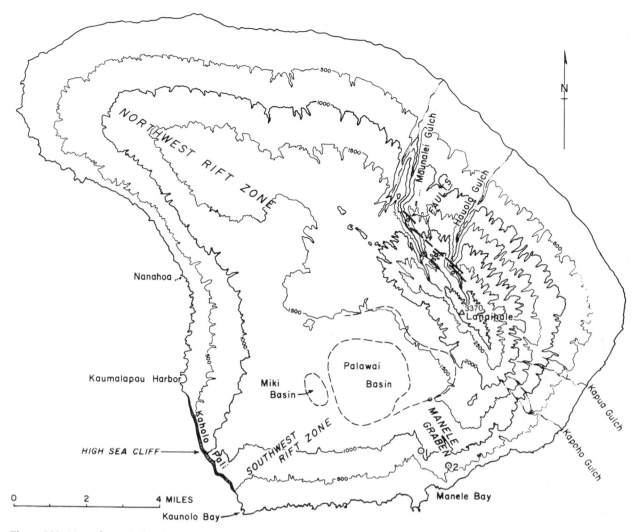

Figure 222. Map of Lanai, showing the bend in the courses of Maunalei and Hauola streams, the Palawai and Miki basins, the high sea cliff of the Kaholo Pali, and the location of fossiliferous deposits indicating former higher sea levels. *1,* Type locality of the Mahana (1,200-foot) stand of the sea; *2,* type locality of the Manele (560-foot) stand.

lava flows in this bench indicates that it is part of the floor of the filled caldera, within which Palawai Basin is an area of still later collapse. Beyond the bench another steep slope rises to Lanaihale (3,370 feet altitude), the highest point on the island. The steep slopes bordering the Palawai and Miki basins are fault scarps, much eroded, and partly buried by taluses.

So far as is known, all of the lavas of Lanai are tholeiitic basalts. They range from olivine-free tholeiites through olivine tholeiites to very olivine-rich oceanites. Some contain phenocrysts of labradorite feldspar, which in a few places are more than an inch long. Some of the feldspar is clear pale yellow, and a little of it has been cut for use as semiprecious gem stones. The lava flows range from about 1 to 100 feet in thickness, but average about 20 feet and seldom exceed 50, except where they have been ponded in depressions. Pahoehoe predominates near the vents, but aa becomes abundant on the lower slopes. At the base of the fault scarps bordering the Manele graben the lava flows have buried ancient talus that accumulated along the escarpments. In the sea cliff on the west coast, 0.65 mile south of Kaumalapau Landing, several inches to 2 feet of red colluvium lies between the lava flows, and other thin layers of colluvium are interbedded with lavas in the walls of Kapoho Gulch, but in general there is very little evidence of erosion or weathering between successive lava flows. A few thin beds of vitric ash, largely altered to palagonite, are intercalated with the lavas.

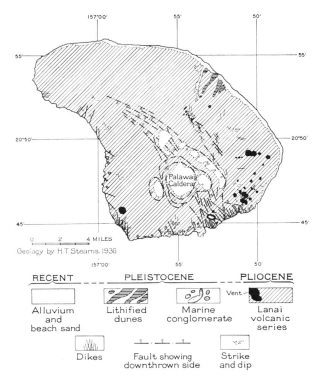

Figure 223. Geologic map of the island of Lanai. (From Stearns, 1940c.)

RECENT --- PLEISTOCENE --- PLIOCENE

Alluvium and beach sand

Lithified dunes

Marine conglomerate

Vent — Lanai volcanic series

Dikes

Fault showing downthrown side

Strike and dip

The subsidiary cones built along the rift zones are either lava cones or spatter cones containing a small proportion of cinder. The cross section of a small shield volcano is visible in the sea cliff 6,000 feet northwest of Kaunolu Bay, where a thin layer of vitric tuff is overlain by a series of thin pahoehoe flows. Another cross section of a similar small shield is exposed at Manele, where layers of pahoehoe 2 to 5 feet thick are associated with spatter. Such very thin flow units generally are found on the slopes of minor cones very close to the vents. The cross sections of several pit craters, filled by talus breccias and later lava flows, have been identified. One is seen at the sharp bend of Maunalei Gulch, about 1,750 feet above sea level (Stearns, 1940c, p. 30). The largest, with a maximum diameter of about 4,000 feet, is Kaluakapo, at the coast half a mile northeast of Manele Bay.

Numerous dikes are exposed in the sea cliffs and canyon walls, the great majority of them parallel to one or another of the rift zones. A

340

Figure 224. In the distance, beyond the deeply dissected shield of West Maui in the foreground, is the island of Lanai, an excellent example of a Hawaiian shield volcano. Caps of cumulus clouds such as that over Lanai, and over West Maui in figure 221, commonly form over the Hawaiian mountains as warm, water-saturated air rises into cooler regions where condensation of the water forms clouds.

total of 124 dikes has been counted by Stearns (1940c, p. 34) in a traverse of 1.5 miles across the southwest rift zone, which is well exposed in the Kaholo sea cliff. Most of the dikes are nearly vertical, and average about a foot in thickness. Much cracking and minor faulting is associated with the dikes in the southwest rift zone, and in places the lava beds are so jumbled by faulting that it is difficult to detect the original layering. A total of 111 dikes and 32 faults is reported in the sea cliff that cuts across the southeast rift zone. One of these dikes is remarkable in that its central part is aa clinker (fig. 225). Dikes exposed in Maunalei Gulch and its tributaries strike northwestward, and are part of the northwest rift zone. This rift zone appears to have shifted with the passage of time, so that the earlier portion of the zone, exposed in Maunalei Gulch, lies farther northeast than the later portion which is associated with the collapse of the caldera.

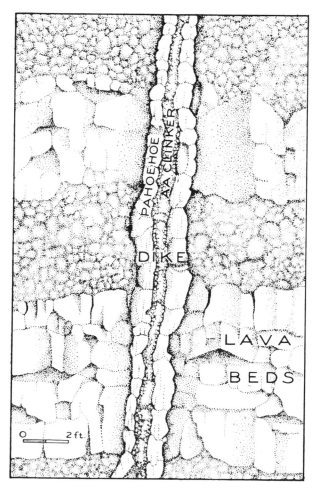

Figure 225. Dike with a center of aa clinker. (From Stearns, 1940c.)

Grabens extend from the caldera southeastward toward Manele Bay and northwestward along the crest of the broad arch of lava flows along the northwest rift zone (fig. 226). Faults are exposed also in the upper parts of Maunalei and Hauola gulches northeast of the summit of the mountain. Displacement on these faults has allowed the central part of the volcano to sink in relation to the outer slopes. Southwest-facing fault scarps probably once existed along these faults, deflecting the streams from their direct radial courses down the northeastern flank of the shield, and thus accounting for the abrupt, nearly right-angle bend in the courses of Maunalei and Hauola gulches (fig. 222).

Three of the localities in the Hawaiian Islands at which evidence of former high stands of the sea were first recognized are on Lanai (Stearns, 1938). Fossiliferous marine conglomerate and remnants of a wave-cut terrace occur 560 feet above sea level in Kaluakapo Crater, just north of Manele Bay (the Manele stand of the sea), and on the south rim of the crater, fossiliferous limestone lies at 625 feet altitude. On the side of a small gulch 0.3 mile northwest of the western tip of Kaluakapo Crater, Stearns found material which he interpreted as fossiliferous marine sediment 1,070 feet above sea level, and stripping of the soil above that level suggested to him that the island had been submerged to a depth of 1,200 feet (the Mahana stand) (fig. 222).

Lanai lies in the rain shadow of West Maui and East Molokai, and consequently the island is very dry. The average rainfall at the summit is only about 38 inches a year. Stream runoff is small and intermittent. As a result, the volcano is little dissected by stream erosion, and what little canyon cutting has occurred is largely confined to the northeastern slope, directly below the zone of maximum rainfall.

The northeastern side of Lanai is also sheltered from wave erosion and is fringed by broad expanses of alluvium and beaches, with no appreciable sea cliff. On the other hand, the southwest side of the island is exposed to the full fury of the waves of southwesterly storms, which have eaten deep into it. Thus we have the apparent anomaly of a low coastline on the windward side of the island and high sea cliffs on the leeward side. This leeward cliff, extend-

Figure 226. A view northeastward over the uplands of Lanai. The line of cliffs in the background is the eroded fault scarp that borders the caldera and extends northwestward along the graben of the northwest rift zone. The nearly level, cultivated surface in the foreground, notched by youthful stream valleys, is on lavas that accumulated in the graben.

ing for 7 miles north of Kaumalapau, reaches a height of 350 feet. Even higher is the spectacular Kaholo Pali (fig. 127), 1,000 feet high at one point 2.2 miles south of Kaumalapau. There is no evidence that these cliffs were formed in any other way than by wave erosion. Present-day wave action is very strong along this coast, and stack rocks are numerous. The Needles, at Nanahoa, 2.5 miles north of Kaumalapau, are among the best examples of stack rocks in Hawaii (fig. 130).

Wind also is a very active erosional agent on Lanai. Along much of the southeastern shore, wind has blown sand inland from the beaches to form a dune ridge 10 to 20 feet high along the inner edge of the coastal flat. These dunes range from slightly consolidated to completely unconsolidated, and they are still being formed. Consolidated calcareous dunes are found on the northeastern slope, extending as far as 2 miles inland and reaching an altitude of 950 feet on the ridge east of Maunalei Gulch. As on Maui, these extensive lithified calcareous dunes are believed to date from a period when sea level was lower than now.

Still another type of dune has resulted from wind erosion and deposition of soil from the weathered basaltic rocks. These dunes are yellowish to reddish brown, a few feet high, and essentially unconsolidated. They can be seen near the head of Kapua Gulch, along the northeastern slope, and on the north end of the island. Most of the gulches on the northeast slope are bordered by conspicuous terraces, as much as 60 feet high, that have resulted from rapid stripping away of the soil by a combination of sheet wash and wind erosion.

The effects of wind action are particularly conspicuous on the north end of the island. In some areas it has swept bare the bedrock surface, cutting long shallow grooves parallel to the predominant northeasterly wind direction, and in places leaving residual pinnacles of weathered rock 10 feet, rarely even 30 feet, high. These residuals are clear testimony that an equal depth of soil has been removed from around them. Elsewhere dunes and sheet deposits of loess several feet thick have accumulated. Whether or not the denudation of the surface and consequent rapid erosion of the soil can be attributed to over-grazing, it is obvious that vast amounts of our mineral wealth, in the form of soil, are rapidly being lost into the sea.

342

# Molokai

The island of Molokai is another volcanic doublet, built primarily by two big shield volcanoes (figs. 227, 228, 229). The older of the two is West Molokai: the lavas of East Molokai overlap its eastern flank, forming the Molokai isthmus. Along Waiahewahewa Gulch, at the west edge of the isthmus, 3 feet of lateritic soil and 6 feet of weathered basalt separate the lavas of the two volcanoes. Extinction of the East Molokai volcano was followed by a long period of volcanic quiet and of erosion before new volcanic eruptions took place. These posterosional eruptions built the small tuff cone of Mokuhooniki, in the ocean just east of Molokai, and the large peninsula on the north shore properly called Makanalua, but familiarly known as Kalaupapa.

## WEST MOLOKAI

The western end of Molokai consists of a very flat shield volcano. In profile, it resembles the great volcano that forms the central-southern part of the island of Hawaii, and like it the mountain is called Mauna Loa. Little is known in detail of the composition of its rocks, but it appears to consist largely of tholeiitic basalts, with only a very thin and discontinuous capping of alkalic olivine basalt and hawaiite. A few thin layers of vitric tuff are interbedded with the lavas, and a bed of explosive breccia containing blocks up to 2 feet across is exposed 0.5 mile south of the summit. Near the new harbor at Hale o Lono (fig. 146), and elsewhere on the south slope, thick platy lava flows of basaltic hawaiite rest on 6 inches to 4 feet of red ashy soil.

The shield was built by eruptions along a principal rift zone that trends east-northeast-ward across its summit, and along a less important rift zone trending northwestward (fig. 230). The principal rift zone is marked only by the broad structural arch built by eruptions along it, and by a swarm of dikes trending parallel to it that have been exposed by erosion in the head of Waiahewahewa Gulch just east of the summit. The northwest rift zone is marked by 16 cinder and spatter cones and by dikes exposed in the sea cliff at the north shore. A bed of dense, fine-grained basalt in Kaeo cone, 2 miles east-southeast of Ilio Point, was quarried by ancient Hawaiians for dense stone used for adzes.

Beneath the sea, West Molokai is prolonged 40 miles west-southwestward by Penguin Bank, the broad, nearly flat surface of which lies 180 feet below sea level (fig. 230). Penguin Bank almost certainly was truncated by wave erosion at a lower stand of the sea, and it may be capped with coral reef formed during the same period. It may be a long extension of the West Molokai shield, but more probably it is an older, independent shield built on the same

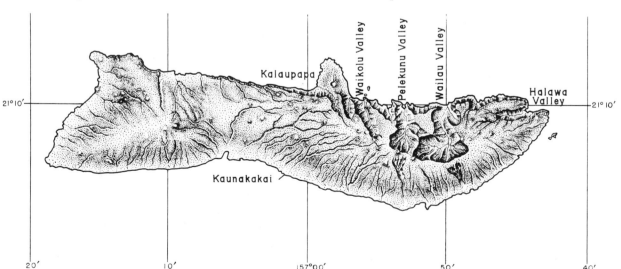

Figure 227. Shaded relief map of the island of Molokai. (From Stearns and Macdonald, 1947.)

343

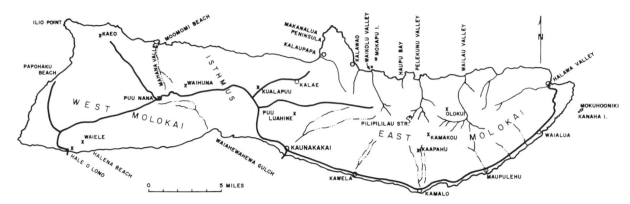

Figure 228. Map of Molokai, showing the location of places mentioned in the text.

trend as the main rift zone of West Molokai. The amount of wave erosion indicated by the low sea cliffs on the west end of Molokai is disproportionately small as compared with that necessary to truncate Penguin Bank.

At the southeast corner of West Molokai, 6 miles from Kaunakakai, two dikes trend northwestward, parallel to the northwest rift zone; they appear to represent an extension of that rift across the summit of the shield. The more westerly of these dikes can be traced up the hillside for nearly a mile, and is one of the few in Hawaii that can be seen merging into a surface lava flow.

There is no indication that a caldera ever was present in the West Molokai shield. Along the northeastern slope, a series of scarps have resulted from faulting which dropped the lower portion of the slope downward in relation to the summit (figs. 229, 230). On most of the faults the downdropped block is to the northeastward, but on two of them the downthrow is to the southwest. The Mahana valley is a graben that lies between the more easterly of these faults, which runs along the western slope of Waihuna hill, and the main escarpment up which the road climbs in a long switchback just west of Mahana. The fault scarps are particu-

344

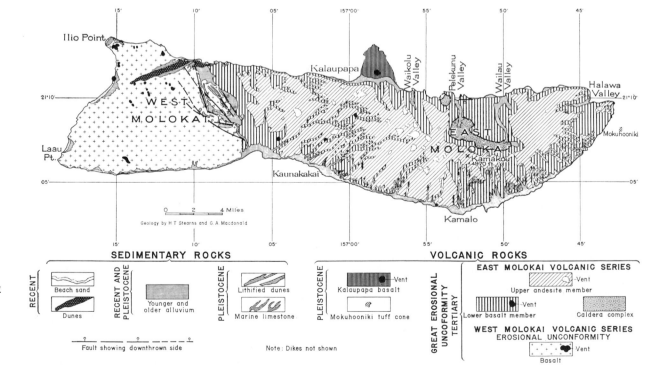

Figure 229. Geologic map of the island of Molokai. (From Stearns, 1946.)

larly conspicuous from the air. Hot water and probably gases rising along the faults have bleached and partly altered the adjacent rocks, and have deposited secondary clay minerals and iron oxides, and masses of calcite, opal, and chalcedony. Fragments of massive and banded chalcedony up to 6 inches across are found scattered along the gulches in that area, and have been cut and polished as semiprecious stones.

Because it is low and partially shielded from the trade winds by the larger mountain to the east, West Molokai is very dry, and streams have cut only shallow gulches. Along most of the coast, wave erosion has been slight, but on the north coast a sea cliff 500 feet high has been cut. A cliff of this height in a shield of such gentle slope represents a considerably greater amount of cutting back by erosion than would be required to cut a similar cliff into a steeper original slope.

Long weathering, accompanied by compara-

tively little stripping by erosion, has left a deep red soil cover over the top of West Molokai. In places recent erosion has cut through the upper horizon exposing the lower horizon, which locally exhibits pronounced iron enrichment in the form of masses of lateritic iron oxide. Particularly on the southern slope, the soil has been largely stripped away at low altitudes, possibly by wave erosion during higher stands of the sea.

Calcareous sand blown inland by the trade winds from Moomomi Beach, at the northwest edge of the isthmus, has formed a belt of dunes as much as 60 feet high and half a mile wide that extends almost completely across the northwestern corner of West Molokai. This belt is sometimes referred to as the Desert Strip. The dunes in it are still active and largely unconsolidated. In the same area, however, dunes of an older generation, which are moderately to well lithified, are now being eroded, and are contributing some sand to the new

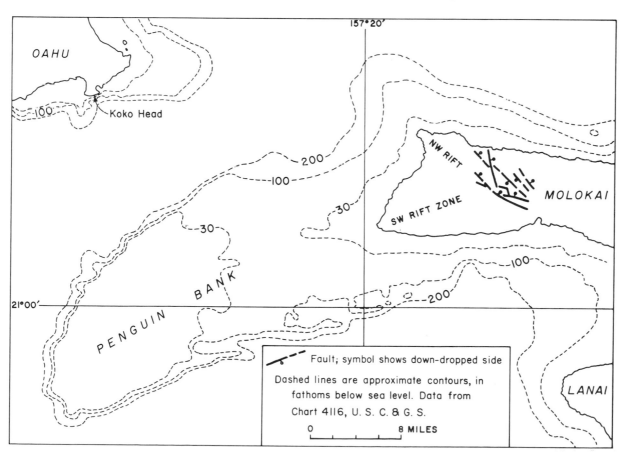

Figure 230. Map showing the relationship of Penguin Bank to the southwest rift zone of West Molokai volcano, faults on the northeastern slope of West Molokai, and the submarine ridge projecting southwestward from Koko Head on Oahu.

345

dunes. A patch of lithified dunes a mile square covers the surface at Ilio Point. The dunes extend below sea level, and are notched by wave erosion 25 and 5 feet above sea level. These, and probably at least part of the lithified dunes of the Desert Strip, were formed from sand blown inland from beaches exposed to the wind at a time of lower sea level.

On the west side of West Molokai, Papohaku Beach is one of the largest white sand beaches in the Islands (fig. 146). During recent years Papohaku and other West Molokai beaches have supplied large amounts of sand for construction purposes on Oahu. Lithified calcareous beach sand ("beach rock") is unusually well developed on some of the Molokai beaches, such as Halena Beach on the south side of West Molokai. The limestone used in building the Honolulu Academy of Arts is beach rock quarried on Molokai. Some beach rock splits readily into thin slabs and has been used as flagstones.

## EAST MOLOKAI

The East Molokai shield was built principally by eruptions along two rift zones, one extending eastward and the other west-northwestward from the summit. On most sides the lava flows slope outward at angles of 6° to 15° from the summit or from points along the rift zones; but on the west, where they were obstructed by the older West Molokai volcano, the slopes flatten to 2° to 3°. The summit of the shield was indented by a caldera 4.5 miles long and 1.5 miles wide that occupied the area of the present heads of Pelekunu and Wailau valleys. A mile and a half to the northwest, half a mile west of the mouth of Pelekunu Valley, was a pit crater nearly a mile in diameter. (See fig. 229.)

The rocks of the East Molokai volcano comprise the East Molokai Volcanic Series, which is divided into lower and upper members. The lower part of the lower member consists of the assemblage of tholeiitic basalt–tholeiitic olivine basalt–oceanite that is typical of the early stages of Hawaiian volcanoes; but in the upper part alkalic olivine basalts appear, and augite phenocrysts begin to accompany the olivine phenocrysts. Quite common is a type of rock containing phenocrysts, as much as half an inch long, of all three principal rock-forming minerals—olivine, augite, and soda-lime feldspar—and sometimes also smaller ones of magnetite. In some places the phenocrysts make up 60 percent or more of the rock. The presence of phenocrysts in such abundance suggests strongly that toward the end of the period of eruption of the lower member the temperature of the underlying magma had dropped to the point at which all of the minerals were crystallizing.

Thin lenticular beds of vitric tuff lie between the lavas. They rarely exceed 3 feet, and are generally less than 2 feet in thickness. In the east wall of Pelekunu Valley three large springs with a total flow of about 1,000,000 gallons a day are perched on tuff beds.

Along the north coast between Waikolu and Wailau valleys there are several faults, on all but one of which the south side has been dropped down. In Mokapu and Okala islands, off the north shore east of the Makanalua (Kalaupapa) peninsula, the lava beds slope southward about 15°. Similar dips are found in the lower part of the sea cliff near the mouth of Waikolu Valley and along the coast as far east as the mouth of Pelekunu Valley. From distant inspection it appears, however, that higher in the cliff they are overlapped by beds that dip northward in conformity with the general form of the East Molokai shield. The cause of these abnormal southward dips is not certain. They have been attributed to faulting (Stearns and Macdonald, 1947, p. 17); but because of the normal dip of the overlying lavas, which themselves are of precaldera age, it is at least possible that the southward-dipping beds are part of the southern flank of a separate, more northerly, shield that was eventually buried by the growing principal shield.

The caldera probably was once completely filled and capped by lava flows, but the upper portion of the fill has been removed by erosion. The caldera-filling lavas differ from those outside it primarily in being essentially horizontal, thicker, and more massive, but also in being somewhat altered. Gases and hot water moving through them commonly have changed the pyroxene to chlorite, and have deposited calcite and chalcedony in vesicles and other openings. At many places talus breccia, in part well consolidated, lies between the former caldera wall and the caldera-filling lavas. On Pilipililau Stream, near the head of Pelekunu Valley, the old caldera wall has been exhumed by erosion, and can be seen to slope 60° to 80° toward the caldera. The rocks in the Haupu Bay crater are much like those in the main caldera. Several stocks of gabbro and coarse-grained basalt, none of them more than half a mile in diameter, intrude the caldera-filling rocks.

The upper member of the East Molokai Volcanic Series (fig. 229) consists predominantly of mugearite, with lesser amounts of hawaiite, trachyte, and intermediate rocks. It is separated from the lower member by only a few inches of soil, indicating that no very long interval of time intervened between the eruption of the two members. Probably the rocks of the upper member originally covered most or all of the top of the shield, but no rocks of the upper member have been found on Olokui Peak, between Wailau and Pelekunu valleys. This peak may have been a high portion of the basaltic shield which was protected from later flows by a high part of the caldera wall. However, it is exceedingly difficult of access and has not been thoroughly explored. Elsewhere, streams have cut gulches and canyons through the upper member, exposing the older basaltic rocks beneath, but leaving it as a cap on the intervening ridges. In the walls of the great valleys near the crest of the mountain the upper member averages about 500 feet thick, and consists of several superimposed flows ranging from about 20 to 100 feet in individual thickness. Downslope the number of flows decreases and the member thins, until, around the periphery of the island, it is only 20 to 50 feet thick. The cap probably never extended far beyond the present shoreline. Ash beds, some of them weathered to soil before they were buried by the succeeding flow, are common between the lavas.

Eruptions of the upper member were more explosive than those of the lower and commonly built large cinder cones, more than 30 of which are scattered across the top of the volcano. Kualapuu and Puu Luahine, on the western slope, are prominent examples. At some vents the erupting magma was so viscous that it formed domes; several lie just west and south of the rim of Waikolu Valley. Kaapahu, the "Camel's Back" on the ridge just east of Kamalo Gulch, is a small dome. A thick flow from the same vent caps the ridge for more than a mile downslope.

The north side of East Molokai is truncated by a great cliff that rises nearly 4,000 feet above sea level (figs. 231, 232). Early workers, unable to visualize such a cliff as the work of wave erosion, attributed it to faulting that supposedly had dropped the northern portion of the shield below sea level, but there is little evidence to support this hypothesis. All but one of the faults actually exposed along and near the north shore show displacement in the opposite direction, with the southern side downthrown. Furthermore, profiles projected from the present surface of the shield to the submarine slopes north of the island do not suggest any large displacement close to shore (fig. 233). Several stack rocks, isolated by marine erosion, lie offshore. The largest of these, Mokapu Island, is nearly a mile north of the general shoreline, and consists of the same complex of thin-bedded lava flows of the lower member cut by numerous dikes as is exposed at the mouth of Waikolu Valley. If any fault exists, it must lie north of Mokapu Island. The only bit of evidence that might support the fault hypothesis is the presence of the later Kalaupapa volcano, built against the base of the cliff (figs. 229, 234). It is possible that the magma of the Kalaupapa volcano was guided to the surface by a fault. In general, however, there is no evidence in Hawaii that magma of the posterosional eruptions was guided to the surface by faults.

There is little question that the prominent submarine escarpment 20 miles northeast of East Molokai (fig. 246) is a fault scarp related to the Molokai fracture zone of the northeast Pacific, but it is wholly unrelated to the cliff that bounds the island.

Thus, although it remains barely possible that the great East Molokai cliff is a fault scarp

Figure 231. A view of East Molokai, looking eastward along the northern sea cliff and across the deeply dissected shield. Pelekunu Valley in the middle distance, and Wailau Valley beyond it, are cutting back into the filled caldera. Note the vertical grooves punctuated by plunge pools on the steep valley walls. A horned spur projects seaward at the near side of Pelekunu Valley.

greatly modified by wave and stream erosion, it is far more probable that it is a sea cliff—an impressive monument to the power of the waves driven by the trade winds across the open sweep of the North Pacific.

The great valleys of the northern slope are clearly the result of stream erosion. Pelekunu and Wailau valleys (fig. 231) extend back into the caldera-filling rocks, and the present valley heads are largely controlled in shape and position by the relatively weak and easily eroded, partly altered rocks of the caldera. The valleys were cut below present sea level during a low stand of the sea, and, with a rise of base level, a thick mass of alluvium was deposited on the valley floors. Halawa Valley, at the east end of the island, is a typical amphitheater-headed valley cut by stream erosion. On the southern slope of the shield the low rainfall has resulted in relatively little stream erosion, but even the resistant mugearite caps between the

gulches are gullied, and their original constructional surface is largely destroyed. The mountain as a whole is in an erosional stage of very late youth or submaturity.

After the northern cliff had been formed and the great valleys cut, volcanic activity resumed. In the ocean a mile east of the east end of the island, hydromagmatic explosions built a cone of vitric tuff (fig. 229), containing many fragments of older basaltic rocks and of reef limestone. The originally glassy tuff has been largely altered to palagonite. Toward the end of the eruption the character of the activity changed: spatter veneered part of the tuff cone, and lava flows poured down its western flank. The change in type of activity probably began when the cone had been built so high that water was excluded from the crater and hydroexplosions no longer occurred. (Exactly similar changes in behavior have been observed during recent years in the eruptions

348

Figure 232. Sea cliffs rise 2,000 to 3,300 feet along the north coast of East Molokai, and great amphitheater-headed valleys cut into this windward side of the shield. Some geologists have suggested a fault origin for the cliff, but projections of the shield profile (fig. 233) indicate that no large fault exists close to shore. Kalaupapa peninsula, in the middle distance, is a small shield volcano built against the base of the sea cliff. Mokapu and Okala islands, on the near side of the Kalaupapa peninsula, are sea stacks.

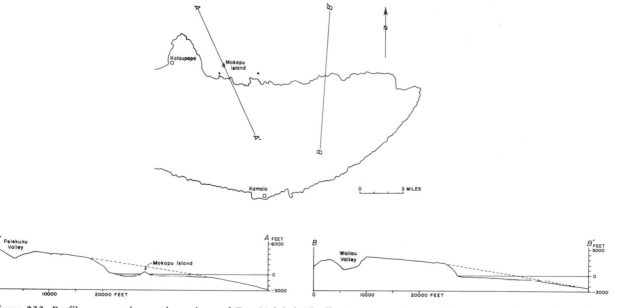

Figure 233. Profiles across the northern slope of East Molokai. The lines on the map at top show the positions of the profiles. (After Stearns and Macdonald, 1947.)

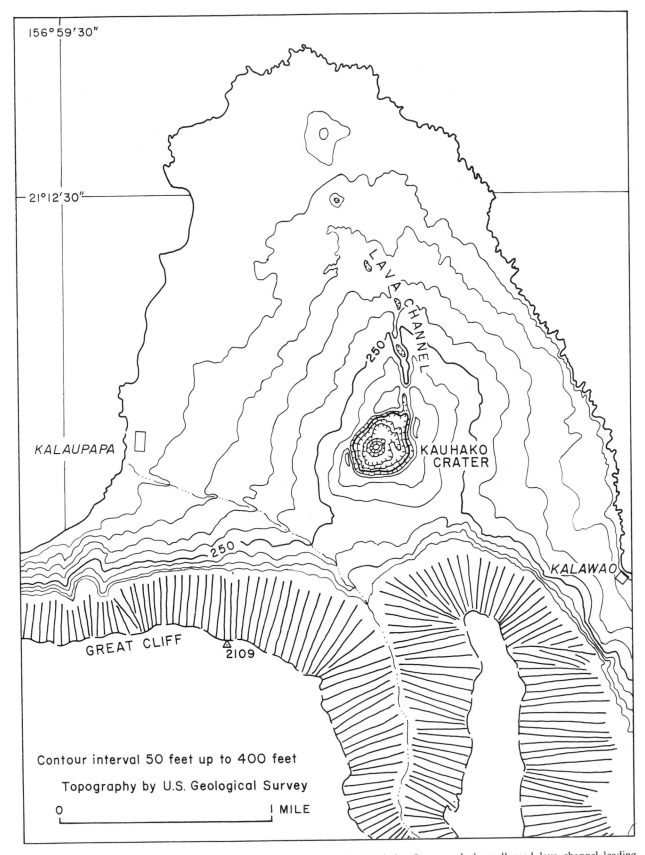

Figure 234. Topographic map of the Kalaupapa peninsula, showing Kauhako Crater and the collapsed lava channel leading northward from it.

of Anak Krakatau, between Java and Sumatra, and Surtsey volcano, south of Iceland.) Wave erosion has cut deeply into the cone, separating it into two islets, Mokuhooniki, and the smaller Kanaha.

At about the same time another eruption built a small shield volcano against the northern cliff, forming the present Kalaupapa peninsula (figs. 229, 235). The Kalaupapa volcano consists of alkalic olivine basalt pahoehoe with abundant phenocrysts of feldspar, olivine, and augite. The shield is about 2 miles across at sea level, and rises to an altitude of 405 feet. Its top is indented by Kauhako Crater, a quarter of a mile across and more than 450 feet deep, containing a pool of brackish water. During the last stage of the eruptions that built the shield,

lava rose in the crater almost to the rim, then drained northward through a large lava tube. The tube has collapsed, but its course is marked by a long trench that extends down the northern slope (fig. 234). Two distinct benches on the sides of the crater mark pauses in the decrease in level of the lava lake as the liquid drained away. The Kalaupapa shield is almost untouched by erosion, and the lava surfaces appear quite fresh. It cannot be very old. However, on the northwestern edge of the peninsula, near the old Federal leprosarium at Kalawao, stream-laid conglomerate forms a terrace, 100 feet above sea level, that must have been graded to a higher stand of the sea. On this basis the Kalaupapa eruption is believed to have taken place during late Pleisto-

351

Figure 235. The basaltic shield volcano of Kalaupapa peninsula, in the foreground, formed late in the history of Molokai, long after the main shield volcano had stopped activity. The settlement of Kalaupapa lies at the far side of the peninsula. The lavas that built the Kalaupapa shield flowed principally from Kauhako Crater, and impinged against the sea cliff cut across the lavas of the main volcano. The crater, which extends below sea level, is more than 450 feet deep, and a quarter of a mile across. The discontinuous trench that extends seaward from the crater is a collapsed lava tube.

cene time (Stearns and Macdonald, 1947, p. 26).

Other evidences of former high stands of the sea are found on East Molokai. East of Kaunakakai massive ledges of coral reef up to about 8 feet thick lie approximately 25 and 100 feet above sea level, and thin fossiliferous deposits in the bottoms of shallow gulches extend up to about 280 feet. Moderately to well consolidated alluvium several hundred feet thick has been deposited in the big valleys and in places forms terraces, now well above the present valley floor, that probably were graded to higher sea levels.

A nearly continuous apron of alluvium lies along the south coast of East Molokai and the isthmus. In part it represents an encroachment of terrestrial sediment, from accelerated erosion of the southern slopes during recent decades, over the shoreward edge of the live fringing reef that borders the south shore of the island. Alluviation has also destroyed many fish ponds. The ancient Hawaiians had built more than 50 mullet ponds around brackish springs along the south shore, the largest of them with an area of about 500 acres, but most of the ponds are now partly or wholly filled with mud and are becoming mangrove swamps.

## Oahu

Oahu is yet another volcanic doublet, formed of the Waianae Range on the west and the younger Koolau Range on the east (figs. 236, 237). Both are the eroded remnants of great shield volcanoes, but the term "range" expresses the fact that they have lost most of the original shield outlines and are now long narrow ridges shaped largely by erosion. Lava flows from the Koolau volcano banked against

352

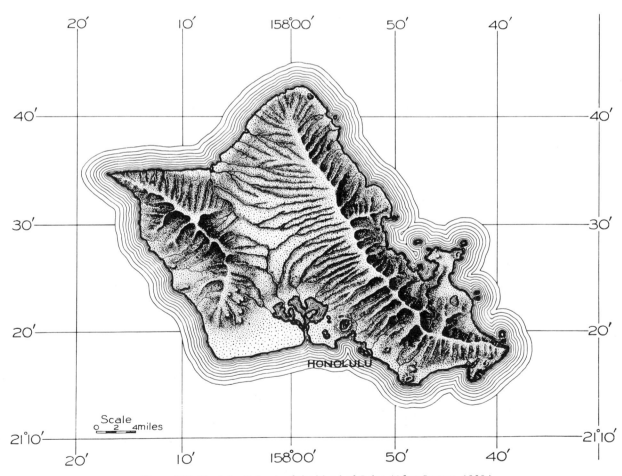

Figure 236. Shaded relief map of the island of Oahu. (After Stearns, 1939.)

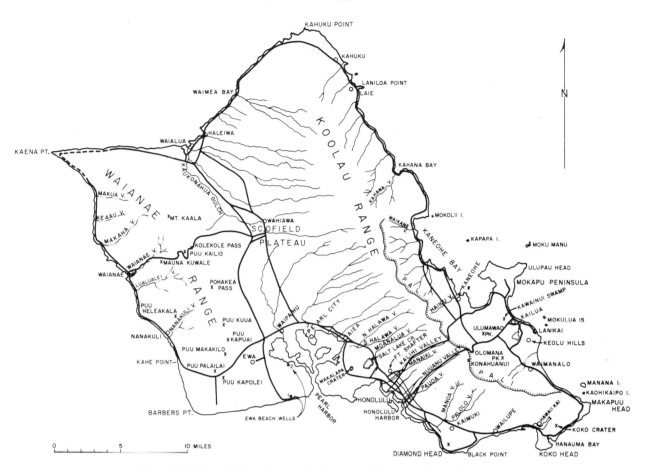

Figure 237. Map of Oahu, showing the location of places mentioned in the text.

the already-eroded slope of the Waianae volcano to form the gently sloping surface of the Schofield Plateau (fig. 238). An erosional unconformity between the rocks of the two volcanoes is visible along Kaukonahua Gulch, at the eastern foot of the Waianae Range, where Waianae lavas slope 10° to 15° northeastward and are overlapped by Koolau lavas dipping 5° northwestward.

After a long period of volcanic quiet during which canyons several thousand feet deep were cut into the Koolau shield, volcanic activity returned, and a series of lava flows, cinder cones, and tuff cones were formed. These, which are very different in composition from the older Koolau rocks, are known as the Honolulu Volcanic Series. The rocks of the older volcanoes are named respectively the Waianae Volcanic Series and the Koolau Volcanic Series.

The Waianae and Koolau volcanoes probably were built on a still older volcanic mass.

Fragments of reef-forming corals and other organisms dredged from the edge of a submarine bank a few miles southeast of Honolulu Harbor have proved to be fossil forms of latest Miocene age, approximately 11 million years old (Menard and others, 1962). Since the oldest known rocks of the Waianae Range appear to be only about 6 million years old (see chap. 17), the shallow marine platform that supported the Miocene coral reef was probably the eroded stub of a still older volcano, perhaps of the same age as the truncated volcanoes that underlie the coral reefs in the Leeward Islands of the Hawaiian group.

CHANGES IN SEA LEVEL

The emerged reefs on Oahu are more extensive than those of any other of the Hawaiian islands and constitute an important aspect of the geology of the island. The Honolulu plain, the

*353*

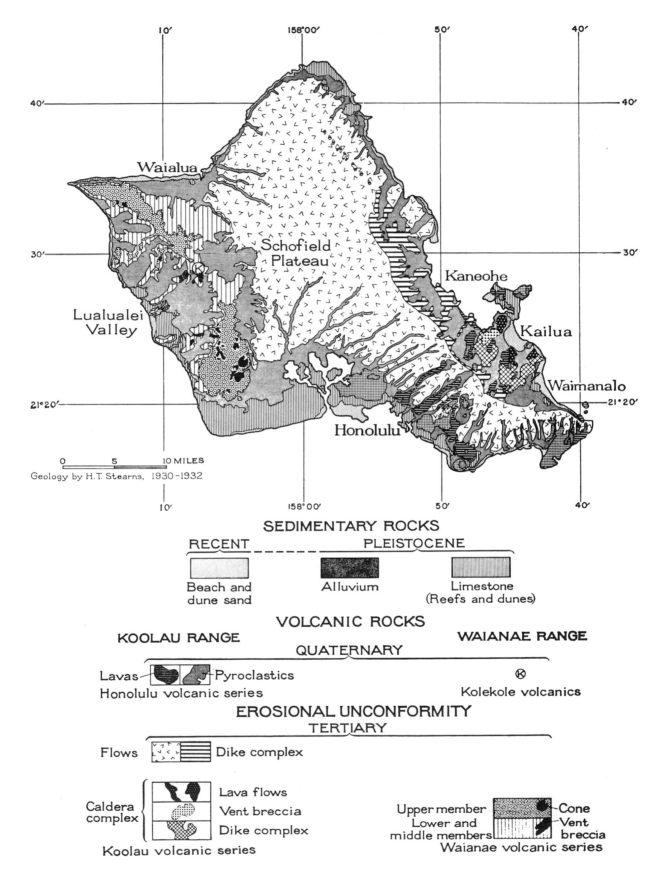

## SEDIMENTARY ROCKS

RECENT ------- PLEISTOCENE

Beach and
dune sand

Alluvium

Limestone
(Reefs and dunes)

## VOLCANIC ROCKS

KOOLAU RANGE           WAIANAE RANGE

QUATERNARY

Lavas ⊢ ▬ ⊣ Pyroclastics           ⊗
Honolulu volcanic series       Kolekole volcanics

### EROSIONAL UNCONFORMITY

TERTIARY

Flows [ ] Dike complex

Caldera   {   Lava flows
complex    Vent breccia
        Dike complex
Koolau volcanic series

Upper member      Cone
Lower and         Vent
middle members   breccia
Waianae volcanic series

354

Figure 238. Geologic map of the island of Oahu. (Modified after Stearns, 1946.)

Ewa plain, and much of the rest of the southern edge of Oahu is underlain by a broad elevated coral reef (fig. 238), partly covered by alluvium carried out from the mountains. Drill holes show that lava flows of the Honolulu Series are interbedded with the reef deposits, and consequently that the two must have been forming, at least in part, at the same time. Fossils in the reef do not place its age closely. Most of the animal species found as fossils still live in the Hawaiian region, and others are still living farther south, which suggests that the reef is not older than late Pleistocene. The nature of the fossil assemblage does, however, indicate that the water in which the animals lived was warmer than that around Oahu today (Ostergaard, 1928), and this in turn suggests that the reef grew during one, or perhaps more than one, of the interglacial stages, when sea level was higher than it is now. Much of the exposed reef surface is close to 25 feet above present sea level and, therefore, it probably grew during the plus-25-foot (Waimanalo) stand of the sea. At other places the reef surface consists of broad terraces, some about 12 feet and others about 5 feet above sea level. Some of these lower parts of the reef may have grown during stands of the sea corresponding to those levels, but others are wave-eroded surfaces cut into the reef that was formed at the higher (plus-25-foot?) stand of the sea. Most of the emerged reef of southern Oahu appears to have formed during the 25-foot stand.

Many other evidences of changes of sea level are found on Oahu. At Laniloa Point, near Laie, lithified dunes of calcareous sand which extend below sea level must have been formed during a lower stand of the sea. At Kailua and Waimanalo, similar dunes formerly bore conspicuous wave-cut nips 22 and 27 feet above present sea level, marking the double Waimanalo (plus-25-foot) stand. These dunes are now being rapidly cut away to furnish sand for construction purposes, and the nips have been destroyed.

There is little evidence on Oahu of the Olowalu (plus-250-foot) shoreline. Blocks of reef limestone weighing up to 500 pounds were found by Stearns in 1932 as high as 200 feet above sea level in the valley just west of Makapuu Head. This was before coral debris had been hauled up the valley to pave roads, and the large blocks of limestone almost surely indicate that the sea was once at least that high on the eastern end of the Koolau Range. Old alluvium in Keaau and Makaha valleys, in the Waianae Range, appears to be graded to about the level of the Olowalu shoreline; and marine muds in Lualualei Valley demonstrate a sea level at least 150 feet above the present one.

Near Waianae, reef limestone extends to 88 feet above sea level and is overlain by almost 10 feet of fossiliferous lithified beach sand. Near Nanakuli, reef limestone reaches an altitude of 95 feet. At Kaena Point fossiliferous conglomerate is 89 feet above sea level, and there are loose coral cobbles up to 100 feet. These outcrops are taken to indicate a stand of the sea, the Kaena stand, about 95 feet above present sea level (Stearns, 1935).

Reef limestone near Kahuku extends to a height of 50 feet above sea level and is overlain by fossiliferous beach sand and conglomerate 10 feet thick, suggesting a stand of the sea about 55 feet above the present one—the Kahuku stand. At Laie, lithified pebbly beach sand overlying marine limestone extends to 76 feet above sea level, and suggests a shoreline at an altitude of about 70 feet—the Laie stand. At Kahe Point, on the Waianae Range, fossiliferous beach conglomerate reaches a similar elevation. Ruhe, Williams, and Hill (1965) have suggested that the deposits attributed by Stearns to the Kahuku and Laie stands of the sea may have been formed during the Kaena stand. In this connection it is certainly justified to emphasize a point made by Easton (1963), that in dropping from any high level, such as 95 feet, to the present or a lower level, the sea surface passes through an infinite number of levels, and a short pause at any one of them may leave distinctive shoreline features.

The terrace of alluvial gravel at Fort Shafter appears to be graded to a stand of the sea about 40 feet above present sea level (Wentworth, 1926), and in the valleys at Hawaii Kai a wave-cut bench at 36 to 41 feet altitude, veneered with marine conglomerate, suggests a shoreline about 40 feet above present sea level

*355*

(Stearns, 1935; Easton, 1963). Stearns named this shoreline the Waialae stand, and placed it about 45 feet above sea level. Evidence for it elsewhere is almost nonexistent, and like the Laie and Kahuku stands it may represent only a short pause in the drop of sea level, with its deposits preserved only at a particularly favorable locality.

No convincing evidence of shorelines higher than 250 feet has been found on Oahu. Evidence for the minus-300, minus-1,700, and minus-3,000-foot stands has been discussed in chapter 11.

A few comments should be made regarding the very conspicuous bench that encircles Hanauma Bay and extends eastward along the coast past Koko Crater. The bench, cut into tuff, is about 12 feet above sea level on the promontories, but descends to sea level in the head of Hanauma Bay. It has been attributed to wave action during the plus-12-foot and/or plus-5-foot stands of the sea, with stronger wave attack on the headlands cutting the bench at a higher level there. However, there appears to be nothing to support this hypothesis. Stronger waves would, indeed, cut to a higher level, but the floor of the nip would still be just below sea level, and the bench would be close to horizontal from the promontories to the bay heads. It would not slope, as does the bench at Hanauma Bay. Recent studies suggest that the Hanauma bench is the result of cutting by the sea at its present level, with lesser amounts of cutting by rainwash and other subaerial processes. The position and slope of the bench seem to be governed by the level of water saturation of the tuff. The heavy surf along the more exposed parts of the coast keeps the rock face of the cliff saturated to a higher level than do the small waves in more protected places. The water-saturated tuff is more resistant to erosion than the unsaturated tuff above it, and the latter is more rapidly cut away, leaving the bench. The surface of the bench is much pitted by "water-level weathering"—loosening of the grains in the tuff by alternate wetting and drying around the edges of water-filled hollows.

Although the Hanauma bench appears to have been formed by the sea at the present level, cut benches in limestone elsewhere on Oahu are clear evidence of stands of the sea at the plus-5 and plus-12 foot levels.

GEOLOGIC HISTORY OF PEARL HARBOR

Pearl Harbor is essentially a series of drowned river valleys (fig. 239), but the area has a complicated history. When the building of the Waianae and Koolau volcanoes came to an end, only a slight embayment existed on the southern coast of Oahu, somewhat seaward of the present shoreline, but as the island sank, 1,200 feet or more, a broad bay developed with a barrier reef across its mouth. The situation was probably much like the present one at Kaneohe Bay. The bay was never very deep; as fast as the island sank, sediment was washed in from the surrounding hills and accumulated in the shallow lagoon behind the reef, forming layers of mud, silt, and sand. The earliest sediments rest on soil formed by weathering of the surface of the basaltic lava rocks. However, as sinking continued, the water did become a little deeper and the sediment changed first to limy muds and then to coral reef. A well, recently drilled about 600 feet inland from the shore at Ewa Beach, passed through this reef at a depth of 786 feet below sea level (Stearns and Chamberlain, 1967). Then the island sank more slowly, and terrigenous (land-derived) silts and sands were laid on top of the reef, eventually to a total thickness of several hundred feet. The same Ewa Beach well passed through the bottom of the sediments into lava rock 1,072 feet below sea level.

The sinking of the island was not uniform and uninterrupted. Several unconformities are found in the cores from the Ewa Beach well and from another drilled 2 miles farther inland. These mark interruptions in the accumulation of the sediment, caused either by slowing of the sinking so that sediment completely filled the lagoon and was transported on to the open sea beyond it, or by an actual reversal of the direction of movement with a slight upheaval of the island. During two of these interruptions, swampy conditions developed and peat was formed. These old swamps are marked by layers of carbonaceous sediment 624 and 358

Figure 239. Pearl Harbor, Oahu. The three prominent arms, or lochs, of Pearl Harbor were formed by drowning of stream valleys cut into the coral reef and other rocks of the coastal plain. The Ewa plain, in the right foreground, is the surface of a coral reef formed during the 25-foot stand of the sea, partly veneered with alluvium.

feet below present sea level. A lesser unconformity is found 406 feet below sea level, and at 203 feet a still-stand is indicated by several feet of calcareous beach rock (Stearns and Chamberlain, 1967).

No evidence of events during the Olowalu (plus-250-foot) stand of the sea has been found in the Pearl Harbor area, but during the Kaena (plus-95-foot) stand a delta of silt and sand grew into the bay near Aiea and Pearl City. The sediments can be seen there in roadcuts. Later, sea level dropped to the Waipio (minus-60-foot) level, and the streams, flowing across the sediments in the old bay, cut valleys into them (fig. 240). The several streams came together into a single master stream. Near the present shoreline the resistant coral reef kept the valley of the master stream narrow, but farther north the soft deltaic sediments were easily eroded and the branching tributary valleys became broad and shallow. At that time eruptions at Salt Lake and Makalapa craters spread ash over the land surface; but erosion continued, cutting into the new ash cover. Then sea level rose again, to the Waimanalo stand, 25 feet above

present sea level. The valleys were drowned, branching embayments were formed, and again sediments were deposited in the head of the bay. At some places the first deposit was a bed of oyster shells, but as the water deepened this was buried by sand or silt. The shell layer, with the beds of detrital sediment covering it, is visible in cuts south of Waipahu. Closer to the open ocean, coral reef grew around the ridges and hillocks left by erosion during the Waipio stand, but the new reefs probably never blocked the entrance to the bay. Again sea level dropped, probably to the Mamala (minus-300-foot) shoreline, with pauses 12 and 5 feet above present sea level, and again erosion cut away part of the sedimentary deposits. Eventually the sea rose to the present level, and again the stream valleys were submerged, forming the present Pearl Harbor. Each of the lochs of Pearl Harbor is the drowned lower part of a tributary valley. The broadly rounded outlines of the individual lochs show the widening of the valleys in the easily eroded, deltaic sediments, while the valley of the master stream at the present

357

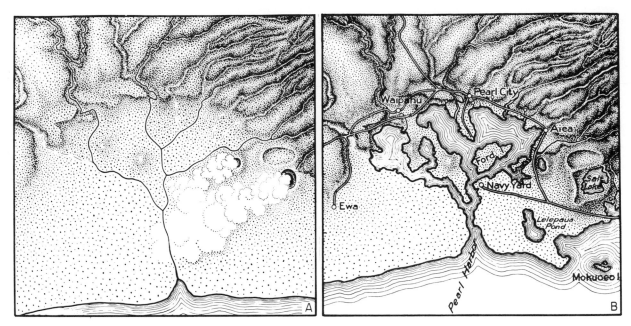

Figure 240. Stages in the development of Pearl Harbor. *A*, Shoreline when the sea stood about 60 feet lower than now; *B*, at present sea level. (From Stearns, 1939.)

harbor entrance was kept narrow by the resistant coral reef.

## WAIANAE RANGE

The Waianae Volcanic Series is divided into lower, middle, and upper members (fig. 238). The lower member comprises the lava flows and associated pyroclastic rocks that built the main mass of the Waianae shield volcano. The middle member consists of rocks that accumulated in the caldera, gradually filling it, and the upper member is the relatively thin cap that appears to have covered the entire top of the shield late in its history. The lower and most of the middle members consist of rocks of the tholeiitic group; but toward the top of the middle member alkalic basalts begin to appear, and the upper member is largely hawaiite with lesser amounts of alkalic olivine basalt. Some very late cones on the south end of the range, which are included in the upper member, are of alkalic olivine basalt. Although in places the lavas of these cones are separated from the older flows of the upper member by a thin layer of ashy soil, there does not appear to have been any prolonged period of erosion. However, in Kolekole Pass a small flow of alkalic olivine basalt rests on alluvium which in turn accumulated in a deep stream-cut valley.

This latter eruption certainly was posterosional; it bears a relation to the Waianae Volcanic Series similar to that between the Koolau Volcanic Series and the posterosional Honolulu volcanics.

Erosion has removed most of the western slope of the Waianae shield and exposed the internal structure of the volcano. The shield was built by eruptions that took place principally along three rift zones, now marked by innumerable dikes. The two principal rift zones trended northwestward and south-southeastward from the summit, and a lesser one trended northeastward. The lavas of the lower member are thin flows of pahoehoe and aa, sloping outward from the summit and from points along the rift zones at angles ranging from about 4 to 14°. No soil layers, and only a very few thin beds of tuff, lie between the lava flows. Explosive activity was slight, and the flows came too frequently to permit any appreciable weathering between them.

The caldera of the Waianae volcano lay in the region just west of Kolekole Pass, and extended for about 9 miles, from the northern side of Makaha Valley to the head of Nanakuli Valley (fig. 174). The lavas that accumulated in the caldera are thick and massive, as compared with the thin flows that poured down the slopes outside the caldera. For the most part

358

they are nearly horizontal, but at Puu Kailio, in the head of Lualualei Valley, they have been deformed into a basin-like structure (the "Kailio syncline"), apparently by sagging as a result of withdrawal of magma from beneath. At many places the lavas of the middle member are separated from those of the lower member by masses of breccia that accumulated as taluses against the cliffs bounding the caldera. The breccia is well exposed in and near Puu Kailio, in the ridge separating Nanakuli and Lualualei valleys, and in the ridge northwest of Makaha Valley. About 1,800 feet of lava flows of the middle member are visible in the valley walls, and they extend on down below the level exposed by erosion.

Eventually the caldera was filled to over-flowing, and lava flows spilled over its western rim. This is recorded in the ridge northwest of Makaha Valley. There, low in the ridge, thick caldera-filling flows can be seen resting against talus breccia, which in turn rests against a buried east-facing cliff that truncates thin westward-dipping flows of the lower member. In the upper part of the ridge, however, the lava flows from within the caldera overtopped the cliff and continued westward as thin flows approximately conformable in slope with the underlying beds of the lower member. The structure is illustrated in figure 241. The outer

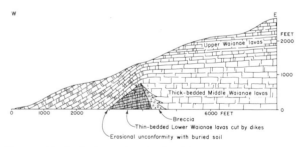

Figure 241. Geologic section in the ridge between Keaau and Makaha valleys, Waianae Range, Oahu, showing the structure at the boundary of the Waianae caldera. (After Stearns and Vaksvik, 1935.)

slope of the volcano must have been protected from lava flows by the caldera wall for a period long enough to permit considerable erosion. The surface separating the lavas of the lower member from those that overflowed from the caldera is marked by a layer of colluvium a few inches to several feet thick, and this rests on

the truncated ends of dikes. Many tens, and perhaps hundreds, of feet of erosion must have been necessary to cut down into the dike complex in this manner. There is a possibility, however, that the buried outer slope of the lower mass of lavas is a tangential fault scarp relatively little changed by erosion.

Mauna Kuwale, in the ridge between Lualualei and Waianae valleys (fig. 243), is a mass of rhyodacite lying between flows of basalt. The first potassium-argon age determinations indicated that the rhyodacite is several million years older than the nearby basalts, and on that basis it was suggested that the rhyodacite was part of an earlier volcano buried by the Waianae volcanics—presumably a part of the same mass that supported the Miocene coral reef already mentioned. However, there is no evidence of erosion between the rhyodacite and the overlying basalt, and later study has indicated that the apparent discrepancy in age is the result of excess argon trapped in the biotite of the rhyodacite (p. 269). The mass appears to be a short thick flow, or possibly a dome and flow, interbedded with the caldera-filling lavas of the Waianae volcano.

Thick massive flows of the upper member form high cliffs of conspicuously light color in the upper part of the valley heads, particularly in the western part of the range. The flows are mostly aa, many of them 50 to 100 feet thick. In Nanakuli Valley and near Mt. Kaala the aggregate thickness of the upper member is about 2,300 feet. From Pohakea Pass southward, lavas of the upper member veneer the entire southern and eastern slopes of the range, but along the northern and northeastern slopes older lavas are exposed beneath them. The upper member was formerly much more extensive, but has been removed over broad areas by erosion. On the western flank it is found only as caps on the Keaau and Kawiwi ridges north of Makaha Valley.

Within the caldera there is no sharp break between the middle and upper members: the beds are conformable, and the passage from tholeiitic to alkalic rock types is transitional. Outside the caldera, however, the transition generally is abrupt and can be recognized both petrographically, from the mineral and chemical composition of the rocks, and by the change in physical character from the thin-bedded tholeiitic basalts of the lower member below to the thick, massive beds of the upper member. From Nanakuli Valley to the south end of the range, the contact is marked by a layer of reddish brown, ashy soil 6 inches to 3 feet thick. Flows of the upper member frequently were separated by a sufficiently long interval of time to allow the cutting of shallow valleys and the formation of thin layers of ashy soil, later buried by the succeeding flows. A valley of this sort is occupied by the massive flow which is being worked at the Kaena Quarry on the north flank of the range.

Along the lower southeastern slope of the range is a row of five very late cones: Puu Kuua, Puu Kapuai, Puu Makakilo, Puu Palailai, and Puu Kapolei. They are composed of a varied mixture of cinder, spatter, and lava flows. At the last two cones the original cinder carapace has been largely stripped away by erosion, leaving only the core of more resistant lava that filled the crater of the cone and the upper part of the underlying conduit. At Puu Palailai, quarrying has removed a large part of the former crater fill. An older, much eroded cinder cone forms Palikea Peak at the head of Nanakuli Valley, and the remains of still another are exposed 2 miles farther south. Masses of cinder are preserved at a few other places, as at Puu Kawiwi, between Waianae and Makaha valleys; and thin beds of vitric tuff are present locally between flows of the upper member. On the south side of Puu Heleakala, between Nanakuli and Lualualei valleys, a cross section of a cinder cone is exposed, enclosed in lava flows of the lower member.

Hundreds of dikes, from a few inches to about 15 feet thick, are exposed in the heads of the broad valleys and in ridges along the western side of the range. Most of them are nearly vertical, and in general they trend parallel to the rift zones. Larger intrusive masses have not been identified, but cobbles of medium- to fine-grained gabbro are found along Nanakuli, Lualualei, and Waianae valleys.

There is a tremendous contrast between the great amount of erosion that has produced the huge valleys on the western side of the Waianae Range (fig. 242) and removed a large portion of the western part of the shield, and the relatively small amount of erosion on the northern and eastern sides (fig. 244). Possible reasons for this have already been discussed in chapter 10.

The valleys of the Waianae Range are choked with enormous accumulations of alluvium and colluvium (fig. 243). Big alluvial fans on the eastern side of the range are explained by the loss of stream water into the permeable lavas of the Schofield Plateau, with consequent reduction in transporting power of the stream. Older, somewhat consolidated alluvium in the western valleys appears to be graded, at least in part, to former higher sea levels, and is being slowly removed by present-day stream erosion. Other alluviation is the result of rise of base level. A well near Waianae passed through about 1,000 feet of alluvium before entering lava rock nearly 1,000 feet below sea level, and another in Lualualei Valley went through 1,200 feet of alluvium, indicating a relative submergence of the island of at least 1,200 feet since the valleys were cut (see chap. 11). As in most other arid regions, the large amounts of modern alluvium are probably related in part to the intermittent nature of stream flow and transportation.

Along the northern edge of the Waianae Range a great cliff rises 750 to 1,000 feet above the coastal lowland. The straightness of the cliff and the absence of any evidence suggesting faulting indicate that it is a sea cliff. It is no longer being attacked by the waves because its base is protected by a mass of coral reef and interbedded detrital sediment, and its lower part is heavily mantled with alluvium. It must have formed before the coastline became partially protected from waves approaching from the northeast by the growth of the Koolau volcano. Above the cliff the ridge tops are approximately parallel to the original surface of the volcano, but appear to have been lowered several tens of feet by erosion.

The top of Mt. Kaala, the highest point on Oahu (fig. 244), is a nearly level plateau, roughly a mile across, which is so little dissected by erosion that drainage is very poor. The area is a swamp, covered with fascinating vegetation, including sphagnum ("reindeer") moss and dwarfed ohia trees only a foot or two

Figure 242. Waianae mountains and valleys. The broad amphitheater-headed valleys are coalescing and reaching maximum size in a late stage of valley development. Only discontinuous thin ridges remain between them. Lualualei and Waianae valleys are in the center, with Makaha Valley behind them. Kaala is the flat-topped mountain on the right skyline. The valleys are filled with alluvium, and near the coast with marine sediments.

361

Figure 243. Kolekole Pass, Waianae Range, in the foreground. In the background the broad alluvium-filled Waianae and Lualualei valleys are separated by a deeply eroded discontinuous ridge. Just to the left of the low gap through the ridge is the rhyodacite flow of Mauna Kuwale, with the town of Waianae behind it.

Figure 244. Kaala, in the Waianae Range, is the highest point on Oahu, rising 4,025 feet above sea level. The flat swampy summit is a little-dissected remnant of the upper surface of the shield. The northern slopes of the shield, on the right side of the photograph, show intense but fairly youthful dissection.

high—a natural bonsai garden. The plateau is held up by nearly horizontal, thick, unusually resistant lava beds of the upper member. It appears to be the little-modified original surface of the volcano.

KOOLAU RANGE

The Koolau volcano was an unusually elongate shield built principally by eruptions along a northwest-trending rift zone. The dike complex formed in this rift zone is exposed in the hills between the crest of the range and Kaneohe Bay and near the heads of the big valleys farther north (fig. 238). The broad ridge extending toward Kahuku marks the northward continuation of the rift zone. Traverses across the dike complex at various places have yielded actual counts of more than 600 dikes per mile, and certainly some were missed because they are not all exposed. The dikes range from a few inches to about 12 feet thick, averaging 2 to 3 feet. Most of them are nearly vertical, but a few dip at angles as low as 60°. Often one dike is directly against another, but

elsewhere masses of the lava flows lie between them.

Excellent exposures of dikes can be seen in the cuts on the Pali Highway near Castle Junction, 3 miles southwest of Kailua (fig. 84). However, this area lies within the caldera of the volcano, and dikes cutting the caldera-filling lavas are far less numerous than in the complex outside the caldera. Therefore, these exposures do not adequately represent the abundance of dikes elsewhere in the complex.

There are indications that another, less prominent rift zone extended southeastward or east-southeastward from the summit of the volcano. Still another minor rift zone is represented by numerous dikes that trend south-southwestward in the general region between Kaimuki, Wailupe, and the crest of the range.

The alteration of the caldera-filling rocks, discussed in a later paragraph, led, in the original systematic study of the geology of Oahu (Stearns and Vaksvik, 1935), to the belief that they represented the top of an older volcano buried by the Koolau volcano. For this reason they were given a separate name—the

Kailua Volcanic Series. Later studies have shown, however, that they are unquestionably a part of the Koolau volcano (Stearns, 1940).

The rocks of the Koolau volcano as a whole are called the Koolau Volcanic Series. Almost all are tholeiitic basalts and olivine basalts with, as usual, small amounts of oceanite. Many of them contain small amounts of the orthorhombic pyroxene, hypersthene. In Moanalua Valley one flow contains an abundance of phenocrysts of labradorite feldspar as much as an inch long. This rock is a slightly alkalic basalt. Similar lava must be exposed somewhere in the face of the Pali on the northeastern side of the ridge, because cobbles of it have been found in the alluvium near Kaneohe. One flow in Kaukonahua Gulch, at the western edge of the Schofield Plateau, appears to be an alkalic basalt, though it is so much altered by weathering that any chemical analysis is unreliable. No other alkalic lavas have yet been found. Thus it appears that the Koolau volcano had barely reached the stage of transition to alkalic lavas when its activity ceased.

Tuff is scarce in the Koolau Series. A few beds of vitric tuff, from a few inches to several feet thick, are found near the crest of the range—for example, in the cliffs near the head of Nuuanu Valley—and some can be seen in cuts along the Pali Highway. In the face of the Pali 1,500 feet above sea level, 1.25 miles south of Waimanalo, the cross section of a cinder-and-spatter cone is exposed. Also in the Pali near the southeast end of the range, a bed of tuff breccia containing many angular fragments of older rocks accompanied by some magmatic debris originated from phreatomagmatic explosions. A bed of similar material, 15 feet thick, is exposed at an altitude of 1,650 feet about 700 feet southeast of the Pali gap, at the head of Nuuanu Valley.

The caldera of the Koolau volcano was about 8 miles long and 4 miles wide, extending from near Waimanalo at the southeast to beyond Kaneohe at the northwest (fig. 174). Its southwestern boundary lies near the base of the Pali, and its eastern boundary is somewhere between the hills at Lanikai and the Mokulua Islands, offshore. In general, the southwestern boundary is difficult to locate precisely because of poor exposures, but it can be placed within a few feet where it crosses the Pali Highway. Lava flows on the northeast side of the small gully just beyond the lower tunnel are thick and nearly horizontal, whereas the flows on the southwest side of the gully are thin and slope toward Honolulu. The boundary runs up the gully.

Within the caldera the rocks have been much affected by rising volcanic gases and hot water. The original pyroxene of the rocks has commonly been changed to chlorite and clay minerals, giving the rock a greenish to greenish gray hue. Silica released during the alteration has been redeposited as one or another of the silica minerals (opal, chalcedony, and quartz), in the form of amygdules filling former vesicles or as irregular masses and veinlets filling other openings in the lavas. Other secondary minerals, such as zeolites and epidote, are present, locally, in abundance. Excellent small crystals of quartz and zeolites have been collected in the Keolu Hills area and around Olomana Peak.

The general horizontal attitude of the caldera-filling lavas was disturbed in some places by sagging, presumably because of removal of magma from underneath. This produced basin-like structures, one of which can be seen in Olomana Peak, and another in the hills just southeast of Kailua (the "Lanikai syncline").

Whether or not the Koolau caldera was ever filled to overflowing we cannot say; the evidence has been destroyed by erosion. However, caldera-filling lavas extend all the way to the top of Olomana Peak (fig. 245)—more than 1,600 feet high. Their total thickness may originally have been more than 3,000 feet. The bottom of the caldera fill lies below sea level, and the southwestern rim must have been higher then the top of Konahuanui Peak (3,105 feet), which is composed of thin lava flows sloping toward Honolulu.

Recent geophysical work indicates that there is a mass of very heavy rock with very high seismic velocities at a depth of only a mile or two beneath the Koolau caldera (Adams and Furumoto, 1965). The mass is centered

363

Figure 245. Olomana Peak, formed by eroding away of the surrounding rocks in the caldera area of the Koolau volcano, Oahu.

approximately beneath the Kawainui Swamp, at the west edge of Kailua (fig. 174). It has been discussed on page 283.

A mass of breccia is exposed at 1,300 feet altitude on the ridge between Kahana and Waikane valleys, and a similar mass, penetrated by a water-development tunnel a mile to the northwest, contains fragments of Koolau rocks, many of them from dikes and ranging up to more than 2 feet in diameter. The breccia is itself cut by a few later dikes. It appears to have resulted from low-temperature explosions (probably phreatic) during the growth of the Koolau shield and to have accumulated in an explosion crater that was buried by later lava flows.

More puzzling is the mass of breccia that caps Ulumawao Peak and forms a large portion of the ridge to the north between Kawainui Swamp and Kaneohe Bay. It is composed of angular fragments of a variety of rocks, all of them resembling flows or dikes of the associated caldera-filling complex, up to 3 feet, but usually only a few inches, in diameter. The breccia lies on a surface that slopes eastward about 15°, approximately parallel to the bedding in the underlying flows, but in places truncating dikes in the underlying rocks. The breccia itself is cut by later dikes, and at the northeast end of the ridge it appears to be overlain by Koolau lava flows which are also cut by dikes. A maximum thickness of 520 feet of breccia is exposed, for the most part quite massive, but locally with a suggestion of bedding. Its origin is uncertain, but most probably it represents a series of mudflows that spread over the floor of a crater more than 2 miles across that lay within the Koolau caldera; the mudflows were later buried by subsequent caldera-filling lava flows, and the entire mass eventually tilted gently eastward by sagging of the caldera floor. However, the truncation of the dikes by the surface beneath the breccia would seem to demand a long period of erosion within the caldera before the upper part of the caldera fill was formed, which in itself seems improbable. The origin of the Ulumawao breccia is still being studied.

Since the caldera boundary lies approximately at the shoreline in the vicinity of Kailua, it is apparent that a large portion of the eastern side of the original shield is now missing, probably having been destroyed by erosion. Northeast of Kailua and Kaneohe a belt of hilly topography lies on the ocean floor, and it has been suggested (Moore, 1964) that this is the result of a giant landslide that was largely responsible for the destruction of the eastern flank of the Koolau shield (fig. 246). However, recent detailed work on the topography of the region (Belshé and others, 1966) does not support this hypothesis. The undersea hills are very large as compared with the bumps on known landslides, and some of them have forms characteristic of volcanic cones. The Tuscaloosa (also called Tuscarora) Seamount, for example, has a height of about 6,000 feet above the surrounding sea floor, and its general form (fig. 247) suggests that it is probably a guyot (Langford, 1969). It appears most likely,

therefore, that the flank of the Koolau shield was removed by a combination of stream and wave erosion.

The origin of the Koolau Pali (figs. 248, 249) through the coalescence of a series of amphitheater-headed valleys, and the cause of the advanced stage of erosion of the lowland at its foot have been discussed in chapter 9.

Great quantities of alluvium and colluvium occupy the large valleys of the Koolau Range. Deeply rotted old colluvium is well exposed in highway cuts along Nuuanu Valley. The causes

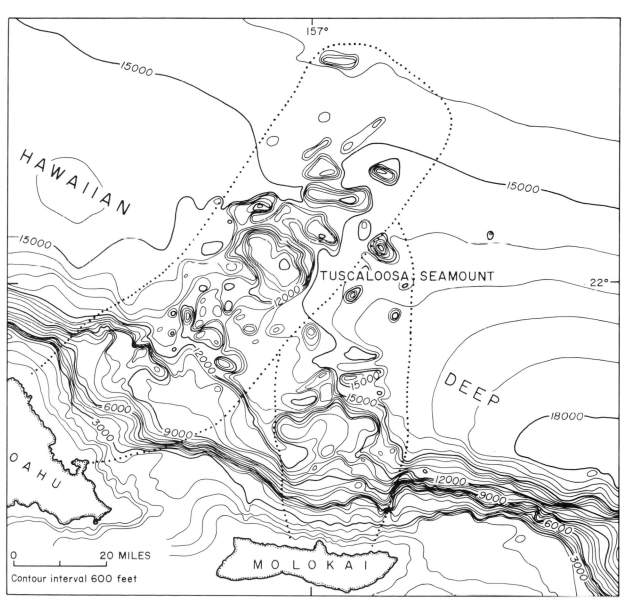

Figure 246. Map showing the submarine topography northeast of Oahu and north of Molokai showing a north-northeast trending belt of seamounts. The areas bounded by heavy dotted lines are believed by J. G. Moore to be huge landslides. (After Moore, 1964.)

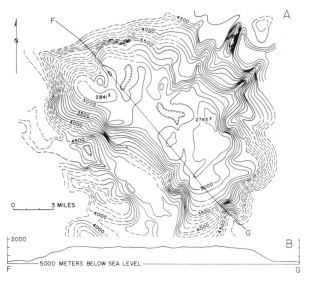

Figure 247. Map of Tuscaloosa Seamount *(A)*, and profile *(B)* along line *F–G* shown on map. (After Langford, 1969.)

of alluviation are probably the same as in the Waianae Range.

## HONOLULU VOLCANIC SERIES

The end of activity of the Koolau volcano was followed by a period of erosion during which much of the eastern flank of the shield was removed, great valleys more than 2,000 feet deep were cut into the rest of the range, the Nuuanu Pali was formed, and alluvium accumulated in the valleys as the island slowly sank at least 1,200 feet. This period of volcanic quiet probably lasted at least 2 million years. Then, on the southeastern end of the Koolau Range, volcanic activity resumed. More than 30 separate eruptions formed cinder, spatter, and ash

Figure 248. The great Pali along the windward side of the Koolau Range extends from the southeastern tip of Oahu behind the community of Waimanalo northward until it disappears from view beyond Kaneohe. The portion of the Pali at the lower edge of the photograph is a sea cliff, but the rest was cut principally by stream erosion.

366

Figure 249. Vertical valleys being cut by plunge-pool action of waterfalls at the head of Haiku Valley, Oahu.

cones, and poured lava flows over the deeply eroded topography and out onto the fringing reef. These constitute the Honolulu Volcanic Series, and have given us some of our best-known landmarks, many of them in or near the city of Honolulu: Diamond Head, Koko Head, Hanauma Bay, Punchbowl, Tantalus, and Salt Lake. The distribution of the vents is shown in figure 250.

The eruptions did not come in rapid succession, but were scattered over a period of hundreds of thousands of years. Their general sequence can be established by (1) the relative amounts of weathering and erosion each has undergone; (2) in some instances by superposition of the products of one eruption over those of another, sometimes with intervening erosion

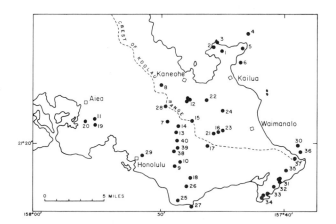

Figure 250. Map of southeastern Oahu, showing the distribution of vents of the Honolulu Volcanic Series. The vents are shown as black dots and are numbered to correspond with the numbered eruptions listed in table 18. (Modified after Stearns and Vaksvik, 1935.)

Table 18. Eruptions of the Honolulu Volcanic Series

| Eruption | Stand of the sea | Resulting structures | Composition of magma |
|---|---|---|---|
| 1. Hawaiiloa | Kahipa (−300 feet) | Cinder cone and lava flow | Melilite nephelinite |
| 2. Pali Kilo | Kahipa | Cinder cone (?) and lava flow | Nephelinite |
| 3. Pyramid Rock | Kahipa | Lava flow | Nephelinite |
| 4. Moku Manu | Kahipa (?) | Tuff cone | Nephelinite |
| 5. Ulupau | Kahipa (?) | Tuff cone (and lava flow?) | Nephelinite |
| 6. Mokolea | Kahipa (?) | Lava flow | Melilite nephelinite |
| 7. Kalihi | Kaena (+95 feet) | Cinder cone and lava flow | Melilite nephelinite |
| 8. Haiku | Kaena | Cinder cone and lava flow | Melilite nephelinite |
| 9. Rocky Hill | Kaena | Spatter cones and lava flow | Nephelinite and melilite nephelinite |
| 10. Manoa | Kaena | Cinder cone and lava flow (?) | Nephelinite |
| 11. Aliamanu | Kaena | Tuff cone with lava flow in crater | Melilite nephelinite |
| 12. Kaneohe | Kaena | Cinder cones and lava flow | Nephelinite and melilite nephelinite |
| 13. Luakaha | Kaena | Cinder cone and lava flow | Nephelinite with a little melilite |
| 14. Makuku | Kaena | Cinder cone and lava flow | Melilite nephelinite |
| 15. Pali | Kaena | Cinder cone and lava flow | Nephelinite |
| 16. Makawao (may be a Koolau vent) | Kaena (?) | Tuff-breccia in vent | Basaltic glass and accidental (Koolau) fragments |
| 17. Kaau | Kaena | Pit crater, lava flows, tuff deposits, and mudflows | Melilite nephelinite |

368

| | | | |
|---|---|---|---|
| 20. Makalapa | Waipio (−40 feet) | Tuff cones | Melilite nephelinite (?) |
| 21. Ainoni | Waipio | Cinder cone and lava flow | Nephelinite |
| 22. Castle | Waipio | Cinder cone and lava flow | Nephelinite with a little melilite |
| 23. Maunavili | Waipio | Cinder cone and lava flow | Nephelinite |
| 24. Training School | Waipio | Cinder cone and lava flow | Nephelinite with a little melilite |
| 25. Diamond Head (Leahi) | Waipio | Tuff cone | Nepheline basanite (?) |
| 26. Kaimuki | Waipio | Shield volcano | Basanitoid |
| 27. Black Point | Waipio | Lava flow | Nepheline basanite |
| 28. Kamanaiki | Waipio | Lava flows | Melilite nephelinite |
| 29. Punchbowl | Waipio | Tuff cone and lava flows | Nephelinite |
| *Koko fissure* | | | |
| 30. Manana | Manana (+5 feet) | Tuff cone | Basanitoid |
| 31. Koko Crater | Manana | Tuff cone | Basanitoid |
| 32. Kahauloa | Manana | Tuff cone with lava flows | Basanitoid |
| 33. Hanauma | Manana | Tuff cone | Basanitoid |
| 34. Koko Head | Manana | Tuff cone with lava flow | Basanitoid |
| 35. Kalama | Mamala (−300 feet) ? | Cinder cone and lava flow | Nepheline basanite |
| 36. Kaohikaipu | Mamala (−300 feet) or Penguin Bank (−180 feet) | Cinder cone and lava flow | Nepheline basanite |
| 37. Kaupo | Mamala or Penguin Bank | Spatter conelet and lava flow | Basanitoid |
| *Tantalus group* | | | |
| 38. Round Top | Manana (+5 feet) to present | Cinder and ash | Melilite nephelinite (?) |
| 39. Sugar Loaf | Manana to present | Cinder cone, ash, and lava flow | Melilite nephelinite |
| 40. Tantalus | Manana to present | Cinder cone, ash, and lava flow | Melilite nephelinite |

*Source:* Modified after Winchell, 1947.

and soil formation; and (3) by their relationships to the deposits and erosional features of various stands of the sea. Careful studies by Stearns and Vaksvik (1935), Wentworth (1926), and Winchell (1947) have arranged the eruptions in the sequence given in table 18. The oldest took place during or before the Kaena (plus-100-foot) stand of the sea; the youngest are almost untouched by erosion and may have occurred within the last 10,000 years. Thus the time span of the eruptions may be taken as roughly from mid-Pleistocene (perhaps half a million years ago) to Recent. The intervals of time between successive eruptions seem to have been as long as, or even longer than, the length of time from the last eruption to the present. Thus we cannot say that the Honolulu volcanic activity has ended forever. We may yet see more eruptions on southeastern Oahu.

The lavas of the Honolulu Volcanic Series include nephelinites, melilite nephelinites, basanites, and alkalic olivine basalts (basanitoids). All are notably rich in magnesium and iron and undersaturated with silica.

The arrangement of the vents of the Honolulu Series in general shows no relationship to the older rift zones of the Koolau volcano, although several lie along a line coincident with the minor south-southwest-trending rift (fig. 250). Most appear to lie along fracture lines that trend northeastward to nearly northward, almost at right angles to the principal rift zones of the Koolau volcano. Thus it is doubtful whether the Honolulu volcanism can properly be regarded as a renewal of activity of the Koolau volcano. Moreover, petrological evidence suggests that the magmas of the Honolulu Series came from a level considerably deeper in the earth than did the Koolau magmas. On the other hand, all of the Honolulu Series vents visible above sea level are close to the Koolau caldera. Perhaps they represent a revival, not of the Koolau volcano itself, but of the same general, deep-seated volcanic hearth—the same "hot spot" within the earth's mantle; or perhaps they tapped still-liquid magma from very deep levels in the slowly cooling magma reservoir that fed the Koolau volcano (Macdonald, 1968).

370

As early as 1840 James D. Dana, the geologist with the U.S. Exploring Expedition, recognized three volcanic vents on the Mokapu peninsula, east of Kaneohe Bay: Ulupau Head, Puu Hawaiiloa, and Pyramid Rock. These vents are now believed to be among the oldest of the Honolulu Series. The evidence, as worked out by H. T. Stearns, is as follows. Ulupau Head is a tuff cone, formed by hydromagmatic eruptions. Fossiliferous marine sediments in its crater, about 70 feet above sea level, apparently were deposited during the Laie (plus-70-foot) stand of the sea, and the outer slopes of the cone appear to have been cliffed by wave erosion at the same level; hence Ulupau Head is older than the Laie stand of the sea. Ulupau tuff rests on a thick flow of nephelinite from the cinder cone, Puu Hawaiiloa, which therefore is older than Ulupau Head. A terrace at an altitude of about 100 feet on the slope of Puu Hawaiiloa suggests that the cone may have been eroded by the Kaena (plus-100-foot) stand of the sea, and the same sea may have benched the tops of adjacent Pali Kilo (which appears to be a high spot on the Hawaiiloa lava flow) and Pyramid Rock. Thus Puu Hawaiiloa probably already existed during the Kaena stand of the sea. However, both the cinder cone and lava flow appear to have formed above sea level, and therefore probably, but not certainly, before the Kaena stand, during the Kahipa (minus-300-foot) stand of the sea. It does seem certain that the eruption was earlier than the Laie stand. Nearby Pyramid Rock appears to be a deeply eroded vent in which the dike feeder has been exposed. It probably erupted even earlier than Puu Hawaiiloa.

An outcrop of nephelinite ("Mokapu basalt") midway between the Ulupau cone and Puu Hawaiiloa may be part of the lava flow from the latter vent. Mokolea Rock, in Kailua Bay, consists of massive columnar-jointed melilite nephelinite lava that must have come from a cone which has been destroyed by erosion. The double island just north of Ulupau Head, Moku Manu, is the eroded remnant of a tuff cone and associated lava flow.

Rocky Hill, just west of the Manoa campus of the University of Hawaii and north of the campus of Punahou School (fig. 253), is a cinder and spatter cone of nephelinite. A well on the south slope of the hill penetrated a layer of calcareous sand and coral fragments 27 feet above sea level, resting on lava that probably is a flow from the Rocky Hill cone. Thus Rocky Hill almost surely is older than the plus-25-foot stand of the sea. Another well, farther south, penetrated what probably is a lava flow intercalated in coral reef that Stearns believes to have formed during the plus-100-foot stand of the sea; on this basis he tentatively assigns the Rocky Hill eruption to the time of the plus-100-foot stand.

Other early eruptions of the Honolulu Series, assigned by Stearns to the Kaena and Laie stands of the sea, took place near the heads of Kalihi, Haiku, and Nuuanu valleys, near Kaneohe, and in upper Palolo Valley. A cinder cone on the divide between Kalihi and Manaiki valleys poured lava flows down both sides of the ridge into the valleys. The Kalihi flow extended all the way to the mouth of the valley. However, the lava exposed in cuts along the Lunalilo Freeway belongs to a later flow that descended Kamanaiki Valley and poured over alluvium that in turn rested on the Kalihi flow.

At least two lava flows descended Nuuanu Valley. The earlier one came from a vent on the east side of the valley near Luakaha Stream, almost directly east of the upper end of the upper reservoir. The cinder cone at the vent is nearly buried by later volcanics. The lava flow extended far beyond the mouth of Nuuanu Valley, at least as far as Iwilei, where wells near the pineapple canneries penetrate 45 feet of the lava beneath two layers of coral limestone. The lava is well exposed along Nuuanu Stream above the cemetery, where it can be seen resting on Koolau lavas and talus from the side of the valley. The upper lava flow is less extensive and has not been positively traced beyond the cemetery. It came from the well-preserved Makuku cinder cone near the west side of the valley a mile from the Pali gap. The Pali Highway cuts through the edge of the cone just south of its junction with the Old Pali Road. The lava is exposed in cuts along the Pali Highway and the Old Pali Road, and its contact

with talus can be seen beneath the link-chain protective grating on the Pali Highway just south of the Wyllie Street intersection. The grating holds back fragments of talus and the basal clinker of the lava flow along the contact which otherwise would cave out onto the highway.

At the head of Nuuanu Valley an eruption occurred at the very top of the Pali, spreading cinder which forms the conspicuous red banks at the southeast edge of the parking area at the Pali lookout. The cinder rests on old alluvium. Looking northeastward from the lookout, one can see steeply inclined layers of the cinder resting against the former face of the Pali. Below the cinder is 15 feet of lithic tuff and tuff-breccia consisting of fragments of Koolau rock, showing that the first events of the eruption were low-temperature steam explosions resulting from contact of the rising magma with ground water in the Koolau dike complex. These were followed by normal magmatic explosions of Strombolian type. The cinder formerly blanketed a more extensive area, but it has been largely removed by erosion, leaving only narrow streaks that filled former gullies in the face of the Pali. The eruption formed a cone that straddled the crest of the Koolau Range, and thin flows of lava also poured down the face of the Pali, interbedded with the cinders. The lava is exposed in a highway cut at the first prominent curve below the tunnels, and at many places along the old road farther downslope. Like many of the flows of the Honolulu Series, it contains many small inclusions of peridotite.

Near the head of Palolo Valley, also, rising magma encountered ground water and generated steam explosions. Kaau Crater (fig. 251), nearly circular and about 1,600 feet in diameter, may be a pit crater, but more probably it was blasted out by the explosions. Its walls are of Koolau lavas. Along the east fork of Pukele Stream, just south of the crater, a melilite nephelinite lava flow rests on Koolau basalt and talus, and is overlain by tuff and mudflow debris. The explosions were violent enough to form a deposit of lithic ash 10 feet thick a mile southwest of the crater. Heavy rains probably accompanied the explosions. The loose ash mantling the hillsides was washed downslope in

*371*

Figure 251. Kaau Crater, and the crest of the Koolau Range at the head of Palolo Valley, Oahu. The crater belongs to the posterosional Honolulu Volcanic Series. The sharp peak beyond the ridge is Olomana Peak, an erosional remnant within the filled caldera of the Koolau volcano.

great volume, and this, together with the ash and blocks that fell directly into the valley bottoms, tended to choke the streams and form temporary dams, which in turn were ruptured, allowing sudden floods of debris-laden water to rush seaward. At any rate, large mudflows moved down the tributaries into the main Palolo Valley. The deposits formed by the mudflows are exposed in cuts near the junction of Waiomao and Palolo valleys, and are most easily seen in a cut behind the Chinese Old Men's Home. They consist predominantly of fragments of Koolau basalt, with less abundant Kaau lava, in a fine earthy matrix. A few tree molds are found. At places the deposit is as much as 60 feet thick. For the most part it is massive, with little or no sign of bedding, but locally it grades into moderately well bedded,

stream-laid conglomerate. Toward the end of the eruption, lava rose in Kaau Crater, probably forming a lava lake, and then poured over the east, south, and west rims and flowed down the valleys of Pukele and Waiomao streams over the tuff and mudflow breccia. Slight recession of the lava in the crater at the end of the eruption left a poorly drained hollow now occupied by a swamp that sometimes contains an open pond.

Kaau Crater (fig. 251) is only one of several vents of the Honolulu Series lying along a nearly straight line that trends north-northeastward across the Koolau Range. Most prominent of these vents is Diamond Head. Others are Kaimuki, Mauumae, and Ainoni, and it requires only a slight bending of the line to extend it through the Training School and

Mokolea vents and several probable submarine cones farther north. The Ulupau and Moku Manu vents lie very close to the line, but slightly to the west. Although the vents lie on the same line, they cover a considerable range of time (table 18).

The sequence of eruptions on the southern end of the line (Mauumae-Diamond Head-Kaimuki) was well shown in the former Kapahulu Quarry. The oldest exposed rock was a pahoehoe flow of nephelinite with a crudely developed pillow structure, showing that it had entered a wet environment—either swamp or the ocean. This lava, once believed to have come from Kaau Crater, probably originated at the Mauumae vent. It was overlain by 4 to 8 feet of Diamond Head tuff containing carbonized plant remains. This, in turn, was overlain by Kaimuki basalt, which is capped by limestone of the Waimanalo stand of the sea.

Mauumae is a small cone of melilite nephelinite cinder and spatter on the lower end of the ridge just east of Palolo Valley. The cinder is visible in cuts along Sierra Drive. The original form of the cone has been largely destroyed by erosion.

Diamond Head (known to the Hawaiians as Leahi) is a tuff cone formed by hydromagmatic explosions. It consists largely of once-vitric ash and lapilli altered to palagonite, but it contains, in addition to the magmatic debris, occasional blocks of Koolau basalt and numerous fragments of coral limestone torn from the underlying reef. The tuff is cemented with calcite and zeolites, and the calcite crystals which

British sailors found there and mistook for diamonds gave the cone its modern name. That the cone was built at least partly, if not wholly, on land is shown by tree molds in the position of growth in the base of the tuff. The water which caused the steam explosions must have been shallow ground water. The layers of tuff wrap over the rim of the cone and dip both outward, and inward into the crater. A considerable width, probably about a quarter of a mile, of the seaward side of the cone has been removed by erosion (fig. 252). The southwestern rim of the cone is higher than the northeastern rim (fig. 11) because trade winds at the time of the eruption blew the ash southwestward. The bedding in the cone is very regular and symmetrical, and no gullying is found between successive layers.

The symmetry of the Diamond Head cone has led some investigators (Bishop, 1901; Wentworth, 1926) to believe that the eruption that formed it lasted only a few hours, or at most a few days. Similar symmetrical cones, however, have been built in historic times by eruptions that continued for weeks or months, such as the recent eruption of Surtsey volcano, south of Iceland. In any case there can be no question that Diamond Head was built by a single, relatively brief volcanic event.

Just southeast of Diamond Head and slightly offset from the Diamond Head-Kaau alignment is Black Point (Kupikipikio). It consists of a lava flow of alkalic olivine basalt (basanitoid) which rests on Diamond Head tuff, and is therefore younger. About 1,000 feet west of

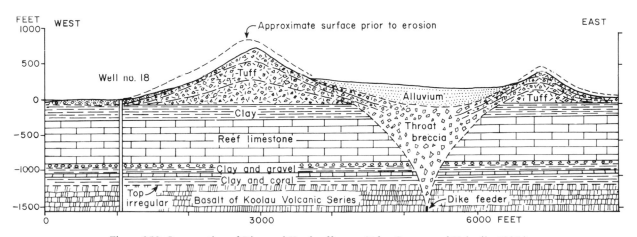

Figure 252. Cross section of Diamond Head tuff cone. (After Stearns and Vaksvik, 1935.)

373

Black Point a basaltic dike cuts limestone that appears to be the same as that which elsewhere underlies the Diamond Head tuff. Another segment of the dike is exposed 700 feet farther northwest. This dike may be the filled feeding fissure of the Black Point eruption. The surface of the Black Point lava flow slopes landward; therefore, the vent from which it was erupted must have been seaward of the present shoreline, and whatever cone may have been built at the vent has been completely removed by erosion. It is interesting to note, however, that a high magnetic anomaly just offshore probably marks the vent (A. Malahoff, personal communication, 1969). Like Diamond Head, Black Point is older than the Waimanalo stand of the sea, and remnants of the Waimanalo reef rest on the lava flow.

In cuts along Diamond Head Road near Black Point black glassy ash, some of it showing dune bedding, lies on talus from Diamond Head and is covered by similar talus. It is later than the Black Point lava and differs from it somewhat in chemical composition. It may be of about the same age as some of the black ash in the area of Round Top and Tantalus, but it also differs in composition from that ash. Its source is unknown.

Kaimuki is a small shield volcano of alkalic olivine basalt, at its center probably about 200 feet thick. Its lava flows are well exposed in cross section along the Lunalilo Freeway. At the summit of the shield a small cinder and spatter cone was built by lava fountains. The summit of the cone, just seaward from the Kaimuki fire station, is indented by a crater about 30 feet deep. As the Kaimuki shield grew it blocked the course of Palolo Stream, causing it to swing westward, around the edge of the shield, and join Manoa Stream.

On the windward side of the Koolau Range, the Ainoni vent is marked by an eroded cinder cone from which a columnar-jointed lava flow 100 feet thick and 0.4 mile wide extends northward for about 0.75 mile. The lava is exposed along Ainoni and Maunawili streams. A short distance northeast of the Ainoni cone, but offset from the Diamond Head-Kaau lineament, is the Maunawili vent, also marked by an eroded cinder cone which sits astride the Aniani Nui Ridge. A deeply weathered lava flow extends northward from the cone. Between the Maunawili and Ainoni vents, on the west side of Makawao Valley, is a mass of breccia half a mile long and a quarter of a mile wide resting unconformably on rocks of the Koolau dike complex. This Makawao breccia is composed of blocks of Koolau basalt as much as 3 feet across, and is generally unbedded. The contacts with the surrounding older rocks appear to be nearly vertical. It is regarded as the filling of a vent blasted out of the Koolau dike complex by steam explosions. Beds of vitric-crystal tuff exposed nearby in the banks of Makawao and Maunawili streams may have come from the Makawao vent during a later phase of the eruption.

The Training School vent is an eroded cinder cone low on the north flank of Olomana Peak, just behind the Girls' Training School. A lava flow from the vent extends northward for more than a mile, spreading out to form the broad, nearly flat surface on which the Olomana and Pohakupu subdivisions are located. In a cut along the Pali Highway just south of the junction with the highway to Waimanalo, the flow is exposed in cross section. There it rests on poorly bedded, unsorted mudflow debris composed largely of pyroclastic material. Apparently the cinder and ash mantling the hillsides near the vent became saturated with water, probably from heavy rains accompanying the early phases of the eruption, and mixed with underlying soil to some extent to form a slurry that flowed off the hillsides and down the valley, to be buried soon afterward by the lava flow. Beneath the mudflow debris red lateritic soil of the pre-eruption land surface can be seen.

About 2 miles to the northwest of the Training School cone, but not on the same line, is the Castle vent—an eroded cinder cone from which a dense lava flow more than 100 feet thick extended northeastward for half a mile. The flow has been largely quarried away for crushed rock. The old Kapaa Quarry was in this lava flow, but the new quarry is working in lava flows and dikes of the Koolau caldera fill.

Just east of Pearl Harbor lies a cluster of overlapping tuff cones including Aliamanu,

Makalapa, and Salt Lake craters. The Aliamanu vent is the oldest. The cones blocked the former courses of Moanalua and Halawa streams and forced the streams to make wide detours to the sea. Well-bedded, water-laid tuff exposed in road cuts on the east side of Halawa Valley is believed to be the same tuff as that forming the rim of Aliamanu Crater. It is overlain by a thick red transported soil, which in turn is covered by air-deposited tuff which can be traced to Salt Lake Crater. Beds of gravel and sand, and mudflow debris are interfingered with the Aliamanu tuff in places, and the tuff contains many leaf imprints, log molds, and fragments of partly petrified wood. The log molds lie parallel to the bedding, showing that the tree fragments had been transported and deposited with the matrix material. Near Aiea the deposits have well-defined, thin, nearly horizontal bedding, and are probably part of a delta that was growing into Pearl Harbor. Along Moanalua Stream, Aliamanu tuff and associated gravel deposits form a terrace that appears to be graded to the Kaena stand of the sea, and the Aliamanu eruption is believed to have taken place during that stand (Stearns and Vaksvik, 1935, p. 109). No lava flow has been found associated with the Aliamanu tuff outside the cone, but within the crater a well penetrated 30 feet of dense melilite nephelinite beneath 62 feet of Salt Lake tuff. The lava was probably poured into the crater in the last stage of the Aliamanu eruption.

The tuffs from Salt Lake and Makalapa craters are similar both in appearance and in their relationships to other rocks, and generally they have not been separated during field studies. The two eruptions probably took place at approximately the same time. In cuts along Puuloa Road near the junction with Moanalua Road the Salt Lake tuff can be seen resting on red soil on the top of the Aliamanu tuff. A small fault offsets the contact and drops the rock to the north several feet in relation to that to the south. The Salt Lake tuff is air laid, and contains in its base many molds of trees that were buried in a standing position. The plants include loulu palm (*Pritchardia* sp.) and tree ferns, which indicates that the local climate at

the time of the eruption was considerably wetter than it is now. The air-laid tuffs extend below present sea level and therefore must have been deposited during a lower stand of the sea, probably the Waipio stand. The tuff is overlain by limestone of the Waimanalo stand.

Salt Lake, once 0.9 miles across, is now being filled and the area will become a golf course. It occupies the crater of the tuff cone, and formerly was fed partly by springs along its northern edge. In 1840 J. D. Dana was told that the lake was very deep, and at his request Hawaiian friends carried a canoe to the lake so that he could sound it. He found that its greatest depth was 16 inches; in 1841 it was only 6 inches! Evaporation was rapid from this broad, very shallow pool, and salt dissolved in the spring water gradually formed a brine so concentrated that salt deposits were precipitated around the shore. In 1910 an artesian well was drilled in the crater and allowed to flow into the lake to increase the depth of the water and reduce its salinity so that it could be used as a mullet pond. The depth of the lake was controlled by means of a drainage tunnel through the southeast rim of the cone.

Punchbowl Hill (Puowaina) is a tuff cone near the center of Honolulu built against the end of a spur of the Koolau Range. Like that of Diamond Head, the tuff is mostly brown and consists largely of palagonitized vitric ash and lapilli with scattered fragments of coral limestone and Koolau basalt, and a few, nearly spherical, dense bombs with rough, spinose surfaces. The contact between the tuff and underlying Koolau basalt is exposed in cuts along Auwaiolimu Street near the Tantalus Drive overpass. Some of the latest ash is gray instead of brown, and close to the crater it grades into cinder and spatter. Most of the cone-building eruptions were phreatomagmatic, as the rising magma encountered ground water close to the surface, but the last eruptions were purely magmatic and did not involve large amounts of steam. Toward the end, lava rose into the crater and formed a pool, but apparently it did not overflow the rim. Near the south and east rims, remnants of the lava formerly could be seen resting on spatter. The lava pool probably was drained through fissures

375

that opened in the flank of the cone. Two dikes of melilite nephelinite, one 3 feet and the other 12 feet thick, cut the southeastern side of the cone. One of them can be seen near the Board of Water Supply reservoir on Alapai Street. A lava flow of the same composition was exposed in cuts along the Lunalilo Freeway, and was also encountered in wells near the Board of Water Supply building on Beretania Street. Punchbowl tuff is overlain by limestone of the Waimanalo stand of the sea and therefore cannot be later than that stand. Other evidence suggests that it probably was formed during the Waipio stand.

Much later than the main mass of Punchbowl are small patches of black glassy ash that lie here and there on its flanks. One of the patches can be seen at the back of the playground at Robert Louis Stevenson School. They fill erosional gullies, and commonly are separated from the older tuff by red soil. Early workers attributed this ash to a new eruption of Punchbowl. Stearns and Vaksvik (1935, p. 147) have pointed out, however, that the ash does not thicken or become coarser toward the crater, as would be the case if Punchbowl were its source. Instead, with little question it is merely a part of the wide-flung blanket of ash that was formed by eruptions of Sugarloaf and Tantalus.

Round Top (Puu Ualakaa), Sugarloaf (Puu Kakea), and Tantalus (Puu Ohia) mark a row of vents along the ridge of the Koolau Range back of the city west of Manoa Valley. All three are cinder cones formed by Strombolian-type eruptions. Round Top may be somewhat older than the other two. No crater is preserved in it, but whether the crater was filled by the last eruptions or was destroyed by erosion is not certain. The cone is largely mantled with black ash from the later vents. Brown to gray, fine, well-bedded cinder is exposed along the road in Ualakaa Park, and, in the quarry at the hairpin bend in Round Top Drive, just north of the park, coarse cinder contains many nearly spherical, dense balls that appear to be spherical bombs smoothed by repeated tossing and milling in the vent. In the valley of Makiki Stream two dikes of nephelinite, found by Horace Winchell, trend toward Round Top and

apparently represent the feeding fissures for the eruption.

Sugarloaf is a double cone built around two craters, and a well-formed crater indents the summit of Tantalus. Both cones appear very young, and their products are intermingled and generally indistinguishable. In addition to ordinary cinder, irregular and ribbon bombs, and lapilli, both contain roughly spherical balls up to about 4 inches in diameter, some with a rough spiny surface and others smoothed like those at Round Top. Both vents produced lava flows. The flow from Sugarloaf issued on the east side of the main cone (fig. 253), about

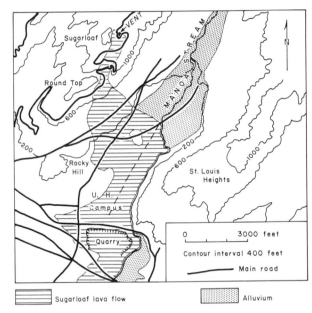

Figure 253. Simplified geologic map of Manoa Valley and surroundings, showing the lava flow from Sugarloaf (also called the Moiliili flow). The lower course of Manoa Stream has been pushed over against the eastern wall of the valley by the lava flow. The approximate former course of the stream is shown by the dashed line. (Adapted from Stearns, 1939.)

10,000 to 20,000 years ago, and cascaded down the wall of Manoa Valley, piling up in great heaps of clinker against the base of the wall. On the valley floor it spread out to form the broad, nearly level surface on which the University of Hawaii is built. The old Moiliili Quarry, at the seaward edge of the University campus, was excavated in the dense center of this 40-foot-thick aa flow of melilite nephelinite. The rock in the quarry walls is unusually coarse grained for a relatively thin lava flow, and contains many veinlets of pegmatoid.

Openings are commonly lined with well-formed crystals of nepheline, augite, apatite, and zeolite, some of them a quarter of an inch long. The lava must have been unusually rich in gas. The general terminus of the flow was formerly marked by a 40-foot escarpment a short distance north of the present Lunalilo Freeway, but beyond this a narrow tongue continued southward another third of a mile through the present Moiliili. Near the mouth of Manoa Valley the lava rests on limestone of the Waimanalo stand of the sea. Although the surface of the flow has now been much modified by construction, in general it has been very little eroded. In the vicinity of Dole Street, however, Manoa Stream has cut a steep-sided gorge about 30 feet deep below the original surface of the flow.

The Sugarloaf flow profoundly altered the drainage of Manoa Valley. In spreading over the valley floor it forced Manoa Stream out of its former channel near the midline of the valley and caused it to swing far eastward, where it now follows the boundary between the flow and the ridge of Koolau rock that bounds the valley on the east (fig. 253). The flow raised the level of the floor of the valley near its mouth, which reduced the gradient of Manoa Stream and caused it to deposit alluvium in the upper part of the valley, building up the valley floor upstream from the lava flow until it is now in essentially continuous slope with the lava-covered portion at the mouth.

The lava flow from Tantalus spilled westward, into upper Pauoa Valley. The abrupt humpy headward termination of the main valley is clearly visible from Punchbowl or from Auwaiolimu Street. This termination is the edge of a mass of lava 500 feet or more thick that filled the upper end of the valley. The flow extends on down the valley for about 1.5 miles. At its mountainward end the top of the flow came within less than 100 feet of the level of a low part of the ridge between it and Nuuanu Valley. The nearly uneroded surface of this upper part of the flow forms the Pauoa Flats. Pauoa Stream has been crowded against the western side of the valley. Like the Sugarloaf flow, the Tantalus flow cannot be more than a few tens of thousands of years old.

The most conspicuous row of vents of the Honolulu Series extends across the eastern end of the island (fig. 254). In order northeastward, the vents are: Koko Head, Hanauma Bay, Kahauloa ("Rifle range") Crater, Koko Crater, the Kalama cinder cone, the Kaupo vent, Kaohikaipu Island, and Manana (Rabbit) Island (fig. 255). Collectively, the cones and lava flows are known as the volcanics of the Koko fissure. A submarine ridge extending 3 miles southwestward from Koko Head shows that this series of eruptions continued beneath the sea (fig. 230).

Manana Island is a cone of palagonite tuff built around two vents, each marked by a crater. The cone is somewhat gullied by erosion and partly mantled by colluvium. A bench, in places more than 150 feet wide, has been cut around it by waves of the Manana (plus-5-foot) stand of the sea. Tuff from the Manana eruption lies on coral reef of the Waimanalo stand on the mainland of Oahu about a mile southwest of the cone, and overlies gravel in a cave cut by the Waimanalo sea now about 20 feet above present sea level in the cliff almost directly below the overlook point where the highway crosses the crest. Thus the Manana eruption occurred between the plus-25 and plus-5-foot stands of the sea.

Kaohikaipu Island is a cone of red to black cinders and spatter, with a pahoehoe lava flow of alkalic olivine basalt on its western side. Since cinder from the Kaohikaipu vent overlies the tuff on Manana Island, the Kaohikaipu eruption must be the younger. It may have occurred at the same time as the Kaupo eruption, which produced a lava flow of similar composition. The fact that the Kaohikaipu eruption built a normal cinder cone instead of a tuff cone suggests that the eruption took place on dry land when sea level was lower than it is now, probably during the Mamala-Kahipa or Penguin Bank stands of the sea (see table 12).

The Kaupo flow issued on the talus at the foot of the cliff, about 200 feet above sea level. It poured downslope into the ocean, building the present Kaupo peninsula, on which Sea Life Park is located. The vent is marked only by a small mound of spatter. The surface of the

Figure 254. Cones of the Honolulu Volcanic Series along the Koko fissure, southeastern Oahu. Beginning at the lower right and extending northeastward across the end of the island are the tuff cones of Koko Head, Hanauma Bay, and Koko Crater, and in the distance are Kaohikaipu and Manana (Rabbit) islands. Kahauloa Crater, between Hanauma Bay and Koko Crater, and the cinder cone in Kalama Valley beyond Koko Crater, are barely visible.

lava flow is almost untouched by erosion and has a very youthful appearance. Although the lack of alteration of the flow surface undoubtedly is partly due to the very dry local climate, the flow may be even younger than the Sugarloaf flow; it is certainly one of the youngest on Oahu.

The Kalama eruption built a cinder cone about 35 feet high with a crater 50 feet deep at its summit. A ridge of spatter and cinder built along the eruptive fissure extended north-northeastward from it. The cone was located on the floor of Kalama Valley a mile from the present shoreline, but it has now been almost entirely quarried away for road material. From the cone a lava flow, largely of aa, extends down the valley to the sea, forming the shoreline just east of Sandy Beach. Its surface is partly covered with alluvium. It is certainly very young, although it looks somewhat older than the Kaupo flow.

The products of the vents along the southern end of the Koko fissure are so intermingled that they are difficult to separate. All of the eruptions were hydromagmatic and built cones of vitric ash, now consolidated into tuff.

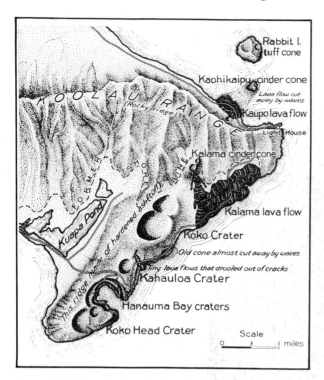

Figure 255. Map showing the location of volcanic vents and lava flows along the Koko fissure, southeastern end of Oahu. (From Stearns, 1939.)

Alteration of the original glass to palagonite has been discussed in chapter 8. Many fragments of coral limestone and Koolau basalt, torn from the underlying rocks by the explosions, are scattered through the tuff. Some of the blocks falling in the soft ash caused bomb sags—downbowing of the ash layers beneath them. Some beds in the cone surrounding Hanauma Bay are especially rich in fragments of limestone and must have come from explosions in the reef itself. One such bed is exposed in the highway cut just east of the junction with the road to the bay. However, at least some of the cones of this group were built on land. Along the southern flank of Koko Crater, 15 feet above sea level, the tuff rests on the soil-covered surface of limestone believed to have been formed during the Waimanalo stand of the sea. Augite phenocrysts are common in the tuff, and are especially abundant in the ridge between Hanauma and Kahauloa craters, where they have been weathered out and can be found scattered loose on the ground.

Koko Head is a tuff cone with two craters on its eastern side. A thin lava flow is interbedded with the Koko Head tuff on the southwest side of Hanauma Bay. The Koko Head cone appears to have overlapped the one surrounding Hanauma Bay. The crater of the latter is now open to the sea. Its southwestern rim appears to have been low originally, perhaps because some of the ash falling on the seaward side during the eruption was carried away by waves and currents, but the present breach in the cone probably is the result of later wave erosion. The Kahauloa cone in turn appears to have been overlapped by the Hanauma cone.

The best preserved, and probably the latest, of the cones is that of Koko Crater (fig. 256). It is horseshoe shaped, open to the northeast, with its highest point (1,204 feet) on the southwestern rim. Like Diamond Head, it was built during a period of trade winds. And like Koko Head, it was built around two vents. A smaller crater lies just to the northeast of the main crater. The prominent dark-colored ledge at the summit of the cone, clearly visible from the parking area at the Halona Blowhole, is the result of erosion cutting through the hard shell of silicified tuff into the softer tuff beneath. In the highway cut just northeast of the blowhole several beds of tuff-breccia consist largely of angular blocks of Koolau basalt, up to a foot in diameter, which were thrown out by hydroexplosions within the Koolau rocks beneath the growing cone. Cracking of the surrounding rocks probably allowed surges of ground water to flow into the hot conduit during pauses in the magmatic explosions. A small remnant of the rim of an older cone forms a detached ridge along the lower southeastern flank of Koko Crater just west of the blowhole.

Probably related to the eruption that built Koko Crater is a series of outbreaks along a northeast-trending fissure that formed a series of small spatter cones and lava flows. One of these is on the northeast side of Koko Crater (fig. 255), and another is on the northeast rim of Kahauloa Crater. The lava from the latter vent flowed seaward over the weathered surface of the tuff. In the highway cut the lava rests on 2 to 4 feet of wind-drifted black sand, and is overlain by several feet of tuff, showing that the flow occurred before the end of the Koko Crater eruption. Another lava flow fol-

*379*

Figure 256. Koko Crater is a compound tuff cone of the Honolulu Volcanic Series formed by eruptions along the northeast-trending Koko fissure near the southeastern tip of Oahu 2 million or more years after the building of the principal Koolau shield volcano. The cone is double, the smaller crater in the foreground being slightly younger than the larger crater. Part of a still earlier cone is hidden behind the ridge on the left. The far side of the cone is highest because ash was blown in that direction during the eruption. The resistant capping on the left rim of the cone has been cemented during palagonitization.

lowed the depression between the Hanauma and Kahauloa cones, and still another poured down the eroded wall of the Hanauma crater into the bay. The path around the eastern side of Hanauma Bay just above water level passes through a natural arch eroded under this lava flow. Erosion of the surrounding tuff has exposed the dike feeder of the flow in the crater wall.

Could eruptions again take place on Oahu? Actually, the period of time which has elapsed since the last eruption is very short in terms of geologic processes—probably shorter than the periods between some of the eruptions in the Honolulu Series. This suggests that future eruptions are indeed possible, although the likelihood that any of us will see one is very small.

On May 22, 1956, a disturbance of some sort took place in the channel between Oahu and Kauai (Macdonald, 1959, pp. 66–68). A patch of yellowish brown material, about 0.25 mile across, appeared in the water, and by the next day it had been drawn out into a streak several miles long. It has generally been interpreted as a discoloration caused by copious flowering of algae. However, pilots flying over at low altitude reported a sulfurous odor, two dead whales were seen floating in it, and the captain of a fishing boat reported floating fragments that looked to him like pumice. Unfortunately, he did not bring any of the pumice-like material to shore; but a few days later pieces of basaltic pumice were found on the beaches at Lanikai and Kahana Bay. The event may have been a submarine eruption.

# Kauai

Kauai consists essentially of a single great shield volcano, deeply eroded, and partly veneered with much later volcanics (fig. 257). The shield has a volume of about 1,000 cubic miles and rises 17,000 feet above the surrounding sea floor. At the top of the shield was a caldera 10 to 12 miles across—the largest in the Hawaiian Islands. The southern flank of the shield collapsed to form a fault-bounded trough, the Makaweli graben, or depression (fig. 260), some 4 miles wide. Lavas erupted in the caldera gradually filled it, except on the

higher northwestern side (fig. 261), and eventually spilled over its low southern rim into the graben, down which they flowed into the sea. Another, much smaller caldera, about 2 miles long, formed on the southeastern flank of the shield. This so-called Haupu caldera also was filled with lava flows. Probably at a somewhat later time another collapse occurred, on the eastern flank of the shield, forming a subcircular caldera 7 to 10 miles wide known as the Lihue basin, or depression (fig. 260). There is no evidence that lavas of the Kauai shield volcano ever poured into the Lihue basin, but at a much later date it was largely filled with posterosional volcanics. The Haupu caldera and Lihue basin are the only flank calderas known on the Hawaiian volcanoes.

Place names mentioned in this section can be found on the location map of Kauai (fig. 258).

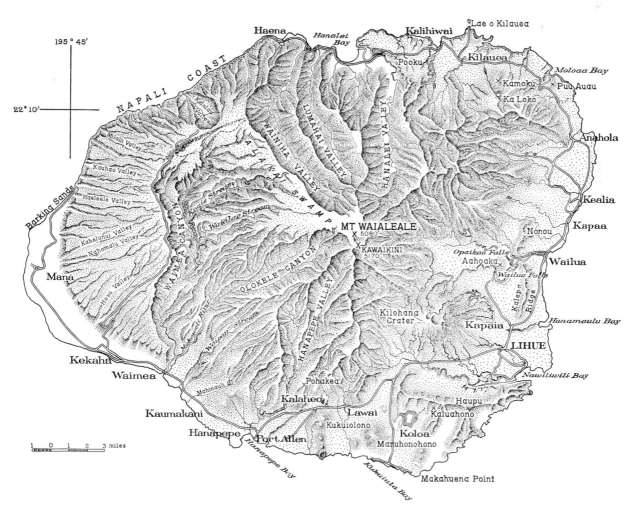

Figure 257. Shaded relief map of the island of Kauai. (From Macdonald, Davis, and Cox, 1960.)

381

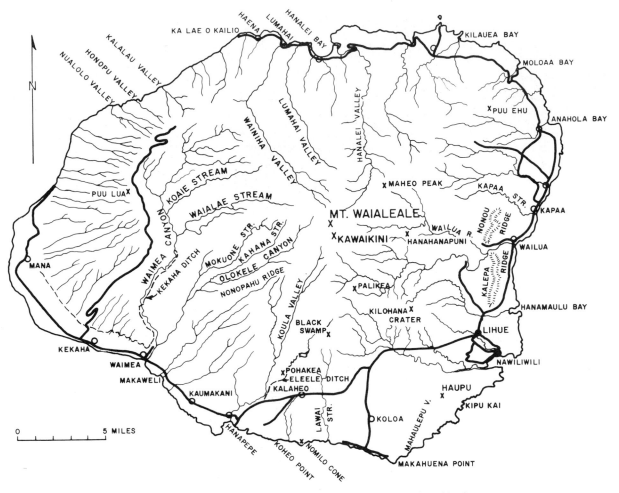

Figure 258. Map of Kauai, showing the location of places mentioned in the text.

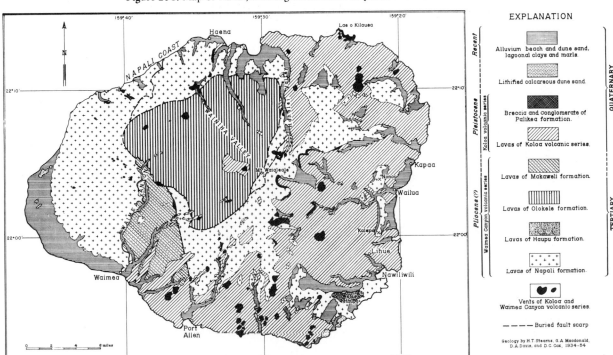

EXPLANATION

Alluvium beach and dune sand, lagoonal clays and marls.

Lithified calcareous dune sand.

Breccia and conglomerate of Palikea formation.

Lavas of Koloa volcanic series.

Lavas of Makaweli formation.

Lavas of Olokele formation.

Lavas of Haupu formation.

Lavas of Napali formation.

Vents of Koloa and Waimea Canyon volcanic series.

– – – – Buried fault scarp

Geology by H.T.Stearns, G.A.Macdonald, D.A.Davis, and D.C.Cox, 1934–54

Figure 259. Geologic map of the island of Kauai. (From Macdonald, Davis, and Cox, 1960.)

Figure 260. Model of the island of Kauai viewed from the south, showing the broad summit area and the relatively little dissected western slope, the manner in which the Waimea Canyon (at the left) cuts diagonally across the slope of the shield, the Makaweli graben to the right of Waimea Canyon and the remarkably circular form of the Lihue basin, or depression, near the eastern (right-hand) edge of the island.

383

Figure 261. Waimea Canyon in the left distance and Kalalau Valley on the right are extending their heads toward each other and will eventually join by cutting through the intervening portion of the Kauai shield. A section of Wainiha Valley appears in the foreground. Of particular interest is the low, saucer-like scarp that swings around the head of Kalalau Valley directly below the tiny single cloud, and may be traced by eye in both directions. This scarp marks the boundary of the caldera, which was not quite filled on this northwestern side. Note the difference in the effects of erosion on the slope lavas to the right of the scarp as compared with those on the caldera-filling lavas to the left.

Figure 262. Napali cliffs, Kauai. The combined erosive power of running water and waves has fashioned a rugged and spectacular coastline on the northwestern perimeter of Kauai. The thin-bedded lavas belong to the Napali Formation, on the outer slopes of the shield.

WAIMEA CANYON VOLCANIC SERIES

The rocks of the Kauai shield volcano are named the Waimea Canyon Volcanic Series (fig. 259), and the portion of them that built the main mass of the shield outside the caldera are called the Napali Formation (Macdonald, Davis, and Cox, 1960) because of their excellent exposures along the Napali Coast of northwestern Kauai (figs. 262, 263). The Napali Formation consists of thin flows of tholeiitic basalt, olivine basalt, and oceanite pahoehoe and aa sloping gently outward in all directions from the summit area. Because of the great amount of erosion of the surface of the shield, very few vents of the Napali Formation are preserved, although the lavas are cut by hundreds of dikes, most of which probably fed flows. Near the top of the west wall of Waimea Canyon, 0.3 mile south of the lookout point, a pit crater a quarter of a mile across and nearly 300 feet deep in the Napali lavas is filled with breccia cut by dikes. It probably was buried by Napali lavas which have been removed by erosion. Masses of breccia near the head of Nualolo Valley, and in the southeast wall of Honopu Valley 0.3 to 0.5 mile above its mouth, also may be fillings of pit craters.

Puu Lua, just west of the rim of Waimea Canyon, is the eroded remnant of a crater filled with talus breccia which is capped by a thick massive lava flow much like that which fills the eastern end of Makaopuhi Crater on Kilauea volcano. The massive crater fill is more resistant to erosion than are the surrounding thin-

384

bedded lavas, and removal of the surrounding rocks has left it standing 200 feet above the ground around it. Puu ka Pele, at the rim of Waimea Canyon a mile southeast of Puu Lua, also appears to be a remnant of a resistant, crater-filling lava flow that has been left standing in high relief by the erosion of the less resistant rocks around it. On the southwest wall of Hoolulu Valley, 3.5 miles southwest of Haena, a small segment of a cinder cone buried by Napali lavas is visible in the cliff above the trail. Thin films of ash and red ashy soil are found here and there between the lava flows, but ash beds more than a few millimeters thick are rare.

The rift zones of the Kauai shield are less clearly marked than those of most Hawaiian volcanoes, and there is a tendency for dikes to radiate in all directions from the summit. Nevertheless, two zones of unusually numerous dikes can be recognized, trending respectively northeastward and west-southwestward (fig. 173). The latter rift zone is the better developed. Dozens of its dikes, from a few inches to 40 feet thick, are well exposed in the walls of Waimea Canyon, but they are far less abundant than those in the dike complexes of Oahu. Along the Napali Coast and in the Kalepa-Nonou Ridge, that bounds the eastern side of the Lihue Basin, most of the dikes are radial, although some trend roughly tangentially to the shield and dip inward toward the caldera at angles averaging as low as 45°. These may be cone sheets, intruded into conical fractures

Figure 263. Napali cliffs, Kauai. This view shows exceptionally well the effects of tropical weathering and erosion by running water, producing the amphitheater-headed valley on the right and the very thin ridges remaining between the giant vertical grooves on the sides of the valleys. The lower portion of Kalalau Valley, partly filled with alluvium, is on the left.

Figure 264. A close view of the walls of Poomau Canyon, a tributary of Waimea Canyon draining from the Alakai Swamp, shows the horizontal flows of the Olokele Formation formed within the caldera. Note the V shape of the youthful canyon.

diverging upward from the top of the magma chamber in the shield, and formed as a result of upward thrust by the magma. However, they cannot be followed far enough to establish their arcuate form. Cone sheets, abundant in some other volcanic districts of the world, such as those of western Scotland, are rare in Hawaii.

The lavas that filled the main caldera make up the Olokele Formation, and those that accumulated in the Makaweli graben are called the Makaweli Formation (Macdonald, Davis, and Cox, 1960). Both of these formations are predominantly tholeiitic, but alkalic olivine basalt and hawaiite appear among the latest flows. The lavas of the Olokele Formation are thick, massive, and essentially horizontal (fig. 264) because they were ponded in the caldera. At several places they can be seen to rest against talus breccia which had accumulated at the foot of the cliffs bounding the caldera. The best such exposure is in Olokele Canyon 2 miles above its junction with Kahana Valley.

The boundary is well exposed also in the wall of Mokuone Valley (fig. 265). The Makaweli lavas are thinner than the caldera-filling flows, but they are thicker and flatter than the Napali lavas because they too were partly confined within a depression with a gently sloping floor. Interbedded occasionally with the Makaweli lavas are lenses of stream-deposited conglom-

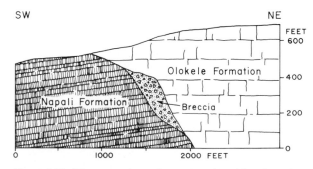

Figure 265. Diagrammatic section across the caldera boundary in the northwest wall of Mokuone Valley, Kauai. The thick horizontal beds of the Olokele Formation, formed in the caldera, rest against the truncated edges of the thin inclined beds of the Napali Formation. (After Macdonald, Davis, and Cox, 1960.)

386

erate, and along the eroded fault scarps at the edges of the graben the lavas rest on talus breccia. This sedimentary material within the Makaweli Formation is known as the Mokuone Member. In places at the upper end of the graben, the Makaweli lavas rest on, or are interfingered with, thick accumulations of cinder and ash blown from lava fountains in the caldera or washed from the caldera region by streams, and banked against the head wall of the graben. Elsewhere, as along the lower portion of Waimea Canyon, layers of ash as much as 3 feet thick lie between the Makaweli lavas, and 0.1 mile west of the intake of the Kekaha ditch, lava rests on 2 to 4 inches of baked red tuffaceous soil. On Nonopahu Ridge, at the south side of Olokele Canyon, a flow of platy basaltic hawaiite rests on 1 to 2 feet of red soil. Clearly, by the time of formation of the Makaweli lavas the tempo of eruption was slowing.

Perhaps at about the same time that the caldera formed at the summit of the shield the smaller Haupu caldera appeared on the southeastern flank. That this lower caldera did not mark the vent of an independent volcano is shown by the fact that the lava beds around it do not slope away from it, but continue to slope away from the center of the island. Lava flows accumulated in the Haupu caldera, burying taluses and forming thick massive beds like those of the main caldera fill. Haupu peak ("Hoary Head") is formed of these resistant lavas (fig. 266), exposed by erosion, and the rocks that filled the caldera are known as the Haupu Formation.

### KOLOA VOLCANIC SERIES

Potassium-argon determinations show the youngest lavas of the Waimea Canyon Volcanic Series to be more than 3 million years old, and the oldest of the posterosional lavas to be less than 1.5 million (table 17). Thus, for more

Figure 266. Haupu peak and the Haupu Ridge, southeastern Kauai, with the upper end of the Menehune Fishpond in the foreground. Haupu consists of massive, nearly horizontal lavas accumulated in a small caldera on the flank of the main Kauai shield volcano, and stands up in relief because it is more resistant to erosion than the lavas around it. The flat, cane-covered surface in the foreground is on posterosional Koloa lavas.

387

Figure 267. Koloa, Kauai. The circular cinder-and-spatter cones mark vents that poured forth lavas of the Koloa Volcanic Series, inundating much of the lower slopes of southeastern Kauai.

than 1.5 million years the Kauai volcano lay quiet, while erosion bit deeply into the shield. But activity returned, and new volcanic rocks were spread over large areas of the eroded surface. The products of this period of renewed volcanism, together with closely associated sedimentary rocks, are known as the Koloa Volcanic Series (Hinds, 1930; Stearns, 1946). Like the posterosional rocks of Oahu, the Koloa volcanics range from alkalic olivine basalt through basanites to nephelinites and melilite nephelinites.

About 40 vents of the Koloa Volcanic Series (fig. 267) have been recognized, scattered widely over the eastern two-thirds of Kauai. Others must have existed to feed the lava flows that at several places poured down from the central uplands, but they either are hidden in the dense vegetation or have been destroyed by erosion. Many of the vents are aligned in roughly north-south rows, and, as on Oahu, the new rifts show no relation to the older rifts of the shield volcano. The distribution of Koloa vents is shown in figure 268.

The cones built at the vents range from small spatter cones characteristic of Hawaiian-type eruptions to large cinder cones formed by Strombolian-type eruptions. Many are of mixed cinder and spatter. Puu Wanawana, 0.75 mile north of Makahuena Point on the southeastern coast, is a spatter cone which contains a small amount of cinder and a few spindle bombs. Puu Hunihuni, half a mile farther inland, consists largely of cinder, which is being quarried for use as road metal. Puu Hi, 2 miles east-northeast of Makahuena Point, is composed largely of highly scoriaceous layers of lava.

Nomilo cone, at the coast south of Kalaheo, is composed principally of cinder, partly cemented by calcium carbonate. Its crater holds a lake fed by springs, the level of the lake being controlled by a man-made tunnel at sea level through the seaward side of the cone. The lake has been used as a fish pond. According to local tradition the water of Nomilo pond becomes roiled and muddy just before volcanic eruptions on the island of Hawaii, but this has not been confirmed by scientific observers.

Puulani, 2.5 miles north of Hanapepe, is a cinder cone partly veneered by a thin flow of lava. Pohakea, 3.7 miles northeast of Hanapepe, is a large cinder cone partly buried by lavas that came from a vent in Black Swamp, to the northeast. Another large cinder cone is Hanahanapuni, in the upper drainage basin of the Wailua River. Kilohana Crater is a collapse crater at the summit of a small shield volcano that fills most of the southern part of the

Lihue basin. Just south of Kalaheo, Kukui o Lono Park lies at the summit of another small shield. Other small shields include Manuhonohono hill, a mile west of Koloa; Puu o Papai and Manieula, northeast of Kaumakani; and Puu Auau, just south of Moloaa Bay.

The northwest side of Kilauea Bay, on the northeastern coast of Kauai, is a much-eroded tuff cone overlain by a cap of thin-bedded, spattery lava. It is the only tuff cone on Kauai. The tuff is palagonitized and contains many bombs and blocks of melilite nephelinite and fragments of reef limestone. Near the head of Kilauea Bay the base of the cone consists of coarse, poorly sorted breccia formed by the first violent steam explosions. Just seaward of the main cone, Makapili rock is a tombolo, a former island tied to the mainland by a sand bar. It is perforated by a sea arch (fig. 269).

Lava flows of the Koloa Series cover about half the surface of the eastern part of the island

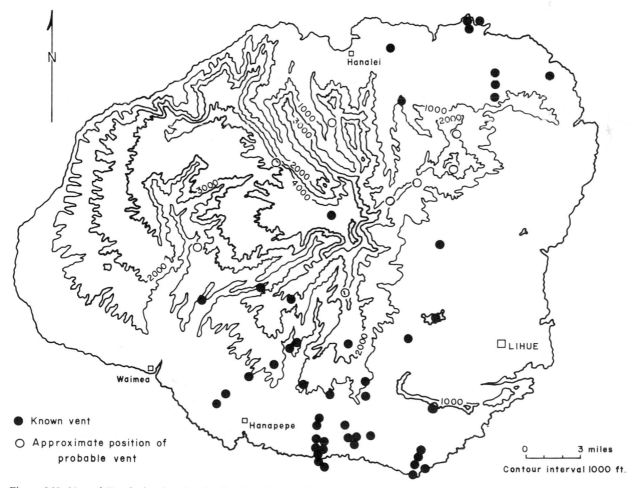

Figure 268. Map of Kauai, showing the distribution of vents (*black dots*) of the Koloa Volcanic Series. The vents show a conspicuous tendency toward a nearly north-south alignment. (After Macdonald, Davis, and Cox, 1960.)

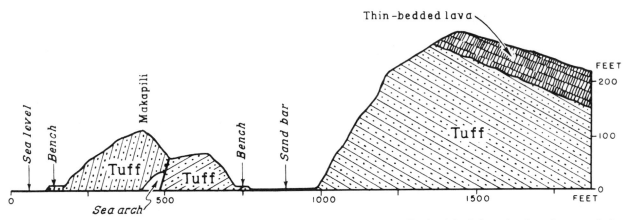

Figure 269. Diagrammatic section through the Kilauea tuff cone, Kauai. Makapili Island is tied to the shore by a tombolo. Note the unconformable relationship of the lavas capping the cone to the underlying tuff. (From Macdonald, Davis, and Cox, 1960.)

(fig. 259). They form the entire floor of the Lihue basin (fig. 270) except for two small kipukas of Waimea Canyon rocks (Aahoaka hill and Puu Pilo) that protrude through them west of the gap through which the Wailua River crosses the Kalepa-Nonou Ridge. The Koloa lavas form broad aprons south of the Waimea Canyon rocks from east of Koloa to Waimea Canyon, and along the northeastern coast of the island. Much of the upper part of Hanalei Valley is floored by these lavas, and they are exposed as remnants of former valley fills along Waimea, Olokele, Hanapepe, and Wainiha can-

yons. A patch of Koloa lavas more than 2 miles long and a mile wide lies almost on the summit of the Kauai shield, just west of Mt. Waialeale.

The greatest exposed thickness of Koloa lavas is 2,100 feet, in the east wall of Hanalei Valley; but they may be even thicker in the Lihue basin and along the southern edge of the island, where their base in not exposed. Individual flows that were confined within valleys may be several hundred feet thick. The residual mass of the latest flow to descend Hanapepe Valley is 600 feet thick, and the flow in Olokele Canyon reaches 1,000 feet.

Figure 270. Haupu Ridge, in the foreground, borders the Lihue plain, which is formed on posterosional Koloa lavas. The Lihue basin appears to be an eroded collapse caldera on the flank of the Kauai shield. Kalepa Ridge, in the middle distance, is a remnant of the old shield surrounded by later Koloa flows.

Both pahoehoe and aa flows are present, and in the dry area near Koloa their surfaces are so well preserved that they appear very recent. Both the Waimanalo shoreline and dunes formed during the Waipio stand of the sea are preserved on them, however, and hence they cannot be more recent than late Pleistocene.

Some of the flows, particularly those accumulated in valleys, have excellent columnar jointing. An especially good exposure can be seen in the bluff just behind the dock at Nawiliwili. Log molds are common in the lavas, though no molds of standing trees have been found. The largest known mold is 30 feet above water level in the north bank of the South Fork of the Wailua River, a mile upstream from its junction with the North Fork. It is 35 feet long, 4 to 5 feet in diameter at its larger end, and 3 feet at its smaller end. Several small branch molds diverge from the trunk. The mold is nearly horizontal, and is enclosed in pillow lava. Hence it probably was formed by lava which surrounded a log lying on the wet bed of the river.

Pillow lavas—spheroidal or ball-like masses which develop when pahoehoe lava flows into water or over very wet ground—are found at several places. In the lowest of the flows that descended Waimea Canyon they are well exposed at the Menehune Ditch, a mile upstream from Waimea village. The pillows there are nearly spherical and range from less than one foot to 10 feet in diameter. Each shows weakly to moderately developed radial jointing, and is enclosed in a skin of black glass 1 to 6 millimeters thick. Angular interstices between the pillows are partly open and partly filled with white to gray marl or marly mud, squeezed up into them from the soft muddy floor of the estuary over which the flow moved. The pillow lava is overlain by a few inches of palagonitized vitric ash ("hyaloclastite") probably formed by littoral explosions where the hot lava came in contact with water. Another excellent exposure is in the base of Wailua Falls, where individual pillows are as much as 15 feet in diameter. The flow occupies a former valley cut into mudflow deposits of the Palikea Formation (described later). Stream gravel in the bottom of the ancient valley is overlain by a few inches of mudstone and peaty material, followed by 2 inches of sandy ash overlain by 1 to 4 feet of thinly laminated fine water-laid ash, and this in turn is overlain by the pillow lava. Commonly the underlying muddy ash has been squeezed up between the pillows. The pillowed zone is 10 to 25 feet thick and passes upward into dense lava without pillows. The upper part of the flow, overriding the earlier pillowed part, was not affected by water.

At any one place Koloa volcanism was not continuous. Quiet periods between eruptions were frequent and of various lengths. Some were short and are marked only by thin layers of weathered ash between the flows. Vitric ash weathers quickly in the warm humid climate of the Kauai lowlands, and a few feet of soil may be formed on ash in only a few hundred years. Other quiet periods were longer, permitting the gullying of the surface by streams and weathering of the rocks to depths of several feet; and still others were so long that canyons several hundred feet deep were cut before the next flow buried the area.

Evidence of pauses in Koloa volcanic activity is most easily seen in highway cuts on the south side of the island. On the ridge 0.3 mile west of Lawai Stream a cut exposes 35 feet of poorly bedded and poorly sorted conglomerate banked against a cliff that truncated Koloa lava flows. Both the lava and the conglomerate are beveled by an erosional surface on which 6 inches to 2 feet of red soil had formed before it was buried by another flow (fig. 271). Unconformities can be seen at the coast also. About 0.3 mile east of Koheo Point a sea cliff 50 feet high was cut across Koloa lavas, and small pockets of soil up to about 2 inches thick were formed on the slightly eroded upper surface of the lavas before later flows buried the soil and cascaded over the sea cliff (fig. 272).

Where the highway crosses Hanamaulu Gulch, a mile north of Lihue, the upper 25 feet of lava rests on an ancient soil zone 6 feet thick which grades downward into much-decomposed Koloa lava at least 30 feet thick. Even in a favorable climate such deep weathering must have required a very long time.

In the Koula branch of Hanapepe Valley a

*391*

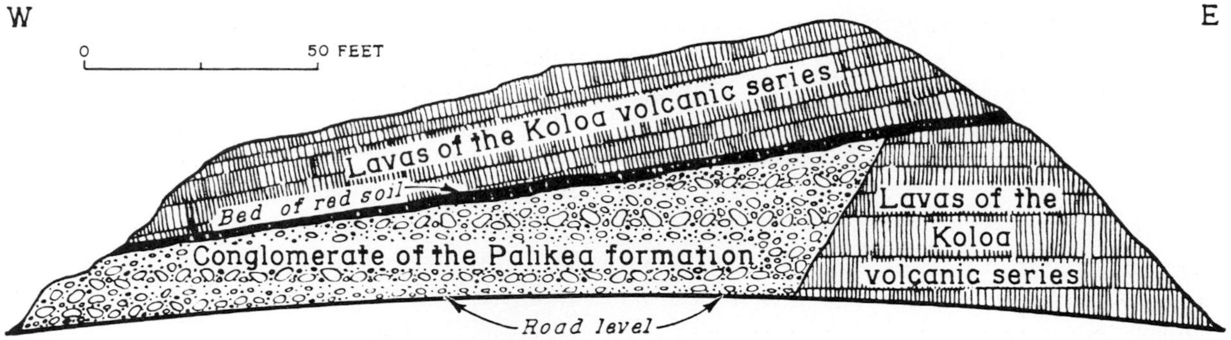

W                                                      E

0        50 FEET

Lavas of the Koloa volcanic series

Bed of red soil

Conglomerate of the Palikea formation

Lavas of the Koloa volcanic series

Road level

Figure 271. Sketch showing the relationships of Koloa lava flows and conglomerate in a highway cut 0.3 mile west of Lawai Stream, Kauai. (From Macdonald, Davis, and Cox, 1960.)

flow of Koloa lava occupies a valley cut through older Koloa flows and into the underlying lavas of the Waimea Canyon Series. The total depth of the lava-filled valley is not known, but its bottom was as much as 350 feet below the base of the earlier Koloa lavas. Even greater amounts of erosion occurred between flows in upper Hanalei Valley. There, early lavas of the Koloa Series filled a valley cut into the Waimea Canyon Series to the level of the divide that separated it from the Wailua drainage basin, and some flows appear to have spilled over the divide. The top of this early lava fill forms a gently sloping, ramp-like surface that extends all the way to the north shore of the island. The base of the fill can be seen on the east side of the present Hanalei Valley 0.4 mile west of Maheo peak, where the lava rests on old colluvium and stream-laid conglomerate. The thickness of the fill in places exceeded 2,000 feet. This epoch of volcanism was followed by a long period of quiet during which Hanalei Stream reexcavated

its canyon along the western side of the fill, cutting down to a level below the bottom of the former valley. This new canyon, 1,000 feet or more deep below the top of the earlier valley fill, was in turn partly filled by later Koloa flows, which have since been trenched by the present canyon.

Presumably the Koloa eruptions were fed by dikes, but very few have been found, probably because erosion has not yet cut deeply enough to expose them. In the west bank of Wahiawa Stream below the intake of the Eleele ditch a 3-foot dike cutting Koloa lavas and intercalated ash beds may be part of the feeder of the nearby Pohakea cinder cone. On the east fork of Lawai Stream a 6-foot dike of nephelinite cuts Koloa cinders, and nearby a 1-foot dike cuts Palikea conglomerate. Others cut Palikea conglomerate in upper Hanalei Valley. Several dikes cut Koloa lavas and Palikea conglomerate and breccia in valleys along the eastern side of the central highland. These may have been the feeders for some of the Koloa flows in the drainage basin of the South Fork of the Wailua River, but if so the cones built at the vents have been eroded away.

At Papapaholahola hill, a mile northeast of Kalaheo, an intrusive stock about a quarter of a mile in diameter is exposed. Most of the stock consists of oligoclase gabbro (known as kauaiite), but its eastern part is nepheline gabbro (ijolite or fasinite). The gabbro contains many inclusions of peridotite (dunite to lherzolite). Boulders of oligoclase gabbro, known locally as lightning stone, are found also along the Waimea River and the lower stretches of Waialae and Koaie streams, but the intrusive bodies from which they were derived have not

392

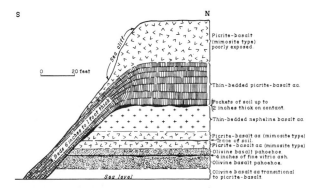

S                          N

0        20 feet

Picrite-basalt (mimosite type) poorly exposed.

Thin-bedded picrite-basalt aa.

Pockets of soil up to 2 inches thick on contact.

Thin-bedded nepheline basalt aa.

Picrite-basalt aa (mimosite type)
Trace of soil.
Picrite-basalt aa (mimosite type)
Olivine basalt pahoehoe.
4 inches of fine vitric ash.
Olivine basalt pahoehoe.

Olivine basalt aa transitional to picrite-basalt.

Sea level

Figure 272. Diagrammatic section of the lava-mantled sea cliff 0.3 mile east of Koheo Point, on the south shore of Kauai. All of the lava flows belong to the Koloa Volcanic Series. (From Macdonald, Davis, and Cox, 1960.)

been located. Lightning stone has also been reported from Olokele Canyon, but it was not found during the geological investigation of the island.

## PALIKEA FORMATION

At the base of the Koloa lavas and interbedded with them are sedimentary breccias and conglomerates known as the Palikea Formation. The breccias consist mostly of angular fragments of rocks of the Waimea Canyon Series, but a few are of Koloa lavas. Generally, bedding is poor and sorting is poor to absent. The fragments range in size from as much as 8 feet across to less than an inch, and in fact grade into the sand and silt of the matrix, which is generally erosional debris, but in places partly tuffaceous. The breccias are usually very well cemented; commonly, when struck by a hammer, they break as readily through the fragments as around them. The breccias are mostly restricted to the base of the Koloa Series and to the vicinity of the steep slopes eroded on the Waimea Canyon Series. They represent the rapid shedding of debris from the steep slopes just before and during the beginning of Koloa time. They are mostly talus and other types of colluvium, of which a large proportion was transported and deposited by mudflows. Some masses of mudflow breccia are found several miles from the steep slopes where they must have originated. In Palikea ridge, at the foot of the central highland 7 miles west-northwest of Lihue, the breccia is 700 feet thick, but nowhere else does it exceed 200 feet. At a few places it is cut by dikes belonging to the Koloa Series.

The breccias grade laterally into stream-laid conglomerates, composed of the same types of materials but better bedded and sorted. Although some of the conglomerate also lies at the base of the Koloa Series, much of it was deposited between Koloa lavas higher in the section. It is distributed all the way from the central highland to the coast, but usually restricted to narrow bands that mark former stream channels. In a few places the conglomerate beds reach thicknesses of more than 100 feet, but they are generally less than 20 feet.

Fragments in the conglomerate range from subangular to well rounded, and up to several feet in diameter. At some places the conglomerate is associated with beds of sandstone and siltstone, largely of detrital origin but in part tuffaceous. Although the well-cemented conglomerates along the Napali Coast are far separated from the Koloa volcanics, they are probably at least in part of the same age as the Palikea Formation.

The great mass of Palikea breccia around the base of the central highland must have been shed from the slopes above at a rate too rapid for the material to be moved seaward by the streams. The cause of the relatively brief period of very abundant landslides and mudflows is an unsolved puzzle, but the fact that it occurred just at the beginning of the Koloa volcanism suggests that the two are related. One possible explanation is the destruction of the heavy vegetation which in the Hawaiian Islands normally binds large amounts of weathered debris to the steep upper walls of the valleys, allowing great volumes of material to move downslope within a short time. Such a destruction of vegetation might have occurred as the climate cooled at the time of onset of glaciation in other regions, but in that case the coincidence with the beginning of Koloa volcanism would be wholly fortuitous. Locally, the vegetation might have been destroyed by volcanic fumes and heavy ash falls, but this would probably not have affected broad areas simultaneously, and, furthermore, ash is not generally present in the Palikea breccias. Another possible cause is a period of numerous strong earthquakes accompanying the advent of Koloa volcanic activity, making the water-soaked weathered material on the upland slopes unstable and precipitating it down the slopes in a series of soil avalanches and mudflows. A similar event, described in chapter 10, brought about the great mudflow of 1868 on the island of Hawaii.

## SEDIMENTARY ROCKS

393

Deposits of old alluvium and colluvium are extensive on Kauai (fig. 273). They are moderately to well cemented, but generally less so than the typical deposits of the Palikea Forma-

Figure 273. Kalalau Valley, Kauai, is a large, amphitheater-headed valley on the Napali Coast. The floor of the valley is composed of moderately consolidated alluvium, which is currently being dissected by streams.

tion. In wet areas the gravels are commonly so decomposed that a pick can be driven easily through both cobbles and matrix. The deposits are not in equilibrium with present conditions, and are being cut away by streams.

Broad aprons of old alluvium lie along the foot of the Kalepa-Nonou and Puu Ehu ridges and along the north side of the Haupu ridge. Smaller patches lie in the heads of Mahaulepu Valley and the Kipu Kai embayment on the south side of the Haupu ridge. Along the Wailua River and Kapaa Stream alluvial terraces probably were graded to the Olowalu stand of the sea. At the mouth of Kalalau Valley the principal alluvial terrace is 110 feet above sea level, and probably was graded to the Kaena stand. A smaller, lower terrace may have been formed during the Waimanalo stand. A mile

northeast of the mouth of Kalalau Valley, gravels have buried an old sea cliff and extend below sea level (fig. 274). They are part of an

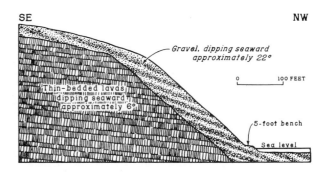

Figure 274. Geologic section on the Napali Coast 1.25 miles northeast of the mouth of Kalalau Stream, showing old alluvium mantling a former sea cliff and extending below sea level. (From Macdonald, Davis, and Cox, 1960.)

alluvial fan that must have been formed during a lower stand of the sea, probably the Waipio stand. At several places old taluses also extend below sea level.

Moderately to well cemented calcareous sand dunes are found in places on the Mana plain and along the southeast coast between Kipu Kai and Makahuena Point; at the last locality they extend below sea level. As elsewhere, they appear to have formed during the Waipio stand of the sea.

Deposits of recent alluvium are widespread along the streams. In part they represent temporary deposits of material in transit to the sea, but a large proportion of the alluvium is the result of aggradation of the lower parts of valleys cut below present sea level in response to a later rise of base level. Deposits of calcareous sand and gravel, marl, and clay on the Mana plain (fig. 275) were formed in a shallow lagoon behind the beach ridge. The proportion of terrigenous sediment increases and the amount of calcareous material decreases toward the inland edge of the plain.

In places the clay beds contain well-formed "buttons" of gypsum, some of them 2 inches in diameter. The buttons consist of a series of wedge-shaped, twinned crystals of gypsum radiating from a common center, forming a disc that tapers from the center to the edge.

Beaches of calcareous sand are beautifully developed on Kauai. The largest are between Haena and Lumahai and at Hanalei Bay on the north side of the island, at Kapaa and Wailua on the east side, at Poipu on the southeast coast, and along the 16 miles of shoreline from Waimea northwestward to the south end of the Napali Coast. In places the sand is cemented into "beach rock." At Ka Lae o Kailio, near Haena, the sand contains enough olivine to give it a greenish color. Just south of the Napali Coast the beach sand is being blown inland into dunes. This area is known as Barking Sands, because some of the sand has a texture such that, with just the right degree of wetness, it makes a peculiar squeaking or yapping noise when it is walked on or squeezed sharply between one's hands.

Figure 275. Mana plain, Kauai, a coastal plain on the southwestern edge of the island composed of earthy and marly lagoon deposits, calcareous beach and dune sand, and alluvium. The cliff at its inner edge is an ancient sea cliff.

# Niihau

Niihau is the deeply eroded remnant of a shield volcano fringed by a platform covered with much later, posterosional volcanics (figs. 276, 277, 278). The surface of the shield remnant slopes gently westward, but is abruptly truncated on the east by a sea cliff 1,200 feet high. All of the lava flows in the shield remnant also slope southwestward, which indicates that only a portion of the southwestern flank of the shield remains. The remnant is notched by valleys as much as 600 feet deep. Between the valleys the spurs have broad, gently sloping tops that are approximately parallel to the underlying lava flows, giving the impression that they are little-dissected remnants of the original shield surface; but dikes crossing them are exposed on their surfaces, showing that a considerable amount of former covering material has been eroded away. Erosion of the spur tops has been by stripping parallel to the bedding, so that the general parallelism of the

surface to that of the original shield has been preserved. The spurs are covered with red lateritic soil up to 5 feet thick, beneath which the lavas are decomposed to a depth as great as 50 feet. The western ends of the spurs are truncated by a sea cliff 200 to 600 feet high, at the base of which lies the plain, floored with later lavas. Projection westward of the gently sloping lava beds in the shield shows that the marine erosion which formed the cliff at the inner edge of the plain cut the edge of the original island back about 2 miles.

## PANIAU VOLCANIC SERIES

The rocks of the shield remnant are known as the Paniau Volcanic Series (Stearns, 1947). They are tholeiitic basalts, olivine basalts, and oceanites. Basaltic hawaiites probably are present, presumably among the latest lavas, but they are very rare. The lavas are cut by numerous dikes, a few inches to 17 feet thick, exposed principally in the eastern and southern sea cliffs and in the stream valleys. The dikes trend southwestward, and represent the southwest rift zone of the volcano. Two small subcircular plugs, less than 500 feet across, are exposed in the walls of Kaailana Valley, near the eastern point of the island. Another plug, 500 feet wide and 1,400 feet long, forms the summit of Kaeo hill, near the center of the island. Kaeo hill is probably the deeply eroded remnant of a former cone on the surface of the shield.

The slope of the lava flows and the trend of the dikes indicate that the summit of the shield lay somewhere northeast of the present island. The valleys on the western flank of the shield remnant radiate in a very general way from a point about 1.5 miles east of the island, and Stearns (1947, p. 9) has suggested that the former summit was located there. If the eastern slope of the shield was similar to the western slope, this would indicate that some 8 miles of the eastern side of the shield has disappeared. If, however, the shield was unsymmetrical, with the eastern slope shorter than the western, the amount that has vanished becomes somewhat smaller, but it can hardly be less than 5

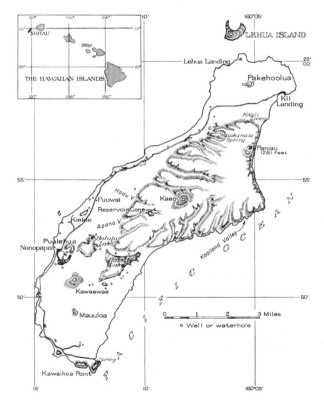

Figure 276. Map of the island of Niihau. (From Stearns, 1947.)

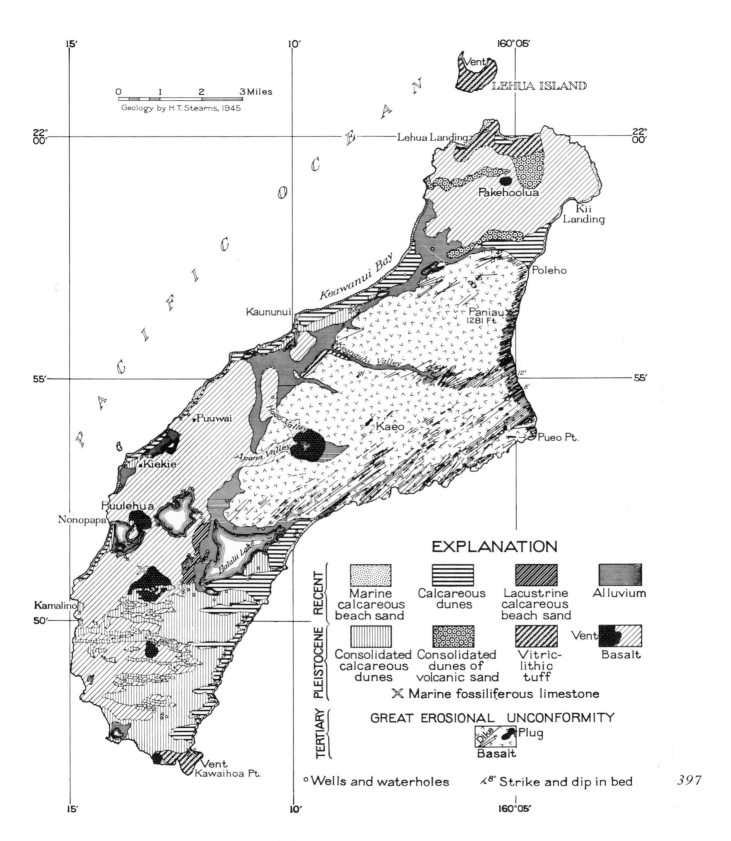

Figure 277. Geologic map of Niihau. (From Stearns, 1946.)

397

Figure 278. Niihau, looking toward Kauai. The island of Niihau is seventh in size of the Hawaiian chain. The high central portion of the island is formed of shield-building, thin-bedded basalt flows. The low plains to the north and south are formed by later volcanics that covered a now-submerged marine platform. Consolidated and unconsolidated sand dunes and beach deposits lie on the late lavas. Three broad playa lakes are visible near the middle of the island.

miles. There is no question that the present eastern cliff is wave cut, but has the whole eastern part of the island been destroyed by erosion? J. D. Dana (1849) suggested that Niihau represents the northwestern edge of Kauai, shifted laterally by faulting from a former position adjacent to the Napali cliffs of Kauai. However, since the slope of the lava flows and trend of the dikes in Niihau do not fit Dana's picture unless the mass was rotated 90° in addition to being transported laterally some 30 miles, all recent workers have rejected Dana's hypothesis. Other geologists have suggested that the eastern side of the shield was dropped below sea level by vertical faulting. It is argued, for instance, that the northeastern side of Niihau is now somewhat sheltered from wave attack by the mass of Kauai to the windward. However, a large part of the wave erosion may have occurred before Kauai was built. Higher rainfall may also have resulted in more active stream erosion, particularly on the windward side, at that time. The only direct evidence is the existence of a shelf 600 feet below sea level which is 1.5 miles wide on the east side of the island. This shelf, which is present also off the western coast, can be interpreted as a wave-cut terrace later submerged nearly 600 feet, and this in turn would indicate that the eastern cliff has been cut back about 1.5 miles by wave erosion (Hinds, 1930, p. 93). This probably can be taken as the minimum of possible wave erosion, but Hinds suggests that the portion of the shield east of the minus-600-foot terrace has been dropped down by faulting. Stearns (1947, p. 10) sees no evidence for faulting, and considers it "at least as probable that a relatively short slope east of the former caldera has been stripped away largely by marine erosion." However, if the eastern part of the island was removed wholly by erosion, why is the minus-600-foot terrace not 5 or 6 miles wide instead of a mile and a half? Marine action is not capable of any appreciable erosion at great depths. Submarine contours indicate a steep, east-facing escarpment trending slightly east of north from a point about 1.5 miles east of the island (fig. 279). This may be the scarp of a fault that has dropped the eastern part of the shield. Faulting seems to us the more probable mechanism for the removal of the part of the shield east of the 600-foot terrace, and it is certainly not ruled out by any present evidence.

398

## KIEKIE VOLCANIC SERIES

The long period of erosion that removed such a large portion of the Niihau shield volcano was terminated by a new series of eruptions, the rocks of which are named the Kiekie Volcanic Series. A terrace several miles wide had been formed around the edges of the shield remnant by wave erosion during the minus-300-foot stands of the sea, and coral reef had grown on the cut platform. The Kiekie lavas were erupted onto this terrace, building it up and widening it. They now form most of the broad low plain that borders the shield remnant on the southwest, west, and north.

Nine vents of the Kiekie Series are visible above sea level. Most of them lie in the coastal plain, but one of them formed a cone nearly a mile across that blocked Apana Valley. Lava from the cone flowed down Apana Valley and the adjacent Haao Valley and spread out on the coastal plain. A breach in the crater wall has been blocked with a dam to form a reservoir. Six of the structures built at other vents are small shield volcanoes, formed of lava flows

with only minor amounts of spatter. The lavas are pahoehoe flows of alkalic olivine basalt. Kawaewae hill, in the south-central part of the plain, is a considerably eroded cinder cone capped by dense massive lava that probably consolidated as a pool in the crater of the cone. The entire end of the island north of the shield remnant consists of a small shield (named by Stearns the Pakehoolua cone), partly overlain by sand dunes and by tuff from the Lehua vent. Lehua Island, just north of Niihau, and Kawaihoa hill at the southern tip of Niihau are tuff cones containing blocks of older basalt and of reef limestone.

The Kiekie lavas on the southern end of Niihau bear remnants of coral reef deposited on them during the Waimanalo (plus-25-foot) and Kaena (plus-95-foot) stands of the sea, and over large areas lateritic soil formed on the lava is covered with lithified calcareous sand dunes. The dunes extend below present sea level and appear to have accumulated during the Waipio negative stand of the sea. Much of the southern and southeastern coast of the island is a sea cliff, 10 to 50 feet high, cut in lithified dunes.

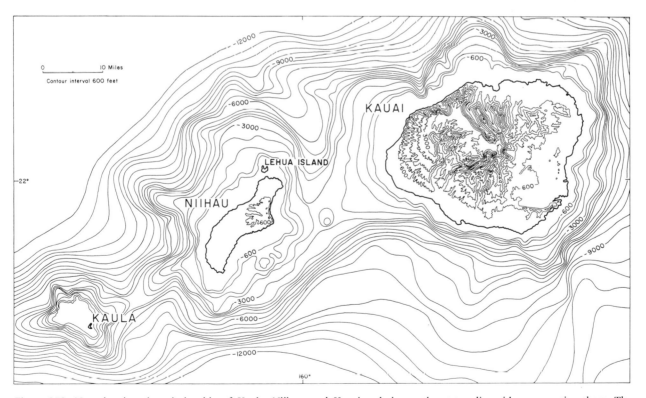

Figure 279. Map showing the relationship of Kaula, Niihau, and Kauai and the northeast-trending ridge connecting them. The submarine contours are generalized. (After Stearns, 1947.)

Traces of both the plus-5- and plus-25-foot shorelines are present on the dunes. The Pakehoolua shield, at the northern end of the island, is partly overlain by lithified dunes of volcanic sand, derived from the Lehua ash, which also extend below sea level and must have formed during a lower stand of the sea. Thus none of the Kiekie volcanics can be more recent than Pleistocene.

# Leeward Islands

The Leeward Islands of the Hawaiian Archipelago form a row of small islands and associated reefs and banks that extends west-northwestward for nearly 1,200 miles beyond Kauai and Niihau. Those closest to Kauai are deeply eroded remnants of volcanoes. Farther northwest the islands are of limestone, but resting on deeply submerged volcanic pedestals. Kaula Island lies 23 miles southwest of Niihau (fig. 279), and belongs with Kauai and Niihau rather than with the rest of the Leeward chain.

In order northwestward, the volcanic islands are: Kaula, Nihoa, Necker, French Frigate Shoal, and Gardner. The last two consist only of small residual pinnacles of volcanic rock surrounded by coral reef, and they constitute a gradation into the wholly limestone islands. Again in order to the northwestward, the limestone islands are: Maro Reef, Laysan, Lisianski, Pearl and Hermes Reef, Midway, and Kure. In addition, more than a dozen other reefs and shoals (fig. 1) mark submarine peaks that do not quite reach the surface of the ocean. Comparatively little geologic work has yet been done on the Leeward Islands.

### KAULA

Kaula Island is a tuff cone resting on a broad base that certainly is a big submerged shield volcano. The cone lies near the southeastern edge of a shoal 8 miles long in a west-northwesterly direction, 4.5 miles wide, and on the average a little more than 200 feet below sea level (figs. 137A, 279). This relatively flat platform must have been cut across the top of the shield volcano by wave erosion, probably during the minus-300-foot stands of the sea. A projection that rises to about 30 feet below sea level 3 miles N 60° W of Kaula is probably an erosional residual, rather than another cone rising from the platform. The general level of the platform probably was raised by a coral reef growing on it, but sinking of the shield became too rapid for reef growth to keep up, and the reef was submerged.

The contours in figure 279 show clearly that the Kaula shield is a part of the Kauai-Niihau massif. The three shields lie on a west-southwest-trending line that is essentially parallel to the southwest rift zones of Kauai and Niihau and appears to reflect one of the fundamental structural trends in the earth's crust in this part of the Pacific. It may have been controlled by one of the splays from the western end of the great Murray Fracture Zone (fig. 170).

Kaula Island is a crescentic erosional remnant of a tuff cone built on the wave-cut platform. It rises about 540 feet above sea level and is just over a mile long from north to south. In composition it resembles Lehua Island north of Niihau and such cones as Diamond Head and Koko Crater on Oahu, and like them it was formed by hydromagmatic explosions. The originally glassy ash is partly altered to palagonite. Many blocks of limestone, torn from the underlying reef, are enclosed in the tuff. Some bombs contain cores of olivine-rich peridotite (dunite and lherzolite). Angular blocks of both nonporphyritic basalt and basalt containing olivine phenocrysts also are present, probably derived from the old shield volcano beneath the reef. An unconformity in the tuff at the north end of the island indicates only a brief pause in the eruptions that built the cone. The composition of the erupting magma has not been determined, but the relationship of the cone to the underlying platform shows clearly that it is a posterosional eruption like those of the Kiekie Series on Niihau. Along much of the shore of Kaula a wave-cut bench lies 4 to 10 feet above sea level. Near the north point a large sea cave is said to extend inward more than 100 feet.

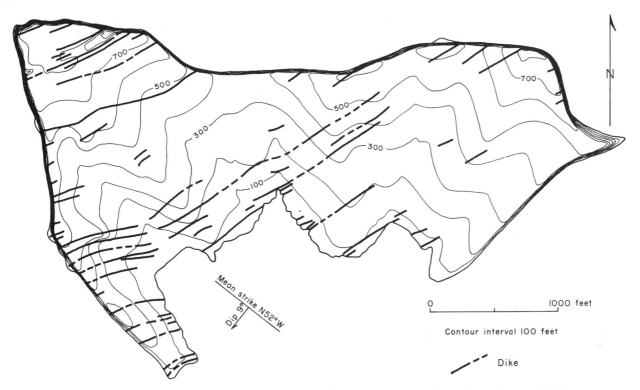

Figure 280. Topographic map of Nihoa Island, showing dikes that cut the lava flows. (After Palmer, 1927.)

## NIHOA

Nihoa Island, 170 miles northwest of Kauai, is approximately 0.85 mile long and averages about 1,500 feet wide. It has two peaks, with a broad swale between them (fig. 280). Miller's Peak, on the west, rises to 895 feet above sea level, and Tanager Peak, on the east, to 852 feet. Bryan (1942) compared it to the southern half of a cowboy's saddle facing west. The northern side of the island is an abrupt sea cliff plunging directly from the crest of the island into the water (fig. 281). On the southern side the surface slopes fairly gently to a line near the coast, where it is truncated by a sea cliff 50 to 100 feet high, with a wave-cut platform at its base 4 to 8 feet above sea level. A tunnel-like sea arch 300 feet long extends all the way through the promontory at the east end of the island.

Nihoa is the eroded remnant of a shield volcano composed of thin lava flows cut by dikes. All of the flows dip southwestward, and thus the island is part of the southwestern side of the shield. Palmer (1927) counted about 25 dikes in the western cliff, and others are

Figure 281. Sketch of the northern side of Nihoa Island. The sea cliff plunges 600 feet into the ocean. (After a photograph in Palmer, 1927.)

*401*

exposed elsewhere. The dikes are nearly vertical, and trend about S 65° W. A platform about 120 to 250 feet deep extends northeastward from the island for about 18 miles. Apparently the whole northeastern portion of a volcanic island some 20 miles across at sea level has been cut away by wave erosion. There is no indication of faulting, but the submarine topography is only poorly known. Another bank, about 17 miles long and 12 miles wide, with depths of 100 to 150 feet, lies 18 miles west-southwest of Nihoa.

The volcanic rocks of Nihoa are predominantly olivine basalts, with less abundant olivine-free basalts and at least one bed of oceanite. They appear to be tholeiitic.

Sea birds nest in great numbers on Nihoa, giving the island its former name, Moku Manu (Bird Island). Thin sheets of guano coat some of the cliffs, and just above the sea cliff on the southern side some of the valleys are partly filled with conglomerate in which the cement is a mixture of clay and guano.

Archeological remains are abundant evidence that the island was formerly occupied by people at least part of the time. Both house sites and terraces used for cultivation have been found, but how the inhabitants obtained their water is a problem. Several seeps are present on the island, the principal one at 270 feet altitude in the large valley on the eastern slope, but the water is somewhat brackish and heavily tainted with guano. Palmer (1927) commented

that the taste was so strong that it seemed impossible that people could have used it, but no other source of water has been found. The bird population, however, probably was much smaller during the period of human habitation.

## NECKER

Necker Island is a narrow bit of land less than 4,000 feet long and averaging about 500 feet wide (fig. 282). Its highest point is 276 feet above sea level. At the east end is an islet about 200 feet long and 75 feet wide which is generally awash. Necker Island lies on a shallow, roughly oval platform about 38 miles long in a northwest-southeast direction and 20 miles wide. The island is an erosional remnant of a shield volcano, composed of thin lava flows cut by nearly vertical dikes. Palmer (1927) reports one dike connecting with a sill that loops around the peak on the northwest peninsula (fig. 282). Another sill, about 2 feet thick, was observed near the eastern end of the island.

The dip of the lava flows is northnortheastward, showing that the original center of the volcano was south-southwestward of the island. The platform on which the island lies unquestionably was cut by wave erosion across a large volcanic massif, but the position on the platform of the island and of the former center of the volcano from which it was carved indicates that the massif was not a single shield, but a group of shields similar to the present

402

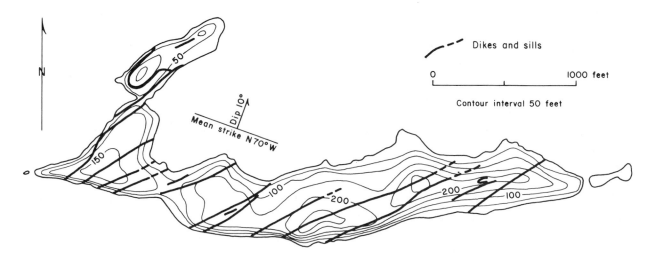

Figure 282. Topographic map of Necker Island, showing dikes. (After Palmer, 1927.)

island of Hawaii. The rocks of Necker Island are mostly olivine basalt, but at least one flow of basaltic hawaiite and several of ankaramite are present. The volcano had entered the late stage. One dike, analyzed by Washington and Keyes (1926), resembles the Hana volcanics of Haleakala and some of the posterosional lavas of Oahu and Kauai in containing more than 19 percent of normative (theoretical) nepheline, although no actual nepheline could be found in the rock (Macdonald, 1949, p. 1554).

The peaks and intervening swales along the crest of Necker Island are remnants of an old stream-eroded topography. The edges of the island are sea cliffs, at the base of which there is usually a wave-cut bench a few feet above sea level.

Long uninhabited except by innumerable sea birds, Necker Island once had a sizeable human population. There are remains of fishponds, ditches, agricultural terraces, and house platforms, and 34 temple platforms have been recognized. Many of the ruins, as well as various artifacts, appear to have been left by an early group of people ("menehunes") who preceded the more recent Hawaiian population. The older remains are overlain by the remains of the work of the later people (Emory, 1928). As on Nihoa, the source of water for these people is a mystery. The present supply appears to be about 10 gallons a day of acrid, guano-tainted water that we would consider undrinkable. Even allowing for the fact that primitive people apparently are able to adjust to very small amounts of very poor water, this quantity hardly seems adequate to support a population large enough to have built all the structures that have been found. Of course, if most of the birds were driven away or killed, the spring water would have been untainted by guano, and this spring supply, in addition to water condensed from the air and rainfall caught and stored in calabashes, may have been sufficient to maintain a small group of primitive people. The irrigation ditches probably operated only during occasional rainstorms. It is interesting to speculate, however, on the possibility of a change in climate. At the time it was discovered by La Perouse, in 1786, the island had not been occupied for a long time. Perhaps 1,500 to 3,000 years ago there was more rainfall than there is now, and later the island had to be abandoned as the climate changed toward the present arid condition.

## FRENCH FRIGATE SHOALS

French Frigate Shoals form a shallow, nearly circular platform about 18 miles across, with a crescent-shaped atoll near its northeastern edge. Within the atoll reef is a broad shallow lagoon containing about 16 small sand islets, none of which reaches more than 12 feet above sea level. The number and form of the islets seems to change from time to time, with changes in wave conditions. A landing strip for aircraft has been built on one of them. Southwest of the atoll the platform generally is between 60 and 150 feet below sea level. A small patch of reef, about 1.5 miles across, lies near its southwestern edge. Near the center of the platform and about 3 miles southwest of the atoll, La Perouse Rock, a mass of lava rock 500 feet long and 80 feet wide, rises 120 feet above the water. About 350 feet to the northwest another projecting mass of lava is 100 feet long, 40 feet wide, and 10 feet high. These tiny sea stacks are the only visible remnants of a great shield volcano.

La Perouse Rock consists of thin lava flows of olivine basalt dipping northwestward at an angle of only 1 or 2 degrees (Palmer, 1927, p. 30). It is largely coated with white guano.

Palmer (1927, p. 31) suggests that the unusual shape of the French Frigate atoll is the result of two generations of reef growth, the leeward reef having grown first, and the windward reef later.

## GARDNER

Gardner Island consists of a pair of stack rocks rising from a shallow wave-cut platform. The larger is a steep-sided pinnacle about 170 feet high, 600 feet long north-south, and averaging about 200 feet wide (fig. 283). Just northwest of it, the smaller pinnacle is about 100 feet high and 250 feet long. Both consist of thin lava flows dipping about 15° westward, cut by a few dikes that strike roughly east-west. The

*403*

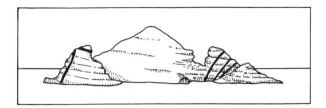

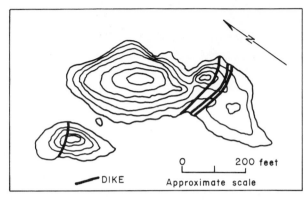

Figure 283. Topographic map and profile of Gardner Island, showing dikes. (After Palmer, 1927.)

dominant rock is fine grained, very dark gray basalt with a few small phenocrysts of olivine, but other specimens of basalt are nonporphyritic. A layer of tuff exposed near the center of the larger islet contains blocks or bombs of basalt and larger gabbroic blocks. A fragment of a cavity lining made up of quartz crystals was also collected. Parts of the island are bordered by a wave-cut bench a few feet above sea level.

Gardner Island is surrounded by a shallow platform extending out from the island 10 or 12 miles to the southwestward, and about 5 miles in all other directions. Undoubtedly it is primarily a wave-cut platform, probably covered to some unknown extent with coral reef. Thus Gardner Island also is a remnant of another big shield volcano truncated by wave erosion.

LAYSAN

Laysan Island (fig. 284) is a little less than 2 miles long and a little more than a mile wide (Bryan, 1942, p. 183). It consists of an oval sandy ridge, reaching just over 40 feet above sea level at the north end, and enclosing a saline lagoon. The lagoon formerly was as deep as 15 feet, but it has been partly filled by

404

drifting sand. Ledges of coral reef rock are exposed in places along the sides of the ridge. The island was once inhabited by great numbers of birds—many of them sea birds, but including also five species of land birds not known elsewhere. Captain John Paty, who annexed the island to the Hawaiian kingdom in 1857, wrote: "The island is literally covered with birds; there is, at a low estimate, 800,000." W. A. Bryan estimated the bird population in 1902 to be nearly 10 million. Phosphates from the large deposits of guano were leached by water that moved downward into limestone, and reacted with it to produce calcium phosphate rock. In 1890 the island was leased for the mining of guano, and by 1904 the deposits were so depleted that the project was discontinued. The unfortunate accompaniment was the wholesale destruction of the bird population, partly by poachers and partly by accidentally introduced rats. Rabbits and guinea pigs also were introduced, and these

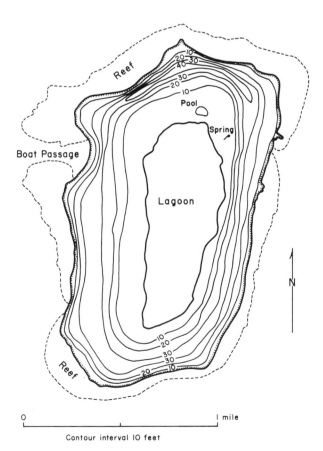

Figure 284. Map of Laysan Island. (After Bryan, 1942.)

animals, together with the mining operations, destroyed most of the vegetation. By 1923 the island had become a desert of sand (Bryan, 1942, p. 188). At the present time the rabbits and guinea pigs have been killed off, and, with help from attempts at revegetation, the island is slowly returning to its former state.

No lava rock has been found in place on Laysan Island, but there are a few boulders of olivine basalt on the reef and beach. They may have come from some part of the basaltic underpinning of the island that is exposed near sea level and have been deposited there by wave action, but more probably they are ballast jettisoned years ago by sailing ships. Such rock ballast, some from as far away as New England, has been found throughout the central Pacific. There can be no question, however, that the Laysan reef rests on a great basaltic volcano.

## MARO REEF

About 80 miles east-southeast of Laysan Island, Maro Reef lies on a shallow bank 31 miles long and 18 miles wide, believed to be the truncated top of another large shield volcano. The reef is about 12 miles long and 5 miles wide, and is generally under breakers. Only one small rock projects as much as 2 feet above sea level. The surrounding bank ranges in depth from about 70 to 120 feet (U.S. Coast Pilot, 1933).

## LISIANSKI

Lisianski Island is a limestone island about a mile long and half a mile wide (fig. 285). A crescentic ridge of sand around the northern end of the island reaches 44 feet above sea level, and a smaller ridge near the southern end reaches 20 feet. In the center of the island a depression, which may once have contained a shallow lake, extends down to about 10 feet above sea level. Thus, in general form the island closely resembles Laysan. Here also, the huge bird population resulted in deposits of guano, particularly in the central depression; and here also rabbits were introduced and destroyed the

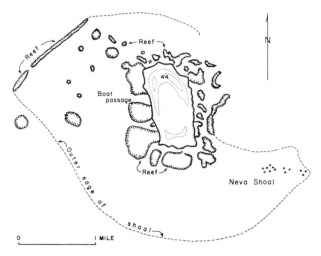

Figure 285. Map of Lisianski Island. (After Bryan, 1942.)

vegetation, and poachers slaughtered the birds for feathers. The rabbits eventually starved to death, and both vegetation and birds are returning.

Lisianski Island is surrounded by a reef, which on the western side encloses a lagoon 2.5 miles wide; and the reef in turn is surrounded by a shallow bank 5 or 6 miles wide on all sides except the southeast, where it extends outward at least 15 miles. Neva Shoal, named after the Russian exploring ship *Neva,* which grounded on the reef east of Lisianski in 1805, is a high part of the reef about a mile southeast of the island. The bank is believed to be the truncated top of a shield volcano, partly covered with reef.

## PEARL AND HERMES REEF

Pearl and Hermes Reef is an atoll about 16 miles long east-west, and about 9 miles wide. The ring reef is lower and less continuous on the northwestern side, which generally is submerged to a depth of 6 to 30 feet. Within the lagoon are 12 small sand islands, the highest of which reaches about 12 feet above sea level. Pearl and Hermes Reef crowns a mountain that rises from more than 15,000 feet below sea level, and undoubtedly is a shield volcano. At the top of a similar mountain 35 miles farther northwest is Gambia Bank, which rises to within about 85 feet of the surface of the water.

*405*

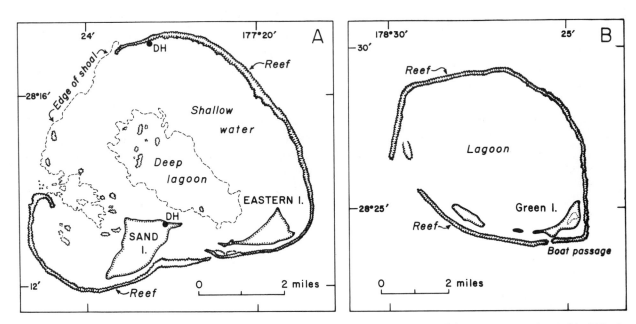

Figure 286. Maps of (A) the Midway Islands and Atoll; (B) the Kure Atoll. (After charts of the U.S. Navy Hydrographic Office.)

## MIDWAY

Midway is a nearly circular atoll 6 miles in diameter enclosing a lagoon 5 miles across (fig. 286A). Within the lagoon are two large and several small islands of calcareous sand. Eastern Island is 1.2 miles long and reaches an elevation of 12 feet. Sand Island is 1.7 miles long and 43 feet high. The reef forms a nearly continuous wall except on the northwestern side. In places the reef rock extends as much as 3 feet above sea level. This has been attributed to a recent downward shift of sea level from a stand of the sea about 5 feet above the present one (Stearns, 1941, 1961; Ladd, Tracey, and Gross, 1967). Considerable portions of the platform at present sea level appear to have been formed by wave erosion of the older reef. A wave-cut nip is well developed on the lagoon side of the high reef remnants.

For more than a century we have believed that the coral reefs of the mid-Pacific rest on volcanic pedestals, but only recently has this belief been proved correct. First, geophysical evidence indicated that the limestone rests on volcanic rocks at relatively shallow depths; more recently, drilling through the limestone has brought up actual samples of the volcanic rock. The latest of these drilling sites is Midway. During 1965 two holes were drilled, one on Sand Island and the other at the reef on the northern edge of the lagoon. From these drill cores can be read the geologic history of Midway (Ladd, Tracey, and Gross, 1967). Both holes went through the limestone into underlying olivine basalt. In the Sand Island hole the top of the basalt was reached at a depth of 516 feet, and in the reef hole at 1,261 feet. Overlying the basalt are conglomerates of basalt cobbles and pebbles and volcanic clays, and these in turn are overlain by limestone. In the reef hole 170 feet of volcanic clays, some of them lignitic, were deposited under swampy conditions, and above these is 500 feet of reef limestone of lower Miocene age (approximately 20 million years ago). The reef then emerged above sea level and was exposed to erosion until late Miocene time (about 12 to 14 million years ago), when it again subsided and more than 100 feet of lagoonal limestone was deposited. The slow accumulation of limestone in the lagoon continued for several million years, into early Pleistocene time, when submergence became more rapid. Another 150 feet of limestone was deposited before the reef again emerged, this time presumably because of lowered sea level caused by the continental glaciation. Again the surface of the reef was eroded, but this was followed by resubmergence and deposition of another 200 feet of reef and lagoon limestone. This is the history read from the Midway drill cores.

## KURE

Kure Island (also known as Ocean Island) is an atoll much like Midway. The ring reef is oval (fig. 286B), with a length of 6 miles and a width of 4 miles. The lagoon has an entrance nearly a mile wide on the southwest side. Three sand islands lie near the southern edge of the lagoon. On the largest of these, Green Island, there are sand dunes as high as 20 feet above sea level. The most westerly island reaches a height of 10 feet.

## JOHNSTON

Neither Wake nor Johnston Island is part of the Hawaiian Archipelago. Johnston Island (fig. 287), about 825 miles west-southwest of Honolulu, rises from deep-ocean depths, but it is directly in line with a southeast-trending ridge on which lie Palmyra, Washington, Fanning, and Christmas islands. It was formerly about 2,800 feet long in a northeast-southwest direc-

tion and averaged about 800 feet wide, but the length has been more than doubled in order to construct an airport large enough to accommodate large airplanes. The highest point, near the northeastern end, was a sand dune that formerly reached 44 feet above sea level, but it has been cut away during construction of the airstrip. The rest of the island is largely sand covered, but in places coral reef crops out close to sea level. About 1.5 miles to the northeast, Sand Island is about 500 feet across and 8 feet high. The two islands lie within the shallow lagoon of an atoll, the bounding reef of which is poorly developed on the southeastern side. The atoll caps a seamount believed to be a large submarine shield volcano.

## WAKE

Wake Atoll also rises from deep-ocean depths, near the middle of a gently-curving arc of seamounts that extends from Necker Island southwestward through the Mid-Pacific Moun-

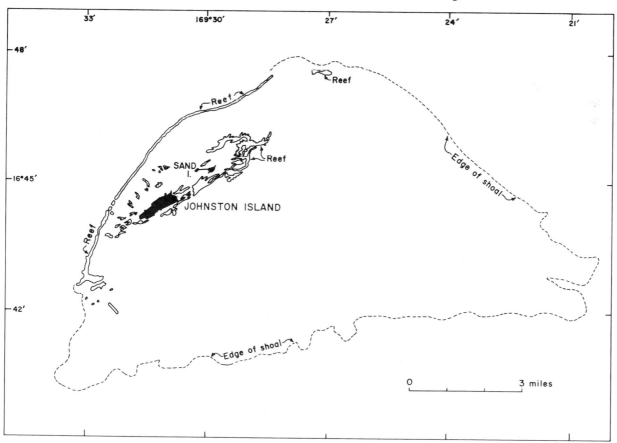

Figure 287. Map of Johnston Island and Atoll. (After Emery, 1956.)

407

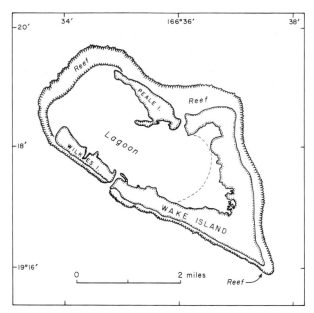

Figure 288. Map of Wake Island and Atoll. (After U.S. Navy Hydrographic Office chart.)

tains (see chap. 11), then curves to the northwest and extends beyond Marcus Island. Wake lies about 2,300 miles N 85° W of Honolulu. A large portion of the atoll has been built above sea level by sand (fig. 288). The main island

(Wake) is V-shaped, each leg of the V being about 3 miles long. Lying northwest of the legs of the V, Wilkes and Peale islands are each about 1.5 miles long. The tops of sand dunes on Wake and Peale islands reach about 21 feet above sea level, and those on Wilkes Island, about 18 feet. The airport is located on Peale Island. Like the atolls of the Hawaiian chain, Wake lies on the summit of a great wave-truncated shield volcano.

## VOLCANIC ACTIVITY

Before leaving the Leeward Islands mention should be made of an apparent volcanic eruption in the area in 1955. On August 20, persons aboard a plane bound from Tokyo to Honolulu sighted what appeared to be a column of smoke rising from the ocean about 55 miles N 85° E of Necker Island. On close approach they saw an oval patch of steaming turbulent water about a mile across, surrounded by a thin line of yellowish surf, with yellowish water drifting away from it. Near one end of the oval was an area of several thousand square yards of

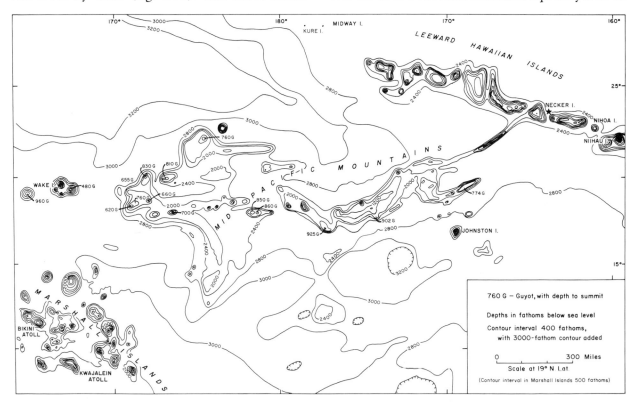

Figure 289. Map of the Mid-Pacific Mountains, showing their junction with the leeward portion of the Hawaiian Ridge. The star just east-northeast of Necker Island marks the site of a submarine volcanic eruption in 1956. (Modified after Hamilton, 1956.)

408

what looked like dry land. This probably was a raft of floating pumice which soon became waterlogged and sank. By the next day, when other planes visited the area, there were no further signs of disturbance other than a slick appearance of the water surface in the formerly turbulent area and a series of long swells sweeping outward from it nearly to Kauai. The locus of the eruption lies just north of the Hawaiian Ridge on the northeastward projection of the Mid-Pacific Mountains and the Necker Ridge (fig. 289), in a depth of about 12,000 feet of water. This is the only volcanic activity which has occurred in the northwestern part of the Hawaiian chain in historic time.

## Suggested Additional Reading

*(principal references for each island are marked with asterisks)*

*General:* Hinds, 1931; Macdonald, 1949*a*, 1968; Macdonald and Katsura, 1964; Malahoff and Woollard, 1966, 1968; Stearns, 1946, 1966*a*

*Hawaii Island:* Dutton, 1884; Macdonald, 1947*a*, 1949*b*; Stearns and Clark, 1930; *Stearns and Macdonald, 1946

*Maui:* *Stearns and Macdonald, 1942

*Lanai:* *Stearns, 1940*c*; Wentworth, 1925*a*

*Kahoolawe:* Macdonald, 1940*b*; *Stearns, 1940*c*

*Molokai:* Lindgren, 1903; *Stearns and Macdonald, 1947

*Oahu:* *Stearns, H. T., 1939, 1940*b*; *Stearns and Vaksvik, 1935; Stearns, N. D., 1935; Wentworth, 1926, 1951; Wentworth and Winchell, 1947; Winchell, 1947

*Kauai:* Hinds, 1930; *Macdonald, Davis, and Cox, 1960

*Niihau:* Hinds, 1930; Macdonald, 1947*b*; *Stearns, 1947

*Leeward Islands:* Bryan, 1942; Ladd, Tracey, and Gross, 1967, 1969; *Palmer, 1927, 1936; Washington and Keyes, 1926

APPENDIX A

# Mineral Localities in the Hawaiian Islands

*Aragonite*—The orthorhombic form of calcium carbonate, aragonite, has been found in limestones at several localities on Oahu. Near the mouth of Lualualei Valley, in the Waianae Range, it occurs in elevated reef limestone that is being quarried for the manufacture of cement.

*Augite*—Very dark green to black crystals of augite are found at several localities, washed out of the loose tephra of cinder and tuff cones, or weathered out of flows of ankaramite. On Oahu, slender crystals of augite up to about 0.3 inch long can be picked up from the ground surface on the slopes of the tuff cones at Koko Head. The most prolific area is above the highway just inland and east of the junction with the road to Hanauma Bay. Many of the crystals are cross-shaped twins.

On Maui, crystals of shorter, blockier shape are found weathered out of an ankaramite flow along the road just below the summit of Haleakala, and just south of the boundary fence of Haleakala National Park near the lower edge of the Park housing area. Some of the crystals are as much as an inch in length. Smaller crystals are scattered over the lower slopes of Namana o ke Akua cinder cone, 0.7 mile northwest of Puu Maile, in Haleakala Crater. Similar crystals are found on several cinder cones on Mauna Kea, Hawaii, particularly at Puu Pa, west of the Kona-Kamuela highway just south of the road to Kawaihae.

*Calcite*—Crystals of calcite, the hexagonal form of calcium carbonate, have been found in cavities in the limestones at several localities, including the above-mentioned quarry in Lualualei Valley, and in lithified calcareous sand dunes at the southeast edge of Kailua, Oahu, that are rapidly being quarried away for sand. Most commonly the crystals are short rhombohedrons, but occasionally they are long slender scalenohedrons, known as dog-tooth spar. Small glistening calcite crystals are common in the tuff at Koko Head, and at Diamond Head, where they are the "diamonds" that gave the hill its name.

Rounded hemispherical masses of calcite are fairly common in vesicles and other openings in lavas throughout the islands, and have received the name "Pele's pearls." One of the best-known localities is the old quarry in Waikakalaua Gulch northwest of Waipahu, Oahu, but they have been found in many road cuts on the islands of Oahu, Maui, and Kauai.

Banded stalactites of calcite formed in shallow caves on the faces of masses of lithified dunes and elevated reefs have been cut and polished as ornaments. The concentric bands range from white to yellow or medium brown in color, and constitute the rock known as travertine, or Mexican onyx. The best-known locality was the dune area at the southeast edge of Kailua, Oahu, mentioned above, but others are found near the highway northwest of Kahuku, and occasionally in cavities in reef limestones exposed by quarrying near Ewa and in Lualualei Valley.

*Chalcedony*—Small masses of chalcedony (cryptocrystalline quartz) are found in the caldera regions of several of the volcanoes, and along the fault zone at the eastern edge of West Molokai. Their color ranges from white to pale blue, green, red, lavender, and brown. Some specimens are banded in different colors, or different shades of the same color. Banded varieties are called agate, and brown or red varieties are called jasper.

Chalcedony is commonly associated with opal—a noncrystalline variety of silica combined with water. The two minerals resemble each other rather closely in appearance, though opal is a little softer than chalcedony and commonly has a somewhat more limpid appearance. Both minerals occur as fillings of vesicles and other cavities in the lavas, as thin veins, and as coatings on rock surfaces.

Chalcedony and opal are quite common in the caldera region of the Koolau volcano on Oahu, particularly in the vicinity of Olomana Peak and in the hills between Kailua and Waimanalo. In the Waianae caldera they have been found in the ridge between Waianae and Lualualei valleys, and in the alluvium in the two valleys. They are found in Iao Valley on West Maui, and in Wailau and Pelekunu valleys on East Molokai. At the east edge of West Molokai they are found along Waiahewahewa Gulch, south of the highway from the airport

to Mauna Loa, and along Kakaaukuu and Kahuuwai gulches north of the highway.

*Feldspar*—Phenocrysts of clear pale yellow to colorless labradorite feldspar occur in hawaiite lava flows exposed in the vicinity of Pohakea Pass, in the Waianae Range on Oahu, and can be found also in the alluvium along Nanakuli Valley. Some fragments can be obtained that are large enough, and free enough of fractures, to be cut as semiprecious gem stones that are sometimes referred to as "Hawaiian topaz." The cut stones are attractive, but the mineral is so soft that great care must be taken to protect it from abrasion. Large clear feldspar crystals occur also in lavas of the Koolau Range in the upper part of Moanalua Valley and adjacent gulches, although there they are rarely, if ever, of gem quality. The same is true of lavas of the Kohala Mountains on Hawaii, where the large feldspars are generally somewhat altered by weathering. Feldspar of gem quality has been found on Lanai in the heads of the big gulches northeast of the Lanaihale ridge.

*Gypsum*—Bladed crystals of gypsum have been found in fractures in lavas around the base of Makapuu Head, on Oahu. Also on Oahu, small masses of crystals and crudely formed rosettes have been found in clays in valleys between Aina Haina and Kuapa Pond, below the level of the shoreline of the 45-foot stand of the sea.

On Kauai, the silts and clays on the Mana plain contain well-formed disk-shaped "buttons" of gypsum.

*Hypersthene*—Small phenocrysts of hypersthene are common in lavas of the Koolau volcano, on Oahu, but have not been found free of the rock. Free crystals of hypersthene, rarely as much as three quarters of an inch long, are present in cinder in a cone 2.5 miles south-southwest of Puu o Keokeo, on the southwest rift zone of Mauna Loa, on Hawaii.

*Obsidian*—Selvedges of obsidian up to about a quarter of an inch thick occur on the edges of many dikes, and as crusts on the surface of some pahoehoe lava flows. The black glass has been cut into small semiprecious gem stones.

412

Because glass alters readily, fresh obsidian generally is found only in the drier parts of the islands. Most of the local obsidian that has been cut came from the western side of the Waianae Range. The Hawaiian basaltic obsidian (tachylite) is duller in luster than many continental obsidians.

Black drop-shaped pellets of obsidian ejected by lava fountains during Hawaiian eruptions are known as Pele's tears. They are especially abundant in the ash deposited in 1959 southwest of Kilauea Iki Crater, and on the slopes of a small prehistoric spatter cone at about 11,200 feet altitude on the Mauna Loa trail.

Lumps of trachyte obsidian are abundant in the pumice cone of Puu Waawaa, on the north slope of Hualalai volcano.

*Olivine*—Crystals of olivine are common in the lavas of the Koolau and Waianae volcanoes, on Oahu, and in the early lavas of all the other Hawaiian volcanoes. As a gem stone, olivine is known as peridot, and some gem olivine has been found in the Hawaiian Islands. It is rare, however, to find a crystal large enough and free enough of fractures to be cut as a gem. Where large crystals are found embedded in the rock it is very difficult to get them out without fracturing them.

Large individual crystals of olivine can be picked up from the ash of the 1924 explosions around Halemaumau Crater, at Kilauea volcano. Near South Point, on Hawaii, large olivine crystals are found in gullies near the top of the Kahuku cliff, where they have been washed out of the layer of yellow ash that covers the ground surface in that region. A few large crystals have been found along the Mauna Loa trail in the 2-mile stretch below Puu Ulaula (Red Hill), where they have been freed from the surface layer of oceanite lava by crushing of the rock beneath the hooves of horses. Well-formed crystals of olivine, up to about 0.3 inch long, are found in some of the cinder cones of Mauna Kea, though they are generally a little decomposed by weathering. One of the best sources is Holoholoku hill, 4 miles south-southeast of the village of Kamuela.

Lumps of olivine (forming the rock dunite), up to several inches across, are found in many of the late-stage and posterosional lavas of the Hawaiian volcanoes, but the individual crystals of olivine in the lumps are generally small. Occasionally, however, grains big enough to cut can be extracted from the lumps. Stearns (1939, p. 68) reports a gem olivine weighing 2 carats obtained from a lump of olivine in cinder from the Tantalus cone in Manoa Valley, Honolulu. Lumps of olivine rock are fairly common in late lavas of the Waianae Range, around the heads of Waianae, Lualualei, and Nanakuli valleys, and in boulders in the alluvium in those valleys, and also in the cinder and lava flows of the late cones, Puu Kapolei, Puu Kapuai, and Puu Makakilo, on the south end of the Waianae Range. In the Honolulu volcanics they are found in the cinder of the Kalama cone, in Kalama Valley, at Puu Hawaiiloa in the Kaneohe Marine Air Station, in the lava and cinder of the eruption that took place at the top of the Pali just east of the Nuuanu gap, in the cinder and ash of the Tantalus cone, and in the tuff at Salt Lake Crater. On Kauai they are found in many of the Koloa lavas. They are especially abundant at the quarry near Halfway Bridge, about 5 miles south of Lihue, and in the lavas along upper Hanalei Valley near the intake of the water tunnel. On Hawaii they have been found in a few late lavas of Mauna Kea and Kohala, and are extremely abundant in the 1801 lava flow of Hualalai. In the latter flow they can be seen in cuts along the highway, but are found by the thousand in the vicinity of the telephone relay station higher up the mountain.

Olivine is concentrated as sand on some Hawaiian beaches. The grains are too small to constitute gem stones, but the green sand is valued by collectors. On Oahu, the best-known localities are the beaches at Diamond Head, Ulupau Head, and Hanauma Bay, where the olivine has been eroded out of the tuff cones and concentrated as placer deposits on the beach. Probably the best locality for olivine sand in the Hawaiian Islands is the beach at Papakolea, on the south coast of Hawaii 3 miles northeast of South Point. There the olivine has been washed out of a littoral cinder cone formed where a lava flow entered the ocean.

*Opal*—See chalcedony.

*Pyrite*—Tiny brassy cube-shaped grains of pyrite have been found in the hills near Kailua, in the caldera area of the Koolau volcano, Oahu, and near the head of Waianae Valley.

*Pyroxene*—See augite and hypersthene.

*Quartz*—Crystals of colorless or white quartz are found in cavities in the lavas in the caldera areas of the Waianae and Koolau volcanoes on Oahu, and less abundantly in those of the West Maui and East Molokai volcanoes. In the latter two areas they have been found along Iao Valley, and in Wailau and Pelekunu valleys. In the Waianae Range they have been found in the vicinity of Kolekole Pass, and in the heads of Waianae and Lualualei valleys. The most prolific locality is in the vicinity of Olomana Peak and in the hills between Kailua and Waimanalo.

The crystals have the shape of hexagonal prisms, commonly broken off at one end and with a six-sided pyramid at the other. A few have been found as much as 1.5 inches long, but most are less than 0.75 inch. They were formed in cavities by gases and hot-water solutions rising through the lavas in the caldera of the volcano. A few have been cut as semiprecious stones. The gems are known as rock crystal, or sometimes as "Hawaiian diamonds" or "Mexican diamonds."

*Zeolite*—Crystals of minerals of the zeolite group are particularly abundant in two areas on Oahu, and are occasionally found in cavities in the lavas in other parts of the Islands. They are common in the hills between Kailua and Waimanalo, and between the Kailua highway and Kaneohe, where, like the quartz, they were deposited by gases or hot solutions moving through the lavas in the caldera. The minerals identified include: radiating spherical aggregates of epistilbite, stubby white glassy crystals of laumontite, pearly white "coffin-shaped" tabular crystals of heulandite, delicate tufts of white needles of ptilolite, and pale yellow nontronite (an iron-bearing clay mineral).

Perhaps the most famous mineral locality in the Hawaiian Islands is the old Moiliili Quarry, now occupied by the athletic fields of the University of Hawaii at the mouth of Manoa Valley. The quarry was opened in a thick lava flow of melilite nephelinite that poured into Manoa Valley from a vent at Sugarloaf, on the ridge west of the valley. During the very last stages of consolidation of the lava, gases deposited zeolites in cavities in the rock. The zeolites include the minerals chabazite, philipsite, and thompsonite. Associated with the zeolites are tiny hexagonal tablets of nepheline ("hydronepheline"), dark green to black needles of augite, and gray needles of apatite. The quarry is no longer active and its walls are overgrown with vegetation, and unfortunately these very interesting miniature mineral specimens are no longer easy to find.

*Suggested Additional Reading*

Dunham, 1933, 1935; Eakle, 1931; Stearns, 1939, pp. 68–71; Winchell, 1947, p. 27

# Geologic Time Scale

| Era | Period | | Epoch | Beginning of interval (years before present) |
|-----|--------|--|-------|------------------------------------------------|
| Cenozoic | Quaternary | | Recent | 10,000–15,000 |
| | | | Pleistocene | 2,000,000–3,000,000 |
| | Tertiary | | Pliocene | 12,000,000 |
| | | | Miocene | 25,000,000 |
| | | | Oligocene | 40,000,000 |
| | | | Eocene | 60,000,000 |
| | | | Paleocene | 70,000,000 |
| Mesozoic | Cretaceous | | | 135,000,000 |
| | Jurassic | | | 180,000,000 |
| | Triassic | | | 225,000,000 |
| Paleozoic | Permian | | | 270,000,000 |
| | Carboniferous | Pennsylvanian | | 325,000,000 |
| | | Mississippian | | 350,000,000 |
| | Devonian | | | 400,000,000 |
| | Silurian | | | 440,000,000 |
| | Ordovician | | | 500,000,000 |
| | Cambrian | | | 600,000,000 |
| Precambrian (oldest rocks of the crust of the earth) | | | | 3,500,000,000– 4,500,000,000 |

*Source:* Modified after Holmes, 1965; and Gilluly, Waters, and Woodford, 1968.

*Note:* The Carboniferous Period of the rest of the world is usually divided into the Pennsylvanian and Mississippian periods in the United States. The oldest exposed rocks of the major islands of the Hawaiian Archipelago are of Pliocene age; thus the entire history of these islands falls within the very latest part of geologic time.

# Glossary

AA. Lava with a rough clinkery surface.

ALLUVIAL FAN. A deposit of rock debris spreading outward in the form of a triangle from the mouth of a valley.

ALLUVIUM. Sedimentary material deposited on the land surface by streams.

AMPHITHEATER-HEADED VALLEY. A valley having a semicircular head and steep walls, with tributary streams frequently entering as waterfalls; sometimes called the "typical Hawaiian valley."

ANDESITE. A lava generally lighter in color than basalt and richer in silicon and sodium.

ANOMALY. A deviation from a general rule or average. In geology and geophysics the term is applied to local deviations in the strength of magnetic or gravitational attraction from the general attraction at the surface of the earth.

ASH. Fine-grained volcanic ejecta, of sand or dust size.

AUGITE. A variety of the dark mineral pyroxene containing calcium, magnesium, iron, aluminum, silicon, and oxygen.

BASAL SPRING. Leakage of fresh ground water at or near sea level.

BASALT. A dark heavy lava rock, rich in iron and magnesium and comparatively poor in silicon. The common lava of Hawaii.

BAUXITE. An aluminum-rich rock consisting of a mixture of gibbsite, allophane, and clay minerals formed by weathering.

BEACH ROCK. A calcareous sandstone formed by lithification of beach sand.

BLACK SAND. Usually sand-sized grains of obsidian formed by littoral explosions as a lava flow enters the sea. The glassy particles wash ashore to form beaches. Other black sands consist of grains of black minerals.

BLOCKS. Angular volcanic ejecta larger than 1.5 inches across, solid when thrown out.

BLOW HOLE (or SPOUTING HORN). Crevice through which seawater and spray issue violently in the form of a fountain.

BLUE ROCK. Very dense, hard, bluish or grayish rock, usually from the interior of an aa flow; quarried for use as concrete aggregate. A local term.

BOMBS. Rounded, flattened, or irregular fragments of volcanic ejecta, molten when thrown out, and larger than 1.5 inches in diameter.

BREACHED CONE. A volcanic cone, one side of which is missing. Such a cone at sea level may be partially filled with seawater.

BRECCIA. A mass of angular rock fragments.

CALDERA. A volcanic crater more than 1 mile in diameter.

CINDER. Irregular, spongy fragments of lava, of bomb or lapillus size, thrown out by volcanic explosions. They are usually solid when they strike the ground.

417

CINDER CONE. A cone-shaped hill of cinder which has piled up around a volcanic vent.

CLINKER. Irregular, spiny fragments composing the top and bottom layers of aa lava flows.

COASTAL PLAIN. An area of low relief, slightly above sea level.

COLLUVIUM. Sedimentary material deposited on the land surface by gravity fall, rain wash, landslide, or mudflow.

COLUMNAR JOINTING. Cracking of lava rock into pencil-shaped fragments (in dikes) or column-like structures (in thick flows or ponded lava lakes).

CRATER. A bowl-shaped depression usually found in the top of a volcanic cone.

CROSS-BEDDING. Thin beds lying at an angle to the thicker bed within which they occur; observed commonly in sand dunes and tuff cones.

DIKE. A sheet-like body of rock formed when magma is forced into a crack and solidifies there.

DIKE COMPLEX. A portion of a volcano containing very numerous dikes cutting earlier rocks.

DOME. A steep-sided hill formed by viscous lava piling up over a vent.

EJECTA. Fragments thrown out, or ejected, by volcanic explosion.

EPICENTER. The point on the earth's surface directly above the place of origin of an earthquake.

EPIMAGMA. Semisolid lava resembling aa in appearance.

FAULT. A fracture in the earth's crust along which one side moves or has moved with respect to the other.

FAULT SCARP. A cliff formed by movement on a fault.

FELDSPAR. A light-colored mineral composed largely of silicon, oxygen, aluminum, and varying proportions of calcium, sodium, and potassium.

FELDSPATHOID. A mineral of a group resembling feldspar in composition, but with a smaller proportion of silica. Nepheline is the commonest feldspathoid.

FISSURE. A crack in the rocks of the earth's crust. Such cracks often guide rising magma to the surface, where it flows out in fissure eruptions.

FUMAROLE. A hole from which volcanic gases issue from the interior of the earth.

GHYBEN-HERZBERG LENS. A lens-shaped body of fresh ground water floating on salt water in underlying rocks.

GIBBSITE. A mineral (aluminum trihydrate) produced by weathering primarily of feldspar or feldspathoid. It is usually the principal constituent of bauxite.

GRABEN. A block of rocks which has dropped down between two or more faults.

HAWAIITE. A variety of andesitic lava rock in which the feldspar is predominantly andesine.

HEADLAND. A cape or promontory; often a high, nearly vertical cliff cut by waves.

HORNITO. A small cone formed on the surface of a lava flow by ejection of spatter from the liquid interior of the flow.

HYDROEXPLOSION. A steam explosion caused by contact of water with magma or hot rock.

IGNEOUS ROCK. Rock formed by solidification of magma or lava.

INCLUSIONS. Fragments of solid rock brought up from depth by rising magma, and left frozen into lava flows or bombs.

INTRUSIVE ROCK. A rock formed by magma solidifying beneath the earth's surface.

KAOLINITE. A mineral (hydrous aluminum silicate) formed by alteration, most commonly by weathering, of aluminum-bearing primary minerals.

KIPUKA. An "island" of land left surrounded by a lava flow.

LACCOLITH. A concordant intrusive body, lens-like in cross section, that has arched up the overlying beds of rock.

LAPILLI. Volcanic ejecta from about 0.25 to 1.5 inches across (singular: lapillus).

LATERITE. Reddish soil high in iron and aluminum oxides. Typical soil of the tropics.

LAVA. Hot liquid rock at or close to the earth's surface, and its solidified products.

LAVA DRAPERY. Cascades of lava that has poured over a scarp and solidified in festoons.

LAVA TREES. Pillars formed by fluid pahoehoe lava surrounding trees and chilling against them.

LAVA TUBE. A natural lava tunnel through which the front of a flow was fed, and left empty when the last of the lava drained out.

LITHIC. Stony.

LITTORAL CONE. A mound of volcanic material at the shoreline formed by steam explosions when a lava flow entered the sea.

MAGMA. Hot liquid rock.

MICROSEISMS. Slight tremors of the earth, particularly the weak vibrations caused by storm waves at sea.

MORAINE. An accumulation of material deposited by a glacier.

MUD ROCK. Local term used by drillers for volcanic tuff.

418

MUGEARITE. A variety of andesitic lava rock in which the feldspar is predominantly oligoclase.

NEPHELINE BASALT. Same as nephelinite.

NEPHELINITE. A variety of basalt-like lava rock containing nepheline instead of feldspar.

OBSIDIAN. Volcanic glass.

OLIVINE. A green mineral composed largely of magnesium, iron, silicon, and oxygen.

PAHOEHOE. Lava with a smooth or ropy-appearing surface.

PALAGONITE. Brownish, waxy substance formed by weathering of volcanic tuff.

PALI. Hawaiian word meaning cliff or scarp.

PELE'S HAIR. Volcanic glass spun out into hairlike form.

PELE'S TEARS. Congealed lava droplets.

PHREATIC. Resulting from the presence of ground water. The steam explosions of Kilauea in 1924 were phreatic explosions.

PISOLITE. In volcanology, a small mud-ball, generally about the size of a pea, formed by raindrops falling through a cloud of volcanic ash.

PIT CRATER. A crater formed by sinking in of the surface; not primarily a vent for lava.

PLAGIOCLASE. A feldspar containing calcium and sodium, but little potassium.

PLANEZE. A triangular erosional remnant of the original slope of the volcano.

PLAYA. A desert basin in which water gathers and evaporates. At times the playa contains a lake; at other times it is dry.

PLUNGE POOL. A basin scoured out at the foot of a waterfall.

PLUTONIC ROCKS. Rocks formed by crystallization of magma, or by solid-state recrystallization, at moderate to great depths within the earth.

POTHOLE. A roughly circular pit formed by abrasion by running water or waves on the bed of a stream or along the shore.

PUMICE. A froth of volcanic glass; very vesicular cinder.

PYROMAGMA. A fluid, gas-charged, very hot lava.

PYROXENE. A group of dark-colored minerals composed largely of silicon, magnesium, iron, and oxygen, with varying amounts of calcium and aluminum.

RETICULITE. Basaltic pumice; also called "thread-lace scoria."

RIFT. A fracture in the earth's crust.

RIFT ZONE. A highly fractured belt on the flank of a volcano along which most of the eruptions take place.

RIPPLE MARKS. Wave or current ripples preserved in rock (usually sandstone).

SCORIA. Slaggy, porous, spongy-appearing ejecta; cinder.

SEA ARCH. An arch on the shoreline formed by waves cutting through the base of a point or promontory.

SEA CAVE. A cave cut out or enlarged by wave action. Some originate as lava tubes.

SEA STACK. An erosional remnant offshore from the main coast line.

SEDIMENTARY ROCK. Rock formed by solidification of particles deposited by water, wind or ice.

SEISMOGRAM. The record of earth tremors made by a seismograph.

SEISMOGRAPH. A device for recording earthquakes.

SHIELD VOLCANO. A volcano having the shape of a very broad, gently sloping dome.

SILL. A sheet of igneous rock intruded between and parallel to older rock layers.

SOLFATARA. A fumarole which liberates sulfur-bearing gases.

SPATTER. Volcanic ejecta thrown out in a very fluid condition. The fragments remain partly fluid when they strike the ground, so that they flatten out and often stick together.

SPATTER CONE. A cone built by the accumulation of spatter.

SPHEROIDAL WEATHERING. A type of weathering which forms onion-like concentric shells, commonly enclosing a rounded, fresh, boulder-like core.

STANDS OF THE SEA. The various elevations on the islands at which former sea levels have been identified.

STREAM CAPTURE. The natural diversion of water from one drainage system to another by encroachment of the stronger stream.

TACHYLITE. Basaltic glass; a variety of obsidian.

TALUS. A heap of rock fragments which have broken off and collected at the foot of a cliff.

TILTMETER. A device for measuring tilting of the ground surface, such as that caused by swelling or shrinking of volcanoes.

TRACHYTE. A light-colored rock, poor in magnesium and iron, and consisting mostly of feldspar rich in potassium and sodium.

TREMOR. A trembling of the ground recorded by seismographs. One variety is caused by movement of magma through volcanic conduits.

TUFF. Consolidated volcanic ash.

TUMESCENCE. Swelling, particularly the swelling of volcanoes before eruption; the opposite of detumescence, or shrinking.

419

UNCONFORMITY. A surface of separation between two rocks or groups of rocks, marking a period of nondeposition and/or erosion. The bedding in the rocks below the unconformity may lie at an angle to that in the rocks above it.

VENT. An opening through which volcanic material reaches the surface.

VERTICAL VALLEY. A giant erosional groove such as those seen on the face of the Oahu Pali.

VESICULAR. Having vesicles.

VESICLE. A bubble hole formed by gas in lava.

VITRIC. Glassy.

WAVE-CUT BENCH. A terrace cut by waves into a sea cliff at the wave base level of erosion.

# References

Abbott, A. T.
>1958 Occurrence of gibbsite on the island of Kauai, Hawaiian Islands. *Economic Geology,* vol. 53, pp. 842–853.

Abbott, A. T., and Pottratz, S. W.
>1969 Marine pothole erosion, Oahu, Hawaii. *Pacific Science,* vol. 23, pp. 276–290.

Adams, W. M., and Furumoto, A. S.
>1965 A seismic refraction study of the Koolau volcanic plug. *Pacific Science,* vol. 19, pp. 296–305.

Agassiz, A.
>1889 The coral reefs of the Hawaiian Islands. *Harvard College Museum of Comparative Zoology Bulletin,* vol. 17, pp. 121–170.

Alexander, W. D.
>1891 *A brief history of the Hawaiian people.* New York: American Book Co.

Allen, E. T.
>1922 Preliminary test of the gases at Sulphur Banks, Hawaii. *Hawaiian Volcano Observatory Bulletin,* vol. 10, pp. 89–93.

Ballard, S. S., and Payne, J. H.
>1940 A chemical study of Kilauea solfataric gases, 1938–40. *Volcano Letter,* no. 469, p. 1.

Belshé, J. C., Cropper, A. G., and Langford, S. A.
>1966 The Tuscarora Seamount province of the Hawaiian Trough. Paper presented at the meeting of the Pacific Science Congress, Tokyo, August 29, 1966.

Bishop, S. E.
>1901 Brevity of tuff cone eruptions. *American Geologist,* vol. 27, pp. 1–5.

Brigham, W. T.
>1909 The volcanoes of Kilauea and Mauna Loa. *B. P. Bishop Museum Memoirs,* vol. 2, no. 4, 222 pp.

Brock, V. E., and Chamberlain, T. C.
>1968 A geological and ecological reconnaissance off western Oahu, Hawaii, principally by means of the research submarine *Ashera, Pacific Science,* vol. 22, pp. 373–394.

Bryan, E. H.
>1942 *American Polynesia and the Hawaiian chain.* Honolulu: Tongg Publishing Co.

Bryan, W. A.
>1915 Evidence of deep subsidence of the Waianae Mountains, Oahu. In *Thrum's Hawaiian Annual for 1916,* pp. 95–125. Honolulu.

Bullard, F. M.
1962 *Volcanoes, in history, in theory, in eruption.* Austin: University of Texas Press.

Bullen, K. E.
1963 *An introduction to the theory of seismology.* 3rd ed. London: Cambridge University Press.

Cathcart, J. B.
1958 Bauxite deposits of Hawaii, Maui, and Kauai, Territory of Hawaii—a preliminary report. U.S. Geological Survey, unpublished open file report, 72 pp.

Chamberlain, T. K.
1968 The littoral sand budget, Hawaiian Islands. *Pacific Science,* vol. 22, pp. 161–183.

Cline, M. G.
1955 *Soil Survey of the Territory of Hawaii* (with sections by A. S. Ayres; William Crosby; P. F. Philipp and Ralph Elliott; O. C. Magistad; C. K. Wentworth; and J. C. Ripperton, E. Y. Hosaka, M. Takahashi, and G. D. Sherman), accompanied by generalized and more detailed soil maps, in color. U.S. Department of Agriculture in cooperation with Hawaii Agricultural Experiment Station.

Cox, A., and Dalrymple, G. B.
1967 Statistical analysis of geomagnetic reversal data and the precision of potassium-argon dating. *Journal of Geophysical Research,* vol. 72, pp. 2603–2614.

Cox, D. C.
1954 Water development for Hawaiian sugar cane irrigation. *Hawaiian Planters' Record,* vol. 54, pp. 175–197.

Daly, R. A.
1911 The nature of volcanic action. *Proceedings of the American Academy of Arts and Sciences,* vol. 47, pp. 47–122.

1933 *Igneous rocks and the depths of the earth.* New York: McGraw-Hill.

Dana, J. D.
1849 Geology. *United States Exploring Expedition,* vol. 10, 756 pp.

1890 *Characteristics of volcanoes, with contributions of facts and principles from the Hawaiian Islands.* New York: Dodd, Mead.

Darwin, Charles
1839 *On the structure and distribution of coral reefs.* Reprint. London: Ward, Lock, Bowden and Co.

Davis, S. N., and DeWiest, R. J. M.
1966 *Hydrogeology.* New York: John Wiley and Sons.

Doell, R. R., and Cox, A.
1965 Paleomagnetism of Hawaiian lava flows. *Journal of Geophysical Research,* vol. 70, pp. 3377–3405.

Dunham, K. C.
1933 Crystal cavities in lavas from the Hawaiian Islands. *American Mineralogist,* vol. 18, pp. 369–385.

1935 Crystal cavities in lavas from the Hawaiian Islands. *American Mineralogist,* vol. 20, pp. 880–882.

Dutton, C. E.
1884 Hawaiian volcanoes. U.S. Geological Survey, Fourth Annual Report, pp. 75–219.

Eakle, A. S.
1931 The minerals of Oahu. *Mid-Pacific Magazine,* vol. 42, pp. 341–343.

Easton, W. H.
1963 New evidence for a 40-foot shore line on Oahu. In *Essays in marine geology in honor of K. O. Emery,* pp. 51–68. Los Angeles: University of Southern California Press.

1965 New Pleistocene shore lines in Hawaii (abstract). Geological Society of America, Cordilleran Section Meeting, Program, p. 21.

1968 Radiocarbon profile of Hanauma reef, Oahu. Geological Society of America, 1968 Annual Meeting, Program, p. 86.

Eaton, J. P.
1962 Crustal structure and volcanism in Hawaii. In *Crust of the Pacific Basin* (G. A. Macdonald and Hisashi Kuno, eds.), pp. 13–29. Geophysical Monograph 6, American Geophysical Union, Washington, D. C.

Eaton, J. P., and Murata, K. J.
1960 How volcanoes grow. *Science,* vol. 132, no. 3432, pp. 925–938.

Eaton, J. P., Richter, D. H., and Ault, W. U.
1961 The tsunami of May 23, 1960, on the island of Hawaii. *Bulletin of the Seismological Society of America,* vol. 51, pp. 135–157.

Eckel, E. B., ed.
1958 *Landslides and engineering practice.* National Research Council, Highway Research Board Special Report 29, 232 pp.

Edmondson, C. H.
1928 *The ecology of an Hawaiian coral reef.* B. P. Bishop Museum, Bulletin 45, 64 pp.

Emery, K. O.
1956 Marine geology of Johnston Island and its surrounding shallows, central Pacific Ocean. *Bulletin of the Geological Society of America,* vol. 67, pp. 1505–1520.

Emery, K. O., and Cox, D. C.
1956 Beachrock in the Hawaiian Islands. *Pacific Science,* vol. 10, pp. 382–402.

Emory, K. P.
1928 *Archaeology of Nihoa and Necker islands.* B. P. Bishop Museum, Bulletin 53, 124 pp.

Finch, R. H.
1940 Engulfment at Kilauea volcano. *Volcano Letter,* no. 470, pp. 1–2.
1941 The filling in of Kilauea Crater. *Volcano Letter,* no. 471, pp. 1–3.

Finch, R. H., and Macdonald, G. A.
1949 Bombing to divert lava flows. *Volcano Letter,* no. 506, pp. 1–3.
1953 *Hawaiian volcanoes during 1950.* U.S. Geological Survey, Bulletin 996-B, pp. 27–89.

Finlayson, J. B., Barnes, I. L., and Naughton, J. J.
1968 Developments in volcanic gas research in Hawaii. In *Crust and upper mantle of the Pacific area* (L. Knopoff, C. L. Drake, and P. J. Hart, eds.), pp. 428–438. Geophysical Monograph 12, American Geophysical Union, Washington, D. C.

Fisher, R. V.
1968 Puu Hou littoral cones, Hawaii. *Geologische Rundschau,* vol. 57, pp. 837–864.

Frankel, J. J.
1967 Forms and structures of intrusive basaltic rocks. In *Basalts: The Poldervaart treatise on rocks of basaltic composition* (H. H. Hess and A. Poldervaart, eds.), vol. 1, pp. 63–102. New York: Interscience Publishers.

Funkhouser, J. G., Barnes, I. L., and Naughton, J. J.
1966 Problems in the dating of volcanic rocks by the potassium-argon method. *Bulletin Volcanologique,* vol. 29, pp. 709–718.
1968 The determination of a series of ages of Hawaiian volcanoes by the potassium argon method. *Pacific Science,* vol. 22, pp. 369–372.

Gilluly, J., Waters, A. C., and Woodford, A. O.
1968 *Principles of geology.* 3rd ed. San Francisco: W. H. Freeman.

Gregory, H. E., and Wentworth, C. K.
1937 General features and glacial geology of Mauna Kea, Hawaii. *Bulletin of the Geological Society of America,* vol. 48, pp. 1719–1742.

Hamilton, E. L.
1956 *Sunken islands of the Mid-Pacific Mountains.* Geological Society of America, Memoir 64, 97 pp.

Hay, R. L., and Iijima, A.
1968 Petrology of palagonite tuffs of Koko Craters, Oahu, Hawaii. *Contributions to Mineralogy and Petrology,* vol. 17, pp. 141–154.

Hess, H. H.
1946 Drowned ancient islands of the Pacific Basin. *American Journal of Science,* vol. 244, pp. 772–791.

Hinds, N. E. A.
1930 *The geology of Kauai and Niihau.* B. P. Bishop Museum, Bulletin 71, 103 pp.
1931 The relative ages of the Hawaiian landscapes. *University of California Publications, Bulletin of the Department of Geological Sciences,* vol. 20, pp. 143–260.

Hitchcock, C. H.
1911 *Hawaii and its volcanoes.* 2nd ed. Honolulu: Hawaiian Gazette.

Hodgson, J. H.
1964 *Earthquakes and earth structure.* Englewood Cliffs, N. J.: Prentice-Hall.

Holmes, Arthur
1965 *Principles of physical geology.* 2nd ed. New York: Ronald Press.

Inman, D. L., Gayman, W. R., and Cox, D. C.
1963 Littoral sedimentary processes on Kauai, a sub-tropical high island. *Pacific Science,* vol. 17, pp. 106–130.

Jackson, E. D.
1966 "Eclogite" in Hawaiian basalts. *U.S. Geological Survey, Professional Paper* 550-D, pp. 151–157.
1968 The character of the lower crust and upper mantle beneath the Hawaiian Islands. *XXIII International Geological Congress Report,* vol. 1, pp. 135–150.

Jackson, M. L., and Sherman, G. D.
1953 Chemical weathering of minerals in soils. *Advances in Agronomy,* vol. 5, pp. 219–318.

Jaggar, T. A.
1925 The Daly glacier on Mauna Kea. *Volcano Letter,* no. 43, p. 1.
1930 The Hualalai earthquake crisis of 1929. *Volcano Letter,* no. 309, pp. 1–2; no. 310, pp. 1–3.
1932 Comparative data about recent eruptions. *Volcano Letter,* no. 370, pp. 1–4.
1936 The bombing operation at Mauna Loa. *Volcano Letter,* no. 431, pp. 4–6.
1940 Magmatic gases. *American Journal of Science,* vol. 238, pp. 313–353.

*423*

1945    Protection of harbors from lava flow. *American Journal of Science,* vol. 243-A, pp. 333–351.

1947    *Origin and development of craters.* Geological Society of America, Memoir 21, 508 pp.

Jaggar, T. A., and Finch, R. H.

1924    The explosive eruption of Kilauea in Hawaii, 1924. *American Journal of Science,* ser. 5, vol. 8, pp. 353–374.

Keller, W. D.

1957    *The principles of chemical weathering.* Revised ed. Columbia, Mo.: Lucas Brothers.

Kinoshita, W. T.

1965    A gravity survey of the island of Hawaii. *Pacific Science,* vol. 19, pp. 339–340.

Kinoshita, W. T., Koyanagi, R. Y., Wright, T. L., and Fiske, R. S.

1969    Kilauea volcano: the 1967–68 summit eruption. *Science,* vol. 166, no. 3904, pp. 459–468.

Kinoshita, W. T., and Okamura, R. T.

1965    A gravity survey of the island of Maui, Hawaii. *Pacific Science,* vol. 19, pp. 341–342.

Kulp, J. L.

1961    Geologic time scale. *Science,* vol. 133, no. 3459, pp. 1105–1114.

Kuno, H., Yamasaki, K., Iida, C., and Nagashima, K.

1957    Differentiation of Hawaiian magmas. *Japanese Journal of Geology and Geography,* vol. 28, pp. 179–218.

Ladd, H. S., Tracey, J. I., and Gross, M. G.

1967    Drilling on Midway Atoll, Hawaii. *Science,* vol. 156, no. 3778, pp. 1088–1094.

1969    Deep drilling on Midway Atoll. *U.S. Geological Survey, Professional Paper* 680-A, 22 pp.

Langford, S. A.

1969    The surface morphology of the Tuscaloosa Seamount. Unpublished M. S. thesis, University of Hawaii, 17 pp.

Leet, L. D., and Judson, S.

1965    *Physical geology.* 3rd ed. Englewood Cliffs, N. J.: Prentice-Hall.

Legget, R. F.

1962    *Geology and engineering.* 2nd ed. New York: McGraw-Hill.

Lindgren, W.

1903    *The water resources of Molokai, Hawaiian Islands.* U.S. Geological Survey, Water Supply Paper 77, 62 pp.

Lyman, C. S.

1851    On the recent condition of Kilauea. *American Journal of Science,* ser. 2, vol. 12, pp. 75–80.

Macdonald, G. A.

1940a   *Petrography of the Waianae Range, Oahu.* Hawaii Division of Hydrography, Bulletin 5, pp. 63–91.

1940b   *Petrography of Kahoolawe.* Hawaii Division of Hydrography, Bulletin 6, pp. 149–173.

1943    The 1942 eruption of Mauna Loa, Hawaii. *American Journal of Science,* vol. 241, pp. 241–256; reprinted in Smithsonian Institution Annual Report for 1943, pp. 199–212.

1945    Ring structures at Mauna Kea, Hawaii. *American Journal of Science,* vol. 243, pp. 210–217.

1947a   *Bibliography of the geology and groundwater resources of the island of Hawaii, annotated and indexed.* Hawaii Division of Hydrography, Bulletin 10, 191 pp.

1947b   *Petrography of Niihau.* Hawaii Division of Hydrography, Bulletin 12, pp. 41–51.

1949a   Hawaiian petrographic province. *Bulletin of the Geological Society of America,* vol. 60, pp. 1541–1596.

1949b   Petrography of the island of Hawaii. *U.S. Geological Survey, Professional Paper* 214-D, pp. 51–96.

1953    Pahoehoe, aa, and block lava. *American Journal of Science,* vol. 251, pp. 169–191.

1954    Activity of Hawaiian volcanoes during the years 1940–1950. *Bulletin Volcanologique,* ser. 2, vol. 15, pp. 120–179.

1955a   *Hawaiian volcanoes during 1952.* U.S. Geological Survey, Bulletin 1021-B, pp. 15–108.

1955b   Distribution of areas of pneumatolytic deposition on the floor of Kilauea caldera. *Volcano Letter,* no. 528, pp. 1–3.

1956    The structure of Hawaiian volcanoes. *Koninklijk Nederlandsch Geologisch-Mijnbouwkundig Genootschap, Verhandelingen,* vol. 16, pp. 274–295.

1958    Barriers to protect Hilo from lava flows. *Pacific Science,* vol. 12, pp. 258–277.

1959    The activity of Hawaiian volcanoes during the years 1951–1956. *Bulletin Volcanologique,* ser. 2, vol. 22, pp. 3–70.

1961    Volcanology. *Science,* vol. 133, no. 3454, pp. 673–679.

1962    The 1959 and 1960 eruptions of Kilauea volcano, Hawaii, and the construction of walls to restrict the spread of the lava flows. *Bulletin Volcanologique,* ser. 2, vol. 24, pp. 249-294.

424

1963a Physical properties of erupting Hawaiian magmas. *Bulletin of the Geological Society of America,* vol. 74, pp. 1071–1078.

1963b Relative abundance of intermediate members of the oceanic basalt-trachyte association—a discussion. *Journal of Geophysical Research,* vol. 86, pp. 5100–5102.

1967 Forms and structures of extrusive basaltic rocks. In *Basalts: The Poldervaart treatise on rocks of basaltic composition* (H. H. Hess and A. Poldervaart, eds.), vol. 1, pp. 1–62. New York: Interscience Publishers.

1968 Composition and origin of Hawaiian lavas. *Geological Society of America, Memoir* 116, pp. 477–522.

1971

Geologic map of the Mauna Loa Quadrangle, Hawaii. U.S., Geological Survey. Geologic Quadrangle Map 897.

1972

*Volcanoes.* Englewood Cliffs, N. J.: Prentice-Hall.

Macdonald, G. A., Davis, D. A., and Cox, D. C.
1960 *Geology and ground-water resources of the island of Kauai, Hawaii.* Hawaii Division of Hydrography, Bulletin 13, 212 pp., colored geologic map.

Macdonald, G. A., and Eaton, J. P.
1955 The 1955 eruption of Kilauea volcano. *Volcano Letter,* no. 529–530, pp. 1–10.

1957 *Hawaiian volcanoes during 1954.* U.S. Geological Survey, Bulletin 1061-B, pp. 17–72.

1964 *Hawaiian volcanoes during 1955.* U.S. Geological Survey, Bulletin 1171, 170 pp.

Macdonald, G. A., and Katsura, T.
1962 Relationship of petrographic suites in Hawaii. In *Crust of the Pacific basin* (G. A. Macdonald and H. Kuno, eds.), pp. 187–195. Geophysical Monograph 6, American Geophysical Union, Washington, D. C.

1964 Chemical composition of Hawaiian lavas. *Journal of Petrology,* vol. 5, pp. 82–133.

Macdonald, G. A., and Orr, J. B.
1950 *The 1949 summit eruption of Mauna Loa, Hawaii.* U.S. Geological Survey, Bulletin 974-A, pp. 1–31.

Macdonald, G. A., Shepard, F. P., and Cox, D. C.
1947 The tsunami of April 1, 1946, in the Hawaiian Islands. *Pacific Science,* vol. 1, pp. 21–37.

Macdonald, G. A., and Wentworth, C. K.
1952 The Kona earthquake of August 21, 1951, and its aftershocks. *Pacific Science,* vol. 6, pp. 269–287.

Malahoff, Alexander, and Woollard, G. P.
1966 Magnetic surveys over the Hawaiian Islands and their geologic implications. *Pacific Science,* vol. 20, pp. 265–311.

1968 Magnetic and tectonic trends over the Hawaiian Ridge. In *Crust and upper mantle of the Pacific area* (L. Knopoff, C. L. Drake, and P. J. Hart, eds.), pp. 241–276. Geophysical Monograph 12, American Geophysical Union, Washington, D.C.

McDougall, Ian
1964 Potassium-argon ages from lavas of the Hawaiian Islands. *Bulletin of the Geological Society of America,* vol. 75, pp. 107–128.

McDougall, I., and Tarling, D. H.
1963 Dating of polarity zones in the Hawaiian Islands. *Nature,* vol. 200, no. 4901, pp. 54–56.

Meinzer, O. E.
1923a *The occurrence of ground water in the United States, with a discussion of principles.* U.S. Geological Survey, Water Supply Paper 489, 321 pp.

1923b *Outline of ground-water hydrology.* U.S. Geological Survey, Water Supply Paper 494, 71 pp.

1942 Occurrence, origin, and discharge of ground water. In *Hydrology* (O. E. Meinzer, ed.), pp. 385–443. New York: McGraw-Hill.

Menard, H. W.
1964 *Marine geology of the Pacific.* New York: McGraw-Hill.

Menard, H. W., Allison, E. C., and Durham, J. W.
1962 A drowned Miocene terrace in the Hawaiian Islands. *Science,* vol. 138, no. 3543, pp. 896–897.

Mink, J. F.
1964 Groundwater temperatures in a tropical island environment. *Journal of Geophysical Research,* vol. 69, pp. 5225–5230.

Moberly, R., Jr.
1963a Amorphous marine muds from tropically weathered basalt. *American Journal of Science,* vol. 261, pp. 767–772.

1963b Rate of denudation in Hawaii. *Journal of Geology,* vol. 71, pp. 371–375.

1968 Loss of Hawaiian littoral sand. *Journal of Sedimentary Petrology,* vol. 38, pp. 17–34.

Moberly, R., Jr., Baver, L. D., Jr., and Morrison, A.
1965 Source and variation of Hawaiian littoral sand. *Journal of Sedimentary Petrology,* vol. 35, pp. 589–598.

*425*

Moberly, R., Jr., and McCoy, F. W., Jr.

1966   The sea floor north of the eastern Hawaiian Islands. *Marine Geology,* vol. 4, pp. 21–48.

Moore, J. G.

1964   Giant submarine landslides on the Hawaiian Ridge. *U.S. Geological Survey, Professional Paper* 501-D, pp. 95–98.

Moore, J. G., and Ault, W. U.

1965   Historic littoral cones in Hawaii. *Pacific Science,* vol. 19, pp. 3–11.

Moore, J. G., and Koyanagi, R. Y.

1969   The October 1963 eruption of Kilauea volcano, Hawaii. *U. S. Geological Survey, Professional Paper* 614-C, 13 pp.

Moore, J. G., and Peck, D. L.

1965   Bathymetric, topographic, and structural map of the south-central flank of Kilauea volcano, Hawaii. U.S. Geological Survey, Map I-456.

Murata, K. J., Ault, W. U., and White, D. E.

1964   Halogen acids in fumarolic gases of Kilauea volcano. *Bulletin Volcanologique,* ser. 2, vol. 27, pp. 367–368.

Murata, K. J., and Richter, D. H.

1966   Chemistry of the lavas of the 1959–60 eruption of Kilauea volcano, Hawaii. *U. S. Geological Survey, Professional Paper* 537-A, 26 pp.

Nakamura, M. T., and Sherman, G. D.

1965   *The genesis of halloysite and gibbsite from mugearite on the island of Maui.* Hawaii Agricultural Experiment Station, Technical Bulletin 62, 36 pp.

Naughton, J. J., and Shaeffer, O. A.

1962   Some preliminary age determinations on Hawaiian lavas. *Symposium of the International Association of Volcanology, Japan,* Abstracts, pp. 47–48.

Oostdam, B. L.

1965   Age of lava flows on Haleakala, Maui, Hawaii. *Bulletin of the Geological Society of America,* vol. 76, pp. 393–394.

Ostergaard, J. M.

1928   *Fossil marine mollusks of Oahu.* B. P. Bishop Museum, Bulletin 51, 32 pp.

Palmer, H. S.

1927   *Geology of Kaula, Nihoa, Necker, and Gardner islands, and French Frigates Shoal.* B. P. Bishop Museum, Bulletin 35, 35 pp.

1936   *Geology of Lehua and Kaula islands. B. P. Bishop Museum Occasional Papers,* vol. 12, no. 13, pp. 3–36.

1946   *The geology of the Honolulu ground water supply.* Board of Water Supply, Honolulu, Hawaii, 55 pp.

1967   Origin and diffusion of the Herzberg principle with especial reference to Hawaii. *Pacific Science,* vol. 11, pp. 181–189.

Patterson, S. H.

1962   Investigation of ferruginous bauxite and plastic clay deposits on Kauai and a reconnaissance of ferruginous bauxite deposits on Maui, Hawaii. U.S. Geological Survey unpublished open-file report; reprinted without plates, 1963, as Hawaii Division of Water and Land Development Circular C-15, 82 pp.

Payne, J. H., and Ballard, S. S.

1940   The incidence of hydrogen sulfide at Kilauea solfatara preceding the 1940 Mauna Loa volcanic activity. *Science,* vol. 93, no. 2384, pp. 218–219.

Peck, D. L., and Minakami, T.

1968   The formation of columnar joints in the upper part of Kilauean lava lakes, Hawaii. *Bulletin of the Geological Society of America,* vol. 79, pp. 1151–1166.

Peterson, D. W.

1967   Geologic map of the Kilauea Crater Quadrangle, Hawaii. U.S. Geological Survey. Geologic Quandrangle Map 667.

Pollock, J. B.

1928   *Fringing and fossil coral reefs of Oahu.* B. P. Bishop Museum, Bulletin 55, 56 pp.

Powers, H. A.

1935   Differentiation of Hawaiian lavas. *American Journal of Science,* ser. 5, vol. 30, pp. 57–71.

1955   Composition and origin of basaltic magma of the Hawaiian Islands. *Geochimica et Cosmochimica Acta,* vol. 7, pp. 77–107.

Richey, J. E., Thomas, H. H., and others

1930   *The geology of Ardnamurchan, north-west Mull, and Coll.* Memoir of the Geological Survey of Scotland, 393 pp.

Richter, C. F.

1958   *Elementary seismology.* San Francisco: W. H. Freeman.

Richter, D. H., Ault, W. U., Eaton, J. P., and Moore, J. G.

1964   The 1961 eruption of Kilauea volcano, Hawaii. *U. S. Geological Survey, Professional Paper* 474-D, 34 pp.

Richter, D. H., and Eaton, J. P.

1960   The 1959–60 eruption of Kilauea volcano. *The New Scientist,* vol. 7, pp. 994–997.

Rittmann, Alfred

1962   *Volcanoes and their activity.* New York: Interscience Publishers.

426

Ruhe, R. V., Williams, J. M., and Hill, E. L.
1965 Shorelines and submarine shelves, Oahu, Hawaii. *Journal of Geology*, vol. 73, pp. 485–497.

Schulz, P. E.
1943 Some characteristics of the summit eruption of Mauna Loa, Hawaii, in 1940. *Bulletin of the Geological Society of America*, vol. 54, pp. 739–746.

Sharpe, C. F. S.
1938 *Landslides and related phenomena.* New York: Columbia University Press.

Shepard, F. P.
1961 Sea level rise during the past 20,000 years. *Zeitschrift für Geomorphologie*, vol. 3, pp. 30–35.

Shepard, F. P., and Dill, R. F.
1966 *Submarine canyons and other sea valleys.* Chicago: Rand McNally.

Shepard, F. P., Macdonald, G. A., and Cox, D. C.
1950 The tsunami of April 1, 1946. *Scripps Institution of Oceanography Bulletin*, vol. 5, pp. 391–528.

Sherman, G. D.
1954 Origin and development of ferruginous concretions. *Soil Science*, vol. 77, pp. 1–8.
1955 *Some of the mineral resources of the Hawaiian Islands.* Hawaii Agricultural Experiment Station, Special Publication 1, pp. 5–28.
1958 *Gibbsite-rich soils of the Hawaiian Islands.* Hawaii Agricultural Experiment Station, Bulletin 116, 23 pp.
1962 Weathering and soil science. International Society of Soil Science, Transactions of Meeting of Commissions IV and V, New Zealand, pp. 24–32.

Sherman, G. D., Cady, J. G., Ikawa, H., and Blomberg, N. E.
1967 *Genesis of the bauxitic Halii soils.* Hawaii Agricultural Experiment Station, Technical Bulletin 56, 46 pp.

Sherman, G. D., and Ikawa, H.
1968 Soil sequences in the Hawaiian Islands. *Pacific Science*, vol. 22, pp. 458–464.

Sherman, G. D., and Uehara, G.
1956 The weathering of olivine basalt in Hawaii and its pedogenic significance. *Soil Science Society of America Proceedings*, vol. 20, pp. 337–340.

Skinner, B. J.
1970 A sulfur lava flow on Mauna Loa. *Pacific Science*, vol. 24, pp. 144–145.

Smith, S. M., and Menard, H. W.
1965 The Molokai fracture zone. *Progress in Oceanography*, vol. 3, pp. 333–345.

Stearns, H. T.
1926 The Keaiwa or 1823 lava flow from Kilauea volcano, Hawaii. *Journal of Geology*, vol. 34, pp. 336–351.
1935 Pleistocene shore lines on the islands of Oahu and Maui, Hawaii. *Bulletin of the Geological Society of America*, vol. 46, pp. 1927–1956.
1938 Ancient shore lines on the island of Lanai, Hawaii. *Bulletin of the Geological Society of America*, vol. 49, pp. 615–628.
1939 *Geologic map and guide of the island of Oahu, Hawaii.* Hawaii Division of Hydrography, Bulletin 2, 75 pp., colored geologic map.
1940a Four-phase volcanism in Hawaii (abstract). *Bulletin of the Geological Society of America*, vol. 51, pp. 1947–1948.
1940b *Supplement to the geology and groundwater resources of the island of Oahu, Hawaii.* Hawaii Division of Hydrography, Bulletin 5, pp. 3–55.
1940c *Geology and ground-water resources of the islands of Lanai and Kahoolawe, Hawaii.* Hawaii Division of Hydrography, Bulletin 6, pp. 3–95, 119–147, colored geologic map.
1941 Shore benches on north Pacific islands. *Bulletin of the Geological Society of America*, vol. 52, pp. 773–780.
1942a Origin of Haleakala Crater, island of Maui, Hawaii. *Bulletin of the Geological Society of America*, vol. 53, pp. 1–14.
1942b Hydrology of lava-rock terranes. In *Hydrology* (O. E. Meinzer, ed.), pp. 678–703. New York: McGraw-Hill.
1945 Glaciation of Mauna Kea, Hawaii. *Bulletin of the Geological Society of America*, vol. 56, pp. 267–274.
1946 *Geology of the Hawaiian Islands.* Hawaii Division of Hydrography, Bulletin 8, 106 pp.; reprinted with supplement, 1967, 112 pp.
1947 *Geology and ground-water resources of the island of Niihau, Hawaii.* Hawaii Division of Hydrography, Bulletin 12, pp. 3–38.
1961 Eustatic shorelines on Pacific islands. In: Pacific island terraces: eustatic? *Zeitschrift für Geomorphologie*, Supplement Volume 3, pp. 1–16.
1966a *Geology of the State of Hawaii.* Palo Alto, Calif.: Pacific Books.

1966b  *Road guide to points of geologic interest in the Hawaiian Islands.* Palo Alto, Calif.: Pacific Books.

Stearns, H. T., and Chamberlain, T. K.
1967  Deep cores of Oahu, Hawaii and their bearing on the geologic history of the central Pacific basin. *Pacific Science,* vol. 21, pp. 153–165.

Stearns, H. T., and Clark, W. O.
1930  *Geology and water resources of the Kau District, Hawaii.* U.S. Geological Survey, Water Supply Paper 616, 194 pp.

Stearns, H. T., and Macdonald, G. A.
1942  *Geology and ground-water resources of the island of Maui, Hawaii.* Hawaii Division of Hydrography, Bulletin 7, 344 pp., colored geologic map.

1946  *Geology and ground-water resources of the island of Hawaii.* Hawaii Division of Hydrography, Bulletin 9, 363 pp., colored geologic map.

1947  *Geology and ground-water resources of the island of Molokai, Hawaii.* Hawaii Division of Hydrography, Bulletin 11, 113 pp., colored geologic map.

Stearns, H. T., and Vaksvik, K. N.
1935  *Geology and ground-water resources of the island of Oahu, Hawaii.* Hawaii Division of Hydrography, Bulletin 1, 479 pp.

Stearns, N. D.
1935  *Annotated bibliography and index of the geology and water supply of the island of Oahu, Hawaii.* Hawaii Division of Hydrography, Bulletin 3, 74 pp.

Strange, W. E., Machesky, L. F., and Woollard, G. P.
1965  A gravity survey of the island of Oahu, Hawaii. *Pacific Science,* vol. 19, pp. 350–353.

Swindale, L. D., and Sherman, G. D.
1964  *Hawaiian soils from volcanic ash.* Food and Agriculture Organization, World Soil Resources Department, pp. 36–49.

Takeuchi, H., Uyeda, S., and Kanamori, H.
1967  *Debate about the earth. Approach to geophysics through analysis of continental drift.* San Francisco: Freeman, Cooper.

Tamura, T., Jackson, M. L., and Sherman, G. D.
1953  Mineral content of low humic and hydrol humic latosols of Hawaii. *Soil Science Society of America Proceedings,* vol. 17, pp. 343–346.

Thornbury, W. D.
1962  *Principles of geomorphology.* New York: John Wiley and Sons.

Tilley, C. E., and Scoon, J. H.
1961  Differentiation of Hawaiian basalts: Trends of Mauna Loa and Kilauea historic magmas. *American Journal of Science,* vol. 259, pp. 60–68.

United States Coast Pilot
1923  *The Hawaiian Islands.* U. S. Coast and Geodetic Survey, 93 pp.

Vaquier, V., Raff, A. D., and Warren, R. E.
1961  Horizontal displacements in the floor of the northeastern Pacific Ocean. *Bulletin of the Geological Society of America,* vol. 72, pp. 1251–1258.

Visher, F. N., and Mink, J. F.
1964  *Ground-water resources in southern Oahu, Hawaii.* U.S. Geological Survey, Water Supply Paper 1778, 133 pp.

Walker, J. L.
1964  *Pedogenesis of some highly ferruginous formations in Hawaii.* Hawaii Institute of Geophysics Contribution 64-10, 405 pp.

Washington, H. S., and Keyes, M. G.
1926  Petrology of the Hawaiian Islands, V. The Leeward Islands. *American Journal of Science,* ser. 5, vol. 12, pp. 336–352.

Watson, L. J.
1955  *The water sources of Honolulu.* Board of Water Supply, Honolulu, Hawaii, 33 pp.

Wentworth, C. K.
1925a  *The geology of Lanai.* B. P. Bishop Museum, Bulletin 24, 72 pp.

1925b  The desert strip of West Molokai. *University of Iowa, Studies in Natural History,* vol. 11, no. 4, pp. 41–56.

1926  *Pyroclastic geology of Oahu.* B. P. Bishop Museum, Bulletin 30, 121 pp.

1927  Estimates of marine and fluvial erosion in Hawaii. *Journal of Geology,* vol. 35, pp. 117–133.

1928  Principles of stream erosion in Hawaii. *Journal of Geology,* vol. 36, pp. 385–410.

1938a  *Ash formations of the island Hawaii.* Hawaiian Volcano Research Association, Hawaiian Volcano Observatory, Special Report 3, 183 pp.

1938b  Marine bench-forming processes: Water-level weathering. *Journal of Geomorphology,* vol. 1, pp. 6–32.

1939  Marine bench-forming processes; II. Solution benching. *Journal of Geomorphology,* vol. 2, pp. 3–25.

1943  Soil avalanches on Oahu, Hawaii. *Bulletin of the Geological Society of America,* vol. 54, pp. 53–64.

428

1944 Potholes, pits, and pans; subaerial and marine (Hawaii). *Journal of Geology*, vol. 52, pp. 117–130.

1951 *Geology and ground-water resources of the Honolulu-Pearl Harbor area, Oahu, Hawaii.* Board of Water Supply, Honolulu, Hawaii, 111 pp.

Wentworth, C. K., and Jones, A. E.

1940 Intrusive rocks of the leeward slope of the Koolau Range, Oahu. *Journal of Geology*, vol. 48, pp. 975–1006.

Wentworth, C. K., and Macdonald, G. A.

1953 *Structures and forms of basaltic rocks in Hawaii.* U.S. Geological Survey, Bulletin 994, 98 pp.

Wentworth, C. K., Powers, H. A., and Eaton, J. P.

1961 Feasibility of a lava-diverting barrier at Hilo, Hawaii. *Pacific Science*, vol. 15, pp. 352–357.

Wentworth, C. K., and Powers, W. E.

1941 Multiple glaciation of Mauna Kea, Hawaii. *Bulletin of the Geological Society of America*, vol. 52, pp. 1193–1218.

1943 Glacial springs on the island of Hawaii. *Journal of Geology*, vol. 51, pp. 542–547.

Wentworth, C. K., Wells, R. C., and Allen, V. T.

1940 Ceramic clay in Hawaii. *American Mineralogist*, vol. 25, pp. 1–33.

Wentworth, C. K., and Winchell, H.

1947 Koolau basalt series, Oahu, Hawaii. *Bulletin of the Geological Society of America*, vol. 58, pp. 49–78.

White, R. W.

1966 Ultramafic inclusions in basaltic rocks from Hawaii. *Contributions to Mineralogy and Petrology*, vol. 12, pp. 245–314.

White, S. E.

1949 Processes of erosion on steep slopes of Oahu, Hawaii. *American Journal of Science*, vol. 247, pp. 168–186; discussion, vol. 248, pp. 511–514.

Wiens, H. J.

1961 The role of mechanical abrasion in the erosion of coral reefs and land areas. *Ninth Pacific Science Congress, 1957, Thailand*, vol. 12, pp. 361–366.

Wilkes, C.

1845 *Narrative of the United States Exploring Expedition during the years 1838–1842.* Vol. 4, pp. 87–231. Philadelphia.

Winchell, H.

1947 Honolulu Series, Oahu, Hawaii. *Bulletin of the Geological Society of America*, vol. 58, pp. 1–48.

Wood, H. O.

1914 On the earthquakes of 1868 in Hawaii. *Bulletin of the Seismological Society of America*, vol. 4, pp. 169–203.

1933 Volcanic earthquakes. In *Physics of the Earth*, vol. 6, *Seismology*, pp. 9–31. National Research Council Bulletin 90.

Wright, T. L., Kinoshita, W. T., and Peck, D. L.

1968 March 1965 eruption of Kilauea volcano and the formation of Makaopuhi lava lake. *Journal of Geophysical Research*, vol. 73, pp. 3181–3205.

Yoder, H. S., Jr., and Tilley, C. E.

1962 Origin of basalt magmas: An experimental study of natural and synthetic rock systems. *Journal of Petrology*, vol. 3, pp. 342–532.

Zetler, B. D.

1948 Seismic sea wave travel times to Honolulu. U. S. Coast and Geodetic Survey chart.

# Photograph Credits

# Index

*433*

*439*